Boundary Value Problems for Linear Evolution
Partial Differential Equations

NATO ADVANCED STUDY INSTITUTES SERIES

Proceedings of the Advanced Study Institute Programme, which aims at the dissemination of advanced knowledge and the formation of contacts among scientists from different countries

The series is published by an international board of publishers in conjunction with NATO Scientific Affairs Division

A	Life Sciences	Plenum Publishing Corporation
B	Physics	London and New York
C	Mathematical and Physical Sciences	D. Reidel Publishing Company Dordrecht and Boston
D	Behavioral and Social Sciences	Sijthoff International Publishing Company Leiden
E	Applied Sciences	Noordhoff International Publishing Leiden

Series C – Mathematical and Physical Sciences

Volume 29 – Boundary Value Problems for Linear Evolution Partial Differential Equations

Boundary Value Problems for Linear Evolution Partial Differential Equations

Proceedings of the NATO Advanced Study Institute held in Liège, Belgium, September 6–17, 1976

edited by

H. G. GARNIR
University of Liège, Liège, Belgium

D. Reidel Publishing Company
Dordrecht-Holland / Boston-U.S.A.

Published in cooperation with NATO Scientific Affairs Division

ISBN 90–277–0788–X

Published by D. Reidel Publishing Company
P.O. Box 17, Dordrecht, Holland

Sold and distributed in the U.S.A., Canada, and Mexico
by D. Reidel Publishing Company, Inc.
Lincoln Building, 160 Old Derby Street, Hingham, Mass. 02043, U.S.A.

Printed in The Netherlands

TABLE OF CONTENTS

PREFACE

Most of the problems posed by Physics to Mathematical Analysis are boundary value problems for partial differential equations and systems.

Among them, the problems concerning linear evolution equations have an outstanding position in the study of the physical world, namely in fluid dynamics, elastodynamics, electromagnetism, plasma physics and so on.

This Institute was devoted to these problems. It developed essentially the new methods inspired by Functional Analysis and specially by the theories of Hilbert spaces, distributions and ultradistributions. The lectures brought a detailed exposition of the novelties in this field by world known specialists.

We held the Institute at the Sart Tilman Campus of the University of Liège from September 6 to 17, 1976. It was attended by 99 participants, 79 from NATO Countries [Belgium (30), Canada (2), Denmark (1), France (15), West Germany (9), Italy (5), Turkey (3), USA (14)] and 20 from non NATO Countries [Algeria (2), Australia (3), Austria (1), Finland (1), Iran (3), Ireland (1), Japan (6), Poland (1), Sweden (1), Zaïr (1)]. There were 5 courses of 6 hours, 1 of 4 hours, 1 of 3 hours, 2 of 2 hours and 1 of 1 hour. Moreover, 30 advanced half an hour seminars were organized by the participants to discuss the last contributions to the field.

I wish to express my warmest thanks to the NATO Organization which was the main sponsor of this meeting and to the University of Liège and the "Fonds National de la Recherche Scientifique" of Belgium who also contributed financially to its achievement.

My special gratitude is due to Professor F. Cerulus, Representative of Belgium to the NATO Science Committee and to Dr. T. Kester, NATO Scientific Officer in charge with the ASI programme.

I have been helped by a Scientific and Organizing Committee constituted by my Liège Colleagues Prof. J. Etienne, J. Gobert, P. Léonard and J. Schmets. The members of our staffs also contributed to the success of the Institute. I am very grateful to all of them.

H.G. GARNIR
Director of the Institute.

LIST OF PARTICIPANTS

GARNIR H.G. : Inst. de Math. Univ. de Liège
15, avenue des Tilleuls / B-4000 LIEGE / BELGIUM

BEALS R. : Dept. of Math. Univ. of Chicago
5734 University avenue / CHICAGO, Illinois 60637 / USA

DUFF G. : Dept. of Math. Univ. of Toronto
TORONTO, Ontario / CANADA

KOMATSU H. : Dept. of Math. Univ. of Tokyo
TOKYO / JAPAN

LIONS J.L. : Collège de France
PARIS VI / FRANCE

MAGENES E. : Istituto di Matematica Universita di Pavia
PAVIA / ITALIE

MUNSTER M. : Inst. de Math. Univ. de Liège
15, avenue des Tilleuls / B-4000 LIEGE / BELGIUM

RAUCH J. : Dept. of Math. Univ. of Michigan
ANN ARBOR, Michigan 48104 / USA

USHIJIMA T. : University of Electro-communications
TOKYO / JAPAN

WAKABAYASHI S. : Dept. of Math. Tokyo Univ. of Education
Tokyo Kyoiku Daigaku Otsuka, Bunkyo-Ku,
TOKYO / JAPAN

WILCOX C. : Dept. of Math. Univ. of Utah
205 Mathematics Building SALT LAKE CITY, Utah / USA

AKTAS Z. : Dept. of Computer Science Middle East Technical Univ.
ANKARA / TURQUIE

ALTIN A. : Matematik Bölümü Ankara Universitesi Fen Facultesi
ANKARA / TURQUIE

ANDREA S. : Division de Matematicas Universidad Simon Bolivar
Apartado Postal n° 5354 / CARACAS / VENEZUELA

BALABAN T. : Instytut Matematyki Université de Varsovie
Palac Kultury i Nauki 9 p / 00-901 WARSZAWA / POLOGNE

BALABANE M. : Dept. de Math. Univ. Paris VI
PARIS / FRANCE

BALAKRISHNAN V.K. : Dept. of Math. Univ. of Maine
Shibles Hall / ORONO, Maine 04473 / USA

BENTOSELA : Dept. Math. Univ. Luminy
MARSEILLE / FRANCE

BESSONNET G. : Laboratoire de Mécanique, Univ. de Poitiers
40, avenue du Recteur Pineau / F-86022 POITIERS / FRANCE

BIERSTEDT K.D. : Gesamthochschule Paderborn, Fachbereich 17 - Mathematik
D-479 PADERBORN / Pohlweg 55 Postfach 1621 / B.R.D.

BRACKX F. : Rijksuniversiteit Gent, Seminarie voor Wiskunkige Analyse
J. Plateaustraat 22 / B-9000 GENT / BELGIUM

BRENNER P. : Dept. of Math. Chalmers Univ. of Technology and Univ. of Göteborg
Fack / Sven Hultins gata 6 / S-402 20 GOTEBORG / SUEDE

CATTABRIGA L. : Istituto Matematico "Salvatore Pincherle"
Piazza di Porta S. Donato, 5 / BOLOGNA / ITALIE

CHAUVEHEID P. : Inst. de Math. Univ. de Liège
15, avenue des Tilleuls / B-4000 LIEGE / BELGIUM

CHEVALIER J. : Inst. de Math. Univ. de Liège
15, avenue des Tilleuls / B-4000 LIEGE / BELGIUM

COIRIER J. : Laboratoire de Mécanique, Univ. de Poitiers
40, avenue du Recteur Pineau / F-86022 POITIERS / FRANCE

COMBES J.M. : Dept. de Math. Centre Universitaire de Toulon
Château Saint-Michel / F-83130 LA GARDE / FRANCE

DAHMEN W. : Inst. für Angedte. Math. und Inf. Univ. Bonn
D-53 BONN / Wegelerstr. 6 / B.R.D.

DASTRANGE N. : Dept. of Math. Pahlavi University-Shiraz
College of Arts and Sciences / SHIRAZ / IRAN

DELANGHE R. : Rijksuniversiteit-Gent Seminarie voor Hogere Analyse
Krijgslaan 271 - Gebouw S 9 / B-9000 GENT / BELGIUM

DE PRIMA C.R. : Dept. of Math. California Institute of Technology
PASADENA, California 91125 /USA

DERMENJIAN Y. : Dept. Math. Univ. Paris XIII
PARIS / FRANCE

DJEDOUR M. : Inst. de Math. Univ. des Sciences et de la Technologie d'Alger
B.P. n° 9 Dar El Beida / ALGER / ALGERIE

DUCLOS P. : Centre de Physique Théorique, Centre National de la Recherche Scientifique
31, Chemin J. Aiguier / F-13274 MARSEILLE CEDEX 2 / FRANCE

EBERLEIN G. : Universität Darmstadt
B.R.D.

EMAMI-RAD H. : Dept. of Math. Univ. of Tehran
TEHRAN / IRAN

ETIENNE J. : Inst. de Math. Univ. de Liège
15, avenue des Tilleuls / B-4000 LIEGE / BELGIUM

FITZGIBBON W.E. : Dept. of Math. Univ. of Houston
Houston, Texas, 77004 / USA

FLORET K. : Mathematisches Seminar, Christiaen-Albrechts-Univ.
KIEL / Olshausenstr. 40-60 / B.R.D.

GARNIR-MONJOIE F. : Inst. de Math. Univ. de Liège
15, avenue des Tilleuls / B-4000 LIEGE / BELGIUM

GENET J. : Dept. de Math. Univ. de Pau
Boîte postale 523 / F-64010 PAU / FRANCE

GERARD P. : Inst. de Math. Univ. de Liège
15, avenue des Tilleuls / B-4000 LIEGE / BELGIUM

GERARD-HOUET C. : Inst. de Math. Univ. de Liège
15, avenue des Tilleuls / B-4000 LIEGE / BELGIUM

GEYMONAT S.: Istituto di Matematica, Polytecnico Torino
TORINO / ITALIE

GOBERT J. : Inst. de Math. Univ. de Liège
15, avenue des Tilleuls / B-4000 LIEGE / BELGIUM

GOSSEZ J.P. : Dept. de Math. Univ. de Bruxelles
Campus de la Plaine / Bd. Triomphe / 1050 BRUXELLES BELGIUM

GOULAOUIC C. : Dept. Math. Centre d'Orsay, Univ. de Paris-Sud
Bâtiment 425 / F-91405 ORSAY / FRANCE

GRUBB G. : Matematiske Institut, Københavns Universitets
Universitetsparken 5 / 2100 KØBENHAVN / DANEMARK

GUILLOT J.C. : Dept. Math. Centre scientifique, Univ. Paris-Nord
Place du 8 mai 1945 / F-93206 SAINT-DENIS / FRANCE

HO-VAN T.S. : Inst. de Math. Univ. de Liège
15, avenue des Tilleuls / B-4000 LIEGE / BELGIUM

INOUE A. : Dept. of Math. Faculty of Sciences, Hiroshima Univ.
HIROSHIMA / JAPAN

JUDGE D. : Mathematical Physics Dept. Univ. College
BELFIELD /DUBLIN 4 / EIRE

KALINDE K. : Université Nationale du Zaïre
ZAIRE / KINSHASA

KARASZOSEN B. : Technical University of West Berlin
D-1 BERLIN 21 / Siegmundshof B 205 / B.R.D.

KESTER T. : NATO / BRUXELLES / BELGIUM

KHOSROVSHAHI G.B. : Dept. of Math. Univ. of Tehran
TEHRAN / IRAN

KIELHOFER H. : Mathematisches Institut A, Univ. Stuttgart
Pfaffenwaldring 57 / D-7 STUTTGART 80 / B.R.D.

LAMI-DOZO E. : Dept. Math. Univ. Bruxelles
Campus de la Plaine / Bd. Triomphe / B-1050 BRUXELLES / BELGIUM

LANGE H. : Gesamthochschule Paderborn, Fachbereich 17 - Mathematik / Pohlweg 55 Postfach 1621 / PADERBORN / B.R.D.

LEONARD P. : Inst. de Math. Univ. de Liège
15, avenue des Tilleuls / B-4000 LIEGE / BELGIUM

LONDEN S.O. : Dept. of Math. Helsinki University of Technology
Otaniemi / FINLANDE

LOUSBERG P. : Inst. de Math. Univ. de Liège
15, avenue des Tilleuls / B-4000 LIEGE / BELGIUM

LUMER G. : Dept. de Math. Univ. de Mons
Avenue Maistriau / B-7000 MONS / BELGIUM

MADAUNE M. : Dept. de Math. Univ. de Pau
Boîte postale 523 / F-64010 PAU / FRANCE

MAHONY J. : Dept. of Math. Univ. of Western Australia
Nedlands, Western Australia, 6009 / AUSTRALIA

MAWHIN J. : Dept. de Math.Univ. de Louvain
2, Chemin du Cyclotron / B-1348 LOUVAIN-LA-NEUVE / BELGIUM

Mc. INTOSH A.G.R. : School of Mathematics and Physics, MacQuarie University
North Ryde-New South Wales 2113 / AUSTRALIA

MEISE R. : Mathematisches Institut, Univ. Düsseldorf
D-4 DUSSELDORF / Universitätstr. 1 / B.R.D.

MIGNOT F. : Dept. Math. Univ. Lille
LILLE / FRANCE

MORAWETZ C. : Courant Institute of Mathematical Sciences
New York University
251 Mercer Street / New York, N.Y. 10012 / USA

MORI Y. : Inst. de Math. Univ. de Liège
15, avenue des Tilleuls / B-4000 LIEGE / BELGIUM

NGUYEN TRI KIEM : Inst. de Math. Univ. de Liège
15, avenue des Tilleuls / B-4000 LIEGE /
BELGIUM

OHWAKI S.I. : University of Kumamoto
KUMAMOTO / JAPAN

ORTON M. : Dept. of Math. Univ. of California, Irvine
IRVINE, California 92664 / USA

OSHER S. : Dept. of Math. Univ. of California at Los Angeles
LOS ANGELES / USA

PAPACOSTAS G. : Univ. Nationale du Gabon
Libreville, BP 911 / GABON

PAQUET L. : Dept. Math. Univ. de Mons
Avenue Maistriau / B-7000 MONS / BELGIUM

PETZSCHE H.J. : Mathematisches Institut Univ. Düsseldorf
D-4000 DUSSELDORF / Universitätstr. 1 / B.R.D.

PUEL : Dept. Math. Univ. Lille
LILLE / FRANCE

RALSTON J.V. : Dept. Math. Univ. of California at Los Angeles
LOS ANGELES / USA

RAY A.K. : Dept. Math. Univ. Ottawa
OTTAWA / CANADA

READ-DERCHAIN C. : Inst. de Math. Univ. de Liège
15, avenue des Tilleuls / B-4000 LIEGE /
BELGIUM

RIFAUT E. : Inst. de Math. Univ. de Liège
15, avenue des Tilleuls / B-4000 LIEGE / BELGIUM

SANTAGATI G. : Seminario Matematico, Univ. de Catania
Corso Italia, 55 / I-95129 CATANIA / ITALIE

SARASON L. : Dept. of Math. Univ. of Washington
SEATTLE, Washington 98195 / USA

SCHAPPACHER W. : Inst. Math. Univ. Graz, Lehrkanzel für angewandte Mathematik
A-8010 GRAZ / Steyergasse 17/5 / AUSTRIA

SCHERER K. : Inst. für Angewandte Math. und Inf. Univ. Bonn
D-53 BONN / Wegelerstr. 6 / B.R.D.

SCHMETS J. : Inst. de Math. Univ. de Liège
15, avenue des Tilleuls / B-4000 LIEGE / BELGIUM

SILBERSTEIN J.P.C. : Dept. Math. Univ. of Western Australia
Nedlands, Western Australia, 6009 /
AUSTRALIA

TAIRA K. : Tokyo Institute of Technology
TOKYO / JAPAN

THYSSEN M. : Inst. de Math. Univ. de Liège
15, avenue des Tilleuls / B-4000 LIEGE / BELGIUM

TRAVIS C. : Dept. of Math. Ayres Hall, Univ. of Tennessee
Knoxville, Tennessee, 37916 / USA

UCHIYAMA K. : Inst. Math. Univ. de Nice
Parc Valrose / F-06034 Nice CEDEX / FRANCE

VANDERMEULEN E. : Inst. de Math. Univ. de Liège
15, avenue des Tilleuls / B-4000 LIEGE /
BELGIUM

VIGNOLI A. : Dipartimento di Matematica,
Universita degli studi della Calabria
C.P. Box 9 / I-87030 ROGES (Cosenza) / ITALIE

WEDER R. : Inst. voor Theoritische Fysica, Univ. Leuven
Celestijnenlaan, 200 D / B-3030 HEVERLEE / BELGIUM

WUIDAR J. : Inst. de Math. Univ. de Liège
15, avenue des Tilleuls / B-4000 LIEGE / BELGIUM

ZIZI K. : Inst. de Math. Univ. d'Oran
ORAN / ALGERIE

LAPLACE TRANSFORM METHODS FOR EVOLUTION EQUATIONS

Richard Beals

University of Chicago

The evolution equations considered here are of the form

$$u'(t) = A(t)u(t) + f(t) \ , \quad t > 0 \ ; \quad u(0) = u_0 \ . \tag{1}$$

The given function f and the unknown function u take values in a complex Banach space X, and the initial condition u_0 is in X. The $A(t)$ are closed linear operators with domains dense in X. The <u>time-independent case</u> is the case $A(t) \equiv A$. As we shall see, taking Laplace transforms (formally) leads very naturally to considering the resolvent operators $R_\lambda = (\lambda I - A)^{-1}$, and what we describe here could as well be titled "resolvent methods for evolution equations." We shall survey some conditions on the R_λ or $R_\lambda(t)$ which (a) bear on the existence, uniqueness, or qualitative behavior of solutions and (b) are verifiable for interesting classes of PDEs. We outline how the conditions are derived for PDEs and how they are used to construct solutions of (1), and indicate some further lines of research.

Garnir (ed.), Boundary Value Problems for Linear Evolution Partial Equations. 1-26.

1. Abstract time-independent problems

If u is a piecewise continuous X-valued function on $[0,\infty)$ $[0,\infty)$ such that $|u(t)| \leq M\exp(-\omega t)$, then its Laplace transform u is the holomorphic X-valued function defined on the half-plane $\mathrm{Re}\,\lambda > \omega$ by

$$\tilde{u}(\lambda) = \int_0^\infty \exp(-\lambda t)u(t)dt.$$

There is a symbolic "inversion formula"

$$u(t) = \frac{1}{2\pi i}\int_{\Gamma_a} e^{\lambda t}\tilde{u}(\lambda)d\lambda \quad , \tag{2}$$

where Γ_a is the oriented vertical line from $a - i\infty$ to $a + i\infty$, and $a > \max\{0, \omega\}$.

Consider the time-independent case of (1), by a purely formal argument. A formal Laplace transformation of (1) and integration by parts gives

$$A\tilde{u}(\lambda) + \tilde{f}(\lambda) = \int_0^\infty u'(t)e^{-\lambda t}dt = -u_0 + \lambda\tilde{u}(\lambda),$$

so

$$\tilde{u}(\lambda) = R_\lambda u_0 + R_\lambda \tilde{f}(\lambda) .$$

We apply (2) and interchange the order of integration:

$$2\pi i u(t) = \int_\Gamma e^{\lambda t}R_\lambda u_0 d\lambda + \int_\Gamma \int_0^\infty e^{\lambda(t-s)}R_\lambda f(s)ds\, d\lambda$$

so

$$u(t) = U(t)u_0 + \int_0^t U(t-s)f(s)ds \tag{3}$$

where

$$U(t) = \frac{1}{2\pi i}\int_\Gamma e^{\lambda t}R_\lambda\, d\lambda \quad , \quad t > 0 . \tag{4}$$

Here Γ is some suitable oriented contour homotopic to Γ_a and having the singularities of R_λ — the spectrum of A — to its left.

All this, we emphasize, is purely formal. We do not need to make this derivation itself rigorous; we only need to show that in certain cases solutions are unique and are expressible by (3) and (4).

Let us begin with the question of uniqueness. Let D_A be the domain of A. It is a Banach space with the graph norm $|u|_A = |Au| + |u|$. Let us suppose $R_\lambda = (\lambda I - A)^{-1}$ exists for real $\lambda > \lambda_0$ and satisfies

(5) $$\lim_{\lambda \to +\infty} e^{-\varepsilon\lambda} \|R_\lambda\| = 0, \quad \text{all} \quad \varepsilon > 0 .$$

<u>Theorem (Lyubich).</u> <u>The problem (1) has at most one solution</u> u <u>continuous to</u> X, <u>such that</u> $u(t) \in D_A$ <u>and the strong derivative</u> $u'(t)$ <u>exists for each</u> $t > 0$.

We sketch the proof. Let u be a solution with data zero, and let $u_1(t) = R_\mu u(t)$ for some $\mu > \lambda_0$. Then u_1 is continuous to D_A, continuously differentiable, and satisfies (1) with data zero. Given $T > 0$, let

(6) $$w(\lambda) = \int_0^T e^{\lambda(T-t)} u_1(t)dt .$$

Then integration by parts shows

(7) $$Aw(\lambda) = u_1(T) + \lambda w(\lambda) ,$$

so

(8) $$w(\lambda) = -R_\lambda u(T) , \quad \lambda > \lambda_0 .$$

Because of (6), w is entire of exponential type and bounded for $\text{Re}\,\lambda < 0$. Because of (8) and (5), $e^{-\varepsilon\lambda} w(\lambda) \to 0$ as $\lambda \to +\infty$. A Phragmen-Lindelof argument shows that w is bounded, hence constant. By (7), the constant must be zero, so $u(T) = (\lambda I - A)u_1(T) = 0$.

The condition (5) will be satisfied in the cases we consider, so we turn to the question of existence and construction of solutions, based on the heuristic formulas (3), (4).

We say that the evolution equation associated to A is strictly parabolic if there are positive M, λ_0, δ such that

(9) R_λ exists and satisfies $\|R_\lambda\| \leq M|\lambda|^{-1}$ whenever $|\lambda| \geq \lambda_0$ and $|\arg \lambda| \leq \frac{1}{2}\pi + \delta$.

In this case we let Γ be the boundary of the region described in (9). Then the integral (4) certainly converges and defines a bounded, infinitely differentiable operator from X to D_A. Moreover,

$$(10) \qquad 2\pi i u'(t) = \int_\Gamma e^{\lambda t} \lambda R_\lambda \, d\lambda = \int_\Gamma e^{\lambda t}(I + AR_\lambda)\, d\lambda = 0 + 2\pi i\, AU(t).$$

Note that $\|U(t)\|$ remains bounded as $t \to 0$. To see this, integrate by parts in (3) to change the integrand to $t^{-1}\exp(\lambda t)R_\lambda^2$, move the contour of integration to the vertical line $\operatorname{Re}\lambda = t^{-1}$, and estimate using (9). It is easy to check that $U(t)u_0 \to u_0$ as $t \to 0$ whenever $u_0 \in D_A$, and the boundedness result just mentioned then yields this convergence for each $u_0 \in X$. We have essentially proved the theorem of Hille, that such an operator A generates a holomorphic semigroup. The following is an easy consequence.

Theorem. If A is strictly parabolic, then for any $u_0 \in X$ and any Holder-continuous f, the time-independent problem (1) has a unique solution u, continuous on $[0, \infty)$ and C^1 on $(0, \infty)$, given by (3), (4).

To see that (4) gives a solution, it is helpful to write

$$\int_0^t U(t-s)f(s)ds = \int_0^t U(t-s)[f(s) - f(t)]ds + V(t)f(t), \tag{11}$$

where

$$V(t) = \int_0^t U(s)ds = \frac{1}{2\pi i}\int_\Gamma (e^{\lambda t} - 1)\lambda^{-1}R_\lambda d\lambda.$$

Now $V(t): X \to D_A$ for $t > 0$. It is easy to see that $U(t)$ has norm $O(t^{-1})$ as a map of X to D_A, so (11) shows that $u(t) \in D_A$ for $t > 0$. The desired differentiability may also be established by using (11) and our other information about $U(t)$.

We say that the evolution equation associated to A is <u>weakly hyperbolic</u> if there are positive M, N, and λ_0, and a such that $0 \leq a < 1$, such that

(12) R_λ exists and satisfies $\|R_\lambda\| \leq M|\lambda|^N$ when $|\lambda| \geq \lambda_0$ and $\text{Re}\,\lambda \geq c_0|\text{Im}\,\lambda|^a$.

(Note that according to our definitions, strictly parabolic implies weakly hyperbolic!) In this case, let Γ be the contour bounding the region (12), and let

$$h_\varepsilon(\lambda) = \exp(-\varepsilon(\lambda_0 - \lambda)^b), \quad \varepsilon > 0,$$

where $a < b < 1$ and we take the principal branch of z^b on the plane slit along the negative real axis. As $\lambda \to \infty$ on or to the left of Γ,

$$|h_\varepsilon(\lambda)| \leq \exp(-\delta|\lambda|^b), \text{ some } \delta > 0.$$

It follows that the integral

$$U(t) = \frac{1}{2\pi i}\int_\Gamma e^{\lambda t}h_\varepsilon(\lambda)R_\lambda d\lambda$$

exists for $\varepsilon > 0$, $t \geq 0$; moreover $U_\varepsilon'(t) = AU_\varepsilon(t)$, $t \geq 0$. Thus if $u_0 \in X$, then $u(t) = U_\varepsilon(t)u_0$ is a solution of $u' = Au$ such that

$u(0) = U_\varepsilon(0)u_0 = J_\varepsilon u_0$. It can be shown [7] that J_ε is injective and has dense range Y_ε. Let us equip Y_ε with the norm which makes J_ε an isometry from X to Y_ε. If $u_0 \in Y_\varepsilon$ and if f is continuous to Y_ε, it follows easily that the (unique) solution of (1) is given by

$$u(t) = U_\varepsilon(t)J_\varepsilon^{-1}u_0 + \int_0^t U_\varepsilon(t-s)J_\varepsilon^{-1} f(s)ds.$$

The requirement of lying in Y_ε may be considered as a smoothness condition, in fact a Gevrey condition, in terms of A [7]. If $v \in Y_\varepsilon$ then v is in the domain of A^n for each n and there is a $\sigma > 0$ such that

$$\sup_n |A^n u| \sigma^n \Gamma(\beta n+1)^{-1} < \infty, \quad \beta = 1/b. \tag{13}$$

Conversely, if (13) is true for large enough σ, then $v \in Y_\varepsilon$.

We have seen that in the strictly parabolic case the homogeneous ($f \equiv 0$) problem has a unique, and well-behaved solution for each $u_0 \in X$; in the weakly hyperbolic case there is a solution for each sufficiently "smooth" u_0. The "natural" initial condition for the homogeneous problem would seem to be $u_0 \in D_A$. By a result of Phillips [27], there is a unique solution u, C^1 on $[0,\infty)$, for each $u_0 \in D_A$ if and only if A generates a C_0 semigroup. The well-known condition for this [17] is that there be constants M and λ_0 such that

(14) R_λ exists and satisfies $\|R_\lambda^n\| \le M(\mathrm{Re}\,\lambda - \lambda_0)^{-n}$ for each λ such that $\mathrm{Re}\,\lambda > \lambda_0$.

Because the constant M is to be independent of n, these inequalities are difficult to verify directly except in the contractive case, and resolvent methods seem of limited value here.

Many other results relating properties of the resolvent operators to properties of solutions of the time-independent problem are known; see in particular the book of S. G. Krein [19], the survey article of Lyubich [24], the extensive paper of Agmon-Nirenberg [2], and also [6], [11], [13], [26], [34]. The "distribution semigroup" approach of Lions [23] also leads to resolvent estimates; see [5], [20] for example. The estimates (12) appear also in the context of distribution semigroups: see Chazarain [10].

2. Abstract time-dependent problems.

One would like to construct a solution of the problem

$$u'(t) = A(t)u(t) + f(t), \quad t > s; \quad u(s) = u_0$$

by a variation-of-constants formula

$$u(t) = U(t, s)u_0 + \int_0^s U(t, r)f(r)dr. \tag{15}$$

The "evolution operator" U should satisfy

$$\frac{\partial}{\partial t} U(t, s) = A(t)U(t, s), \quad t > s \; ; \; U(s, s) = I. \tag{16}$$

In (16), s is merely a parameter. In particular, if $U(t) = U(t,0)$ then

$$U'(t) = A(t)U(t) , \quad t > 0 \; ; \; U(0) = I. \tag{17}$$

Formally, we may attempt to solve (17) by writing it as an integral equation and solving by the Picard iterative method:

$$\begin{aligned} U(t) &= I + \int_0^t A(s)U(s)ds , \\ U(t) &= \sum_0^\infty V_n(t) , \quad V_0 = I , \quad V_{n+1}(t) = \int_0^t A(s)V_n(s)ds. \end{aligned} \tag{18}$$

There is a case discovered by Ovcyannikov and Treves, among others, in which (18) makes sense even for unbounded $A(t)$.

Suppose (X_σ) is a family of Banach spaces, $0 \leq \sigma \leq 1$. Suppose $X_\sigma \subset X_\tau$ if $\tau < \sigma$, and $|u|_\tau \leq |u|_\sigma$. Suppose $A(t)$ is bounded from X_σ to X_τ if $\tau < \sigma$, with norm $\|A(t)\|_{\tau,\sigma} \leq C(\tau-\sigma)^{-1}$, and suppose $t \mapsto A(t)$ is continuous to $\mathcal{L}(X_\sigma, X_\tau)$. Then it is not difficult to show by induction that the operators V_n in (18) satisfy

$$\|V_n\|_{\tau,\sigma} \leq (n!)^{-1}[Cn(\tau-\sigma)^{-1}t]^n .$$

Therefore the series (18) converges in $\mathcal{L}(X_\sigma, X_\tau)$ if $|t| < (\tau-\sigma)(Ce)^{-1}$, and solutions exist for small time. The classical Cauchy-Kowalewski theorem can be proved in this way: suppose $A(t)$ is a first order system of PDEs in $x = (x_1, \ldots, x_n)$ with coefficients which are continuous in (t, x) and analytic in x for x near 0. Let X_σ consist of those functions of x analytic near 0, such that

$$|u|_\sigma = \sup_\alpha (\alpha!)^{-1} |u^{(\alpha)}(0)| \sigma^{|\alpha|} < \infty .$$

See [33] for details. For a recent version in which the $A(t)$ are analytic pseudodifferential operators, see Baouendi-Goulaouic [4].

We consider now a <u>time-dependent strictly parabolic case</u>. Suppose that the $A(t)$ have a common domain D, with fixed norm $|\ |_D$, and that $t \mapsto A(t)$ is Lipschitz-continuous to $\mathcal{L}(D, X)$. We assume a uniform version of (9):

(19) $R_\lambda(t) = (\lambda I - A(t))^{-1}$ exists and satisfies $\|R_\lambda(t)\| \leq M|\lambda|^{-1}$ when $|\lambda| \geq \lambda_0$ and $|\arg \lambda| \leq \frac{1}{2}\pi + \delta$.

Finally, we assume that for some fixed μ, $t \mapsto R_\mu(t)$ is continuous to $\mathcal{L}(X, D)$.

Theorem (Tanabe [32] - Sobolevskii [29]). *There are operators* $U(t,s)$, $t \geq s$, *such that* $(t,s) \mapsto U(t,s)$ *is continuous in the strong operator topology for* $t \geq s$, *continuous in norm for* $t > s$, *and such that (16) holds.*

Start with an approximate solution $U_0(t,s)$:

$$U_0(t,s) = \frac{1}{2\pi i}\int_\Gamma e^{\lambda(t-s)}R_\lambda(s)d\lambda. \tag{20}$$

Then

$$\frac{\partial}{\partial t} U_0(t,s) = A(t)U_0(t,s) - \Phi_1(t,s), \tag{21}$$

where

$$\Phi_1(t,s) = [A(t) - A(s)]U_0(t,s).$$

From (21) and (15) we get (formally)

$$U_0(t,s) = U(t,s) - \int_s^t U(t,r)\Phi_1(r,s)dr .$$

One may hope, therefore, to find U in terms of U_0 by an integral equation of the form

$$U(t,s) = U_0(t,s) + \int_s^t U_0(t,r)\Phi(r,s)ds . \tag{22}$$

If (16) and (22) were true, we should expect

$$0 = (\frac{\partial}{\partial t} - A(t))U(t,s) = -\Phi_1(t,s) + \Phi(t,s) - \int_s^t \Phi_1(t,r)\Phi(r,s)dr.$$

Thus we obtain an integral equation for Φ and solve by iteration:

$$\Phi(t,s) = \Phi_1(t,s) + \int_s^t \Phi_1(t,r)\Phi(r,s)dr , \tag{23}$$

$$\Phi = \sum_1^\infty \Phi_n , \quad \Phi_{n+1}(t,s) = \int_s^t \Phi_1(t,r)\Phi_n(r,s)dr . \tag{24}$$

Now it follows from (20) that $U_0(t,s)$ has norm $\leq C(t-s)^{-1}$ as operator from X to D, so the Φ_1 are uniformly bounded as operators in X (locally), and the series in (24) converges to a solution of (23). With enough more work, it can be shown that

U given by (22) has the stated properties. (The estimates (19) are stronger than necessary; see [30], for example.)

Next we consider an abstract time-dependent weakly hyperbolic case. We assume a uniform version of (12), together with some strengthening of it:

(25) $R_\lambda(t)$ exists and satisfies $\|R_\lambda(t)\| \leq M|\lambda|^{\eta-1}$ when $|\lambda| \geq \lambda_0$ and $\operatorname{Re}\lambda \geq c_0|\operatorname{Im}\lambda|^a$, where $0 \leq a, \eta < 1$;

(26) when λ is real and $> \lambda_0$, $\|R_\lambda(t)\| \leq M|\lambda|^{-1}$.

Choose β such that $1 < \beta < a^{-1}$. As noted in section 1, the operator defined formally by (20) makes sense on the Gevrey space $X_{\tau,t} \subset X$, where

$$|u|_{\tau,t} = \sup_m |A(t)^m u| \tau^m \Gamma(\beta m+1)^{-1} .$$

Our aim here is to combine the Ovcyannikov-Treves and Tanabe-Sobolevskii methods. A careful analysis of $U_0(t,s)$ shows that in a fixed interval $\tau, \sigma \in [\tau_0, \sigma_0] \subset (0,\infty)$, $U_0(t,s)$ maps $X_{\sigma,s}$ to $X_{\tau,s}$ with norm

$$\|U_0(t,s)\|_{\tau,\sigma;s} \leq C(\sigma-\tau)^{-\rho}$$

provided $\sigma > \tau$ and $|t-s| \leq T = T(\tau_0)$. Here $0 \leq \rho < 1$. Reasonable (i.e. verifiable) assumptions on the $A(t)$ imply

$$|u|_{\tau,t} \leq |u|_{\sigma,s} \quad \text{if } |t-s| \leq c(\sigma-\tau),$$

$$|(A(t) - A(s))u|_{\tau,t} \leq C|t-s||\tau-\sigma|^{-\beta} \quad \text{if } |t-s| \leq c(\sigma-\tau).$$

Let Φ_1 be as before. Combing the last three inequalities,

(27) $|\Phi_1(t,s)u|_{\tau,t} \leq C|t-s||\tau-\sigma|^{-\rho-\beta}|u|_{\tau,s}$ if $|t-s| \leq C_1(\sigma-\tau)$.

Let $\delta = 2-\beta+\rho$. If $\beta < \eta^{-1}$, then $\delta > 0$ and by induction we get

(28) $|\Phi_n(t,s)u|_{\tau,t} \leq \Gamma(n\delta)^{-1}[C\Gamma(\delta)|t-s|(\sigma-\tau)^{\delta-2}]^n$.

Thus Φ may be defined by (24), and U by (22); see [9].

The (explicit or implicit) assumption above that the A(t) have the same domain is essential, since otherwise Φ_1 may not be defined. This assumption is natural for the Cauchy problem, but not for general mixed problems (where the boundary conditions are incorporated in the domain of A). Kato and Tanabe [18] removed the assumption by taking

$$U_0(t,s) = \frac{1}{2\pi i}\int_\Gamma e^{\lambda(t-s)}R_\lambda(t)dt$$

and looking again for a solution of the form (22). The formal solution is again given by (24), with

$$\Phi_1(t,s) = -\left(\frac{\partial}{\partial t} + \frac{\partial}{\partial s}\right)U_0(t,s) .$$

(For another approach, see Da Prato [12]).

It is possible to consider abstract mixed problems more directly. We begin with the time-independent case. Suppose W is a Banach space dense in X and

$$B: W \to Y, \ \mathcal{A}: W \to X, \quad \text{bounded.}$$

Suppose B is onto, and let A be the restriction of $\mathcal{A}$ to ker(B). Let $\mathcal{A}_\lambda = \lambda I - \mathcal{A}$ and

$$S_\lambda = \mathcal{A}_\lambda \oplus B: W \to X \oplus Y .$$

Then S_λ has inverse T_λ if and only if $\lambda I - A$ has inverse R_λ. A formal Laplace transform shows that the mixed problem

(29) $u' = \mathcal{A}u + f, \quad Bu = g, \quad u(0) = u_0$

would have a solution of the form

(30) $u(t) = U(t)J_X u_0 + \int_0^t U(t-s)h(s)ds$.

Here J_X is the injection $X \to X \oplus Y$, $h(s) = [f(s), g(s)]$, and $U(t): X \oplus Y \to X$,

$$U(t) = \frac{1}{2\pi i} \int_\Gamma e^{\lambda t} T_\lambda \, d\lambda \ . \tag{31}$$

It is too much to expect that $\|T_\lambda\| \leq M|\lambda|^{-1}$ in the region described by (19), say, but one may have

$$\|T_\lambda\| \leq M|\lambda|^{-\rho} \ , \quad \|T_\lambda J_X\| \leq M|\lambda|^{-1} \ .$$

If so, the integral (31) exists and it can be shown that for Lipschitz functions f and g, (30) is a solution of (29).

In the time-dependent version, the boundary operators $B(t)$ and the operators $\mathcal{A}(t)$ vary, while W is fixed; of course the domain of $A(t) = \ker B(t)$ may vary within W. We look for solutions of the time-dependent problem (29) in the form

$$u(t) = U(t, 0)J_X u_0 + \int_0^t U(t, s)h(s)ds \ ,$$

where

$$\frac{\partial U}{\partial t}(t, s) = A(t)U(t, s), \ BU(t, s) = 0, \ t > s, \ U(s, s)J_X = I.$$

A formal argument suggests that U can be constructed by (22), (24), where

$$U_0(t, s) = \frac{1}{2\pi i} \int_\Gamma e^{\lambda(t-s)} T_\lambda(s) ds,$$

$$\Phi_1(t, s) = (\mathcal{A}(t) - \mathcal{A}(s))U_0(t, s) \oplus (B(s) - B(t))U_0(t, s).$$

This approach does not seem to have been carried through, even for the strictly parabolic case.

3. The Cauchy problem for PDEs.

We begin with a system of first order in t, having constant coefficients. Let $\mathbf{x} = (x_1, \ldots, x_n)$, $D_j = -i\partial_j$, $u = (u_1, \ldots, u_m)^t = u(\mathbf{x}, t)$, and

$$A(u) = a(D)u = \sum a_\alpha D^\alpha u ,$$

where the a_α are $m \times m$ matrices. Let

$$X = (L^2)^m = L^2 \oplus L^2 \oplus \ldots \oplus L^2 ,$$

and let the domain D_A consist of those $u \in X$ such that $a(\xi)\hat{u}(\xi) \in (L^2)^m$, where $\hat{u}$ is the Fourier transform and $a(\xi) = \sum \xi^\alpha a_\alpha$. (For a more general treatment of constant coefficient problems, see the book of Gelfand-Shilov [15].) A Fourier transform-ODE argument shows uniqueness of solutions of the Cauchy problem for $\partial_t - A$ (with X as space of initial conditions!) with no further assumptions, but let us consider the criterion of Lyubich anyway. It is easy to see that $\lambda I - A$ is invertible if and only if the matrix $\lambda I - a(\xi)$ is invertible for every $\xi \in R^n$, and also

$$\|(\lambda I - A)^{-1}\| = \sup_\xi \|(\lambda I - a(\xi))^{-1}\| < \infty .$$

Suppose $R_\lambda = (\lambda I - A)^{-1}$ exists for large real λ. The Seidenberg-Tarski Theorem [28] implies that $r(\lambda) = \|R_\lambda\|$ is an algebraic function of λ for large real λ, so the Lyubich growth condition (5) is automatically satisfied when the resolvent exists for large real λ.

If $-A$ is strongly elliptic, i.e., $\partial_t - A$ is parabolic in the classical sense, it is readily seen that (9) is satisfied and the equation is strictly parabolic in the sense above. Ellipticity is not necessary: consider

$$A = \begin{bmatrix} 0 & I \\ \Delta & \Delta \end{bmatrix}$$

This operator is not elliptic, but satisfies (9). The operator

$\partial_t - A$ is not parabolic in the classical sense, though it is in our sense (or that of Gelfand-Shilov [15]).

It should also be noted that the choice of space X is important. The evolution equation for

$$A = \begin{bmatrix} 0 & I \\ -\Delta^2 & 2\Delta \end{bmatrix}$$

obtained by reducing $(\partial_t - \Delta)^2$ to first order in t, is not strictly parabolic in X, but it is in $H^2 \oplus L^2$. (As usual, H^s denotes the Sobolev space consisting of $u \in \mathcal{S}'$ such that $(1+|\xi|)^s \hat{u}(\xi) \in L^2$.)

Consider now the "weakly hyperbolic case." Again there is some conflict of terminology — we have already noted that "strictly parabolic" implies "weakly parabolic.") Consider the operators A_1, A_2, A_3 given respectively by

$$(32) \qquad \begin{pmatrix} D_1 & 0 \\ D_2 & D_1 \end{pmatrix}, \begin{pmatrix} 0 & 1 \\ D_2^3 & i\Delta + D_1 \end{pmatrix}, \begin{pmatrix} 0 & 1 \\ -\Delta & 0 \end{pmatrix}.$$

Then $\partial_t - A_1$ has characteristic polynomial $p(\tau, \xi) = \det(i\tau I - a(\xi)) = -(\tau - \xi_1)^2$, so is (weakly) hyperbolic in the classical sense, but the spectrum of A_1 is all of $\mathbb{C}$. While $\partial_t - A_2$ is not hyperbolic in the classical sense, it is weakly hyperbolic in our sense. Finally, $\partial_t - A_3$, which is obtained by reducing the wave equation to first order in t, is weakly hyperbolic in $H^1 \oplus L^2$ but not in $L^2 \oplus L^2$ in the stronger sense of (25), since $\|R_\lambda\| \to 1$ as $\lambda \to +\infty$.

To exhibit a general class of weakly hyperbolic equations, we suppose that $A_0 = a_0(\Delta)$ is homogeneous of order 1, and that the roots τ of $\det(i\tau I - a_0(\xi))$ are real and $\neq 0$ for $\xi \in R^n \setminus (0)$; thus A_0 itself is elliptic. An easy argument based in part on homogeneity shows [8]:

$$\|\lambda I - a_0(\xi))^{-1}\| \leq C|\mathrm{Re}\,\lambda|^{-k}|\lambda|^k(|\lambda| + |\xi|)^{-1} \tag{33}$$

when $\mathrm{Re}\,\lambda \neq 0$, where k is the maximum multiplicity of the roots τ. If $A = A_0 + B$, where B is a constant matrix, then $R_\lambda = (\lambda I - A)^{-1}$ exists whenever $(\lambda I - A_0)^{-1}$ exists and $\|B(\lambda I - A_0)^{-1}\| \leq 1/2$; moreover in such a case

$$\|R_\lambda\| \leq 2\|(\lambda I - A_0)^{-1}\| \leq 2c(\mathrm{Re}\,\lambda)^{-k}|\lambda|^{k-1}.$$

This is true when

$$|\mathrm{Re}\,\lambda| \geq c_0|\lambda|^a, \quad a = 1 - k^{-1},$$

so $\partial_t - A$ is weakly hyperbolic. The condition here that A_0 be elliptic is unfortunate — it rules out Maxwell's equations — though A_1 above shows what may otherwise occur. When there is an associated "divergence", a homogeneous system $B = b(D)$ such that $a_0(\xi)$ is invertible on $\ker b(\xi)$ for $\xi \in R^n \setminus (0)$, then we may restrict A to the closure of $\ker(B)$ in $(L^2)^m$ and obtain a weakly hyperbolic problem.

In discussing operators with variable coefficients, we shall consider single equations of parabolic and weakly hyperbolic type, starting with a time-independent parabolic equation. Suppose

$$P = p(x, \partial_t, D) = \sum_{qk + |\alpha| \leq M} p_{k\alpha}(x)\,\partial_t^k D^\alpha$$

where the coefficient of ∂_t^m ($m = M/q$) is 1; and all coefficients are smooth with bounded derivatives of all orders. Consider the associated <u>homogeneous</u> operators

$$Q_z = q_z(x, D_0, D) = \sum_{qk+|\alpha|=M} p_{k\alpha}(x)(zD_0)^{qk} D^\alpha .$$

We suppose that the Q_z are uniformly elliptic, uniformly for z such that $|z| = 1$ and $q|\arg z| \leq \frac{1}{2}\pi + \delta$. Reduce to a system of first order in ∂_t by the standard method: $v_j = \partial_t^{j-1} u$, $1 \leq j \leq m$, and

$$(Av)_j = v_{j+1} , \ j < m, \quad (Av)_m = - \sum_{k<m} p_{k\alpha}(x) D^\alpha v_{k+1} .$$

Let $X = H^{M-q} \oplus H^{M-2q} \oplus \ldots \oplus H^0$. Considering A in X is equivalent to considering $A^\#$ in $(L^2)^m$, where $A^\# = SAS^{-1}$ and $S = s(D)$ is the matrix of pseudodifferential operators with symbols

(34) $\quad s(\xi) = \mathrm{diag}(<\xi>^{M-q}, <\xi>^{M-2q}, \ldots, 1), \quad <\xi>^2 = 1 + |\xi|^2.$

Then $A^\#$ is a matrix of ψdo's of order 1. Our assumptions imply that $P_z = z^q D_0 - A^\#$ is elliptic for z in the above set. Choose $\varphi \neq 0$ in $\mathcal{D}(R)$. Given $u \in \mathcal{D}(R^n)^m$, let

$$u_\lambda(x_0, x) = \varphi(x_0) \exp(i\mu \cdot \xi_0) u(x) , \quad \mu = \lambda^{1/q} .$$

The standard elliptic estimate for P_z :

(35) $\qquad \|v\|_1 \leq C(\|P_z v\| + \|v\|)$

shows (taking $z = \mu|\mu|^{-1}$ and letting $v = u_\lambda$) that

$$|\lambda| \|u\| + \|u\|_1 \leq C_1(\|\lambda I - A^\#)u\|)$$

when $|\arg \lambda| \leq \frac{1}{2}\pi + \delta$ and $|\lambda|$ is large. Thus $\partial_t - A^\#$ is strictly **parabolic** in X. (The trick of deriving estimates with a parameter directly from known elliptic estimates seems to be due to Agmon.) The passage to the time-dependent case via the

Tanabe-Sobolevskii method is clear: assume the coefficients are Lipschitz in t.

We turn to weakly hyperbolic problems. Suppose

$$P = p(x, \partial_t, D) = \sum_{k+|\alpha| \leq m} p_{k\alpha} \partial_t^k D^\alpha ,$$

and $p_{m0} = 1$. Suppose the coefficients are smooth with bounded derivatives and suppose the roots τ of

$$p_0(x, i\tau, \xi) = \sum_{k+|\alpha| = m} p_{k\alpha}(i\tau)^k \xi^\alpha$$

are real and $\neq 0$ when $\xi \in R^n$ (0). Reduce to a first order system $\partial_t - A$ as above, and take $X = H^{m-1} \oplus H^{m-2} \oplus \ldots \oplus H^0$. Assume for the moment that the coefficients of P are constant and that P is homogeneous. Let $A^\# = SAS^{-1}$, where S has symbol (34) with $M = m$, $q = 1$. Then $a^\#(\xi)$ satisfies an estimate (33), and it follows that

$$(36) \quad |\lambda| \sum \|u_j\|_{m-j} + \sum \|u_j\|_{m-j+1} \leq C(\mathrm{Re}\,\lambda)^{-k} |\lambda|^k \sum \|(v_\lambda)_j\|_{m-j} ,$$

where $v_\lambda = (\lambda I - A^\#)u$. Now estimates like (36) with "frozen" coefficients can be patched together by a partition of unity (as in the old elliptic theory), or a pseudodifferential parametrix can be constructed, to carry (35) over to the variable coefficient case. In that case (35) holds with $2k$ in place of k and λ large, $|\mathrm{Re}\,\lambda| \geq c_0 |\lambda|^a$; see [8].

To pass to the time-dependent weakly hyperbolic case by the methods of the preceding section, we need the strong estimates relating the abstract Gevrey spaces associated with the operators $A(t)$. It is enough to assume that the coefficients are

Lipschitz in t and lie in a sufficiently small classical Gevrey class in x (uniformly in t). Results for such problems were first obtained by Ohya [25] and Leray-Ohya [22], using methods very different from those here; see also Steinberg [31].

4. Mixed problems for PDEs.

Consideration of mixed problems leads rapidly to ODEs, so we begin with a quick discussion of constant coefficient ODEs. Let $d = d/dy$ and consider the problem

$$(37) \qquad Pu(y) = f(y), \quad y \geq 0 \; ; \quad B_k u(0) = g_k \,, \quad 1 \leq k \leq r \,,$$

where

$$P = p(d) = \sum_0^m p_j d^j \,, \quad B_k = b_k(d) = \sum_0^{m_j} b_{kj} d^j \,.$$

We assume that $p_m = 1$. If the B_k are of different orders and $r \leq m$, then (37) will have an $(m-r)$-dimensional affine space of solutions. Suppose that f has compact support, and we look for a <u>bounded</u> (as $y \to \infty$) solution. If $p(z)$ has no pure imaginary roots z, then any solution of the homogeneous ($f = 0$) problem grows or decays exponentially. There is a base of decaying solutions

$$(38) \qquad \varphi_j(y) = \frac{1}{2\pi i} \int_{\Gamma_-} e^{zy} z^{j-1} p(z)^{-1} dz \,, \quad 1 \leq j \leq m_- \,,$$

where m_- is the number of roots of p with negative real part, and Γ_- is a curve in $\operatorname{Re} z < 0$ enclosing those roots. The problem (37) with $f = 0$ will have a unique bounded solution for each $g = (g_1, \ldots, g_r) \in \mathbb{C}^r$ if and only if the matrix Q with entries

$$(39) \qquad Q_{kj} = \frac{1}{2\pi i} \int_{\Gamma_-} b_k(z) z^{j-1} p(z)^{-1} dz$$

is non-singular.

Now let Γ_+ be a curve in $\operatorname{Re} z > 0$ enclosing the remaining roots of p, and let

$$G(y) = \pm \frac{1}{2\pi i} \int_{\Gamma_{\mp}} e^{zy} p(z)^{-1} dz \ , \quad \pm y > 0, \tag{40}$$

$G(0) = 0$. Then $PG = 0$ for $y \neq 0$, G has continuous derivatives of orders up to $m-2$, and

$$G^{(m-1)}(0+) - G^{(m-1)}(0-) = 1.$$

It follows that if f has compact support, then $u = G*f$ is a bounded solution of $Pu = f$. The L^2-norm of u is bounded by the L^2-norm of f multiplied by the L^1-norm of G, and similarly for derivatives of u in terms of derivatives of G.

Under the above assumptions, the unique bounded solution of (37) is $u = u_0 + u_1$, where

$$u_1 = G*f \ , \quad u_0 = \sum c_j \varphi_j \ , \quad c = Q^{-1}[g - Bu_1] \ . \tag{41}$$

The condition that Q be non-singular is equivalent to the algebraic condition that no non-trivial linear combination of the $b_k(z)$ be a multiple of $p_-(z) = \prod (z - z_j)$, the z_j's being the roots of p with negative real part.

Consider now a constant coefficient problem in the quarter space $R_+ \times R^n \times R_+ = \{(t, x, y)\}$, of the parabolic kind considered in the previous section:

$$Pu = f \ , \quad B_k u(t, x, 0) = 0, \quad k \leq r \ , \quad \partial_t^j u(0, x, y) = v_j(x, y), \quad j < m, \tag{42}$$

where

$$P = p(\partial_t, D_x, \partial_y) = \sum_{qi + |\alpha| + j = M} p_{i\alpha j} \partial_t^i D_x^\alpha \partial_y^j \ ,$$

$$B_k = b_k(\partial_t, D_x, \partial_y) = \sum_{qi + |\alpha| + j = M_j} \partial_t^i D_x^\alpha \partial_y^j \ ,$$

and $M_j < M$, $P_{m00} = 1$ $(m = Mq^{-1})$. Let us examine the corresponding family of ODEs with

$$(43) \quad p_{\lambda,\xi}(d) = p(\lambda, \xi, d), \quad b_{k,\lambda,\xi}(d) = b_k(\lambda, \xi, d),$$

$\xi \in R^n$, $|\arg \lambda| \leq \frac{1}{2}\pi + \delta$. Under the ellipticity assumption of the preceding section, $p_{\lambda,\xi}(z)$ will have no imaginary roots. We assume that each matrix $Q(\lambda, \xi)$ for (43) is non-singular. Let $\rho = \rho(\lambda, \xi) = |\lambda|^{1/q} + |\xi|$. Homogeneity considerations show that

$$(44) \quad |Q(\lambda, \xi)^{-1}_{jk}| \leq C\rho^{M-M_k-j}$$

and that the roots z of $p_{\lambda,\xi}$ are of size $O(\rho)$ and at distance $O(\rho)$ from the imaginary axis. It follows that

$$(45) \quad |\partial_y^k G(\lambda, \xi; y)| \leq C\rho^{1-M+k} \exp(-\delta\rho|y|),$$

$$(46) \quad |\partial_y^k \varphi_j(\lambda, \xi; y)| \leq \rho^{j-M+k} \exp(-\delta\rho|y|),$$

Now reduce (42) to a system of first order in t, with the boundary conditions $Bu = 0$ incorporated in the domain of the corresponding A. To solve $(\lambda I - A)v = h$, we take the Fourier transform in the x-variables, and get a problem equivalent to the boundary value problem for (43). The representation (41) for the solution, and the estimates (44)-(46) lead to the estimates (35) of the last section. The passage to coefficients variable in x (and to operators with lower-order terms) is now made in the way which is standard for elliptic boundary conditions, and one may use the Kato-Tanabe methods to allow variation in t.

The results we have outlined for strictly parabolic problems are due to Agranovich-Vishik [3]. The authors obtain L^2-results by the Laplace transform approach, but do not reduce to a system of first order in t and so do not use the Kato-

Tanabe method for the time-dependent case. The latter approach was suggested by Browder and carried out by Lau [21] to get L^p-results.

Finally, we consider weakly hyperbolic problems. Consider (41), where P and B_k are of the above form with $q = 1$, so $m = M$ and the operators are homogeneous. We suppose that the roots τ of $p(\tau, \xi, i\eta)$ are real and $\neq 0$ for non-zero $(\xi, \eta) \in R^{n+1}$. Then $p(\lambda, \xi, z)$ has no imaginary roots z when $\xi \in R^n$ and $\text{Re}\,\lambda \neq 0$. We assume that the matrices $Q(\lambda, \xi)$ associated with (43) are non-singular for each such λ, ξ. Homogeneity considerations and the Seidenberg-Tarski theorem show that in place of (44)-(46) we get estimates

$$|Q(\lambda, \xi)^{-1}_{jk}| \leq C\sigma^N \rho^{M-M_k-j},$$

$$|\partial_y^k G(\lambda, \xi; y)| \leq C\sigma^N \rho^{1-M+k} \exp(-\delta\sigma^N \rho |y|),$$

$$|\partial_y^k \varphi_j(\lambda, \xi; y)| \leq C\sigma^N \rho^{j-M+k} \exp(-\delta\sigma^N \rho |y|),$$

where $\rho = |\lambda| + |\xi|$ and $\sigma = |\lambda^{-1} \text{Re}\,\lambda|$. These estimates give estimates of the form (36) for the system of first order in time. Thus the problem is weakly hyperbolic. The passage to coefficients varying with x and problems on a cylindrical domain is carried out once more by a patching argument; see [8].

5. Further remarks.

We have emphasized the "strictly parabolic" and "weakly hyperbolic" problems in this summary because for such problems the qualitative features one expects can be derived, ultimately, from resolvent estimates; moreover the resolvent estimates obtained are sufficiently stable to carry over to operators with variable coefficients. For strictly hyperbolic equations, for example, semigroup or constructive (Fourier integral operator) methods seem to be necessary for optimal results.

Boundary conditions for mixed problems have as ancestors the Lopatinskii-Shapiro conditions of elliptic theory. Conditions for constant coefficient strictly hyperbolic equations were given by Agmon [1] and for general constant-coefficient Petrowskii-correct problems by Hersh [16], both in a quarter space.

The Agranovich-Vishik-Lau treatment of parabolic problems described here obviously relies heavily on the (mixed) homogeneity of the principal terms of the operators, and it is the corresponding (weighted) ellipticity which makes the passage from constant to variable coefficients relatively straightforward. There are problems which are "strictly parabolic" in the present sense which are not in the Agranovich-Vishik class, for example

$$(\partial_t^2 - \Delta\partial_t - \Delta)u = f \ , \quad t > 0 \ , \quad x_n > 0 \ ,$$

$$u\big|_{x_n = 0} = 0 \ , \quad u\big|_{t=0} = v_0 \ , \quad \partial_t u\big|_{t=0} = v_1 \ .$$

This example is due to Donaldson, who has made an extensive study of the type of non-homogeneous problems for which the corresponding estimates are stable enough to pass to variable coefficients [14].

The study of weakly hyperbolic problems in [8], outlined here, also relies heavily on homogeneity considerations. Again there are problems weakly hyperbolic in the abstract sense but not of this type, but there seems to be no general study of such problems.

The work of Leray-Ohya for the Cauchy problem shows that the "ellipticity" assumption (non-vanishing of the characteristic roots) in our discussion should not be necessary; it would be interesting to remove it from the present approach.

Results for the time-dependent weakly hyperbolic Cauchy problem can be obtained by the methods outlined here, but the time-dependent mixed problem remains to be done. Some refinement of the Kato-Tanabe method might suffice; an alternative (and possibly more promising) approach is to refine the method outlined at the end of section 2 along the lines of the refinement of the Tanabe-Sobolevskii method as given in section 2.

Bibliography

1. Agmon, S., Problemes mixtes pour les équations hyperboliques d'ordre supérieur, Les Équations aux Dérivées Partielles, ed. CNRS, Paris 1962.

2. Agmon, S., and L. Nirenberg, Properties of solutions of ordinary differential equations in a Banach space, Comm. Pure Appl. Math. 16 (1963), 121-139.

3. Agranovich, M.S., and M.I. Vishik, Elliptic problems with a parameter and parabolic problems of general type, Uspehi Mat. Nauk 19 (1963), 53-161; Russian Math. Surveys 19 (1963), 53-159.

4. Baouendi, M.S., and C. Goulaouic, paper appeared in Communications in Part. Diff. Equ., 1 (1976).

5. Barbu, V., Les semi-groupes distributions differentiables, C.R. Acad. Sci. Paris 267 (1968), A875-A878.

6. Beals, R., On the abstract Cauchy problem, J. Functional Analysis 10 (1972), 281-299.

7. ________, Semigroups and abstract Gevrey spaces, J. Functional Analysis 10 (1972), 300-308.

8. ________, Hyperbolic equations and systems with multiple characteristics, Arch. Rat. Mech. Anal. 48 (1972), 123-152.

9. ________, Abstract evolution equations of weakly hyperbolic type, to appear.

10. Chazarain, J., Problèmes de Cauchy abstraits et applications à quelques problèmes mixtes, J. Functional Analysis 7 (1971), 386-446.

11. Da Prato, G., Semigruppi regolarizzabili, Richerche Mat. 15 (1966), 223-248.

12. ________, Somme de génerateurs infinitésimaux de classe C_0, Rend. Acad. Lincei 45(1968), 14-21.

13. ________, and E. Giusti, Equazioni di evoluzione in L^p, Ann. Scuola Norm. Sup. Pisa 21(1967), 485-505.

14. Donaldson, T., A Laplace transform calculus for partial differential operators, Memoirs Amer. Math. Soc. 143, Providence 1974.

15. Gelfand, I. M., and G. E. Shilov, Generalized Functions, vol. 3, Gos. Izdat Fiz.-Mat. Lit., Moskva 1958; Academic Press, New York 1967.

16. Hersh, R., Boundary conditions for equations of evolution, Arch. Rat. Mech. Anal. 16 (1964), 243-264.

17. Hille, E., and R. S. Phillips, Functional Analysis and Semigroups, Amer. Math. Soc., Providence 1956.

18. Kato, T., and H. Tanabe, On the abstract evolution equations, Osaka Math. J. 14(1962), 107-133.

19. Krein, S. G., Differential Equations in a Banach Space, Izdat. Nauka, Moscow 1963; Transl. Math. Monographs, vol. 29, Amer. Math. Soc., Providence 1971.

20. Larsson, E., Generalized distribution semigroups of bounded linear operators, Ann. Scuola Norm. Sup. Pisa 21 (1967), 137-159.

21. Lau, R., Elliptic equations with a parameter and applications to parabolic problems, thesis, Yale University 1967.

22. Leray, J., and Y. Ohya, Systèmes linéaires, hyperboliques non-stricts, Deuxième Colloq. l'Anal. Fonct., Centre Belg. Rech. Math., Louvain 1964.

23. Lions, L. L., Les semigroupes distributions, Portug. Math. 19 (1960), 141-164.

24. Lyubich, Ya. I., The classical and local Laplace transformations in an abstract Cauchy problem, Uspehi Mat. Nauk 21 (1966), 1-52; Russian Math. Surveys 21 (1966), 1-52.

25. Ohya, Y., Le problème de Cauchy pour les équations hyperboliques à caractéristiques multiples, J. Math. Soc. Japan 16 (1964), 268-286.

26. Pazy, A., On the differentiability and compactness of semigroups of linear operators, J. Math. Mech. 17 (1968), 1131-1141.

27. Phillips, R.S., A note on the abstract Cauchy problem, Proc. Nat. Acad. Sci. U.S.A. 40 (1954), 244-248.

28. Seidenberg, A., A new decision method for elementary algebra, Ann. Math. 60 (1954), 365-374.

29. Sobolevskii, P.E., Equations of parabolic type in a Banach space, Trudii Mosk. Mat. Obshch. 10 (1961), 297-350; Amer. Math. Soc. Trans. (2) 48 (1965), 1-62.

30. ________, A certain type of differential equation in a Banach space, Differ. Uravn. 4 (1968), 2278-2280.

31. Steinberg, S., Existence and uniqueness of solutions of hyperbolic equations which are not necessarily strictly hyperbolic,

32. Tanabe, H., On the equations of evolution in a Banach space, Osaka Math. J. 12 (1960), 363-376.

33. Treves, F., Basic Linear Partial Differential Equations, Academic Press, New York 1975.

34. Zaidman, S., Some asymptotic theorems for abstract differential equations, Proc. Amer. Math. Soc. 25 (1970), 521-5.

HYPERBOLIC DIFFERENTIAL EQUATIONS AND WAVES

G.F.D. Duff

Department of Mathematics, University of Toronto,
Toronto, Canada

CONTENTS

Garnir (ed.), Boundary Value Problems for Linear Evolution Partial Equations. 27-155.

PREFACE. These notes on Hyperbolic Differential Equations and Waves are centred about the existence and properties of wave solutions for the full space problem and for the half space or mixed problem. Thus the emphasis is on elementary solutions, the geometry of wave surfaces, singularities and lacunas in the case of propagation in R^n . The mixed initial boundary problem is studied from several points of view in the constant coefficient case followed in Chapter 6 by a description of the extensive recent work on the variable coefficient problem. The exposition is intended to be self-contained as far as possible, but in a topic that has attracted so much interest in recent years some references to the current literature have been necessary for reasons of space. By its nature the topic of hyperbolic equations spans the range between theoretical analysis and applied techniques so an attempt has been made to show the interest of both as well as their mutual interaction.

My thanks are due to Professor Garnir and the sponsors of the NATO Advanced Study Conference on hyperbolic differential equations and wave propagation for the invitation to give these lectures and for their cooperation in preparing these notes for publication.

G.F.D. Duff

CHAPTER 1. HYPERBOLIC EQUATIONS AND CAUCHY'S PROBLEM

1.1 Historical introduction

The classification of linear partial differential equations into the three well known types, elliptic, parabolic and hyperbolic, corresponds to basic properties and problems which are widely different for the equations of the various types. In these notes the solutions of hyperbolic equations, often known as waves, will be studied in the context of the initial value problem, or Cauchy problem, while recent developments in the theory and application of the mixed initial and boundary value problems will be described. The study of mixed problems has received much attention during the past fifteen years and the initiation of this emphasis was perhaps stimulated by the remarks of L. Gårding in his review of linear partial differential equations at the Edinburgh International Mathematical Congress in 1958.

We begin with the Cauchy problem in the case of constant coefficients. Historically, the d'Alembert solution of the initial value problem for the one space dimensional wave equation,

$$\frac{\partial^2 u}{\partial t^2} = c^2 \frac{\partial^2 u}{\partial x^2} ,$$

with $u(x,0) = \phi(x)$, $\frac{\partial u}{\partial t}(x,0) = \psi(x)$, was the first "wave formula"; it is

$$u(x,t) = \frac{1}{2}\{\phi(x+ct)+\phi(x-ct)\} + \frac{1}{2c}\int_{x-ct}^{x+ct} \psi(x)ds .$$

Already the properties of continuous dependence on data, propagation of singularities along characteristics, and finite domain of dependence are visible in this d'Alembert formula.

Wave propagation in higher dimensions was studied in detail by Hadamard with his method of the 'finite part' of a divergent integral, and using also the method of descent which is related to the "clean cut' wave propagation property for an odd number of space dimensions, which we study below under 'lacunas'. Formulas for the 'elementary solution' were given by Hadamard and Herglotz, but with the advent of distributions according to L. Schwartz, a more convenient and comprehensive notation became available, and we make use of it as follows. Let the Laplacian Δ be defined in R^n as

$$\Delta = \Delta_n = \frac{\partial^2}{\partial x_1^2} + \frac{\partial^2}{\partial x_2^2} + \cdots + \frac{\partial^2}{\partial x_n^2} ,$$

where $\{x_1, x_2, \ldots, x_n\}$ are Cartesian coordinates. Let $\delta(t)$ denote the symbolic Dirac delta distribution, defined by

$$\int f(t)\delta(t)dt = f(0) ,$$

and $\delta_n(x)$ the corresponding n dimensional Dirac distribution. Then the wave equation in R^n :

$$\frac{\partial^2 u}{\partial t^2} - \Delta u = \delta_n(x)\delta(t)$$

has the elementary solution

$$\delta^{(n-3)/2}(t^2 - r^2)H(t) ,$$

where $H(t) = 0(x < 0)$ and $1(x \geq 0)$ is the Heaviside function. Also the fractional order of differentiation is defined using the Riemann - Liouville fractional integral of order $\alpha \in \mathbb{C}$;

$$I^{\alpha}f(x) = \frac{1}{\Gamma(\alpha)} \int_a^x (x-t)^{\alpha-1} f(t)dt .$$

Indeed an integration theory for the wave equation depending on a generalization of this fractional integral has been put forward by M. Riesz, but we shall use only the more adaptable method of distributions and Fourier transforms as described in Shilov (1), or Gelfand and Shilov (1, vols. 1 and 3).

In the general theory of linear partial differential operators, as described by Hörmander and others, hyperbolic polynomials play a special rôle as the operators which permit a well posed Cauchy problem. A condition of real and distinct roots for homogeneous hyperbolic polynomials generalizes the algebraic character of the wave operator $\tau^2 - \rho^2$. We study here the elementary solutions and their singularities which lie on well defined wave fronts, separating regions of analytically distinct character that are related to lacunas.

Symmetric hyperbolic systems of first order equations are related by elimination processes to hyperbolic higher order equations and we shall examine the Riemann matrix for such a system with constant coefficients. The Riemann matrix or elementary solution also shares the wave front geometry and the analytic character of the elementary solution of the higher order equation. We also study in this chapter hyperbolic systems of second order equations and anisotropic waves of which anisotropic elastic waves are a prominent example. Then we take up the properties of certain 'regular' equations and systems which are not hyperbolic and may therefore exhibit such 'parabolic' characteristics as wave propagation with infinite speed. We postpone to the next chapter consideration of mixed initial boundary value problems in which the spatial domain is bounded by a hyperplane boundary, so that boundary conditions and reflection phenomena arise.

1.2 Elementary properties and solutions

For equations with constant coefficients, we make repeated use of the Fourier transform in R^n :

$$Ff(\xi) = \hat{f}(\xi) = \frac{1}{(2\pi)^{\frac{n}{2}}} \int_{R^n} e^{ix\cdot\xi} f(x)\,dx$$

and its inverse

$$F^{-1}f(x) = f(x) = \frac{1}{(2\pi)^{\frac{n}{2}}} \int_{R^n} e^{-ix\cdot\xi} f(\xi)d\xi ,$$

where $x = (x_1, \ldots, x_n) \in R^n$, $\xi = (\xi_1, \ldots, \xi_n) \in R^n$. Fourier transforms of distributions may be defined by considering the space S: of all $\phi \in C^\infty(R^n)$ such that $\sup_x |x^\beta D^\alpha \phi(x)| < \infty$, where $\alpha = (\alpha_1, \ldots, \alpha_n$, $\beta = (\beta_1, \ldots, \beta_n)$ are multi indices (Hormander, Gelfand and Shilov , vol. 1). The set S' of temperate distributions may be defined as the dual of S , i.e. the set of continuous linear forms on S . Then the Fourier transform for $u \in S'$ is defined as $\hat{u}(\phi) = u(\hat{\phi})$ for all $\phi \in S$. The Fourier transform maps S onto S and S' onto S' , isomorphically, and is continuous in the weak topology on S . By Parseval's theorem, if $u \in L^2(R^n)$, then $\hat{u} \in L^2(R^n)$ and $\|u\|_2 = \|\hat{u}\|_2$.

Note also that if $u \in \mathcal{E}'$ then its Fourier transform is the function $\hat{u}(\xi) = u_x(e^{-ix\cdot\xi})$, which defines an entire analytic function of ξ called the Fourier - Laplace transform. By the Paley - Wiener theorem (Hörmander 1, p. 21) it follows that an entire analytic function $u(\zeta)$ is the Fourier Laplace transform of a distribution with support in the sphere $|x| \leq A$ if and only if there are constants C , N such that $|u(\zeta)| \leq C(1 + |\zeta|)^N e^{A|\mathrm{Im}\,\zeta|}$. We shall employ this theorem to estimate the domain of dependence, or the support, of the distributions representing waves propagated with finite velocity.

As the Fourier transform of the Dirac distribution is a constant, the Fourier transform of $-i\frac{du}{dx}$ is $\xi\hat{u}(\xi)$, and the Fourier transform of a convolution $u_1 * u_2$ is the product $\hat{u}_1 \hat{u}_2$, it follows that the smoothness of a function u is reflected in the polynomial smallness of $\hat{u}$ at infinity, and vice versa. We shall frequently use Fourier transforms of polynomials which thus represent distributions of finite order.

Example 1. The elementary solution of the wave equation $u_{tt} = \Delta u + \delta_n(x)\delta(t)$. Here

$$\hat{u}_{tt}(\xi, t) = -\xi^2\hat{u} + \delta(t) .$$

Assuming $u \equiv 0$ for $t < 0$, we find after one integration over t that

$$\hat{u}(\xi\,,0) = 0 \quad , \quad \hat{u}_t(\xi\,,0+) = 1 \,.$$

Hence we require for $t > 0$ this solution of the ordinary differential equation $\hat{u}_{tt} + \xi^2\hat{u} = 0$, namely

$$\hat{u}(\xi\,,t) = \frac{\sin|\xi|t}{|\xi|} \,.$$

Consequently,

$$\hat{u}(x\,,t) = \frac{1}{(2\pi)^n}\int_{R^n} e^{ix\cdot\xi}\,\frac{\sin|\xi|t}{|\xi|}\,d\xi$$

$$= \frac{1}{(2\pi)^n}\int_{R^n} e^{ix\cdot\xi}\,\frac{\sin|\xi|t}{|\xi|}\,|\xi|^{n-1}\,d|\xi|d\Omega_n \,,$$

where $\quad d\Omega_n = \sin^{n-2}\theta d\theta d\Omega_{n-1}$

$$= -(1-\mu^2)^{\frac{n-3}{2}}\,d\mu d\Omega_{n-1} \,,$$

where $0 \le \theta \le \pi$ and hence $-1 \le \mu = \cos\theta \le 1$. Here $d\Omega_n$ is the solid angle element in n space dimensions, $d\Omega_{n-1}$ in $n-1$ dimensions while ω_n , ω_{n-1} denote the total solid angles. Thus we find, formally,

$$u(x\,,t) = \omega_{n-1}\int_0^\infty M(|x||\xi|)\sin(|\xi|t)|\xi|^{n-2}\,d|\xi| \,,$$

where

$$M(s) = \frac{1}{(2\pi)^n}\int_{-1}^{1} e^{is\mu}(1-\mu^2)^{\frac{n-3}{2}}\,d\mu \,.$$

As the factor $|\xi|^{n-2}$ in the expression above may lead to divergence of the integral, in the classical though not in the distribution sense, we consider the following device which applies for n odd. We write

$$u(x\,,t) = \omega_{n-1}(-1)^{\frac{n-1}{2}}\left(\frac{\partial}{\partial t}\right)^{n-2}\int_0^\infty M(|x||\xi|)\cos(|\xi|t)d|\xi|$$

and observe that the inner integral can be written in the form of a Fourier integral

$$\frac{1}{2}\int_{-\infty}^{\infty} M(|x||\xi|)e^{i|\xi|t}\,d|\xi| = \frac{1}{2|x|}\int_{-\infty}^{\infty} M(s)e^{ist/|x|}\,ds \,.$$

From the definition of $M(s)$ it is clear that $(2\pi)^{n/2} M(s)$ and $(1-\mu^2)^{(n-3)/2} H(1-\mu^2)$ are a pair of Fourier transforms, so that the above Fourier integral equals

$$\frac{1}{(2\pi)^n} \frac{1}{2|x|} \left(1 - \frac{t^2}{|x|^2}\right)^{\frac{n-3}{2}} H\left(1 - \frac{t^2}{r^2}\right) .$$

Thus the elementary solution u now takes the form

$$u(x\,,t) = \frac{\omega_{n-1}}{(2\pi)^n}(-1)^{\frac{n-1}{2}} \left(\frac{\partial}{\partial t}\right)^{n-2} \left\{ \frac{1}{2|x|} \left(1 - \frac{t^2}{|x|^2}\right)^{\frac{n-3}{2}} H\left(1 - \frac{t^2}{|x|^2}\right) \right\} .$$

Since the polynomial in t being differentiated has degree $n-3$, the coefficient or term that accompanies H in the Leibnitz expansion will be zero. Since $H' = \delta$, the support of this distribution lies on the light cone $t = |x|$.

Thus if we consider the initial value problem with data $u(x\,,0) = 0$, $u_t(x\,,0) = \psi(x)$, the solution can be shown to be (Courant 1, vol. II, p. 686), (Shilov 1, pp. 154, 288)

$$u(x\,,t) = \frac{1}{(n-2)!} \frac{\partial^{n-2}}{\partial t^{n-2}} \int_0^t (t^2 - r^2)^{\frac{n-3}{2}} rQ(x\,,r)dr ,$$

where the mean value

$$Q(x\,,r) \equiv \frac{1}{\omega_n} \int_{\Omega_n} \psi(x + \alpha r) d\Omega_\alpha , \quad |\alpha| = 1 .$$

It can be shown that the same formula gives the solution for n even either by direct calculation or by the method of descent from n dimensions to $n-1$ dimensions.

For comparison we include here the elementary solution of the heat flow equation in n dimensions, and that of the Stokes equation describing viscous flow.

Example 2. Let $u(x\,,t) \equiv 0$ for $t < 0$, and

$$\frac{\partial u}{\partial t} - \Delta_n u = \delta(x)\delta(t) .$$

Then

$\hat{u}(\xi\,,t)$ satisfies the ordinary differential equation

$$\frac{\partial \hat{u}}{\partial t} + |\xi|^2 \hat{u} = \frac{\delta(t)}{(2\pi)^{\frac{n}{2}}}, \quad \hat{u}(\xi, t) \equiv 0, \quad t < 0.$$

Thus

$$\tilde{u}(\xi, t) = \frac{e^{-|\xi|^2 t}}{(2\pi)^{\frac{n}{2}}}$$

so that

$$u(x, t) = \frac{1}{(2\pi)^n} \int_{R^n} e^{-ix\cdot\xi - |\xi|^2 t} d\xi$$

$$= \frac{1}{(2\sqrt{\pi t})^n} e^{-|x|^2/4t} H(t).$$

Example 3. The Stokes equation, which we shall consider for one space dimension only, combines the properties of wave propagation and diffusion and it occurs in many branches of fluid mechanics and vibration theory (Lagerstrom et al 1), (Ewing et al 1, p. 272). The equation is

$$u_{tt} = u_{xx} + u_{xxt} + \delta(x)\delta(t).$$

It is more convenient to find this elementary solution by using the Laplace transform

$$\hat{u}(t, y) = \int_{-\infty}^{\infty} e^{-xy} u(t, x) dx.$$

Thus we obtain for $\tilde{u}(t, y)$ the ordinary differential equation

$$\tilde{u}_{tt} = y^2 \tilde{u} + y^2 \tilde{u}_t + \delta(t)$$

which has solution

$$u(t, y) = Ae^{\lambda_1 t} + Be^{\lambda_2 t}.$$

Here $\lambda_1 \lambda_2$ are the roots of the characteristic equation

$$\lambda^2 - y^2\lambda - y^2 = 0,$$

so $\lambda_1 = \frac{y}{2}[y - \sqrt{y^2 + 4}]$

$$\lambda_2 = \frac{y}{2}[y + \sqrt{y^2 + 4}]$$

The initial conditions $\tilde{u}(0,y) = 0$ and $u_t(0,y) = 1$ suffice to determine A and B so that

$$\tilde{u}(t,y) = \frac{e^{\lambda_1 t} - e^{\lambda_2 t}}{\lambda_1 - \lambda_2}$$

$$= e^{\frac{y^2 t}{2}} \frac{\left(e^{\frac{1}{2}y\sqrt{y^2+4}t} - e^{-\frac{1}{2}y\sqrt{y^2+4}t}\right)}{y\sqrt{y^2+4}} .$$

The inversion integral is

$$u(t,x) = \frac{1}{2\pi i}\int_{c-i\infty}^{c+i\infty} \tilde{u}(t,y)e^{xy}\,dy$$

and we note that the expression for $\tilde{u}(t,y)$ is unchanged by a permutation of the square root $\sqrt{y^2+4}$ so that this contour of integration can be deformed past the branch points at $\pm 2i$ without a contribution arising there. However a two sheeted Riemann surface for the square root function is necessary to evaluate the integral asymptotically. We shall use the method of steepest descents (Erdelyi,1) for which we must examine each of the two exponential terms separately.

We set $\alpha = \frac{x}{t}$ and observe that the first exponent is then

$$g_1(\eta)t = (\alpha y + \lambda_1(y)t) .$$

The zeros of $g_1(y)$ are found from the cubic equation $y^3 + \frac{1}{2}\alpha y^2 + 4y + 2(\alpha - \frac{1}{2}) = 0$ and thus a real root exists. For t positive, the second exponential term does not contribute to the leading asymptotic order of magnitude, and so is omitted. A detailed calculation may be found in (Duff and Ross, 1) and we quote here only some results for $\alpha = \frac{x}{t} \sim 0$, $\alpha \sim \infty$ and $\alpha \sim 1$.

For $\alpha \sim 0$ we find

$$u(t,x) \approx \frac{1}{2\sqrt{3\pi t}} \exp\left(-t + 3.2^{-2/3} t^{1/3} x^{2/3}\right)$$

For $\alpha \sim \infty$,

$$u(t,x) \approx \frac{1}{2\sqrt{3\pi t}} \exp\left(t - \frac{x^2}{4t}\right)$$

and for $\alpha \sim 1$, that is $x \sim t$, we find

$$u(t\,,x) \approx \frac{1}{\sqrt{2\pi t}} \exp\left[-0.579t\left(\frac{x}{t}-1\right)^2\right] .$$

We may observe that along the line $x = t$, or $\alpha = 1$, the decrease of this elementary solution is algebraic only, not exponential, just like the behaviour of the heat flow solution for $x = 0$. The lines $x = \pm t$ are called subcharacteristics for the Stokes equation; they are determined by the linear terms in the expansions of $\lambda_1(y)$ and $\lambda_2(y)$ for y small. Along each subcharacteristic $x = \pm t$ travels a slowly diffusing wave so Stokes equation displays the properties of both hyperbolic and parabolic types.

1.3 Finite propagation speed

Because the property of wave propagation with finite speed is fundamental for hyperbolic equations, we include here the uniqueness proof of this property. Let

$$L(u) = u_{tt} - \Delta u = u_{tt} - \sum_{j=1}^{n} u_{x_j x_j} = 0$$

and let $S\ \phi(x\,,t) = 0$ denote an initial surface which is spacelike:

$$\phi_t^2 - (\nabla\phi)^2 = \phi_t^2 - \sum_{j=1}^{n} (\phi_{x_j})^2 > 0 .$$

Suppose that u and u_ϕ vanish on S (zero Cauchy data) and let P be a point such that the retrograde characteristic cone with vertex P meets the interior of S and is cut off by S from the "past". Then we assert that $u(P) = 0$. (Figure 1).

For the proof we require the identity

$$2u_t L(u) = -2\sum_{j=1}^{n} \left(u_t u_{x_j}\right)_{x_j} + \sum_{j=1}^{n} \left(u_{x_j}\right)^2_t + \left(u_t^2\right)_t ,$$

which we integrate over region R bounded by the retrograde cone and S. Since the expression on the right is a divergence,

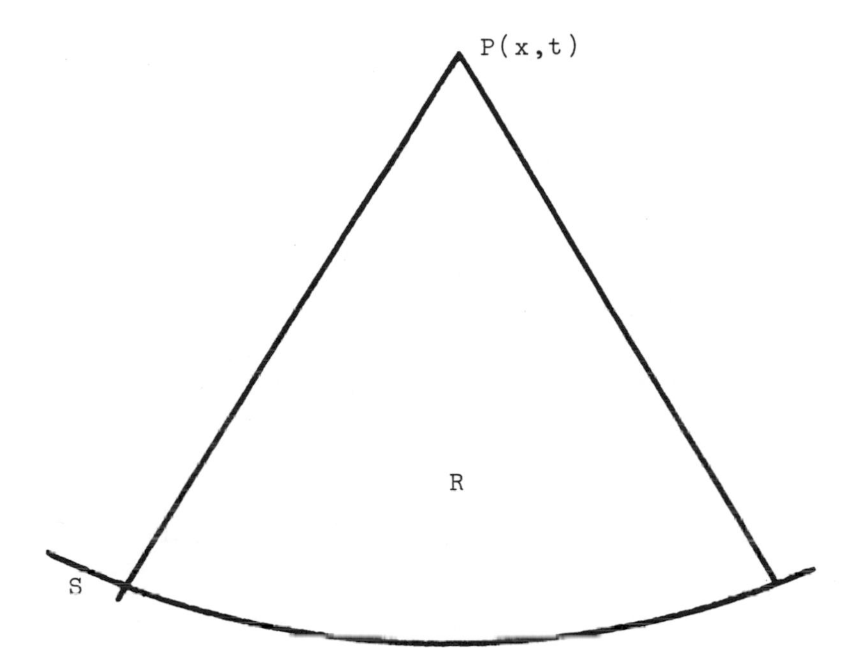

Fig. 1. Domain of dependence of P(x,t).

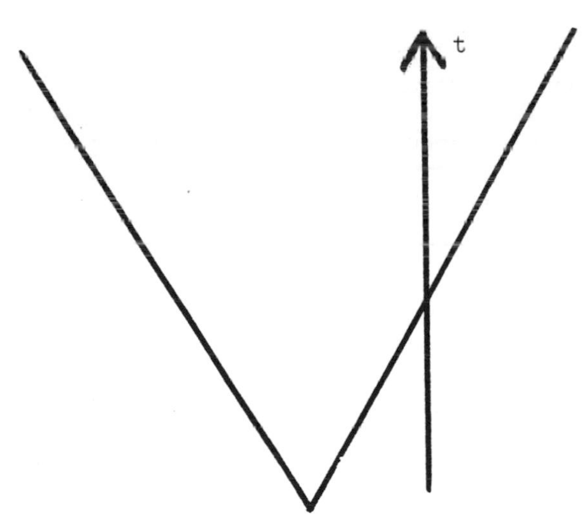

Fig. 2. Intersection of a sharp or diffuse wave front with a world line.

we can apply Gauss' Theorem, and obtain

$$0 = \int_{\text{cone}} \left(u_t^2 t_n + \sum_{j=1}^{n} u_{x_j}^2 t_n - 2 \sum_{j=1}^{n} u_t u_{x_j} x_{jn} \right) dS$$

$$= \int_{\text{cone}} \frac{1}{t_n} \sum_{j=1}^{n} \left(u_{x_j} t_n - u_t x_{jn} \right)^2 dS ,$$

where (x_{jn}, t_n) denotes the unit normal to the cone, with $t_n^2 = \sum_{j=1}^{n} x_{jn}^2$ which expresses the characteristic or null property of the cone. Note also that the vanishing of Cauchy data removes the integral over S from the equation. From the positive definiteness of the last expression above we conclude that $u_{x_j} t_n - u_t x_{jn} = 0$, $j = 1 \ldots n$. As each of these expressions is a derivative tangential to the n-dimensional cone, it follows that all tangential derivatives vanish so that u is constant on the cone. Finally, therefore, being zero on S , u must also be zero at P .

The foregoing proof of uniqueness can easily be extended to second order equations with variable coefficients which can be interpreted as defining an indefinite Riemannian metric. Again, the null cone or characteristic cone is defined by the geodesic lines of this metric which are null (of zero length) in the metric, and which pass through the given point P (Courant, 1, p. 564).

The domain of dependence of u on the Cauchy data is now clearly defined by the retrograde cone with vertex P . Any change of data on S , which does not affect values within or on the cone, does not change $u(P)$, so we note that the domain of dependence of P is defined by the surface and interior of the cone. For odd space dimensions, the actual domain of dependence is the surface of the cone only, while for even space dimensions the interior of the cone, and not the surface, is the actual domain.

Another description of this difference in behaviour of the wave equation in even and odd dimension is as follows. In the odd dimensional case, a short signal emitted is received as an equally short signal. Thus we speak of sharp wave propagation, and sharp wave fronts which change instantaneously at a given point of space signalling the arrival of an abrupt signal. The elementary solution in these cases contains Dirac distributions or their derivatives. We say that in odd space dimensions the wave equation satisfies Huyghens' Principle (Courant, 1, p. 208) of sharp wave propagation. (Figure 2). We also speak of the interior of the wave cone as a <u>lacuna</u>, or gap, in this case. In contrast, the wave equation in even space dimensions has an elementary solution with support the interior of the wave cone. Thus the arrival or onset of a signal is sharp, but its ending trails on forever at a given point of space, even if the emitted signal terminates. We say that this wave propagation is diffuse.

For the initial value problem there is an extended form of Huyghens' Principle, due to Lax (Courant, 1, p. 735) which describes the singularities of the solution. Since the elementary solution is singular only on the wave cone, it follows that singularities (lack of smoothness) of the data are propagated only along wave cones.

1.4 Mixed initial and boundary value problems

The existence proof for solutions of Cauchy's problem for a second order hyperbolic differential equation in n variables was given by Sobelev and by Kryzanski and Schauder in 1936. The latter authors also were able to prove existence for the mixed problem in which one boundary condition is specified as well as two initial conditions. We shall discuss their method and particularly the L^2 estimates used.

We begin with the problem in a cylinder, $D = R \times [0, T]$, where $R \subset R^n$. Let S be the boundary surface $S = \partial R$, and let $\Sigma = S \times [0, T]$, while R_0 and R_T denote initial and final positions of the space region R. (Figure 3). Consider the hyperbolic equation

$$L(u) = u_{tt} - \Delta u + b\cdot\nabla u + cu = f \quad \text{in} \quad D = R \times [0, T] \tag{1.4.1}$$

We impose Cauchy data

$$u(x, t) = u_0(x), \quad u_t(x, t) = u_1(x), \quad x \in R \tag{1.4.2}$$

and a boundary condition of the Dirichlet type

$$u(x, t) = g(x, t), \quad (x, t) \in S \times [0, T] \tag{1.4.3}$$

or of the Neumann type or more general Robin type

$$\frac{\partial u}{\partial n} + hu = g(x, t), \tag{1.4.4}$$

where g, h are given functions on $S \times [0, T]$. We assume for the moment sufficient smoothness of R, f, g, h. By subtracting if necessary a suitable function from u, we can assume that the boundary conditions are homogeneous — that is that $g(x, t) \equiv 0$ on $S \times [0, T]$. (This may involve some minor adjustment of smoothness conditions at a later stage).

Multiply (1.4.1) by u_t and integrate over the domain $D = R \times [0, T]$. We find

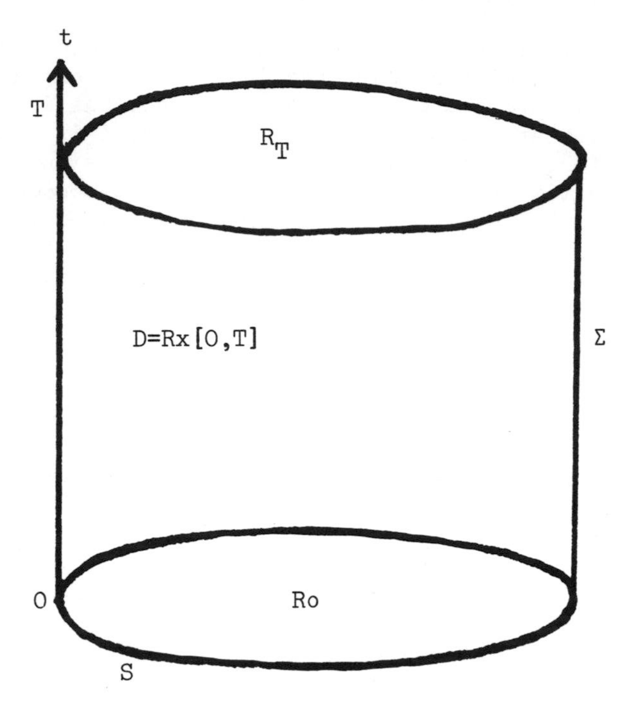

Fig. 3. Mixed problem on a space-time cylinder.

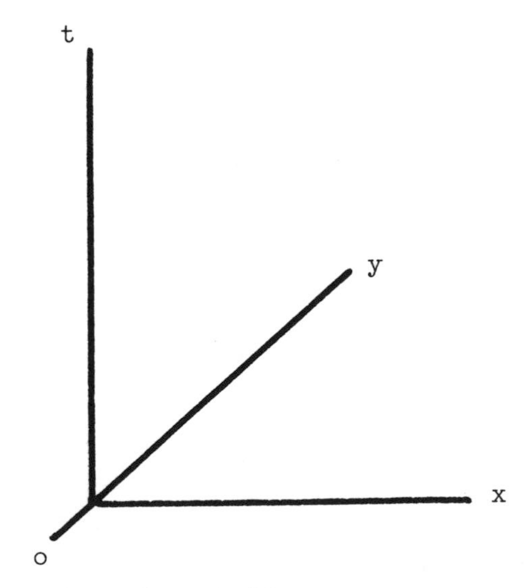

Fig. 4. Mixed problem in a half-space x>o.

$$\iint_D u_t\{u_{tt} - \Delta u + b\cdot\nabla u + cu - f\}dxdt$$

$$= \iint_D \{\tfrac{1}{2}\frac{d}{dt}(u_t^2 + (\nabla u)^2) + b\cdot\nabla u u_t + cuu_t - fu_t\}dxdt$$

$$- \iint_{S\times[0,T]} u_t \frac{\partial u}{\partial n}\, dSdt \qquad (1.4.5)$$

$$= \frac{1}{2}E(T) - \frac{1}{2}E(0) + \iint_D (b\cdot\nabla u + cu - f)u_t\, dxdt$$

$$- \iint_{S\times[0,T]} u_t \frac{\partial u}{\partial n}\, dSdt ,$$

where

$$E(t) = \int_{R_t} \{u_t^2 + (\nabla u)^2\}\, dx ,$$

and ∇u denotes the n-dimensional gradient vector with components u_{x_i}.

From the homogeneous boundary conditions we see that for the Dirichlet condition u_t vanishes since $u \equiv 0$ for all t, on S. For the Neumann condition we require a further transformation of the dependent variable. Let n denote a space coordinate normal to S, and let h be defined in a neighbourhood of S as a function of $x \in S$ and n. Then set $v = u\exp(-\int h dn)$ and observe that v satisfies the condition $\frac{\partial v}{\partial n} = 0$ on S. Since the above transformation leads to a different equation for v of the same type, we can assume this done beforehand. Then reverting to u as variable, we have $\frac{\partial u}{\partial n} = 0$ on $S\times[0,T]$. In both of these cases, the surface integral will vanish.

We now estimate $E(t)$ for $0 < t < T$ in terms of the data and the integral

$$H(t) = \int_{R_t} u^2\, dx . \qquad (1.4.6)$$

We note that for $t = 0$ we have $H(0) = \int u_0^2\, dx$ which is given, while

$$\frac{dH(t)}{dt} = 2\int_{R_t} uu_t\, dx$$

$$\leq 2\, H(t)^{\frac{1}{2}}E(t)^{\frac{1}{2}} .$$

Thus

$$\frac{d}{dt} H(t)^{\frac{1}{2}} \leq E(t)^{\frac{1}{2}}$$

whence on integration with respect to t,

$$H(t)^{\frac{1}{2}} \leq H(0)^{\frac{1}{2}} + \int E(t)^{\frac{1}{2}} \, dt .$$

Applying Schwarz inequality to the integral term we easily find

$$H(t) \leq 2H(0) + 2t\, E(t) . \tag{1.4.7}$$

All terms in the integral over D on the right of the main estimate can now be majorized by integrals of $E(t)$ or $H(t)$, with coefficients depending on the data. Thus

$$\iint_D b \cdot \nabla u \cdot u_t \, dxdt \leq \max b \int_0^T dt \int_R u_t \cdot \nabla u \, dx$$
$$\leq \max b \int_0^T \frac{1}{2} E(t) dt$$

while

$$\iint_D cuu_t \, dxdt \leq \max c \int_0^T H^{\frac{1}{2}}(t) E^{\frac{1}{2}}(t) dt$$
$$\leq \max c \left(\int_0^T H(t) dt \right)^{\frac{1}{2}} \left(\int_0^T E(t) dt \right)^{\frac{1}{2}}$$
$$\leq \max c \left(\int_0^T \left(2H(0) + TE(t) \right) dt \right)^{\frac{1}{2}} \left(\int_0^T E(t) dt \right)^{\frac{1}{2}}$$
$$\leq \operatorname{Max} c \left(2T\, H(0) + (2T+1) \int_0^T E(t) dt \right)$$

and

$$\iint_D u_t f \, dxdt \leq \int_0^T E(t)^{\frac{1}{2}} \| f \|_D \, dt$$
$$\leq \left(\int_0^T E(t) dt \right)^{\frac{1}{2}} \left(\int_0^T \| f \|_D^2 \, dt \right)^{\frac{1}{2}} ,$$

where

$$\| f \|_D^2 = \int_D f^2 \, dx .$$

Supposing then that b, c are bounded, we obtain an estimate for $E(t)$ of the form

$$\begin{aligned} E(T) &\le E(0) + C_1 H(0) + C_2 \int_0^T E(t)dt \\ &= C_0 + C_2 \int_0^T E(t)dt . \end{aligned} \tag{1.4.8}$$

This inequality of Gronwall type is easily "integrated" to yield

$$E(T) \le C_0' \, e^{C_2 T} .$$

(Set $Y(T) = \int_0^T E(t)dt$, and employ a comparison theorem for the first order differential inequality satisfied by Y .)

Thus $E(T)$ is bounded by a number depending only on T , the data and coefficients of the equation.

To complete the uniqueness theorem is now very easy, for this estimate applied to the difference of two solutions gives $E(T) \equiv 0$, whence $u \equiv 0$.

To complete the existence theorem several methods are available but we shall describe the original method using analytic approximation (Kryzanski - Schauder, 1, Courant, 1, vol. II, p. 670). Given data, coefficients and domain of k degrees of differentiability, we first construct a sequence of analytic problems that approximate the given one uniformly for $0 \le t \le T_1$ in the sense of the uniform norm for all derivatives up to and including k . Analytic solutions exist (except that a "corner condition" must hold, see below) and these are defined in a uniform t - interval. The foregoing estimates are now applied to the solution and its derivatives up to and including order k , and they show uniform convergence in the H norm and E norm to a limit; that is, convergence in the space W_2^k . By the Sobolev lemma, all derivatives of order less than $k - \frac{n}{2}$ will converge uniformly, and it will follow that a solution with derivatives of orders less than $k - \frac{n}{2}$ exists in a certain time interval $0 \le t \le T_1$. Repeating the entire procedure, we can extend the time interval for which the solution is defined as long as the smoothness hypotheses continue valid. The solution remains in the Hilbert space $W_2^k(D)$. These results can also be established for equations with variable coefficients $a_{ik}(x , t)$ of $\partial^2 u/\partial x_i \partial x$

Consider now more general linear boundary conditions. For simplicity we now work locally, with initial hyperplane $t = 0$, and boundary hyperplane $x = 0$. (Figure 4). Let y_i denote coordinates of a space variable in the boundary, $i = 1 , \dots , n-1$, and set $y_1 = y$. Then the most general linear boundary condition of first order is

$$Bu = pu_t + qu_x + ru_y + wu = g .$$

By solving $Bu = g$ as a first order partial differential equation we can arrange that $g = 0$, and by solving

$$Bv = pv_t + qv_x + rv_y + wv = 0 ,$$

we can formally reduce the boundary condition by setting $u = vz$. Then z satisfies $pz_t + qz_x + rz_y = 0$, so we shall assume that the coefficient $w = 0$ henceforth.

Let the hyperbolic equation considered be the wave equation

$$Lu = u_{tt} - u_{xx} - u_{yy} - \sum_{j=2}^{n-1} u_{y_j y_j} = f ,$$

and let us choose a first order <u>multiplier</u>

$$Mu = \alpha u_t + \beta u_x + \gamma u_y ,$$

noting that $\alpha = 1$, $\beta = \gamma = 0$ in the previous result. After calculation we obtain an integral identity

$$\iint Mu\, Lu\, dV dt = \int_{R_T} \left[\frac{\alpha}{2}\left(u_t^2 + u_x^2 + u_y^2 + \sum_j u_{y_j}^2\right) + \beta u_x u_t + \gamma u_y u_t\right] dV$$
$$+ \int_S \int_0^T \left[-\frac{\beta}{2}\left(u_t^2 + u_x^2 - u_y^2 - \sum_{j=1}^{n-1} u_{y_j}^2\right) - \alpha u_x u_t - \gamma u_x u_y\right] dS dt$$
$$+ \cdots ,$$

where the terms omitted are a quadratic form over the spacetime domain in derivatives of u . We see that the new energy integral $E(t)$ contains the quadratic form

$$\frac{\alpha}{2}\left(u_t^2 + u_x^2 + u_y^2 + \sum_{i=2}^{n-1} u_{y_i}^2\right) + \beta u_x u_t + \gamma u_y u_t$$

and this form is positive definite only if $\alpha^2 > \beta^2 + \gamma^2$, the condition for the vector (α, β, γ) defining the multiplier Mu to be timelike. This we now assume.

We observe that the integrated surface terms contain the quadratic form

$$Q(0) = -\frac{\beta}{2} u_t^2 - \frac{\beta}{2} u_x^2 + \frac{\beta}{2} u_y^2 + \frac{\beta}{2} \sum_{i=2}^{n-1} u_{y_i}^2 - \alpha u_x u_t - \gamma u_x u_y .$$

As the surface integral terms are evaluated at the lower limit $x = 0$, this form must be negative definite (or at least, bounded above, if an estimate including $E(t)$ is to be found.

Algebraically, the problem becomes: for what values of p, q, r can a spacelike multiplier (α, β, γ) be found, so that $Q(0)$ is bounded above independently of u_t, u_x, u_j, u_{y_i} ?

Clearly this form of the problem does not in general have solutions — that is, nonempty sets of boundary coefficients leading to quadratic estimates.

Consider the boundary condition $Bu = pu_t + qu_x + ru_y = 0$ where $p \neq 0$, and without loss of generality $p > 0$. In the quadratic estimates choose

$$\alpha = p \quad , \quad \beta = 0 \quad , \quad \gamma = r ,$$

which is compatible with $\alpha^2 > \beta^2 + \gamma^2$ provided $|r| < p$. Then

$$\begin{aligned} -Q(0) &= 2\alpha u_x u_t + 2\gamma u_x u_y \\ &= 2u_x(pu_t + ru_y) \\ &= -2u_x(qu_x) \\ &= -2qu_x^2 \geq 0 \end{aligned}$$

provided $q \leq 0$. We conclude that <u>the estimates and existence theorem will hold if</u> $p > 0$, $q \leq 0$, $|r| < p$.

However the limitations of this method are evident, and the mixed problem has therefore had to be approached by other and more penetrating means. First among these has been consideration of mixed problems with constant coefficients, wherein the stability or well posedness can be studied using exponential solutions.

1.5 Elementary solution in a fixed region

Let R be a fixed region of space, with boundary surface S, and consider the solution of the wave equation

$$u_{tt} = c^2 \Delta u + F(P, t)$$

in $R \times [0, T] = D$ with given initial conditions

$$u(P, 0) = f(P)$$

$$u_t(P, 0) = g(P)$$

and a boundary condition on $S \times [0, T]$ of one of the three classical forms:

a) Dirichlet: $u(P,t) = h(P,t)$, $P \in S$

b) Neumann: $u_n(P,t) = h(P,t)$, $P \in S$

c) Robin: $u_n + \ell(P)u = h(P,t)$, $P \in S$,

where $\ell(P) > 0$.

These problems can be treated by analogy with the theory of domain functionals for elliptic or parabolic equations, with the one difference that the Green's functions now obtained are distributions and the convergence of their eigenfunction expansions is in the distribution sense.

For simplicity consider the Dirichlet boundary condition and let all data but $u_t(P,0) = g(P)$ be zero. Then we make use of the eigenfunctions $u_n(P)$ and eigenvalues $\lambda_n = k_n^2$ of the domain R :

$$\Delta u_n + k_n^2 u_n = 0 \; ; \quad u_n(P) = 0 \quad \text{on} \quad S \; .$$

Here we assume the existence of a complete orthonormal set of eigenfunctions and we expand the solution $u(P)$ in a Fourier series:

$$u(P) = \sum c_n(t) u_n(P) \; ,$$

where

$$c_n^{(t)} = \int_R u(P) u_n(P) dV_P \; .$$

Now

$$\begin{aligned} c_n''(t) &= \int_R u_{tt}(P) u_n(P) dV_P \\ &= c^2 \int_R \Delta u \, u_n(P) dV_P \\ &= c^2 \int_R u \cdot \Delta u_n(P) dV_P + c^2 \int_S \left(u_n \frac{\partial u}{\partial n} - u \frac{\partial u_n}{\partial n}\right) dS \\ &= -c^2 k_n^2 \int_R u(P) u_n(P) dV_P \\ &= -c^2 k_n^2 c_n(t) \; . \end{aligned}$$

Here we have noted $u_n = u = 0$ on S , the eigenvalue equation for u_n and the definition of $c_n(t)$. Also initial conditions for $c_n(t)$ follow if we note that $c_n(0) = 0$ and

$$c_n'(0) = \int_R u_t(P\,,0)u_n(P)dV_P$$
$$= \int_R g(P)u_n(P)dV_P$$
$$= g_n\,,$$

where the Fourier coefficient g_n is thus defined.

Since

$$c_n''(t) + c^2k_n^2c_n(t) = 0$$

we now have

$$c_n(t) = g_n \frac{\sin(k_n ct)}{k_n c}\,.$$

The solution function is

$$u(P\,,t) = \sum_{n=1}^{\infty} c_n(t)u_n(P)$$
$$= \sum_{n=1}^{\infty} \int_R g(Q)u_n(Q)dV_Q u_n(P) \frac{\sin(k_n ct)}{k_n c}$$
$$= \int_R K(P\,,Q\,,t)g(Q)dV_Q\,,$$

where

$$K(P\,,Q\,,t) = \sum_{n=1}^{\infty} u_n(P)u_n(Q) \frac{\sin k_n ct}{k_n c}$$

is the Green's function or elementary solution of the wave equation for this region and boundary condition.

A study of its derivation shows that the series for K converges in the distribution sense. Because of conceptual difficulties with this type of convergence, Green's function treatments of the wave equation were seldom used for general domains. However there is no lack of clarity or rigor in their use provided that the limitations of the distribution convergence are observed. In practice it is sufficient to assume that the data functions are test functions with a finite degree of differentiability.

If the series can be summed then the Green's function becomes known, and this has been done for R^n when the elementary solution is found. Outside the wave cone, the Green's function vanishes, but the multiple reflections within a region must be taken into account to find the singular set or wave front set. For a half space this leads to reflection problems to be considered in Chapters 3 and 4 below.

Setting $Lu = \frac{1}{c^2} u_{tt} - \Delta u$, we find that K has the following properties:

$$L_P K(P, Q, t) = \delta(t)\delta(P, Q),$$

where the Dirac distribution is given by the dyadic expansion

$$\delta(P, Q) = \sum_{n=1}^{\infty} u_n(P)u_n(Q).$$

Also

$$\lim_{t\to 0+} K(P, Q, t) = 0$$

$$\lim_{t\to 0+} K_t(P, Q, t) = \delta(P, Q)$$

while

$$K(P, Q, t) = 0 \quad \text{for} \quad P \in S.$$

A general representation formula for the solution of the nonhomogeneous problem can be found by applying Green's formula to the product region $D = R \times [0, T]$, and letting the functions within be $u(Q, T)$ and $K(P, Q, T+\varepsilon-\tau)$. To prove such a result strictly it is necessary to remove the singularities of K, say by convolving K with a smooth test function having support in a neighbourhood, which is then made to approximate to the Dirac distribution. Details are left to the reader. We have, with $\varepsilon > 0$

$$\iint_D [u(Q, \tau)L_Q K(P, Q, T+\varepsilon-\tau) - Lu(Q, \tau)K(P, Q, T+\varepsilon-\tau)]dVd\tau$$

$$= \frac{1}{c^2}\int_R \{u(Q, T)K_t(P, Q, \varepsilon) - u_t(Q, T)K(P, Q, \varepsilon)\}dV$$

$$- \frac{1}{c^2}\int_R u(Q, 0)K_t(P, Q, T+\varepsilon) - u_t(Q, 0)K(P, Q, T+\varepsilon\}dV$$

$$- \int_0^T d\tau \int_S \left\{u(Q, \tau)\frac{K(P, Q, T+\varepsilon-\tau)}{\partial n_Q} - \frac{\partial u(Q, \tau)}{\partial n} K(P, Q, T+\varepsilon-\tau)\right\}dS.$$

Let $\varepsilon \to 0$ and also replace $u(Q, 0)$, $u_t(Q, 0)$ by Cauchy data, similarly for other data. In the integral over D the first term vanishes as $T+\varepsilon-\tau > 0$. In the first integral on the right we use $K_t \to \delta$ as $\varepsilon \to 0$ while the second term vanishes in the limit. In the surface integral the first term brings in Dirichlet data while the second vanishes as K satisfies the homogeneous Dirichlet condition.

The result can be written

$$u(P,t) = -c^2 \int_D K(P,Q,t-\tau)F(Q,\tau)dV_Q\, d\tau$$
$$+ \int_R K(P,Q,t)g(Q)dV_Q - \int_R K_t(P,Q,t)f(Q)dV_Q$$
$$+ c^2 \int_0^t d\tau \int_S \frac{\partial K}{\partial n_Q}(P,Q,t-\tau)h(Q,\tau)dS_Q .$$

Similar representation formulas can be derived for the solutions of the Neumann or Robin boundary conditions, using the eigenfunctions and eigenvalues for the corresponding Sturm - Lionville problem to define the appropriate wave kernel K. However this method will clearly not work for boundary conditions involving time derivatives and these must be discussed as mixed initial boundary value problems.

CHAPTER 2. HYPERBOLIC EQUATIONS OF HIGHER ORDER AND SYSTEMS

2.1 Hyperbolic equations of higher order. The slowness and wave surfaces

Just as the wave equation has a real characteristic cone $\tau^2 = \xi^2$ with two real roots $\tau = |\xi|$, $\tau = -|\xi|$ for all real ξ, so a hyperbolic equation of higher order is defined by this same property. Let $P(\xi)$ be a homogeneous polynomial of degree m in $\xi = (\xi_1, \dots, \xi_n, \xi_{n+1} = \tau)$, let $D_k = -i\partial/\partial_{x_k}$, and let $\alpha = (\alpha_1, \dots, \alpha_{n+1})$ be a multi-index with $|\alpha| = \alpha_1 + \dots + \alpha_{n+1}$. Then P is called hyperbolic with respect to $\tau = \xi_{n+1}$ if

$$P(\xi + \tau N) = 0 \qquad N = (0, 0, \dots, 0, 1)$$

has only real roots τ for real ξ. Thus the normal cone of P, with equation $P(\xi) = 0$, has m real sheets. (Figure 5). (There is a definition of hyperbolicity for a nonhomogeneous polynomial (Hörmander, 1, p. 130), but we shall confine attention to the homogeneous case.)

The characteristic surfaces of P are those surfaces across which a solution can have a singularity, in particular a jump of a derivative. If a plane wave

$$u = f(\xi_1 x_1 + \dots + \xi_n x_n + t\tau)$$

has a singularity (or is a solution in an open set) then, since $u_{x_k} = \xi_k f'$, $u_t = \tau f'$, and so on, we have $P(\xi_1, \dots, \xi_n, \tau) = 0$ assuming only that $f^{(m)} \neq 0$. When f has a singularity, it will lie on a characteristic plane $\xi_1 x_1 + \cdots + \xi_n x_n + t\tau = \text{const.}$ In the $x_1 \cdots x_n$ space R^n, this singularity appears as a progressing plane wave front, and we may set $\tau = 1$ without loss of generality. Thus $x \cdot \xi = -t + \text{const.}$ and the velocity of propagation in R^n is $|\xi|^{-1}$, while the direction of the normal to the plane wave front is the direction of $\xi = (\xi_1, \dots, \xi_n)$.

Thus the set of possible normals is given by $P(\xi_1, \dots, \xi_n, 1) \equiv P(\xi, 1) = 0$, while the velocity of propagation is inversely proportional to $|\xi|$. The surface $S : P(\xi, 1) = 0$ is called the normal surface, or sometimes the slowness surface, of P. (Courant, 1, vol. 2, Ch. 6). The normal surface S is the point set of plane wave slowness vectors in the dual space of ξ. For the wave equation S is the sphere $|\xi|^2 = 1$.

Polynomial P is called strictly hyperbolic if all roots τ of $P(\xi, \tau) = 0$ are real and distinct. Thus $P_1 = (\tau^2 - \xi_1^2 - \xi_2^2) \cdot (\tau^2 - 4\xi_1^2 - 4\xi_2^2)$ is strictly hyperbolic with normal surface two concentric circles, but $P_2 = (\tau^2 - 4\xi_1^2 - \xi_2^2)(\tau^2 - \xi_1^2 - 4\xi_2^2)$ is not strictly hyperbolic because it has double roots at the intersections of the ellipses. The normal surface of P_2 consists of two ellipses with orthogonal major axes and four points of intersection giving rise to double roots of $P_2(\xi, \tau) = 0$.

A characteristic surface $\phi(x, t) = \text{const.}$ of $P(\xi, \tau)$ satisfies the first order partial differential equation $P(\nabla\phi, \phi_t) = 0$. As is known from the theory of first order equations, the most general characteristic surface is an envelope of plane characteristics. The envelope of all characteristics through the origin is the characteristic cone $C = \{(x, t) | x \cdot \xi + t = 0, P(\xi, 1) = 0\}$. To every sheet of S corresponds a sheet of C which thus also contains m real sheets. To each point of S corresponds a tangent to C.

Again, it is convenient to take $t = \text{const.}$ and define the wave surface $W(t)$ as the intersection of C with the plane $t = \text{const.}$ Thus $W(t)$ is the t-fold magnified image of $W = W(1)$; if a point source disturbance occurs at the origin at the instant $t = 0$, such as $\delta_n(x)\delta(t)$, then $W(t)$ is the wave front locus of the propagating disturbance.

To each point of S corresponds a tangent hyperplane of W, which is the polar of the given point as pole, the reference conic of the polarity being the unit sphere. It follows that W is obtained geometrically from C by duality and is the polar reciprocal of C with respect to the unit sphere. This relationship is involutory and C is obtained from W by polar inversion as well.

The degree of the algebraic surface W is however

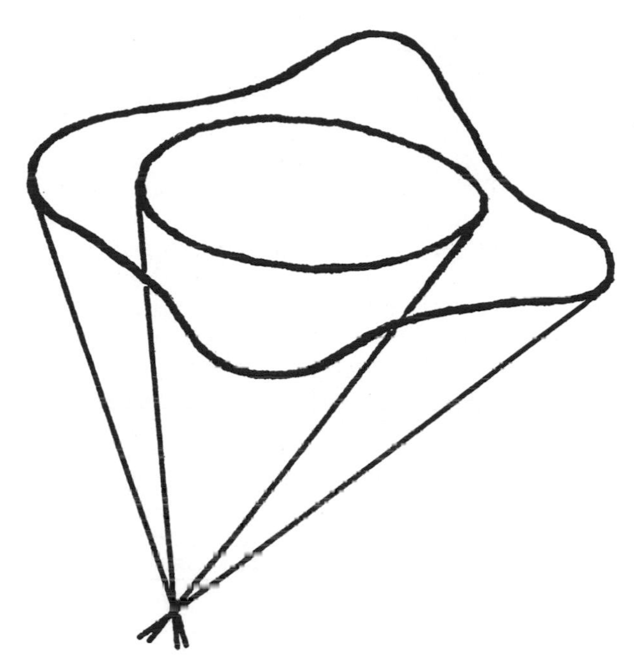

Fig. 5. Normal cone $P(\xi)=0$.

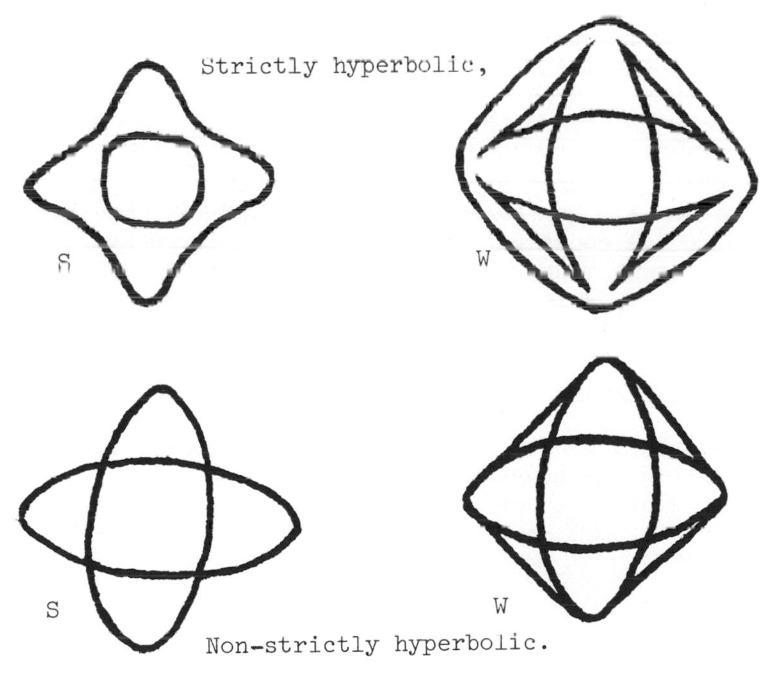

Fig. 6. Normal and Wave Surfaces of Fourth Order.

considerably higher in general than the degree of S ; it is rather the class number of S , the greatest number of tangent planes to S passing through a given n-l-plane. The degree of W is at most $m(m-1)^{n-1}$, and will be less if S has certain point singularities that may appear in the non-strictly hyperbolic case.

If P is strictly hyperbolic, S consists of m concentrically enclosed sheets, of which the outermost will extend to infinity for m odd. The wave surface W reverses the order of containment and is bounded for hyperbolic P . As the degree of P is m , any straight line can meet the innermost sheet of S at most twice, the next at most four times, and so on. Thus the innermost sheet of S is convex while dually the outermost or leading sheet of W must also be convex having at most two parallel tangent hyperplanes. The next sheet of W can have at most four hyperplane tangents parallel, which allows simple cusp singularities, and the successive interior sheets of W can have more complicated point singularities. Thus the wave surface of P_1 in the above example consists of two concentric circles. But the intersecting ellipses of the normal surface P_2 give rise to a wave surface of two intersecting dual ellipses together with ruled surface components forming the convex hull of the set of ellipses. Figure 6. A suitable perturbation such as $P_2(\xi,T) + \epsilon\frac{\partial P_2}{\partial T}(\xi,T)$ can separate the double points leaving a convex inner sheet and class 4 outer sheet for the perturbed normal surface. The perturbed wave surface then has a simple convex outer sheet and an inner sheet with four large cusp formations each containing two cusp points.

To summarize, the relationship between S and W can be stated in several different ways. Each is the polar reciprocal with respect to the unit circle, of the other. The point equation of S is the tangential equation of W , and vice versa. Given S , we can construct W as follows: let π be a tangent plane to S at P , and Q the foot of the normal from the origin to π . Then let R be the point inverse to Q with respect to the unit sphere; R lies on W . Reciprocally, the same construction yields P on S given R on W . We see also that the position vector on S is a normal at the corresponding point of W , and vice versa. To a point of inflection on S corresponds a cusp on W , to a double point on S a double ruled surface or planar sheet on W .

Further discussion of the algebraic geometry of reciprocal surfaces may be found in Courant (1), Duff (3), and in Musgrave (1-3).

2.2 Elementary solutions and estimates.

Consider now the construction of an elementary solution of the equation

$$P(D_x,D_t)u = \delta_n(x)\delta(t) ,$$

that is, a solution with $u \equiv 0$ for $t < 0$. Expressions involving the n+1 dimensional Fourier transform have been studied by Atiyah, Bott and Gårding (1,2), and Hörmander (1, p.137), but we shall follow the n-dimensional Fourier transform method of Petrowsky (1) and Gelfand-Shilov, (vol. 1). Setting

$$\hat{u}(\xi,t) = \int_{\infty}^{\infty} e^{ix\circ\xi} u(x,t)\,dx ,$$

we find

$$P(\xi,D_t)\,\hat{u}\,(\xi,t) = \delta(t) , \qquad \hat{u} \equiv 0,\ t < 0$$

and this ordinary differential equation has the Green's function Duff-Ross (1)

$$G(\xi,t) = i^{m+1} \sum_{k=1}^{m} \frac{e^{i\lambda_k(\xi)t}}{\frac{\partial P}{\partial \lambda}\left(\xi,\lambda_k(\xi)\right)}$$

$$= \frac{1}{2\pi i^m} \oint \frac{e^{i\lambda t}d\lambda}{P(\xi,\lambda)} ,$$

where the contour indicated encircles all roots λ of $P(\xi,\lambda)=0$.

Then the elementary solution is

$$K(x,t) = \hat{G}(\xi,t) = \frac{1}{(2\pi)^{n+1} i^m} \int_{R^n} e^{ix\cdot\xi} d\xi \oint \frac{e^{i\lambda t}d\lambda}{P(\xi,\lambda)} .$$

It will be noted that a sum of terms each referring to one root $\lambda_k(\xi)$ are implied on the right side here. Thus it is possible to transform this expression into an integral over the normal surface S and thus to obtain the following formulae of Herglotz (1) and Petrowsky (1). As we shall give the corresponding formulae for first order systems in detail, we omit the calculation here and quote the results, (see for example Gelfand-Shilov, vol. 1, p.139). For odd n, we have

$$K(x,t) = \frac{-(-1)^{\frac{1}{2}(n-1)}}{2(2\pi)^{n-1}(m-n-1)!} \int_S (x\cdot\xi+t)^{m-n-1} \operatorname{sgn}(x\cdot\xi+t) \frac{dS}{|\nabla P|\operatorname{sgn}(\xi\cdot\nabla P)}$$

where ∇P denotes the gradient with respect to ξ of $P(\xi,t)$ at a point of S. For even n, we have

$$K(x,t) = \frac{2(-1)^{\frac{1}{2}n}}{(2\pi)^n(m-n-1)!}\int_S (x\cdot\xi+t)^{m-n-1}\ell n\left|\frac{x\cdot\xi+t}{x\cdot\xi}\right|\frac{dS}{|\nabla P|\,\mathrm{sgn}(\xi\cdot\nabla P)}$$

The existence theorems for Cauchy's problem for higher order hyperbolic equations have been proved by Leray (1), Gårding (3), Hörmander (1) and others using estimates which generalize the second order case. The essence of these estimates is an algebraic proof of positive definiteness of the energy integral that generalizes $E(t)$ in the second order case. We shall give here a sketch of this proof, due to Leray (1).

A suitable multiplier of order $m-1$ is needed for quadratic integral estimates, and Leray observed that the necessary algebraic property of the multiplier Q is that it should be strictly hyperbolic and that the sheets of its normal cone should separate those of P, the strictly hyperbolic operator under study. Let

$$P(\xi,\lambda) = \sum_{k=1}^{m}\left(\lambda - \lambda_k(\xi)\right)$$

where the $\lambda_k(\xi)$ are real and distinct for ξ real. Then let

$$Q(\xi,\lambda) = \sum_{k=1}^{m-1}\left(\lambda - \mu_k(\xi)\right) ,$$

where $\lambda_1 < \mu_1 < \lambda_2 < \mu_2 < \ldots < \mu_{m-1} < \lambda_m$ holds for $\xi \in R^n$. Such operators exist since the polynomial $\frac{\partial P}{\partial \lambda}$ has this property by Rolle's Theorem, and any hyperbolic polynomial sufficiently close to it does also. Let

$$P_k(\xi,\lambda) = \prod_{j\neq k}\left(\lambda - \lambda_j(\xi)\right) \qquad k = 1,\ldots,m$$

and observe that P_k has degree m-1 in λ so that the m independent polynomials P_k form a basis for polynomials of degree m-1 in λ. Hence

$$Q(\xi,\lambda) = \sum_{k=1}^{m}\gamma_k(\xi)\,P_k(\xi,\lambda) ,$$

where the coefficients γ_k are easily evaluated by setting $\lambda = \lambda_\ell$; as $P_k(\xi,\lambda_\ell) = 0$, $\ell \neq k$. Thus

$$\gamma_\ell = \frac{Q(\xi,\lambda_\ell)}{P_\ell(\xi,\lambda_\ell)} = \frac{\prod_{\ell\neq k}(\lambda_\ell-\mu_k)}{\prod_{\ell\neq k}(\lambda_\ell-\lambda_k)} > 0$$

by the separation property.

The energy integral will be defined as

$$E(t) = \int \sum_{k=1}^{m} \gamma_k |P_k(D_x,D_t)\, u|^2\, dx$$

and it will be noted that $E(t)$ contains pseudodifferential operators specified by the P_k . However, by Parseval's Theorem

$$E(t) = \int_{R^n} \sum_{k=1}^{m} \gamma_k(\xi) |P_k(\xi,D_t)\,\hat{u}|^2 d\xi$$

and we shall work with Fourier transforms $\hat{u}(\xi,t)$ in ξ to estimate derivatives later.

Since $D_t = -i\,\partial/\partial t$, we have

$$\frac{\partial}{\partial t}\sum_{k=1}^{m} \gamma_k(\xi)|P_k(\xi,D_t)\,\hat{u}|^2$$

$$= \sum_{k=1}^{m} \gamma_k(\xi)\; i[D_t P_k(D_t)\hat{u}\cdot P_k(\bar{D})\bar{\hat{u}} - P_k(D_t)\,\hat{u}\,\bar{D}_t P_k(\bar{D}_t)\,\bar{\hat{u}}]$$

$$= \sum_{k=1}^{m} \gamma_k(\xi)\; i[(D_t-\lambda_k)\,P_k\,(D_t)\,\hat{u}\cdot P_k(\bar{D}_t)\,\bar{\hat{u}}$$

$$- P_k(D_t)\,\hat{u}\,(\bar{D}_t-\lambda_k)\,P_k\,(\bar{D}_t)\,\bar{\hat{u}}]$$

$$= i\sum_{k=1}^{m}\left\{P(D_t)\,\hat{u}\cdot\gamma_k\,P_k(D_t)\,\hat{u} - \gamma_k\,P_k(D_t)\,\hat{u}\,P(\bar{D}_t)\,\hat{u}\right\}$$

$$= i[P(D_t)\,\hat{u}\,Q(\bar{D}_t)\,\bar{\hat{u}} - Q(D_t)\,\hat{u}\,P(\bar{D}_t)\,\bar{\hat{u}}]$$

In these expressions the conjugate operator $\bar{D}_t = i\,\partial/\partial t$ operates only on $\bar{\hat{u}}$. Integrating over ξ , we obtain

$$\frac{\partial}{\partial t} E(t) = -2\int \mathrm{Im}[P(D_t)\,\hat{u}\cdot Q(D_t)\,\hat{u}]\, d\xi$$

$$= -2\int \mathrm{Im}[P(D_x,D_t)\,u\cdot Q(D_x,D_t)\,\bar{u}]\,dx$$

again using Parseval's theorem. To obtain estimates, we set $P(D_x,D_t)u = f$ and observe that the right side now contains derivatives of u up to order m-1 only. Thus, integrating over t, we have

$$E(t) = E(0) - 2\int_0^t\int \mathrm{Im}[f \quad Q(D)\,\bar{u}]\,dx\,dt'$$

$$\leq E(0) + 2\int_0^t \|f\|_2 \|D^{m-1}u\|_2 dt'$$

$$\leq E(0) + c\int_0^t \|f\|_2 \; E(t')\,dt' \quad .$$

This inequality of Gronwall type can be integrated to yield a bound for $E(t)$ that depends only on t, the region and equation, and the data. We shall not pursue this aspect here, referring the reader for example to Hormander, Chapter 9.2. For variable coefficients, further terms will appear on the right side of these estimates.

However we now wish to show that all derivatives of order up to m-1 can be estimated in the L^2 norm in terms of $E(t)$. In view of Parseval's relation, we have

$$\int |D_x^\alpha D_t^\ell u|^2 \,dx = \int |\xi^\alpha \lambda^\ell|^2 |\hat{u}|^2 \,d\xi$$

where $|\alpha| + \ell \leq m-1$. We now consider $\xi^\alpha\lambda^\ell$ as a polynomial in λ and write

$$\xi^\alpha\lambda^\ell = \sum_{k=1}^{m-1} \rho_k(\xi)P_k(\xi,\lambda) \; .$$

Here the coefficients $\rho_k(\xi)$ can be evaluated by setting $\lambda = \lambda_k(\xi)$, with the result that

$$\rho_k(\xi) = \frac{\xi^\alpha \lambda_k(\xi)^\ell}{P_k(\xi,\lambda_k(\xi))} \quad .$$

Since all $\lambda_k(\xi)$ are homogeneous of degree 1, $\rho_k(\xi)$ is homogeneous of degree zero in ξ, and can be estimated by considering values of ξ with $|\xi| = 1$. Because all $\lambda_k(\xi)$ are distinct, it now follows readily that $\rho_k(\xi)$ is bounded. Hence

$$\int |D_x^\alpha D_t^\ell u|^2 dx \le \sum_{k,l} \int \rho_k(\xi)\bar{\rho}_\ell(\xi)\, P_k(\xi,D_t)\,\hat{u}\,\overline{P_\ell(\xi,D_t)\,\hat{u}}\, d\xi$$

$$\le C \sum_k \int |P_k(\xi,D_t)\,\hat{u}\,|^2 d\xi$$

$$\le \quad C \quad \int \sum_k |P_k(D_x,D_t)\,u\,|^2 dx$$

$$= K\,E(t) \;.$$

This completes the demonstration of these estimates for higher order homogeneous hyperbolic differential equations.

A treatment of the existence theorem using a reduction to a hyperbolic system of first order equations is given by Mizohata (2 , Chapter 6). We shall next turn to the study of estimates for first order systems.

Consider next a first order symmetric hyperbolic system of the form

$$\frac{\partial u_r}{\partial t} = \sum_{\nu=1}^{n} A^\nu_{rs} \frac{\partial u_s}{\partial x_r} + B_{rs} u_s \quad ,$$

with initial conditions $u_r(x,0) = f_r(x)$ in a domain R , and boundary conditions to be specified below. Here $r,s = 1,\ldots,m$ and $x = 1,\ldots,n$ while $A^\nu = (A^\nu_{rs})$ and $B = (B_{rs})$ are $m \times m$ matrices with components suitably smooth functions of x and t , where $x = (x_1,\ldots,x_n)$. The coefficient matrices of the first derivative terms are assumed symmetric: $A^\nu_{rs} = A^\nu_{sr}$ and unless otherwise specified all coefficients and functions will be assumed real. (In the complex case, the A^ν are Hermitean). Symmetric hyperbolic systems were introduced by Friedrichs (1).

In the above normal form the hyperbolic character of the system is expressed by the symmetry of the A^ν and L^2 estimates for the n_r can be derived as follows. Let

$$E(t) = \int_R u_r u_r \, dV$$

where summation over r is understood. Then

$$\frac{dE(t)}{dt} = 2\int_R u_r \frac{\partial u_r}{\partial t}\, dV$$

$$= 2\int_R u_r \left(\sum_{r=1}^{n} A^{\nu}_{rs} \frac{\partial u_s}{\partial x_r} + B_{rs} u_s\right) dV$$

$$= 2\left(\int_R \frac{\partial}{\partial x_r}\left(A^{\nu}_{rs} u_r u_s\right) + \left(B_{rs} - \sum_r \frac{\partial A_{rs}}{\partial x_r}\right)\right) dV$$

$$= \int_S A^{\nu}_{rs} u_r u_s n_r \, dS + \int_R \left(B_{rs} - \sum_r \frac{\partial A_{rs}}{\partial x_r}\right) dV \; .$$

Here the symmetry property has been utilized to express the quadratic form as a perfect differential. Also n_r denotes the outward normal to S the bounding surface of R .

Boundary conditions will be specified, to control the surface integral term. Consider the signature of the matrix

$$A \cdot n = \sum_r A^{\nu} n_{\nu}$$

at a typical point of S . By symmetry the eigenvalues are real and we suppose k are positive and $m-k$ negative. Performing if necessary a rotation in the u_r vector space, we may write

$$(A \cdot n)_{rs} u_r u_s = A^{I}_{rs} u_r u_s + A^{II}_{rs} u_r u_s$$

where $A^{I} = \mathrm{diag}(\lambda_1, \ldots, \lambda_k)$ is positive (or non-negative) and $A^{II} = \mathrm{diag}(\lambda_{k+1}, \ldots, \lambda_m)$ is negative. The quadratic form as a whole is therefore bounded above if the boundary conditions are taken as

$$u_j = g_j \qquad\qquad j = 1, \ldots, k$$

with datum functions g_j . More generally, however, we may take linear boundary conditions

$$u_j = \sum_{\ell=k+1}^{m} S_{j\ell} u_{\ell} + g_j$$

provided that substitution in the quadratic form again yields an expression bounded above. The condition for this is seen to be that the matrix

$$C_{\ell p} = \left(\sum_{j=1}^{k} \lambda_j s_{j\ell} s_{jp} + \lambda_\ell \delta_{\ell p}\right) \qquad \ell , p = k+1 , \ldots , m$$

is negative, say $C_{\ell p} << - \delta I_{II}$, where $\delta > 0$.

Under these conditions the surface integral is bounded above by a quantity $G(t)$ dependent on the data, and

$$\frac{dE(t)}{dt} \delta \int_{\partial R} u_{II}^2 \, dS \le G(t) + KE(t) ,$$

where K is an upper bound for the matrix $B - \sum \frac{\partial A^\nu}{\partial x_\nu}$ over the region and valid for the time period considered. Supposing also that $G(t) \le G$, we easily find for $E(t)$ a bound of the form

$$E(t) + \delta \int_S u_{II}^2 \, dS\, dt \le \left(E(0) + \frac{G}{K}\right) e^{Kt} .$$

This bound, and similar estimates for the derivatives of u_r with respect to x_ν and t up to any desired order, can be used as in §1.4 to establish an existence theorem for the symmetric hyperbolic system. Uniqueness of the solution is evident as $E(0)$ and G can be taken as zero in the event that all data are zero. A slight modification covers the presence of a non-homogeneous term on the right side of the differential equation. We return in §6.3 to consideration of the types of boundary conditions that lead to well posed initial and boundary value problems.

2.3 The Riemann matrix of a hyperbolic system

Consider the first order system

$$\frac{\partial \vec{u}}{\partial t} = \sum_{\nu=1}^{n} A_\nu \frac{\partial \vec{u}}{\partial x_\nu} \qquad \nu = 1 , 2 , \ldots , n , \tag{2.2.1}$$

where $\vec{u} = (u_1 , \ldots , u_m)$ is an m component vector, and $A_\nu = (a_{rs\nu})$ are $m \times m$ constant matrices. We assume that (2.2.1) is hyperbolic, that is, the characteristic roots of the matrix $A(\eta) = \sum_{\nu=1}^{n} A_\nu \eta_\nu$ are all real. For a hyperbolic system, the initial value problem of Cauchy is correctly set (well-posed), the initial condition being $\vec{u}(x , 0) = \vec{g}(x)$. The solution $\vec{u}$ can be expressed as a convolution

$$\vec{u}(x , t) = \int_{R^n} R(x-z , t)\vec{g}(z)dz ,$$

where $R(x, t)$ denotes the Riemann matrix or elementary solution: thus

$$\frac{\partial R}{\partial t} = \sum_{\nu=1}^{n} A_\nu \frac{\partial R}{\partial x_\nu} \qquad \text{for} \quad t > 0$$

and initially,

$$R(x, 0) = \delta(x)E ,$$

where E denotes the $m \times m$ unit matrix. A solution $\vec{u}(x, t)$ of the nonhomogeneous system

$$\frac{\partial \vec{u}}{\partial t} = \sum_{\nu=1}^{n} A_\nu \frac{\partial \vec{u}}{\partial x_\nu} + \vec{f}(x, t)$$

can easily be expressed in the convolution form

$$\vec{u}(x, t) = \int_0^t \int_{R^n} R(\vec{x}-\vec{y}, t-\tau) f(\vec{y}, \tau) d\vec{y} d\tau .$$

To calculate the Fourier transform of R, we note that

$$\hat{R}(\xi, t) = \frac{1}{(2\pi)^{\frac{n}{2}}} \int_{R^n} R(x, t) e^{ix\cdot\xi} dx$$

and

$$\frac{\partial \hat{R}(\xi, t)}{\partial t} = -i \, \Sigma A_r \xi_\nu \hat{R}(\xi, t)$$

with $\hat{R}(\xi, 0) = (2\pi)^{-n/2} E$. The matrix solution of this matrix ordinary differential equation is

$$\hat{R}(\xi, t) = \frac{1}{(2\pi)^{\frac{n}{2}}} \exp\left(-i \sum_{\nu=1}^{n} A_\nu \xi_\nu\right)$$

$$= \frac{1}{(2\pi)^{\frac{n}{2}}} \exp\left(-iA(\xi)\right) .$$

Hence the inverse Fourier transform leads to the formula

$$R(x, t) = \frac{1}{(2\pi)^n} \int_{R^n} \exp(-i[x\cdot\xi E + A(\xi)t]) d\xi .$$

We now study this matrix in its dependence on x and t, its support, singularities, and asymptotic behaviour. Duff (4).

Let $\lambda_k(\xi)$, $k=1,\dots,m$ be the characteristic roots of $A(\xi) = \sum_\nu A_\nu \xi_\nu$, that is, roots of

$$\det(A(\xi) - \lambda E)) = |A(\xi) - \lambda E| = 0 .$$

The $\lambda_k(\xi)$ are assumed real, and they are of first degree in ξ, so that $\lambda_k(s\xi) = s\lambda_k(\xi)$. There exists a nonsingular matrix $T = (t_{ik}(\xi))$ which reduces $A(\xi)$ to its Jordan canonical form: $A = TJT^{-1}$; $J = T^{-1}AT$. When A has distinct roots,or is symmetric, then $J = \mathrm{diag}(\lambda_j(\xi))$. By permanence of matrix functional relations,

$$\exp(-iA(\xi)t) = T\exp(-iJ(\xi))T^{-1} ,$$

and supposing J diagonal we see that

$$R(x,t) = \frac{1}{(2\pi)^n} \int_{|\eta|=1} T\,\mathrm{diag} \int_0^\infty \exp(-i(x\cdot\eta + t\lambda_k(\eta))|\xi|) \cdot |\xi|^{n-1}\, d|\xi| T^{-1}\, d\Omega_\eta ,$$

where $\xi = |\xi|\eta$, $|\eta| = 1$. Setting $s = x\cdot\eta + t\lambda_k(\eta)$ we consider the inner integral

$$\int_0^\infty \exp(-is|\xi|)|\xi|^{n-1}\, d|\xi|$$

$$= (-i)^n (s-i0)^{-n}(n-1)!$$

This is the Fourier transform of a distribution as set forth in Gelfand-Shilov (1, vol. 1, p. 172). Thus

$$R(x,t) = \frac{(-i)^n(n-1)!}{(2\pi)^n} \int_{|\eta|=1} T(\eta)\mathrm{diag}(x\cdot\eta + t\lambda_k(\eta) - i0)^{-n} T^{-1} d\Omega_\eta$$

$$= \frac{(n-1)!}{(2\pi i)^n} \int_{|\eta|=1} (x\cdot\eta E + tA(\eta) - i0)^{-n}\, d\Omega_\eta .$$

Note that for $t=0$ this form reduces to the distributional plane wave representation of the δ function (Gelfand-Shilov, 1, vol. 1, p. 77), namely

$$\delta_n(x)E = \frac{(n-1)!}{(2\pi i)^n} \int_{|\eta|=1} (x\cdot\eta - i0)^{-n}\, d\Omega_\eta \cdot E .$$

The formula shows incidentally that $R(x,t)$ is homogeneous of degree $-n$ in $\vec{x}$ and t together. The distributional character of R is now explicit as the singularities have been gathered into the integrand. We shall study below the extent to which the $n-1$ remaining integrations smooth out this singularity of order n.

Example 1. If $n = 1$, then $R(x, t) = \delta(xE + tA)$.

Example 2. Let $m = 1$ so the system becomes the scalar equation

$$\frac{\partial u}{\partial t} = \sum_{\nu=1}^{n} a_\nu \frac{\partial u}{\partial x_\nu} .$$

Then $R(x, t) = \delta(x_\nu + ta_\nu)$ a scalar n-dimensional delta function. The domain of dependence of the solution is in this case a single point, $x_\nu = -ta_\nu$, $\nu = 1, \dots, n$.

Example 3. The 1 dimensional wave equation,

$$u_t = v_x , \quad v_t = u_x , \quad A = \begin{pmatrix} 0 & 1 \\ 1 & 0 \end{pmatrix} .$$

Then

$$R(x, t) = \delta\begin{pmatrix} x & t \\ t & x \end{pmatrix} = \frac{1}{2}\begin{pmatrix} \delta(x+t) + \delta(x-t) , & \delta(x+t) - \delta(x-t) \\ \delta(x+t) - \delta(x-t) , & \delta(x+t) + \delta(x-t) \end{pmatrix} .$$

Returning to the general system in n space dimensions, we see that a plane wave

$$\vec{u}(x, t) = \vec{a} f(x \cdot \xi + t\tau)$$

will satisfy the system if and only if

$$\sum \left(A_{pq}(\xi) - \tau\delta_{pq}\right) a_q = 0 .$$

That is, τ must be a characteristic root $\lambda_k(\xi)$:

$$\det\left(A(\xi) - \tau E\right) = 0 ,$$

and $\vec{Q}$ must be an associated eigenvector. The normal cone is the above determinant locus, while again the normal surface S is defined as the intersection of the normal cone with the hyperplane $\tau = -1$. The jth sheet of the wave cone is found from the envelope relations

$$\frac{\partial s_j}{\partial \xi_r} = x_r + t \frac{\partial \lambda_j}{\partial \xi_r} = 0 , \qquad \begin{matrix} j = 1, \dots, m \\ r = 1, \dots, n . \end{matrix}$$

Setting $t = 1$ we see that W is found by eliminating $\xi_1, \dots, \xi_n$ from the n relations with j fixed.

We now show that the domain of dependence is contained in the convex closure of the wave surface $W(t)$. The energy integral is

$$E(t) = \int_{R(t)} u_p u_p \, dV ,$$

where $R(t)$ is a suitable time varying space domain. Let σ denote the inward velocity of the boundary of R normal to itself. Then

$$\frac{dE(t)}{dt} = 2 \int_{R(t)} u_p \frac{du_p}{dt} dV - \int_{\partial R} u_p u_p \sigma \, dS$$

$$= 2 \int_{R(t)} u_p \sum_\nu A_{pq\nu} \frac{\partial u_q}{\partial x_\nu} dV - \int_{\partial R} u_p u_p \sigma \, dS$$

$$= \int_{R(t)} \frac{\partial}{\partial x_\nu} (A_{pq\nu} u_p u_q) dV - \int_{\partial R} u_p u_p \sigma \, dS$$

$$= \int_{\partial R} (\sum_\nu A_{pq\nu} n_\nu - \sigma\delta_{pq}) u_p u_q \, dS .$$

Here we have used the symmetry property $A_{pq\nu} = A_{qp\nu}$ to form the derivative of the quadratic expression in u_p, u_q. To make this last expression non-positive we must choose σ at least as large as the largest eigenvalue of $A(n) = \sum_\nu A_\nu n_\nu$. But this condition is exactly fulfilled if we choose $R(t)$ to be the convex closure of the wave surface $W(t_1 - t)$, where t_1 is a given positive number. (Figure 7). If now $u_p(x , 0) = 0$ in $W(t_1)$ then $E(t) \equiv 0$ for $0 \leq t \quad t_1$ and it follows that $u_p(0 , t_1 - 0)$ is zero. If $u_p(x , t)$ is continuous at $(0 , t)$ then $u_p(0 , t_1)$ will also be zero, and this completes the proof.

Conversely, we see that the support of the Riemann matrix $R(x , t)$ is the convex closure of $W(t)$.

2.4 The order of the singularities

Returning to the representation formula for $R(x , t)$, we note that the integrand is singular when one of the phases $s_j = x \cdot \eta + t\lambda_j(\eta)$ vanishes. Since the wave surface $W(t)$ is the envelope of these loci, the principle of stationary phase suggests that any singularities of $R(x , t)$ will lie on $W(t)$. The singular parts of these integrals can be regarded as the sum of all singular contributions from the characteristic planes through the point. The density of these, weighted by solid angle in the η space, is itself singular precisely on the envelope $W(t)$, and the order of singularity of this density determines the order of the singularity of $R(x,t)$ on $W(t)$.

Denote the Riemann - Liouville fractional integral of order α by

$$I_+^\alpha f(x) = \frac{1}{\Gamma(\alpha)} \int_a^x (x-t)^{\alpha-1} f(t)\,dt = \frac{1}{\Gamma(\alpha)} \int_0^b u^{\alpha-1} f(x-u)\,du ,$$

where the upper limit b on the right may be replaced by $x-a$ if necessary. Denote the Weyl fractional integral of order α by

$$I_-^\alpha f(x) = \frac{1}{\Gamma(\alpha)} \int_x^{a_1} (t-x)^{\alpha-1} f(t)\,dt = \frac{1}{\Gamma(\alpha)} \int_0^{b_1} u^{\alpha-1} f(x+u)\,du .$$

These fractional integrals satisfy

$$\frac{d}{dx} I_\pm^{\alpha+1} f(x) = I_\pm^\alpha f(x)$$

and the composition law

$$I_\pm^\alpha I_\pm^\beta f(x) = I_\pm^{\alpha+\beta} f(x)$$

holds. We note also that

$$I_\pm^{\frac{\alpha}{b}} f(x) = \frac{K^{\frac{\alpha}{b}}}{\Gamma\left(\frac{\alpha}{b}\right)} \int_0 u^{\alpha-1} f\left(x \mp Ku^b\right) du .$$

where $K > 0$.

We require the following stationary phase lemma proofs of which can be found in Duff (4) or Ludwig (2).

<u>Lemma</u>. Let

$$I = \int f(S) g(\eta)\, d\eta_1 \cdots d\eta_{n-1} ,$$

where f is a distribution singular at the origin and

$$S = s + tP(\eta) = s + tQ_2(\eta) + \cdots$$

is a power series in $\eta_1 \cdots \eta_{n-1}$ with nonsingular quadratic leading term having n_+ positive and n_- negative coefficients in principal axis form. Then I has an asymptotic expansion in s with leading term

$$\frac{\pi^{\frac{n-1}{2}}}{2^{n-1}} \frac{t^{-(n-1)/2}}{\sqrt{\det(Q)}} g(0) I_-^{n_+/2} I_+^{n_-/2} f(s)$$

which proceeds in steps of a half order integration in successive terms.

We now calculate the singularity of $R(x,t)$ at a point of the sheet of the wave surface dual to a given root, say $\lambda_k(\eta)$. Choose a fixed time t, and a point x close to W_k. Let x_0 be the foot of the perpendicular from x to W_k, and η_0 the corresponding value of η. Then the tangent plane to W_k at x_0 has the equation $x \cdot \eta_0 + t\lambda_k(\eta_0) = 0$. Since also η_0 is normal to W_k at x_0, we have $s = (x - x_0)\cdot\eta_0$ where s is the distance $|x - x_0|$ of x from W_k.

Now write

$$\begin{aligned} S_k &= x\cdot\eta + t\lambda_k(\eta) \\ &= s + x\cdot(\eta - \eta_0) + x_0\cdot\eta_0 + t\lambda_k(\eta) \\ &= s + x\cdot(\eta - \eta_0) + t\big(\lambda_k(\eta) - \lambda_k(\eta_0)\big) . \end{aligned}$$

Let $\eta_0 = (0, 0, 0, \ldots, 0, 1)$ and expand $\lambda_k(\eta)$ in powers of $\eta_1, \ldots, \eta_{n-1}$. We find

$$S_k = s + \sum_j \left(x_j + t\frac{\partial\lambda_k}{\partial\eta_{0j}}\right)\eta_j + \frac{t}{2}\sum_{jk}\frac{\partial^2\lambda_k}{\partial\eta_{0j}\partial\eta_{0j}}\eta_j\eta_\ell .$$

Using the envelope relations $x_{0j} + t\frac{\partial\lambda_k}{\partial\eta_{0j}} = 0$, and $x_j = x_{0j} + s\eta_{0j}$, we find

$$\begin{aligned} S_k &= s + s\eta_0\cdot(\eta - \eta_0) + \frac{t}{2}\sum_{j,\ell=1}^{n-1}\frac{\partial^2\lambda_k}{\partial\eta_{0j}\partial\eta_{0\ell}}\eta_j\eta_\ell \\ &= \cos\theta\cdot\left(s + \frac{t}{2}\sum_{j,\ell=1}^{n-1}\frac{\partial^2\lambda_k}{\partial\eta_{0j}\partial\eta_{0\ell}}\eta_j\eta_\ell + \cdots\right) , \end{aligned}$$

where the terms omitted are of third or higher order in η_j, and where $\cos\theta = \eta - \eta_0$.

We shall apply the lemma for stationary phase to the diagonalized form of the Riemann matrix, and therefore it is convenient to express the components of the diagonalizing matrix T in a particular algebraic form. The columns of T are eigenvectors $\vec{t}_k$ that satisfy

$$A(\eta)\vec{t} = \lambda\vec{t}$$

for $\lambda = \lambda_k$. Thus the components $t_i(\lambda, \eta)$ can be computed as polynomial functions of λ and η. We shall also normalize the $t_i(\lambda, \eta)$ to unit length which involves a division by a square root of a polynomial. Then the normalized $t_i(\lambda, \eta)$, $i = 1, \ldots, n$ are homogeneous functions of λ and η of degree zero.

The transforming matrix $T = \big(t_i(\lambda_k, \eta)\big)$ is orthogonal (or unitary) so that $T^{-1} = T'$ (or $\bar{T}'$). For any matrix function $f(A)$, we have an expression for the (i, k) component as follows:

$$\begin{aligned} f(A)_{ik} &= \big(Tf(J)T^{-1}\big)_{ik} \\ &= \big(Tg(J)\bar{T}'\big)_{ik} \\ &= \big(T \operatorname{diag} f(\lambda_j)\bar{T}'\big)_{ik} \\ &= \sum_{j=1}^{m} t_i(\lambda_j, \eta) f(\lambda_j) \bar{t}_k(\lambda_j, \eta) . \end{aligned}$$

Therefore the components of the Riemann matrix are

$$R_{ik}(x, t) = \frac{(n-1)!}{(2\pi i)^n} \int_{|\eta|=1} \sum_{j=1}^{m} \frac{t_i(\lambda_j, \eta)\bar{t}_k(\lambda_j, \eta)}{\big(x\cdot\eta + t\lambda_j(\eta) - i0\big)^n} \, d\Omega_\eta .$$

Taking the term singular on the jth sheet W_j, we have to consider

$$\frac{(n-1)!}{(2\pi i)^n} \int_{|\eta|=1} \frac{t_i(\lambda_j, \eta)\bar{t}_k(\lambda_j, \eta)}{(S_j - i0)^n} \, d\Omega_\eta ,$$

where

$$S_j = \cos\theta\big(s + \tfrac{t}{2} c_{k\ell}\, \eta_k \eta_\ell + \cdots\big) .$$

Noting that $\cos\theta = \eta\cdot\eta_0$ is equal to unity when $\eta = \eta_0$, we find the leading singular term to be

$$\frac{(n-1)!}{2^{\frac{3n-1}{2}} \pi^{\frac{n+1}{2}} i^n t^{\frac{n-1}{2}}} \frac{t_i(\lambda_j, \eta) t_k(\lambda_j, \eta)}{\sqrt{\det c_{k\ell}}} I_-^{n_+/2} I_-^{n_-/2} (s - i0)^{-n} .$$

Here the $\det(c_{k\ell})$ can be shown to be the Gaussian curvature of

S at the value of η involved. If the surface S is strictly convex then either n_+ or n_- equals zero and in certain cases where the delta function portion of $(s-i0)^{-n}$ appears, the support of the singular term will be a point, or will lie entirely on one side of the wave sheet W_j.

In certain directions η the Gaussian curvature of S may vanish; such an inflection on S corresponds to a cusp on W and a generalized form of the stationary phase lemma is needed. Duff (4):

If

$$S = s + t\,\Sigma\, c_\alpha \eta^\alpha ,$$

where the leading terms in the series are of degree n_j in η_j, then the integral I has the asymptotic expansion

$$c \cdot t^{-n+d_1+d_2}\, I_-^{d_1} I_+^{d_2} (s-i0)^{-n} ,$$

where

$$d_1 = \sum_+ \frac{1}{n_j} , \quad d_2 = \sum_- \frac{1}{n_j} ,$$

the sums denoting terms with positive or negative values in the series expansion. The expansion proceeds in steps of $1/n_j$ integrations in each variable.

In certain cases the numbers n_j must be defined with reference to a Newton diagram for the power series. Also note that an odd order term such as η^3 gives rise to both positive and negative values and hence a term in $I_+^{1/3}$ and another in $I_-^{1/3}$. An integral of a distribution $(x-i0)^{-n}$ is to be understood as being regularized in the sense of Gelfand - Shilov (1, vol. 1, p. 24), which means that certain terms are introduced to avoid divergence of the integrals over test functions giving the values of the integrated distribution.

Example

$$\int_{-\infty}^{\infty} f(s + t\eta^3)\,d\eta = \frac{t^{-1/3}}{3}\left[I_-^{1/3} f(s) + I_+^{1/3} f(s) \right] .$$

The vanishing of one principal curvature on S, increasing the first term from order 2 to order 3, increases the order of the spatial cusp singularity by $\frac{1}{6}$. Also the power of decay with time t is reduced by $\frac{1}{6}$ so that at large distances the wave front remains more concentrated along the corresponding rays.

Consider two sheets of S that approach each other by a deformation leading to a double point. While still separate, there will be on one sheet an inflection locus giving rise to a cusp locus as above. Within the inflection locus a large curvature on S develops leading to a widely extended near flat wave front bounded by the cusp locus and having correspondingly weak amplitude. In the limit a double point of S forms, and the plane or ruled surface wave front drops to a weaker singularity. Ludwig and Granoff (1) have derived a system of propagation equations for this wave front, and estimate its singularity as ½ unit smoother for each dimension of planarity.

We conclude this section by giving a further formula for $R(x, t)$ expressed as an integral over the normal surface S. The equation of the kth sheet of S can be written as $\lambda_k(\xi) = |\xi|\lambda_k(\eta) = 1$. We note that the surface element on S is given by

$$\eta\cdot\zeta dS = |\xi|^{n-1}\, d\Omega_\eta ,$$

where ζ is the unit normal to S so that $\eta\cdot\zeta = \cos\theta$, the angle θ is that between the radius vector and the normal on S or on W. Also we observe that

$$\begin{aligned}
&(x\cdot\eta E + tA(\eta) - i0)^{-n}\, d\Omega_\eta \\
&= |\xi|^n (x\cdot\xi + tA(\xi) - i0)^{-n}\, d\Omega_\eta \\
&= (x\cdot\xi + tA(\xi) - i0)^{-n}\, \xi\cdot\eta\, dS .
\end{aligned}$$

Therefore, since $\lambda = 1$ on S, and $t_i(\lambda, \eta)$ are homogeneous of degree zero,

$$R_{ik}(x, t) = \frac{(n-1)!}{(2\pi i)^n} \int_S \frac{t_i(1, \xi)\bar{t}_k(1, \xi)}{(x\cdot\xi + t - i0)^n}\, \xi\cdot\eta\, dS$$

This formula is the closest analogue of the Herglotz - Petrowsky formula for higher order hyperbolic equations.

The singular locus of this integrand is the section of S by the hyperplane $x\cdot\xi + t = 0$, so the singularities of the solution arise from points on this locus or in its neighbourhood. As t increases, the hyperplane moves away from the origin, and an instant t' at which it is tangent to a sheet S_j of S is the moment the corresponding wave sheet W_j of $W(t)$ reaches x. As $t \to t'_j$ the intersection of the variable hyperplane with S_j, which is a "vanishing cycle" in the sense of Petrowski (1, p. 327), shrinks to a point and disappears. Thus the singularity of the solution involves the eigenvectors attached to that point.

To summarize: the leading singular term in R_{ik} is a distribution multiple of the eigenvector $t_i(1, \xi)$, with leading term polarized parallel to it:

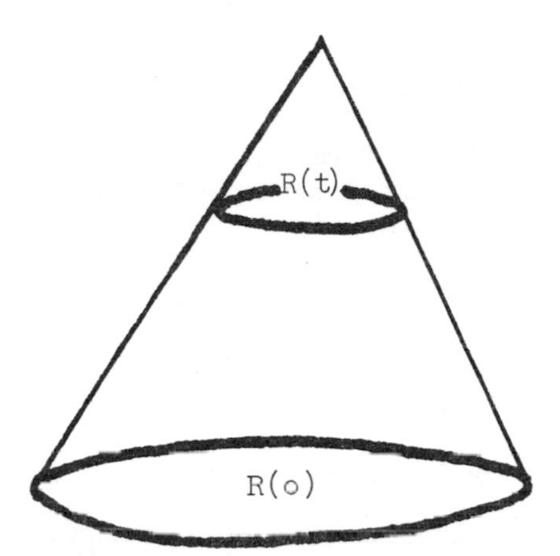

Fig. 7. Region of exclusive influence.

Fig. 8. Singularity at an ordinary point.

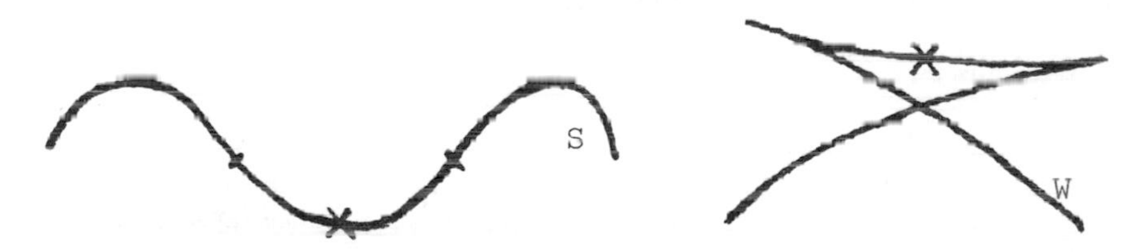

Fig. 9. Singularity at cusp locus of W corresponding to an inflection point of S .

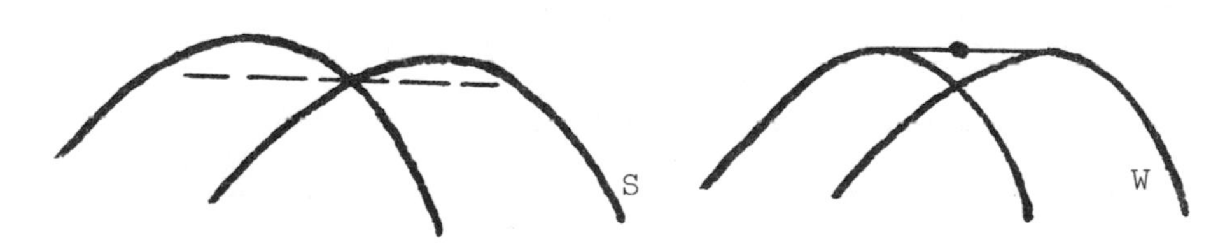

Fig. 10. Singularity on a ruled surface of W corresponding to a double point of S .

$$R_{ik}(x, t) \sim c t_i(1, \xi)\bar{t}_k(1, \xi) t^{-\Sigma\frac{1}{n_s}} I_{\pm}^{\Sigma\frac{1}{n_s}} (x \cdot y + t - i0)^{-n} .$$

Further studies of the precise forms of this singularity in the highly singular cases are still desirable. However, the leading cases may be indicated as follows (Duff 4, Ludwig 1, 4).

1. An ordinary point of S ($\det Q = K$ the Gaussian curvature of $S \neq 0$). Then all $n_s = 2$, the order of the singularity is $n - \frac{1}{2}(n-1) = \frac{1}{2}(n+1)$ and the corresponding point of W is an ordinary point. (Figure 8).

2. An Inflection point or locus on S ($\det Q = 0$) and corresponding on W a cusp locus. Then one or more $n_s \geq 3$, giving a singularity $\frac{1}{6}$ or more higher on W, with attenuation in time reduced by the same power of t. (Figure 9).

3. A double point or locus of S, and a ruled surface portion on W. Then to a point of the ruled surface corresponds a dual line $x \cdot \xi + t = 0$ through the double point but not tangent to S. Hence the n_s is reduced to 1, and the singularity is $\frac{1}{2}$ degree lower for each such dimension or index n_s, while attenuation in t is as much higher. This case is related to 2. above when 2 sheets of S meet in a limiting case. (Figure 10).

CHAPTER 3. PROPERTIES OF ELEMENTARY SOLUTIONS

3.1 Lacunas

The name 'lacuna' (or gap) was first used by Petrowski whose famous paper (1) can be regarded as the foundation of the modern theory. However the basic idea of lacunas, namely the vanishing of a wave solution in certain space regions determined by wave fronts, goes much further back in the case of the wave equation for which the bounded domain of dependence was known by the early 19th century. Also the name of "Huyghens Principle" was given to the sharp or clean cut wave propagation in 1 or 3 (or odd) space dimensions, and this can be subsumed under the theory of lacunas merely by noting that the interior of the wave surface is a lacuna for n odd. For n even diffusion of waves occurs and the interior of W is not a lacuna.

We shall give here a brief description of the results of Petrowsky and of some results of Atiyah, Gårding, and Bott (1, 2) who have recently extended and consolidated Petrowsky's work. These studies involve extensive work in the algebraic topology of the surfaces S and W, which we shall not attempt to describe in detail, referring the reader to the original papers instead.

Petrowsky considers a hyperbolic polynomial $P(D)$,

homogeneous of degree m, and the elementary solution $K(t,x)$ such that

$$P(D)K(t,x) = \delta_n(x)\delta(t),$$

while $K(t,x) \equiv 0$ for $t < 0$. The formulas given in a previous section for $K(t,x)$ are then expressed by means of the function

$$\phi(x,s) = \int_{P=0} \frac{d\Omega}{\Sigma x_k \xi_k + s},$$

where s is complex and the integration is carried out over the normal surface S. Then, for n odd,

$$K(t,x) = \frac{-2i}{(2\pi i)^n (m-n-1)!} \int_0^t (t-s)^{m-n-1} v(s,x)dx,$$

and for n even

$$K(t,x) = \frac{2}{(2\pi i)^n (m-n-1)!} \int_0^t (t-s)^{m-n-1} u(s,x)dx,$$

where $\phi(s+0,x) = u(s,x) + iv(s,x)$.

Thus $K(t,x)$ is expressed by means of a distribution very similar to the power $(x-i0)^{-n}$ used for the Riemann matrix. However a lacuna in the sense of Petrowsky will occur for those regions determined by $W(x,t)$ where $\phi(x,s) = 0$. Petrowsky expresses ϕ as an integral over certain topological cycles in the $n-2$ dimensional algebraic surface $P(\xi,1) = 0$ and $x \cdot \xi + t = 0$. This surface has complex dimension $n-2$.

Petrowsky defines $p-2$ dimensional cycles C_{real} and C_{imag} for the cases n odd or even, respectively, as follows. In the hyperplane $x \cdot \xi + t = 0$ choose a coordinate ξ_n so that $x \cdot \xi = x_n \xi_n$ and the hyperplane becomes $x_n \xi_n + t = 0$. Then let ξ_{n-1} be any other ξ coordinate (in this particular system). Project parallel to ξ_{n-1} in the hyperplane $x_n \xi_n + t = 0$, that is draw the straight lines parallel to ξ_{n-1} in it. For C_{real} and n odd, take the real intersection points of these lines with the normal surface S. For C_{imag} and n even, take the complex intersection points, of these lines with S, together with the real points that are limits of the complex points. Both cycles formed in this way have dimension $p-2$, and lie in the algebraic surface $S \cap \{x \cdot \xi + t = 0\}$ of complex dimension $p-2$.

Then Petrowsky shows that a lacuna occurs if these cycles are homologous to zero on the complex surface. More precisely, in the case $m \le n$, there is a lacuna if and only if the cycle C_{real} (n odd) or C_{imag} (n even) corresponding to an inner point $(-x_k)$ of the region is homologous to zero in the above complex algebraic surface. In the case $m > n$, a further construction is given for a cycle Σ formed by drawing certain "films" on C_{real} or C_{imag}. The above necessary and sufficient condition then applies to Σ in this case $m > n$.

These Petrowsky lacunas are stable, that is, they do not disappear under sufficiently small variations of the coefficients, if P is regularly hyperbolic so that S has no double points.

In the case $n = 2$, Petrowsky shows by calculating a sum of residues that if all points of intersection of S with $x \cdot \xi + t = 0$ are real, then there is a lacuna. Dually, this means that if m real tangents to $W(t)$ pass through (x), then x lies in a lacuna. Examples of this occur in two dimensional magnetohydrodynamic wave propagation, Weitzner (1), Bazerand Yen (1). Petrowsky also shows in general that if a system of hyperbolic equations with constant coefficients gives rise to a single higher order equation satisfied by each component of the system, then a lacuna of the higher order equation is necessarily a lacuna for the system. However the converse proposition is not in general true, as the system may have other lacunas as well.

Atiyah, Bott and Gårding have extended and refined Petrowski's results. They define a polynomial P to be hyperbolic if and only if its principal part a satisfies $a(\theta) \neq 0$ for a given $\theta \in R^n$, $\theta \neq 0$, and also $P(\xi + t\theta) \neq 0$ for all real ξ when $|\mathrm{Im}\ t|$ is large enough. Then $P(D)$ is said to be hyperbolic with respect to $\theta \left(P \in hyp(\theta), a \in Hyp(\theta)\right)$. Thus when $P = a \in Hyp(\theta)$ the condition implies that $a(\xi + t\theta) = 0$ has m real roots for every $\xi \in R^n$. Also $a(\xi)$ is real, apart from a possible constant complex factor which can be discarded. If the roots of $a(\xi + t\theta) = 0$ are all distinct, $a \in Hyp^0(\theta)$ and is called strongly hyperbolic. Indeed every P with principal part a is hyperbolic only if a is strongly hyperbolic; these P are denoted as the class $hyp^0(\theta)$. For simplicity we shall assume that P and a are complete polynomials in R^n, that is, that they are not polynomials on any proper linear subspace of R^n.

Supposing that $a(\xi)$ is a homogeneous polynomial, let $A = \{a(\xi) = 0\}$ be the associated hypersurface, the complex normal cone. Then its real part $\mathrm{Re}\ A$ meets every real straight line parallel to θ in m points. If A is strongly hyperbolic and the line does not meet the origin, these m points are distinct. The component $\Gamma(A, \theta)$ of $\mathrm{Re}\ Z - \mathrm{Re}\ A$, where $Z = (x_1, \ldots, x_n) \subseteq \mathbb{C}^n$, that contains θ is an open convex cone. Now $P \in hyp(\theta)$ will imply $P \in hyp(\eta)$ for every $\eta \in \Gamma(A, \theta) = \Gamma(P, \theta)$.

The fundamental solution $E(P, x) = E(P, \theta, x)$ of P with support in $x \cdot \theta \geq 0$ is just the Fourier Laplace inverse transform of P^{-1}, that is

$$E(P, x) = (2\pi)^{-n} \int_{R^n} P(\xi - i\eta)^{-1} e^{ix(\xi - i\eta)} \, d\xi ,$$

where $\eta \in \Gamma$, such that $P(\xi - i\eta) \neq 0$. The expression is independent of η and by the Paley-Wiener-Schwartz theorem, $E(P, x) = 0$ unless $x \cdot \eta \geq 0$ for all $x \in \Gamma$. This condition can be taken to define the dual wave-cone $K = K(P, \theta) = K(A, \theta)$ of P. This cone of wave propagation is closed and convex, and meets every half space $x \cdot \eta = \text{const.}$ where $\eta \in \Gamma$, in a compact set. K is also the closed convex hull of the support of E. Also if $P = a + b$ where b has lower degree then it can be shown that $ba^{-1}(\xi - is\theta)$ tends to zero uniformly in ξ when $s \to \infty$. Consequently the relation

$$E(P, \theta, x) = \sum_{k=0}^{\infty} (-1)^k b(D)^k E(a^{k+1}, \theta, x)$$

is valid and this reduces the problem to the case of homogeneous a.

When a is strongly hyperbolic it can be shown by choice of a suitable path of integration that E is holomorphic in x outside the wave cone. If double points or other singularities occur, the wave cone must be interpreted as the convex hull as in earlier sections, and the same holomorphic behaviour of E can then be established.

Supposing a is homogeneous, one can perform a radial integration in the integral. Atiyah, Bott and Gårding give the following result: For $x \notin W(A, \theta)$ the wave cone, then

$$D^\nu E(a, x) = i(2\pi)^{1-n} \int_{\alpha^*} \chi_q^0(ix \cdot \xi) \xi^\nu a(\xi)^{-1} \omega(\xi)$$

when $q = m - n - |\nu| \geq 0$, and

$$D^\nu E(a, x) = (2\pi)^{-n} \int_{t_x \partial\alpha^*} \chi_q^0(ix \cdot \zeta) \zeta^\nu a(\xi)^{-1} \omega(\xi) ,$$

when $q < 0$. Here

$$\chi_r^0(t) = \begin{cases} t^r/r! & \text{for } r \geq 0 \\ (-1)^{r+1}(-r-1)! \, tr & \text{for } r < 0 , \end{cases}$$

and the Leray cycle

$\alpha^* = (A, x, \theta)^* \in H_{n-1}(Z^* - A^*, X^*)_*$ is the homology class of the images of a certain $\beta(x)^+$ in Z^* under the maps

$$\xi \to \xi - i v(\xi) \qquad \xi \in R \qquad u \neq 0 .$$

Also $\omega(\xi) = \Sigma_k \xi_k \tau_k(\xi)$, where $\tau_k(\xi)$ is the right cofactor of $d\xi_k$ in $\tau(\xi) = d\xi_1 \wedge \cdots \wedge d\xi_n$.

The tube operation $t_x : H_{n-2}(X^* - X^* \cap A^*) \to H_{n-1}(Z^* - A^* \cup X^*)$ is generated by the boundary of a small 2-disc in the normal bundle of X^* when its centre moves on X^*. Because of the orientation $x \cdot \xi \omega(x \cdot \xi) > 0$, the homology class α^* depends strongly on the parity of n and is the counterpart of Petrowsky cycles C_{real} and C_{imag} for the cases n odd or even. Petrowsky's formulas can be obtained for $q \geq 0$ by taking one residue onto A^*, and for $q < 0$ by taking two successive residues onto $A^* \cap X^*$.

The definition of lacuna is also extended; thus a component L of the complement of the singular support of E is called a lacuna if E has a C^∞ extension from L to $\bar{L}$. Then a lacuna still has the property that it is bounded by sharp wave fronts — fronts that arrive "without advance warning". When E vanishes in L, L is said to be a strong lacuna. If $n > 4$, there are homogeneous hyperbolic a such that W is not all of $SSE(a)$ (singular support of $E(a)$); however if k is large enough, $W = SSE(a^k)$. A lacunary component L of the complement of W such that $E(P)$ has a C^∞ extension from L to $\bar{L}$ is called regular. In a regular lacuna, $E(a)$ is a polynomial of degree $m-n$ in (x, t), so that for $m < n$ the lacuna is strong as $E(a)$ must vanish. If L is a regular lacuna for all powers a^k, $k = 1, 2, \ldots$, and $P \in hyp(\theta)$ has principal part a, then $E(P, \theta)$ will be an entire function in L and hence L is a lacuna in the extended sense. This will occur if $\partial\alpha^*(A, x, \theta) = 0$ in $H_{n-2}(X^* - X^* \cap A^*)$ which is essentially Petrowsky's condition. There are examples for every m, n of such stable Petrowsky lacunas.

The main result of Atiyah, Bott and Gårding is stated as follows: For k sufficiently large, all regular lacunas for a^k are Petrowsky lacunas, with the support of $E(a^k)$ being the propagation cone $K(A, \theta)$ and singular support the wave front surface $W(A, \theta)$. Also, for k large enough, the trivial (exterior) lacuna is the only strong lacuna for a^k. Their two papers contain a variety of other results based on a detailed study of the cohomology of algebraic varieties.

3.2 Second order systems

A hyperbolic system represents wave propagation without frictional dissipation and it is therefore natural to consider systems of the type

$$\frac{\partial^2 u_r}{\partial t^2} = \sum_{p,q,s} C_{pqrs} \frac{\partial^2 u_s}{\partial x_p \partial x_q},$$

where $p, q = 1, \ldots, n$ and $r, s = 1, \ldots, m$, the order of the system being therefore $2m$. For this system we derive an expression for the Green's matrix, the elementary solution in R^n with a point source at $x = 0$, $t = 0$.

An exponential solution

$$u_r = a_r e^{i(\lambda t + \xi_p x_p)}$$

will satisfy the system provided

$$(\lambda^2 \delta_{rs} - \sum_{p,q} c_{pqrs} \xi_p \xi_q) a_s = 0 .$$

Thus λ^2 must be a characteristic root of the matrix $C(\xi) = (\Sigma c_{pqrs} \xi_p \xi_q)$, which we shall assume is positive definite so that λ remains real. We shall also assume without loss of generality that $c_{pqrs} = c_{qprs}$, and we assume further that $C(\xi)$ is symmetric: $c_{pqrs} = c_{pqsr}$. It then follows that $\lambda_N(\xi)$, $N = 1, \ldots, m$, is real and of degree one in ξ: $\lambda_N(\tau\xi) = |\tau| \lambda_N(\xi)$.

As $C(\xi)$ is symmetric, it may be diagonalized by an orthogonal unitary matrix T:

$$C(\xi) = TJT^{-1} ,$$

where $J = \text{diag}\big(\lambda_N^2(\xi)\big)$. The columns of T are the normalized eigenvectors of $C(\xi)$, constructible as homogeneous functions of degree zero in λ, ξ from the linear equations for a_s as above. Set

$$a_s = t_s(\lambda, \zeta) ,$$

then

$$T_{rs} = t_s\big(\lambda_r(\xi), \xi\big)$$

is homogeneous of degree zero in ξ. The orthogonal or unitary relations $T\tilde{T} = E$ and $\tilde{T}T = E$ become

$$\sum_s t_p(\lambda_s, \xi) t_q(\lambda_\xi, \xi) = \delta_{pq}$$

and

$$\sum_r t_r(\lambda_M, \xi) t_r(\lambda_N, \xi) = \delta_{MN} .$$

We seek a solution of the homogeneous equations with initial conditions

$$u_p(x,0) = 0 \quad , \quad \frac{\partial u_p}{\partial t}(x,0) = g_p(x) \ ,$$

where $g_p(x)$ will later be taken as a delta function. An appropriate form of solution is given by the Fourier transform

$$\hat{u}_p(\xi,t) = \sum_N C^N(\xi) t_p(\lambda_N(\xi),\xi) \sin(\lambda_N(\xi)t) \ .$$

At $t = 0$ we have

$$g_p(\xi) = \sum_N C^N(\xi) t_p(\lambda_N(\xi),\xi) \lambda_N(\xi) \ ,$$

whereas on multiplication by $\bar{t}_p(\lambda_M(\xi),\xi)$ and summation over p we find

$$C^M(\xi) = \frac{1}{\lambda_M(\xi)} \sum_p \hat{g}_p(\xi) \bar{t}_p(\lambda_M(\xi),\xi)$$

so that

$$\hat{u}_p(\xi,t) = \sum_{N,q} t_p(\lambda_N,\xi) t_q(\lambda_N,\xi) \hat{g}_q(\xi) \frac{\sin(\lambda_N(\xi)t)}{\lambda_N(\xi)} \ .$$

If now we specialize $g_q(x)$ to be a delta function we have $g_q(\xi) = \delta_{qq}$, say, and it is evident that the Green's matrix has the form

$$\hat{G}_{pq}(\xi,t) = \sum_{N=1}^{m} t_p(\lambda_N,\xi) t_q(\lambda_N,\xi) \frac{\sin(\lambda_N(\xi)t)}{\lambda_N(\xi)} \ .$$

Taking the inverse Fourier transform we find

$$G_{pq}(x,t) = \frac{1}{(2\pi)^n} \int_{R^n} e^{-ix\cdot\xi} \sum_N t_p(\lambda_N,\xi) \bar{t}_q(\lambda_N,\xi) \frac{\sin(\lambda_N(\xi)t)}{\lambda_N(\xi)} d\xi.$$

Setting $\xi = |\xi|\eta$, $d\xi = |\xi|^{n-1} d|\xi| d\Omega_\eta$, we have

$$G_{pq}(x,t) = \int_{\Omega_\eta} \sum \frac{t_p(\lambda_N,\eta) \bar{t}_q(\lambda_N,\eta)}{(2\pi)^n \lambda_N(\eta)} d\Omega$$

$$\cdot \int_0^\infty e^{-ix\cdot\eta|\xi|} \sin(\lambda_N(\eta)|\xi|t) |\xi|^{n-2} d|\xi| \ .$$

The radial integral may be expressed as a difference of two

integrals each of the form

$$\int_0^\infty e^{-ia|\xi|}|\xi|^{n-2}d|\xi| = F(|\xi|_-^{n-2})$$

$$= i^{n-1}(-1)^{n-1}(n-2)!\left[a^{-n+1} - \frac{(-1)^{n-1}}{(n-2)!}\, i\pi\delta_{(a)}^{(n-2)}\right]$$

$$= i^{n-1}(-1)^{n-1}(n-2)!(a+i0)^{-n+1} .$$

Thus we have

$$G_{pq}(x, t) = \frac{(n-2)!}{2(2\pi i)^n}\int_{\Omega_\eta}\sum \frac{t_p(\lambda_N, \eta)\bar{t}_q(\lambda_N, \eta)}{\lambda_N(\eta)}$$

$$\cdot \left[(x\cdot\eta + t\lambda_N(\eta) + i0)^{-n+1}\right.$$

$$\left. - (x\cdot\eta - t\lambda_N(\eta) + i0)^{-n+1}\right]d\Omega_\eta .$$

Again, using the relation $\xi\cdot\zeta dS = |\xi|^n\, d\Omega_\eta$, we can transform this expression to an integral over the m-sheeted normal surface S. On the Nth sheet $\lambda_N(\zeta) = 1$, so

$$G_{pq}(x, t) = \frac{(n-2)!}{2(2\pi i)^n}\int_S t_p(1, \xi)t_q(1, \xi)\left[(x\cdot\eta + t + i0)^{-n+1}\right.$$

$$\left. - (x\cdot\eta - t + i0)^{-n+1}\right]\xi\cdot\zeta dS .$$

Here again ζ is the unit normal to S.

An interesting example of a second order system is the set of equations of motion for elastic waves in a 3-dimensional anisotropic medium (Duff, 3). Here $m = n = 3$ and a detailed analysis of the normal and wave surfaces has been made by Musgrave (1, 2, 3, 4). Section diagrams for zinc (hexagonal symmetry) and nickel (cubic symmetry) are shown in Figures 11,12,13. A detailed analysis of the solution formulas shows for this type of system that: there is a sharp wave (of degree equal to that of the data) on each sheet of the wave surface, (so that for certain directions in the examples shown five sharp waves cross each point). There is also a continuous wave, or volume wave again with the smoothness of the initial data, in the regions between the outer and innermost wave sheets. But the innermost region is a lacuna so there is no diffusion of waves just as in the case of the wave equation in three space dimensions

The elastic equations for an isotropic medium form a particularly simple case which was actually solved by Stokes (1) in 1849. The number of elastic constants is reduced to two and

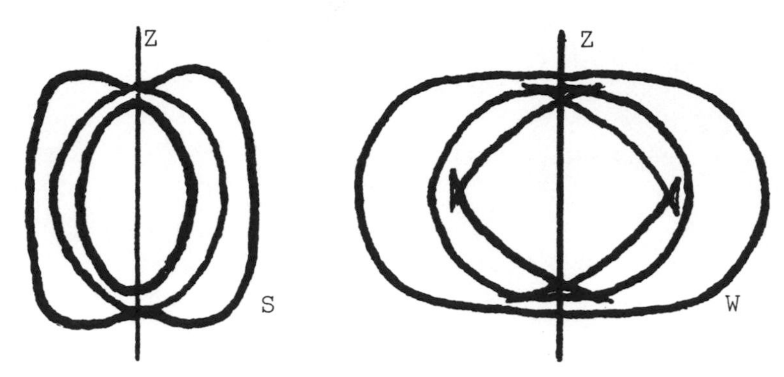

Fig. 11. Inverse surface S , and wave surface W , for elastic waves in zinc. Rotational symmetry about Z axis (after Musgrave).

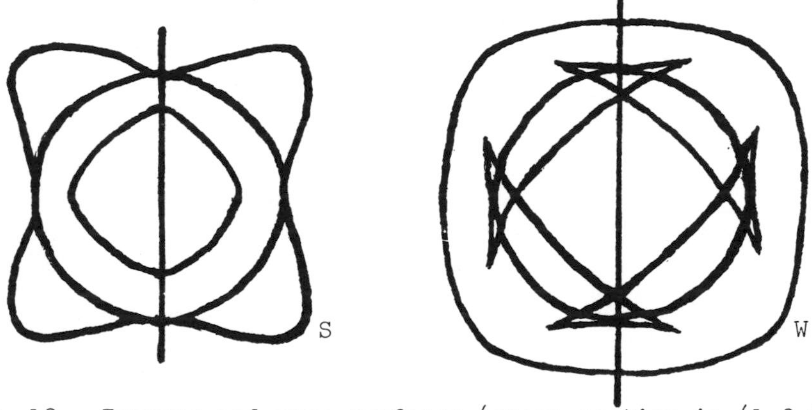

Fig. 12. Inverse and wave surfaces (cross section in (1,0,0) plane) of nickel, (cubic symmetry).

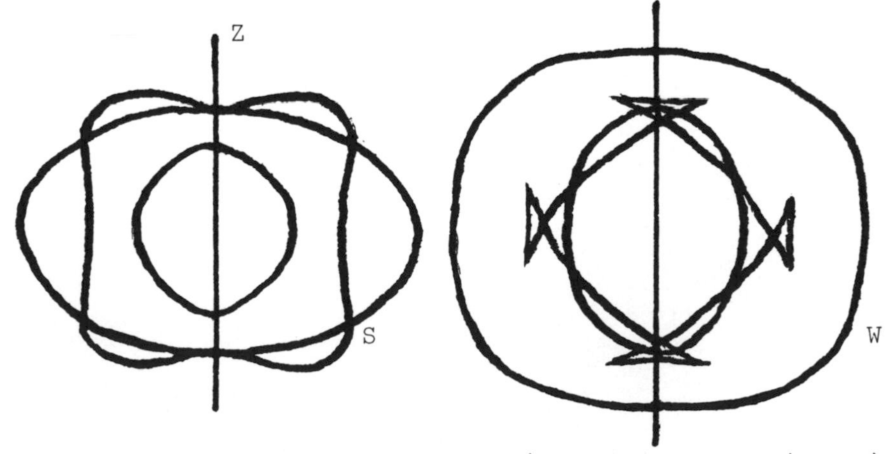

Fig. 13. Inverse and wave surfaces (cross section in (1,1,0) plane) of nickel, (cubic symmetry).

the equations take the form

$$\rho \frac{\partial^2 u_p}{\partial t^2} = (\lambda+\mu) \frac{\partial \theta}{\partial x_p} + \mu \nabla^2 u_p ,$$

with

$$\theta = \sum_p \frac{\partial u_p}{\partial x_p} .$$

Taking density $\rho = 1$ for simplicity, we may note that the normal surface consists of two spheres. The inner sheet is single, and the outer sheet has multiplicity two. Thus the outer wave sheet corresponds to P-waves of pressure, or compression, with one eigenvector oriented radially. The double inner spherical wave sheet represents two tangential modes of oscillation, called S waves or shear waves. The two eigenvectors are tangential and the S wave velocity is $\sqrt{\mu}$, whereas the P wave velocity is $\sqrt{\lambda+2\mu}$.

The solution formula for the initial value problem is

$$u_p(x,t) = \frac{t}{4\pi} \int_{\Omega_\zeta} \zeta_p \zeta_q g_q(\sqrt{\lambda+2\mu}\, t\zeta + x) d\Omega_\zeta$$

$$+ \frac{t}{4\pi} \int_{\Omega_\zeta} (\delta_{pq} - \zeta_p \zeta_q) g_q(\sqrt{\mu}\, t\zeta + x) d\Omega_\zeta$$

$$+ \frac{t}{4\pi} \iint_{\sqrt{\mu}\, t < |z| < \sqrt{\lambda+2\mu}\, t} (3\zeta_p \zeta_q - \delta_{pq}) g_q(z+x) d\Omega_\zeta \frac{d|z|}{|z|} .$$

Here $z = |z|\zeta$, $|\zeta| = 1$. The central lacuna is evident. Stokes solution underlies much applied work in seismology which in turn contributes to the understanding of the structure of the Earth, and, recently of the Moon. Boundary waves for this system in the half space problem are called Rayleigh waves, and will be studied later. Detailed study of both direct and reflected waves may be found in Cagniard (1) and Ewing, Jardetsky and Press (1).

Another famous example which shows interesting wave propagation properties is the set of equations of crystal optics (Courant, 1, p. 602). These have the form

$$\sigma_p \frac{\partial^2 u_p}{\partial t^2} = \Delta u_p - \frac{\partial}{\partial x_p} \theta , \qquad p = 1,2,3 ,$$

where $\theta = \operatorname{div} \vec{u} = \Sigma \frac{\partial u_p}{\partial x_p}$.

Here (u_1, u_2, u_3) are components of the electric vector E, and $\sigma_1, \sigma_2, \sigma_3$ are numbers related to the three principal dielectric constants of the crystalline optic medium.

The normal surface is the two sheeted Fresnel surface

$$F(\xi) = \sum_{p=1}^{3} \frac{\xi_p^2}{|\xi|^2 - \sigma_p} = 1 ,$$

which has four double "conical" points in the coordinate plane of the largest and smallest σ_p. The wave surface turns out to be a Fresnel surface also:

$$F_1(x) = \sum_{p=1}^{3} \frac{x_p^2}{|x|^2 - \sigma_p^{-1}} = 1 .$$

The numbers σ_p^{-1} are the reciprocals of the dielectric constants. The wave surface also includes four plane "lids" dual to the conical points.

3.3 Localization

The more complicated singularities of elementary solutions are related to higher order zeros of the hyperbolic polynomial, that is, to multiplicities of the normal surface. Atiyah, Bott and Gårding (1) have developed a general approach to the study of singularities based on a process of localization of P, and adapting to hyperbolic polynomials a method of Hörmander (1). Given a hyperbolic polynomial P of degree m and a point $\xi = (\xi_1, \ldots, \xi_n, \tau)$ in the vicinity of which P is to be studied, consider the expansion of $t^m P(t^{-1}\xi + \zeta)$ in ascending powers of t. Let $P_\xi(\zeta)$ be the first nonzero term, of degree say p in t:

$$t^m P(t^{-1}\xi + \zeta) = t^p P_\xi(\zeta) + O(t^{p+1}) .$$

Call $p = m_\xi(P)$ the multiplicity of P at ξ, ordinarily equal to the number of sheets of the normal cone passing through ξ. The polynomial $P_\xi(\zeta)$ is called the localization of P at ξ. If $a(\xi)$ is the principal part of P, and $a(\xi) \neq 0$, then $P_\xi(\zeta) = a(\xi) = \text{const.}$ If $a(\xi) \neq 0$ but $\text{grad}\, a(\xi) \not\equiv 0$, then $p = 1$ and $P_\xi(\zeta) = \text{grad}\, a(\zeta) \cdot \zeta + \text{const.}$ has degree 1. If P is not strongly hyperbolic, higher orders may occur at multiple points of $P(\xi) = 0$. If $P = a$ is homogeneous, then

$$t^m P(t^{-1}\xi + \zeta) = a(\xi + t\zeta) = t^p a_\xi(\zeta) + O(t^{p+1}) .$$

It can be shown that $m_\xi(P) = m_\xi(a)$, that a_ξ is the principal part of P_ξ and is hyperbolic, and that the local cone $\Gamma(a_\xi, \theta)$ contains the normal cone $\Gamma(a, \theta)$. The wave cone, or propagation cone $K(P, \theta)$ is the geometric dual of $\Gamma(P, \theta)$:

$$K(P, \theta) = K(a, \theta) = \{x \mid x \in R^{n+1}, x\Gamma(a, \theta) \geq 0\} .$$

Generally, the convex boundary of $K(P, \theta)$ is the outermost wave surface. However the singular support of $E(P, \theta, x)$ consists of other, interior, sheets as well. These can be described using the local propagation cones $K_\xi(a, \theta) = K(a_\xi, \theta)$. When $a(\xi) \neq 0$, $\xi \in R^{n+1}$, then $P_\xi(\zeta) = a(\xi) = \text{const.}$ and $\Gamma_\xi = R^{n+1}$, $K_\xi = \{0\}$. If $a(\xi) = 0$ but $\text{grad } a(\xi) \neq 0$, then $a_\xi(\zeta) = \Sigma\zeta_k \partial a/\partial\xi_k$ so that Γ_ξ is the half space $a_\xi(\theta)^{-1}a_\xi(\zeta) > 0$ and K_ξ is the half ray spanned by $a_\xi(\theta)^{-1}\text{grad } a(\xi)$. If P is not strongly hyperbolic, Γ_ξ may be smaller than a half space and K_ξ larger than a half ray — they may both be dual proper cones. We can now define the wave front surface $W(a, \theta)$ as the union of all the K_ξ:

$$W = W(P, \theta) = W(a, \theta) = \bigcup_\xi K_\xi(P, \theta) , \quad \xi \neq 0 .$$

The fundamental solution is

$$E(P, x) = (2\pi)^{-n} \int_{R^{n+1}} \frac{e^{ix(\xi + i\eta)}}{P(\xi + i\eta)} d\xi ,$$

where $\eta \in - s\theta - \Gamma(P, \theta)$ with s large enough, has as its support $K(P, \theta)$. For the principal part a, the formula can be written as a distribution integral,

$$E(a, x) = \frac{1}{(2\pi)^n} \int_{R^{n+1}} a_-(\xi)^{-1} e^{ix\xi} d\xi ,$$

where

$$a_-(\xi)^{-1} = \lim_{t\to 0_+} a(\xi + it\eta)^{-1} , \quad \eta \in -\Gamma(a, \theta) .$$

For a nonhomogeneous hyperbolic polynomial P, we may write $P = a + b$, and

$$P^{-1}(\xi + i\eta) = \sum_{k=0}^{\infty} (-1)^k b(\xi + i\eta)^k a(\xi + i\eta)^{-k-1} .$$

Then it follows that

$$E(P, \theta, x) = \sum_{k=0}^{\infty} (-1)^k b(D)^k E(a^{k+1}, \theta, x) ,$$

where the series converges in the distribution sense. For proofs we refer to Atiyah, Bott and Gårding (1, p. 143).

The localized elementary solution $E_\xi(P, \theta, x) = E(P_\xi, \theta, x)$ has its support in the local cone: $S(E_\xi) \subset K_\xi = K(P_\xi, \theta) \subset K(P, \theta)$. Hörmander (2) has shown that in a certain sense E_ξ is just a localization of E, as in the following localization theorem of Atiyah, Bott and Gårding.

Let p be the multiplicity of P at ξ. Then

$$t^{m-p} e^{-itx\xi} E(x) \to E_\xi(x) \quad \text{as} \quad t \to \infty ,$$

where the limit is taken in the distribution sense. Further,

$$S(E_\xi) \subset SS(E)$$

for $\xi \neq 0$, where SS denotes singular support.

We shall give the extremely direct and instructive proof by Atiyah, Bott and Gårding. Let

$$E_t(x) = e^{-itx} E(x) = (2\pi)^{-n} \int_{R^{n+1}} P(t\xi + \zeta + i\eta)^{-1} e^{ix(\zeta + i\eta)} d\zeta .$$

Then

$$\left(t^{m-p} E_t(x), g(-x)\right) = (2\pi)^{-n} \int_{R^{n+1}} t^{p-m} P(t\xi + \zeta + i\eta)^{-1} \hat{g}(\zeta + i\eta) d\zeta.$$

With $\eta = s\theta$ for sufficiently large s, the right hand side approaches (for $t \to \infty$)

$$(2\pi)^{-n} \int_{R^{n+1}} P_\xi(\zeta + i\eta)^{-1} \hat{g}(\zeta + i\eta) d\zeta = \left(E_\xi(x), g(-x)\right) .$$

This shows that $t^{m-p} e^{-itx\xi} E(x) \to E_\xi(x)$. Now let $g \in C_0(V)$, where V is the complement of SSE. With $\xi \neq 0$, then by the Riemann - Lebesgue lemma,

$$\int t^{m-p} e^{-itx\xi} E(P, x) g(x) dx$$

will tend to zero as t tends to infinity. Thus $\left(E_\xi(x), g(x)\right)_V = 0$ and so $S(E_\xi)$ lies in the complement of V which is $SS(E)$.

This result yields the generalized envelope relation $SS(E) \supset \bigcup_{\xi \neq 0} S(E_\xi)$. Whether equality holds in this inclusion is not known. The theorem also extends to derivatives of $E(P, x)$: if $F = Q(D)E$, then $F_\xi = Q_\xi(D)E(P_\xi)$ for ξ real and the multiplicity factor m is that of $Q/P = f$. Then $t^{m_\zeta(f) - m(f)} e^{-itx\xi} F(x) \to F_\xi(x)$ as $t \to \infty$, in the distribution sense, while $S(F_\xi) \subset SS(F)$, for $\xi \neq 0$.

Analytic continuation with respect to a complex power parameter s is also defined, thus let

$$E_s(a, x) = (2\pi)^{-n} \int a(\xi + i\eta)^{-s} e^{ix(\xi+i\eta)} d\xi , \quad \eta \in -\Gamma(a, \theta) .$$

Then E_s is of degree $ms - n$ in x and $a(D)E_s = E_{s-1}$, while $E_0(x) = \delta(x)$. This construction generalizes the work of M. Riesz, (1) and is related to results of Gelfand - Shilov (1).

In their second paper (2), Atiyah, Bott and Gårding study behaviour of $E(a, x)$ close to the wave front surface $W(P, \theta)$ which is a cone in $R^{n+1} = \{x_1, \ldots, x_n, x_{n+1} = t\}$ if $\theta = \{0, 0, \ldots 0, 1\}$. Let L be a component of the complement of $W(a, \theta)$ and y a point of ∂L. Then E is said to be C^∞ (or holomorphically) sharp if y has a neighbourhood N such that E has a C^∞ (or holomorphic) extension from L to $\bar{L} \cap N$. There is a "local Petrowski condition" for holomorphic sharpness from L at y, namely $\beta(a, x, \theta)^* \in H_{n-2}(Y^* - Y^* \cap A^*)$ when $x \in L$ is close enough to y. (Compare with the global Petrowsky condition $\beta(a, x, \theta)^* = 0$ in $H_{n-2}(X^* - X^* \cap A^*)$, which implies x belongs to a lacuna for all powers of a.) It is shown that the local Petrowsky condition is necessary for sharpness at any point of W with non-degenerate curvature. For such ordinary or non-degenerate points, the Petrowsky condition is therefore both necessary and sufficient for sharpness.

The remaining singular points of E are those arising from multiple points of $a(\xi) = 0$, and forming ruled surface or "plane" components of the wave fronts. A hyperplane wave front of $W(a, \theta)$ corresponds to a conical point of $\mathrm{Re}\, A \equiv \{a(\xi) = 0\}$ the normal cone, and the main result is given for this case. At a conical point the localized polynomial $a_\xi(\zeta)$ has lineality with minimal dimension for a cone, this implies $L(a_\xi) = C\zeta$. (The lineality $L(Q)$ is the largest linear space such that Q is a polynomial on R^{n+1}/L). The plane portion of $W(a, \theta)$ is then just the local propagation cone $K(a_\xi, \theta) \subset R^{n+1}$ which spans the hyperplane $x \cdot \xi = 0$. If, close to y, $W(a, \theta)$ coincides with $K(a_\xi, \theta)$, then we say that $y \in K(a_\xi, \theta)$ is a simple point of $W(a, \theta)$.

We also need the concept of a reduced wave surface $\hat{W}$ for the localized polynomial. The reduced wave front $\hat{W}(a, \theta)$ is the union of the local propagation cones $K(a_\xi, \theta)$ when ξ is real and does not lie in $L(A)$. We note that for a complete polynomial the reduced wave cone coincides with the ordinary wave cone. Thus the reduced wave cone $W(a_\eta, \theta)$ is distinct in this situation because a_η has nonzero lineality. After these preliminaries we can state the result for plane wave fronts. It is that if $y \in K(a_\xi, \theta) - \hat{W}(a_\xi, \theta)$ is a simple point of $W(a, \theta)$, then for all $k > 0$, $E(a^k, \theta_x)$ is holomorphically sharp at y from both sides of $K(a_\xi, \theta)$. We omit the proof, which depends on showing that the local Petrowsky condition holds, and instead refer the reader to the papers cited by Atiyah, Bott and Gårding.

When a function or distribution has different asymptotic expansions on the two sides of a hyperplane $x_1 = 0$, then it is said to be weakly sharp there. The asymptotic expansion of its jump can then be defined, and takes the form

$$J(f) \sim \sum_j \theta_j(x_1)f_j(x') \qquad x' = (x_2, \dots, x_n),$$

where $f_j(x') \in C^\infty(M : x_1 = 0)$. Here also

$$\theta_k(t) = \frac{\theta(t)t^k}{k!}, \quad \theta_{-k-1}(t) = \delta^{(k)}(t)$$

are integrals and derivatives of the Heaviside function $\theta(t) = H(t) = 1$ for $t > 0$, and 0 for $t < 0$. $J(f)$ is essentially defined by $f = f_1 + J(f)$, where $f_1 \in C^\infty$.

If $\xi = (1, 0, 0, 0, \dots)$ is a conical point of $a(\xi) = 0$ with multiplicity p, then $a(\zeta) = \zeta_1^{m-p}a_\xi(\zeta') \bmod \zeta_1^{m-p-1}$, where $\zeta' = (\zeta_2, \dots, \zeta_n)$ and $a_\xi(\zeta')$ is a complete polynomial of degree p. The dual cone $K(a_\xi, \theta)$ is then contained in the hyperplane $x_1 = 0$. Then $E(a, \theta, x)$ is weakly sharp across $x_1 = 0$, and the asymptotic expansion of its jump is

$$\sum_{j \geq 0} \theta_{m-p-1+j}(x_1)H_{p+1-j-n}(x'),$$

where the indices specify homogeneity. Here also

$$H_{p+1-j-n}(x') = \sum_{0 \leq \ell \leq j} Q_{j\ell}(D')E(a_\eta^{\ell+1}, x'),$$

with $\theta_{j\ell}$ polynomials defined as follows by the formal expansion of $a(\zeta)^{-1}$: in terms of rational functions with denominators powers of ζ_1 and a_ξ:

$$a(\zeta)^{-1} = \sum_{0 \leq \ell \leq j} \zeta_1^{p-m-j} a_\xi(\zeta')^{-\ell-1} Q_{j\ell}(\zeta').$$

This expansion relates the singularities of $E(a, \theta, x)$ across the plane face to the behaviour of the elementary solutions of the localized powers. If the elementary solutions $E(a^k, \theta, x)$ are themselves holomorphically sharp for large k and some $y \neq 0$ with $y_1 = 0$, then the above series converges for small $x-y$ to the jump function of $E(a, \theta, x)$, which is a locally bounded function as $p < m$, and is continuous if $p < m-1$. Indeed, the homogeneity of the terms in the series expansion show that the jump then has a form that can be given for any power a^k as

$$J\big(E(a^k, \theta, x)\big) = \Theta_{k(m-p)-1}(\xi x)H_{kp+1-n}(x),$$

where the indices again indicate homogeneity, and where H is holomorphic.

Also it can be shown that if the Petrowsky condition for a_ξ holds at y, that is, $\beta(a_\xi, y, \theta)^* = 0$ in $H_{n-3}(Y^* - Y^* \cap A_\xi^*)$, with a_ξ being considered as a polynomial in R^n mod C_η, then $H_{kp+1-n}(x)$ is a polynomial. In particular, if $kp < n-1$, it must be zero, so that $E(a^k, \theta, x)$ is holomorphic across $W(a, \theta)$ at y. Thus if y is in a lacuna for all powers of a_ξ (considered as a polynomial on R^n/C_ξ) then $H(x)$ must be a polynomial and so must itself or in its derivatives vanish for $kp < n-1$. Then, as stated above, $E(a^k, \theta, x)$ is holomorphic across $W(a, \theta)$ at y so that no singularity is carried on this sheet of the wave surface.

As an application of this powerful result, it is easily seen that for $n = 3$, the wave front of a first order symmetric hyperbolic system carries no singularity on the relatively open plane parts associated with multiple points of its normal surface. For $n > 3$, the same will be true for hyperplane wave front parts if the appropriate Petrowsky condition holds.

Thus the singular support of $E(a, \theta, x)$ does not contain lacunas of the localization provided that certain homogeneity conditions hold. The local Petrowsky condition is thereby related to the global Petrowsky condition in the next lower dimension.

Employing the localization method, Tsuji (5) has recently studied the case where $P(\xi) = P(\tau, \xi)$ is a product of strictly hyperbolic polynomials. Here τ is dual to $D_t = i\partial/\partial t$, t a fixed time coordinate, and ξ_k is dual to $D_{x_k} = i\partial/\partial x_k$, $k = 1, \ldots, n$. THus

$$P(\tau, \xi) = \prod_{i=1}^{\kappa} P_i(\tau, \xi)^{\alpha_i},$$

where P_i has order m_i and $\Sigma \alpha_i m_i = m$. Then for localization about (σ^0, ξ^0) consider

$$\begin{aligned}
& e^{-is(t\sigma^0 + x\xi^0)} E(t, x) \\
&= (2\pi)^{-n-1} \int_{R^{n+1}} \frac{e^{it(\tau - s\sigma^0) + ix(\xi - s\xi^0)}}{P(\tau, \xi)} \, d\sigma d\xi \\
&= (2\pi)^{-n-1} \int_{R^{n+1}} \frac{e^{i(t\tau + x\xi)} \, d\sigma d\xi}{P\big(s(\sigma^0, \xi^0) + (\tau, \xi)\big)} \\
&= s^{-m} (2\pi)^{-m-1} \int_{R^{m+1}} \frac{e^{i(t\tau + x\xi)} \, d\sigma d\xi}{P\big((\sigma^0, \xi^0) + s^{-1}(\tau, \xi)\big)} .
\end{aligned}$$

Expanding in powers of s^{-1}, we have

$$\frac{1}{P\big((\sigma^0\xi^0 + s^{-1}(\tau , \xi)\big)} = s^{-p}\left\{ \sum_{k=0}^{N} p^{(k)}(\sigma^0 , \xi^0 , \tau , \xi)s^{-k} + \frac{s^{-N-1}}{(N+1)!} R_{N+1}(\sigma^0 , \xi^0, s , \tau , \xi)\right\} ,$$

where p is the local multiplicity of P at (σ^0 , ξ^0), and

$$p^{(k)}(\sigma^0\xi^0 , \tau , \xi) = \sum_{j=1}^{k} \frac{Q_{kj}(\sigma^0\xi^0 , \tau , \xi)}{P_{(\sigma^0\xi^0)}(\tau , \xi)^{j+1}}$$

and Q_{kj} is a polynomial of degree $\le mj$.

Define

$$E_k(t , x) = (2\pi)^{n-1} \int_{R^{n+1}} p^{(k)}(\sigma^0\xi^0 , \tau , \xi)e^{i(t\tau+x\xi)}\, d\sigma d\xi$$

amd set $e_k = -m-k+p$. Then the elementary solution has an asymptotic expansion:

$$e^{-is(t\sigma^0+x\xi^0)}E(t , x) \sim \sum_{j=0}^{\infty} E_j(\sigma^0 , \xi^0 , \tau , \xi)s^{e_j}$$

for which each remainder term

$$s^{-e_N}\left(e^{-is(t\sigma^0+x\xi^0)}E(t , x) - \sum_{j=0}^{N-1} E_j(\sigma^0 , \xi^0 , \tau , \xi)s^{e_j}\right)$$

tends to E_N in $D'(R^{N+1})$ as $s \to \infty$. Also the singular support contains the union of all localized supports:

$$SSE \supseteq \bigcup_{(\sigma^0,\xi^0)\in S^n} \bigcup_{j=0}^{\infty} \operatorname{supp} E_j(\sigma^0 , \xi^0 , \tau , \xi) .$$

Denoting this union set on the right by $WF(E)$, the wave front set, Tsuji shows that it is closed, and gives a special uniformity argument which shows that equality holds above, i.e. that $WF(E) = SSE$.

The order of singularity of E on $WF(E)$ is then studied using a Hilbert space definition of the order which gives results ½ step different from our earlier definition based on homogeneity. It is shown by estimating orders of the E_k that

1) For strictly hyperbolic P, the order agrees with our earlier value of $-m + \frac{1}{2}n + \frac{3}{2}$ (homogeneous definition).

2) If the multiplicity is constant and k-ple, the singularity becomes $k-1$ steps sharper than in 1), and

3) If the multiplicity is not constant (multiple points), the singularity at a point of the ruled surface is $\frac{1}{2}(k-1)$ steps sharper than in 1).

CHAPTER 4. THE HALF-SPACE PROBLEM, WITH CONSTANT COEFFICIENTS

4.1 Boundary conditions for the wave equation

We study here our first "mixed" problem in which boundary conditions as well as initial conditions are given. For the wave equation

$$Lu = \frac{\partial^2 u}{c^2 \partial t^2} - \Delta u$$

it is appropriate to give Cauchy initial conditions $u(x, 0) = f(x)$, $u_t(x, 0) = g(x)$, for $x > 0$, and a single boundary condition

$$B(D_t, D_x, D_y)u = h(t, y),$$

for $x = 0$, $t > 0$ and arbitrary $y_2, \ldots, y_n$. Thus $u = u(t, x, y_2, \ldots, y_n)$ is sought in the quarter space $t > 0$, $x > 0$.

Since waves propagate with the maximum velocity c, it follows that any point more distant than ct from the boundary is not affected by the presence of the boundary condition, and we shall therefore not need to treat this region further. On its boundary $x = ct$ there arise possible discontinuities or singularities due to the onset of a wave emitted from the boundary at time $t = 0$. Certain compatibility conditions connecting the initial and boundary data will determine the presence and magnitude of this wave. We shall chiefly treat the region $ct > x$ in which the boundary conditions take effect.

Consider the problem of the reflection of waves from the boundary; given that $Lu = \delta(x - x_0)\delta_{n-1}(y)\delta(t)$ with zero Cauchy data and $Bu = 0$. Waves spread with velocity c from the source and first encounter the boundary when $t = x_0/c$. Their reflection or absorption is determined by $B(\tau, \xi, \eta)$.

<u>Example 1.</u>

$$Bu = u = 0 \quad \text{for} \quad x = 0 .$$

The solution can be found by placing an opposite "image" at the image point $(-x_0, 0)$ and the reflected wave has just the opposite sign. The entire incident and reflected waves remain within the portion of the sphere $r \leq ct$, where $r^2 = (x - x_0)^2 + \Sigma y^2$, where $x > 0$. These results also hold if B contains only even powers of d/dx.

Example 2.

$$Bu = \frac{\partial u}{\partial x} = 0 \quad \text{for} \quad x = 0 .$$

This Neumann condition can be satisfied by a positive image at $(-x_0, 0)$ and the waves are reflected with no change of sign — they are even functions of x. Velocity of propagation at most c is again observed. The same is seen to hold if B contains only odd powers of $\partial/\partial x$.

If however $Bu = 0$ gives rise to wave propagation in the boundary, then we can only assert that the disturbance cannot reach any point with $r > ct$ and $x > ct - x_0$. We now examine the region $r > ct$ and $x \leq ct - x_0$ which may be reached by a wave travelling on the boundary for a part of its path, and note that this will occur only if B contains both even and odd powers of d/dx. This problem was first studied by Bondi (1).

Let

$$v = Bu ;$$

since we assume B has constant coefficients also, we have

$$Lv = LBu = BLu = 0 ,$$

in the region $r > 0$. Since $v = 0$ on B, reflection with change of sign only holds for v and hence v is propagated with velocity at most c. Thus $v = 0$ for $r > ct$. That is, for $x \geq 0$ and $r > ct$ we have

$$Bu = v = 0 .$$

Since u vanishes for $r > ct$, $x > ct - x_0$, it follows that there is a characteristic surface of the operator B separating the support of u from this outer region. Since u satisfies both $Lu = 0$ and $Bu = 0$ in $r > ct$, $x \leq ct - x_0$, the boundary of its support there must be a characteristic of L and of B. Thus we conclude that u vanishes in this region unless L and B have a common characteristic in $r > ct > h$, $0 < x \leq ct - h$. This gives a second condition necessary for "faster than sound" boundary wave propagation.

Bondi has shown that such "ultrasonic" propagation does occur for capillary surface waves. The basic equation is the wave equation, $Lu = u_{tt} - \Delta u = 0$ in three space dimensions. Let the surface be z = const. then the boundary condition is

$$B_1 u = \Delta u + \left(\frac{\partial^2}{\partial x^2} + \frac{\partial^2}{\partial y^2}\right)\frac{\partial u}{\partial z} = 0 \ .$$

As $\Delta u = u_{tt}$ we may write

$$Bu = u_{tt} + \left(\frac{\partial^2}{\partial t^2} - \frac{\partial^2}{\partial z^2}\right)u_z = 0 \ ;$$

and now L and B have common characteristics namely $z \pm t = \text{const.}$ Hence "infinitely fast" boundary propagation may occur.

To construct a solution, we place the origin at the image point 0 of S as shown in Figure 14.

Let $r^2 = x^2 + y^2 + z^2$,
$r_0^2 = x^2 + y^2 + (z+2h)^2$
and $\rho^2 = x^2 + y^2$.

We set

$$u = u_0 + u_1 + u_2 \ ,$$

where

$$u_0 = \frac{H(t - r_0)}{r_0}$$

is the emitted wave, and

$$u_1 = \frac{H(t - r)}{r}$$

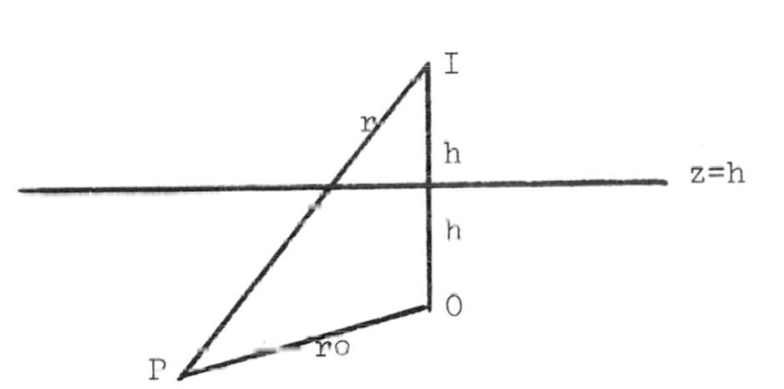

Fg. 14. Field point P source 0, and image I in plane.

is the first approximation by even reflection to the reflected wave. Then the remaining "diffracted" wave u_2 satisfies

$$Bu_2 = \left[\frac{\partial^2}{\partial t^2} + \frac{\partial}{\partial z}\left(\frac{\partial^2}{\partial t^2} - \frac{\partial^2}{\partial z^2}\right)\right]u_2 = -\frac{2\partial^2}{\partial t^2}\frac{H(t - r)}{r}$$

$$= -\frac{2}{r}\,\delta'(t - r) \ .$$

Also $Lu_2 = 0$ and $u_2 = 0$ whenever $z + h > t$.

As the equation for u_2 contains no derivatives with respect to x or y , a solution of the form

$$\Phi(t , z , \rho) = \int_{-\infty}^{\infty} d\zeta \int_0^t d\tau\, H(\tau , \zeta , \rho) F(t-\tau , z-\zeta)$$

may be tried, where $H(t , z , \rho) = -2r^{-1}\delta'(t - r)$. This will satisfy all conditions provided the kernel F satisfies

$$BF = \left[\frac{\partial^2}{\partial t^2} + \left(\frac{\partial^2}{\partial t^2} - \frac{\partial^2}{\partial z^2}\right)\frac{\partial}{\partial z}\right]F = 0$$

$$F(0\,,z) = 0$$

$$\left[\left(1 + \frac{\partial}{\partial z}\right)\frac{\partial F(t\,,z)}{\partial t}\right]_{t=0} = \delta(z)\ .$$

A typical exponential solution for $BF = 0$ is

$$\exp\left[iuz \pm iut\sqrt{\frac{iu}{1+iu}}\right]\ .$$

Combining these to satisfy the other two conditions, we are led to

$$F(t\,,z) = \frac{1}{2\pi}\int_{-\infty}^{\infty} du\,\frac{e^{iuz}}{1+iu}\,\frac{\sin\left(ut\sqrt{\frac{iu}{1+iu}}\right)}{u\sqrt{\frac{iu}{1+iu}}}\ .$$

In the integrand, the only singularity is the essential singularity at $u = i$, and we choose the path of integration to pass above $u = i$. Then $F(t\,,z)$ will vanish whenever the path of integration can be closed by the upper infinite semicircle. Hence it follows that $F(t\,,z) = 0$ for $z > t$.

Thus in the expression for Φ we need only consider values of ζ and τ satisfying $t-\tau \geq z-\zeta$, or $\zeta > z+\tau-t$. As $H(\tau\,,\zeta\,,\rho) = 0$ for $\zeta > \tau$, we find $\Phi(t\,,z\,,\rho) = 0$ for $z > t$. Thus ϕ satisfies the correct boundary relation and vanishes for $z > t$. The consolidated expression for Φ is now

$$\Phi(z\,,t)$$

$$= \frac{1}{\pi}\int_{-\sqrt{t^2-\rho^2}}^{\sqrt{t^2-\rho^2}} \frac{d\zeta}{\sqrt{\zeta^2+\rho^2}}\int_{-\infty}^{\infty} du\,\frac{e^{iu(z-\zeta)}}{1+iu}\cos\left[u\sqrt{\frac{iu}{1+iu}}\left(t-\sqrt{\zeta^2+\rho^2}\right)\right]$$

for $t \geq \rho$, while $\Phi = 0$ for $t < \rho$. In the permitted interval for ζ given by

$$z + \sqrt{t^2-\rho^2} \geq z - \zeta \pm \left\{t - \sqrt{\zeta^2+\rho^2}\right\} \geq z - \sqrt{t^2-\rho^2}\ ,$$

we see that the path of integration can again be closed if $z > \sqrt{t^2-\rho^2}$. That is, $r > t$, $z > 0$ imply $\Phi = 0$ as well.

Hence Φ is confined to regions that can be reached with the velocity of sound. But Φ cannot coincide with u_2, for $\Phi \to \infty$ as $\rho \to 0$. Also Φ turns out not to satisfy the wave equation at $\rho = 0$.

A new complementary integral Ψ is thus required, satisfying the wave equation and boundary condition, and cancelling the above singularity. Bondi shows that it is

$$\Psi(t, z, \rho) = -\frac{2}{\pi}\int_{-\infty}^{\infty} du \frac{e^{iuz}}{1+iu} \cos\left[ut\sqrt{\frac{iu}{1+iu}}\right] K_0\left(\sqrt{\frac{\rho u}{(1+iu)}}\right),$$

where the path of integration also passes above $z = i$, and $\mathrm{Re}\, u/\sqrt{1+iu} > 0$ along the path. By closing the path of integration one finds that $\psi = 0$ for $z > t$. A detailed calculation is necessary to show that the singularities for $\rho = 0$ cancel. Then we can take $u_2 = \Phi + \Psi$, and Bondi shows that this solution is unique.

By the steepest descent method it can be shown that for $\rho > 5t$,

$$\Psi \sim \frac{4t}{\pi\rho^2} \exp\left[-\left(\frac{\rho}{2t}\right)^2 z - \frac{\rho^2}{4t}\right],$$

so there is a very small signal at large distances close to the boundary. For $1.05 \leq \rho/t \leq 4.98$ the values of Ψ oscillate and decrease exponentially, while for $\rho/t \leq 1.05$ they oscillate with damping as $\rho^{-\frac{1}{2}}$ only.

This capillary wave problem is the first known example of a well posed and physically meaningful ultrasonic boundary wave propagation.

4.2 The Oblique derivative problem

As a further example of a boundary condition for the wave equation, let us take the oblique derivative boundary condition

$$Bu = \frac{\partial u}{\partial t} - \alpha\frac{\partial u}{\partial z} - \beta\frac{\partial u}{\partial x} = 0,$$

where the boundary is $z = 0$ and the domain the half space $z > 0$ in three dimensions.

For the source solution take

$$u_1(x, y, z, t) = \delta(c^2t^2 - r^2),$$

where $r^2 = x^2 + y^2 + (z - \ell)^2$, the source being thus located at $(0, 0, \ell)$. The first reflected wave front may be represented by

$$u_2(x, y, z, t) = \delta(c^2t^2 - r_1^2),$$

where r_1 is the distance from the image point: $r_1^2 = x^2 + y^2 + (z+\ell)^2$. We choose to represent the full solution u as follows:

$$u = u_1 - u_2 + u_3,$$

so that u_3 satisfies the wave equation

$$u_{tt} - c^2\Delta u = 0$$

and the boundary condition

$$\begin{aligned} Bu_3 &= B(u_2 - u_1) \\ &= \left(\frac{\partial}{\partial t} - \beta\frac{\partial}{\partial x}\right)(u_2 - u_1) - \alpha\frac{\partial}{\partial z}(u_2 - u_1). \end{aligned}$$

As the first terms cancel by symmetry, and the second terms in $\partial/\partial z$ are equal and opposite on $z = 0$, we find

$$Bu_3 = -2\alpha\frac{\partial u_2}{\partial z},$$

Thus

$$Bu_3 = 4\alpha(z+\ell)\delta'(c^2t^2 - r_1^2).$$

We treat this condition as an inhomogeneous first order partial differential equation for u_3, of the form

$$u_t - \alpha u_z - \beta u_x = g(x, z, t).$$

The characteristic equations are

$$\frac{dt}{1} = \frac{dz}{-\alpha} = \frac{dx}{-\beta},$$

with first integrals $z + \alpha t = z_1$, $x + \beta t = x_1$. Setting also $t = t_1$, we have

$$x = x_1 - \beta t_1, \quad z = z_1 - \alpha t_1, \quad t = t_1$$

and

$$\frac{\partial u}{\partial t_1}(x, z, t) = u_t - \alpha u_z - \beta u_x = g(x_1 - \beta t_1, z_1 - \alpha t_1, t_1).$$

The solution of the first order equation is therefore

$$u_3(x\,,z\,,t) = \int_0^{t_1} g(x_1 - \beta\tau\,,z_1 - \alpha\tau\,,\tau)d\tau$$

$$= 4\alpha \int_0^{t_1} (z - \alpha\tau + \ell)\delta'(c^2\tau^2 - r_1^2)d\tau\ ,$$

where

$$r_1^2 = (x_1 - \beta\tau)^2 + y^2 + (z + \ell - \alpha\tau)^2$$

$$= A\tau^2 + B\tau + C\ ,$$

where

$$A = \alpha^2 + \beta^2$$

$$B = -2\big(\beta x_1 + \alpha(z_1 + \ell)\big)$$

$$C = x_1^2 + y^2 + (z_1 + \ell)^2\ .$$

In the integral for u_3 we see that the contribution will be zero unless $c\tau = r_1$ within the interval of integration. That is, the linear characteristic must encounter a point of the reflected wave front. Hence we see that the contribution is confined to those points which lie in the "shadow" of the reflected wave front as defined by the linear characteristic rays. (Figure 15).

Noting that

$$\frac{d}{d\tau}(c^2\tau^2 - r_1^2) = 2(c^2 - A)\tau - B\ ,$$

we find

$$u_3(x\,,z\,,t) = 4\alpha \int_0^{t_1} \frac{z - \alpha\tau + \ell}{2(c^2 - A)\tau - B}\ \delta'(c^2\tau^2 - r_1^2)d(c^2\tau^2 - r_1^2)$$

$$= -4\alpha\ H(ct_1 - r_1)\ \frac{d}{d(c^2\tau_1^2 - r_1^2)}\left(\frac{z_1 - \alpha\tau_1 + \ell}{2(c^2 - A)\tau_1 - B}\right)$$

$$= -4\alpha\ \frac{H(ct_1 - r_1)\big(\alpha B - 2(c^2 - A)(z_1 + \ell)\big)}{\big(2(c^2 - A)\tau_1 - B\big)^3}\ .$$

Here τ_1 denotes the root of $c^2\tau^2 - r_1^2 = 0$, and we see that the denominator of the expression for u_3 vanishes only when the reflected wave front is tangent to the characteristic line through the field point. That is, the singularity of the wave lies on the characteristic surface formed by the shadow lines tangent to the reflected wave surface. (Figure 15). Thus an ultrasonic wave appears, if and when the direction $(-\beta, 0, -\alpha, 1)$ becomes tangent to the reflected wave cone at a point of the physical half space $z > 0$. This is possible only if $\alpha^2 + \beta^2 > c^2$.

4.3 The line source earthquake

The propagation of elastic "earth quake" waves in a half space, a fundamental problem in seismology, has been studied by many authors including Rayleigh (1), Lamb (1), Lapwood (1), Nakano (1, 2), and Sobolev (2, 3). Here we shall describe only the simplest case of a surface line source parallel to the y axis, referring the reader to Ewing, Jardetsky, and Press (1, Ch. 2) for more complete details.

For the displacement vector $\vec{u}$ in the elastic wave equations take scalar and vector potentials ϕ, ψ so that

$$\vec{u} = \text{grad}\,\phi + \text{curl}\,\psi .$$

It then follows easily that ϕ and ψ satisfy wave equations with the pressure and shear wave velocities

$$\phi_{tt} = \alpha^2\Delta\phi \quad , \quad \psi_{tt} = \beta^2\Delta\psi \, ,$$

where $\rho\alpha^2 = \lambda + 2\mu$, $\rho\beta^2 = \mu$ with ρ density, λ and μ the isotropic elastic constants.

For the two dimensional problem with z representing depth below the surface, x horizontal distance from a source point, the potentials satisfy

$$u = \phi_x - \psi_z \quad , \quad w = \phi_z + \psi_x \, .$$

The surface or boundary conditions of vanishing stress are

$$p_{zx}\Big|_{z=0} = \mu(w_x + u_z) = \mu(2\phi_{xz} + \psi_{xx} - \psi_{zz}) = 0$$

$$p_{zz}\Big|_{z=0} = \lambda\theta + 2\mu w_z = \lambda\Delta\phi + 2\mu(\phi_{zz} + \psi_{zx}) = 0 \, .$$

The plane wave expressions

$$\phi = Ae^{-\nu z - ikx + i\omega t} \quad , \quad \psi = Be^{-\nu' z - ikx + i\omega t}$$

satisfy the wave equations provided

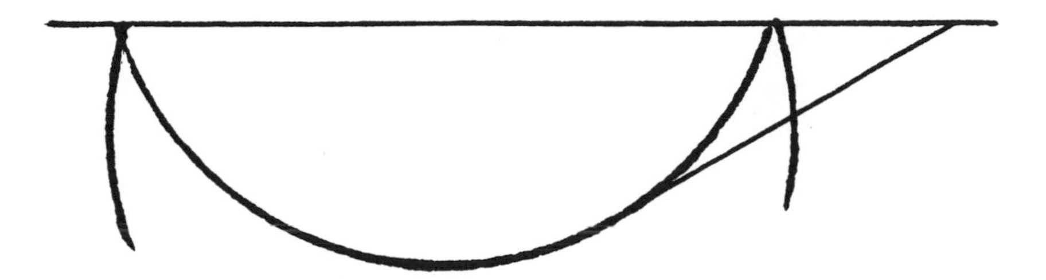

Fig. 15. Oblique ultrasonic wave front.

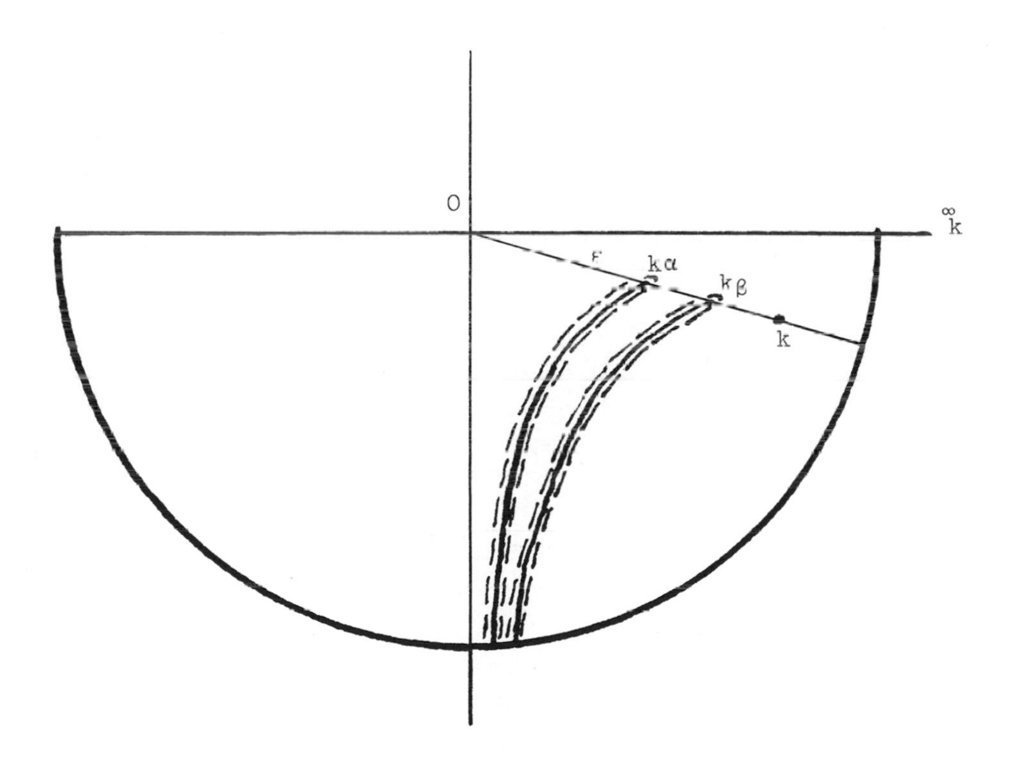

Fig. 16. Contours of integration.

$$\nu^2 = k^2 - k_\alpha^2 \ , \ k_\alpha = \frac{\omega}{\alpha} \quad ; \quad \nu'^2 = k^2 - k_\beta^2 \ , \ k_\beta = \frac{\omega}{\beta}$$

and we choose the coefficients A , B to satisfy boundary conditions representing a force applied vertically to the surface:

$$p_{xz}\Big|_{z=0} = 0 \ , \quad p_{zz}\Big|_{z=0} = Ze^{-ikx+i\omega t} \ .$$

Inserting the plane wave expressions in these conditions, we find the linear system

$$2Ai\nu k - (2k^2 - k_\beta^2)B = 0$$

$$(2k^2 - k_\beta^2)A + 2ik\nu' B = \frac{Z(k)}{\mu} \ ,$$

leading to the expressions

$$A = \frac{2k^2 - k_\beta^2}{F(k)} \cdot \frac{Z(k)}{\mu} = \frac{A_1}{F(k)} \ , \quad B = \frac{2ik\nu}{F(k)} \cdot \frac{Z(k)}{\mu} = \frac{B_1}{F(k)} \ ,$$

where the determinant

$$F(k) = (2k^2 - k_\beta^2) - 4k^2\nu\nu'$$

is known as Rayleigh's function. Apart from its branch points, $F(k)$ has a zero K at a point greater than k_β . Numerical calculation shows that the corresponding (reciprocal) wave velocity is 92% - 95% of the shear wave velocity $(\mu/\rho)^{\frac{1}{2}}$, depending on the ratio of λ to μ . Choosing values for the roots ν , ν' with $\mathrm{Re}\ \nu \geq 0$, $\mathrm{Re}\ \nu' \geq 0$, we see that the free waves arising from this zero, diminish exponentially as depth z increases. They are known as surface waves, or Rayleigh waves.

To represent a line source of strength Q at $x = 0$, we set $Z(k) = -Q dk/2\pi$ and integrate over k , and then over ω . Thus

$$\begin{pmatrix} \phi \\ \psi \end{pmatrix} = \frac{1}{2\pi} \int_{-\infty}^{\infty} \int_{-\infty}^{\infty} \begin{pmatrix} A_1 e^{-\nu z} \\ B_1 e^{-\nu' z} \end{pmatrix} \frac{e^{-i(kx - \omega t)}}{F(k)} \, dk d\omega \ .$$

Considering first the inner integral over k , we select branch cuts in the k plane so that $\mathrm{Re}\ \nu \geq 0$, $\mathrm{Re}\ \nu' \geq 0$.

For complex $\omega = s - i\sigma$, $\zeta = k + i\tau$, we have $k_\alpha = (s - i\sigma)/\alpha$ and $\nu^2 = k^2 - k_\alpha^2 = (k + i\tau)^2 - (s - i\sigma)^2/\alpha^2 = (\zeta^2 + i\tau)^2 - (s - i\sigma)^2/\alpha^2$ so that $\mathrm{Re}\ \nu = 0$ requires that

$$k^2 - \tau^2 + 2ik\tau - (s^2 - \sigma^2 - 2is\sigma)/\alpha^2$$

be real and negative. Then

$$k\tau = -s\sigma/\alpha^2 \ , \ k^2 - \tau^2 < (s^2 - \sigma^2)/\alpha^2 \ .$$

Thus the branch cut will lie on a hyperbola defined by the first of these conditions, and on the part of it described by the second. For $s > 0$ the branch cuts and pole are situated as in Figure 16, and the integral along the real axis can be expressed as a sum of two branch cut integrals, a residue at κ , and an evanescent large semicircle term.

The residue at $\kappa = \kappa_1\omega$ yields the Rayleigh wave terms

$$\begin{pmatrix}\phi_1\\ \psi_1\end{pmatrix} = i\int_{-\infty}^{\infty}\begin{pmatrix}A_1(K_1)e^{-\nu z}\\ B_1(K_1)e^{-\nu' z}\end{pmatrix}\frac{e^{-i\omega(\kappa_1 x - t)}}{F'(\kappa_1)}\frac{d\omega}{\omega} ,$$

where the homogeneity of A , B and F in k has been used to factor out all terms in ω. Writing $\nu = \nu_1\omega$, $\nu' = \nu_1\omega$, where $\nu_1^2 = \kappa_1^2 - \alpha^{-2}$, $\nu_1'^2 = \kappa_1^2 - \beta^{-2}$, we see that the contributions to the displacements from the poles are, in vector matrix form,

$$\begin{pmatrix}u_1\\ w_1\end{pmatrix} = \begin{pmatrix}\phi_{1x} - \psi_{1z}\\ \phi_{1z} + \psi_{1x}\end{pmatrix}$$

$$= i\int_{-\infty}^{\infty}\begin{pmatrix}-i\kappa_1 \ , & \nu_1'\\ \nu_1 \ , & -i\kappa_1\end{pmatrix}\begin{pmatrix}A_1(\kappa_1)e^{-\nu_1\omega z}\\ B_1(\kappa_1)e^{-\nu_1'\omega z}\end{pmatrix}\frac{e^{-i\omega(\kappa_1 x - t)}}{F'(\kappa_1)}\,d\omega .$$

But

$$\int_{-\infty}^{\infty} e^{-\nu_1\omega z - i\omega(\kappa_1 x - t)}\,d\omega$$

$$= \int_0^{\infty} e^{-\omega(\nu_1 z + i(\kappa_1 x - t))}\,d\omega + \int_0^{\infty} e^{\omega(\nu_1 z - i(\kappa_1 x - t))}\,d\omega$$

$$= \frac{1}{\nu_1 z + i(\kappa_1 x - t)} + \frac{1}{\nu_1 z - i(\kappa_1 x - t)}$$

$$= \frac{2}{\nu_1 z\left[1 + \left(\frac{k_1 x - t}{\nu_1 z}\right)^2\right]} .$$

Hence the Rayleigh waves are smooth, within the wave fronts, and are an approximation to the delta function for small z. Indeed,

$$\begin{pmatrix} u_1 \\ w_1 \end{pmatrix} = 2i \begin{pmatrix} -i\kappa_1 & \nu_1' \\ \nu_1 & -i\kappa_1 \end{pmatrix} \begin{pmatrix} A_1(\kappa_1')\Big/\nu_1 z\left(1+\left(\dfrac{\kappa_1 x - t}{\nu_1 z}\right)^2\right) \\ B_1(\kappa_1)\Big/\nu_1' z\left(1+\left(\dfrac{\kappa_1 x - t}{\nu_1' z}\right)^2\right) \end{pmatrix} \frac{1}{F'(\kappa_1)} .$$

The wave maxima follow the boundary with velocity κ_1 which is slightly less than the compressional wave velocity. Note that the amplitude of this wave does not diminish with time or distance.

Returning to the expressions for the potentials ϕ and ψ, we note that $F(k)$ contains only the product $\nu\nu'$ and is therefore single valued if a cut is made along the real axis from k_α to k_β. It is possible to express $\frac{\partial\phi}{\partial z}$ and $\frac{\partial\psi}{\partial t}$ in terms of integrals having this same single-valued property except on this cut.

Thus

$$\frac{\partial\phi}{\partial z} = \frac{-1}{2\pi}\int_{-\infty}^{\infty}\int_{-\infty}^{\infty} \frac{(2k^2 - k_\beta^2)}{F(k)} \nu e^{-\nu z + i(kx-\omega t)} dk d\omega .$$

Set $k = \kappa\omega$, so that $dk d\omega = \omega d\kappa d\omega$, and use the homogeneity of k_β, ν and $F(k)$. We find, with $\nu = \nu_1\omega$,

$$\frac{\partial\phi}{\partial z} = \frac{-1}{2\pi}\int_{-\infty}^{\infty}\int_{-\infty}^{\infty} \frac{\kappa^2 + \frac{1}{\beta^2}}{F(\kappa)} \nu_1 e^{-\omega(\nu_1 z + i(\kappa z - t))} d\kappa d\omega$$

$$= \frac{-1}{2\pi z}\int_{-\infty}^{\infty} \frac{\kappa^2 + \frac{1}{\beta^2}}{F(\kappa)} \cdot \frac{d\kappa}{\left(\left(1 + \frac{\kappa x - t}{\nu_1 z}\right)^2\right)},$$

where the inner integral is evaluated as above. The new integral in κ has integrand single valued except on the cut. Deform the contour using a large lower half plane semicircle, which gives zero contribution as the integrand is $O(\kappa^{-2})$ for large κ. The contour now becomes a loop about the positive real axis. As the integrand is single valued for $0 < \kappa < \kappa_\alpha = \alpha^{-1}$ and for $\kappa > \beta^{-1}$, these parts of the contour cancel, except for the

Rayleigh pole earlier evaluated. We are left with a loop contour C about the cut $\alpha^{-1} \le \kappa \le \beta^{-1}$, and therefore

$$\frac{\partial\phi_2}{\partial z} = \frac{-1}{2\pi z}\int_C \frac{\kappa^2 + \frac{1}{\beta^2}}{F(\kappa)} \cdot \frac{d\kappa}{\left[1 + \left(\frac{\kappa x - t}{\nu_1 z}\right)^2\right]} ,$$

where $\phi = \phi_1 + \phi_2$ the first term being the Rayleigh wave. An integration over z from ∞ to z then yields ϕ_2 which can be taken to vanish for z large.

Similarly,

$$\frac{\partial\psi}{\partial t} = \frac{-1}{2\pi}\int_{-\infty}^{\infty}\!\!\int \frac{2ik\nu\omega e^{-\nu' z - i(kx-\omega t)}}{F(k)}\, dk d\omega$$

$$= \frac{-1}{2\pi}\int_{-\infty}^{\infty}\!\!\int \frac{2i\kappa\nu_1 e^{-\omega(\nu_1' z + i(\kappa x - t))}}{F(\S)}\, d\kappa d\omega$$

$$- \frac{-1}{2\pi z}\int_{-\infty}^{\infty} \frac{2i\kappa\nu_1\, d\kappa}{F(\kappa)\nu_1'\left[1 + \left(\frac{\kappa x - t}{\nu_1' z}\right)^2\right]} ,$$

where $k = \kappa\omega$ and ω is again integrated out. Since ν_1 has branch point $\kappa = \alpha^{-1}$ and ν_1' has branch point $\kappa = \beta^{-1}$, it follows that the quotient ν_1/ν_1' is single valued except on the cut. Thus $\partial\psi/\partial t$ can also be expressed as a sum of a Rayleigh pole contribution and a loop contour C integral. The latter is

$$\frac{\partial\psi_2}{\partial t} = \frac{-1}{2\pi z}\int_C \frac{2i\kappa\nu}{F(\kappa)\nu} \cdot \frac{d\kappa}{\left[1 + \left(\frac{\kappa x - t}{\nu_1' z}\right)^2\right]} ,$$

and ψ_2 can be determined by integration over time from 0 to t . We shall omit further details.

Observe that the loop integrals represent a bundle or packet of waves that propagate with velocities from α to β . There are two sharp wave fronts represented by the leading and trailing edges of this packet. By the Paley Wiener theorem it can be shown that ϕ and ψ vanish outside the leading wave front, and this is an instance of a general result for hyperbolic boundary conditons discussed in Chapter 4. The Rayleigh wave trails the inner wave front, and at large distances becomes the largest term.

4.4 The General Mixed Boundary and Initial Value Problem

Consider a hyperbolic operator $P(D)$ of higher order m, an initial manifold $t = x_{n+1} = 0$, and a boundary hyperplane $x_1 = x = 0$. We now study the construction of a formal solution for appropriate mixed boundary conditions. In the following section we take up the deeper existential problems for what boundary conditions is such a problem well posed. Here we consider a source point at $(\ell, 0, 0, 0, \ldots 0)$ and construct a reflected wave solution for the waves generated by this point source. Thus $Pu = \delta(x-\ell)\delta_{n-1}(x)\delta(t)$, and $u \equiv 0$ for $t < 0$.

Let $\lambda_k(\xi, \xi_j)$, $k = 1, \ldots, m$ be the roots of $P(\xi) = P(\xi_1, \xi_j, \lambda) = 0$ where $\xi_{n+1} = \lambda$. A plane wave solution is

$$e^{i(x\xi + x_j\xi_j + t\lambda_k)}$$

and on the boundary $x = 0$ it induces a disturbance with tangential wave numbers $(\xi_j, \lambda_k(\xi, \xi_j))$. Reflection in the form of waves travelling towards positive values of x can take the form

$$e^{i\{x_i\mu_{k\ell} + x_j\xi_j + t\lambda_k(\xi,\xi_j)\}},$$

where $P(\mu_{k\ell}, \xi_j, \lambda_k) = 0$ with ξ_j, λ given. We select those roots $\mu_{\kappa\ell}$ with $\mathrm{Re}(\mu_{\kappa\ell})\lambda_k < 0$ which represent wave propagation in the reversed direction of increasing x. Thus we have $\ell = 1, \ldots, k_1$, where k_1 is the number of characteristic surfaces issuing from $x = 0$ into the domain $x > 0$, and is equal to the number of boundary conditions.

Let the boundary conditions be

$$B_h(D_x, D_{x\cdot}, D_t)u = 0 \qquad h = 1, \ldots, k_1$$

for $x = 0$. The distinct polynomials B_h shall be linearly independent and may involve high orders of differentiation. Thus a plane wave with phase

$$\Xi_k = (x-\ell)\xi + x_j\xi_j + t\lambda_\kappa(\xi, \xi_j)$$

gives rise to a trial solution

$$u = e^{i((x-\ell)\xi + x_j\xi_j + t\lambda_k)} + \sum_{k=1}^{k_1} c_h(\xi, \xi_j, \lambda_k) e^{i(x\mu_{\kappa\ell} + x_j\xi_j + t\lambda_k)}$$

Applying the boundary conditions, we find that the coefficients

c_l are to be determined by the relations

$$0 = B_h u = e^{i(x_j\xi_j+t\lambda_k)}\left[e^{-il\xi}B_h(\xi , \xi_j , \lambda_k) + \sum_{m=1}^{k_1} c_m \cdot B_m(\mu_{k\ell} , \xi_j , \lambda_k)\right].$$

Now let

$$\Delta(\xi_j , \lambda_k) = \det[B_h(\mu_{k\ell} , \xi_j , \lambda_k)] \qquad h , \ell = 1 , \dots , k .$$

We also write

$$\Delta_\ell(\xi , \xi_j , \lambda_k) = \det[B_h(\mu'_{k\ell} , \xi_j , \lambda_k)] ,$$

where $\mu'_{k\ell} = \mu_{k\ell}$ for $\ell \neq h$, $\mu_{kh} = \xi$. Then

$$c_\ell(\xi , \xi_j , \lambda_k) = -e^{i\ell\xi}\frac{\Delta_\ell(\xi , \xi_j , \lambda_k)}{\Delta(\xi_j , \lambda_k)} .$$

With $\lambda_k = \lambda_k(\xi , \xi_j)$, this reflection coefficient prescribes the amplitude and phase of a reflected wave of the ℓth mode produced by an incident kth mode. The above quantities are all algebraic functions of (ξ , ξ_j) and Δ is a symmetric function of the $\mu_{k\ell}$, $\ell = 1 , \dots , k_1$. The "boundary discriminant" Δ can be regarded as a pseudo-differential operator governing the propagation of waves on the boundary. We can construct a normal surface S_B and wave surface W_B for Δ ; it is convenient to regard S_B as a cylinder in R^n with generators parallel to the ζ_1 axis. A zero of $\Delta(\zeta_j , \lambda_k)$ gives a set of wave numbers for which the boundary conditions are not independent with respect to P . These characteristic or resonance frequencies of the B_h will give rise to surface waves. Real zeros of Δ , that is, zeros for which $\mu_{k\ell}$ is real, will give rise to new wave fronts that may be of the ultrasonic or supersonic type. Zeros corresponding to complex values of the $\mu_{k\ell}$ give rise to exponentially attenuated waves within the space region, which are called Rayleigh waves in seismology. Branch points are possible in the $\mu_{k\ell}(\xi_j , \lambda_k)$ and these also give rise to "branch waves" or "head waves" with ruled surface wave fronts having geometry related to the intersection of reflected wave fronts with the boundary. To calculate these waves we must form the full expression for the reflected elementary solution (Duff, 5 , p. 204).

We take the incident elementary solution in the form

$$K(t\,,x) = \frac{i^{-m}}{(2\pi)^n}\int_{R^n}\sum_{k=1}^{m}\frac{e^{i\left(x\cdot\xi + x_j\xi_j + t\lambda_k(\xi)\right)}}{\left(\frac{\partial P}{\partial\lambda}\right)_{\lambda=\lambda_k(\xi)}}\,d\xi$$

and the reflected term then becomes

$$K_2(t\,,x\,,x_j) = \frac{i^{m-1}}{(2\pi)^n}\int_{R^n}\sum_{k=1}^{m}\sum_{\ell=1}^{k_1}\frac{\Delta_\ell(\xi\,,\xi_j\,,\lambda_k)}{\Delta(\xi_j\,,\lambda_k)}$$

$$\cdot\ \frac{e^{i(x\mu_{k\ell} - \ell\xi + x_j\xi_j + t\lambda_k)}}{P_\lambda(\xi\,,\xi_j\,,\lambda_k)}\,d\xi\ .$$

Upon carrying out the usual radial integration over $|\xi|$, we find the following distributional expression for K_2 :

$$K_2(x\,,x_j\,,t) = \frac{i^n(-1)^m}{(2\pi)^n}\int_{\Omega_\eta}\sum_{k=1}^{m}\sum_{\ell=1}^{k_1}\frac{\Delta_\ell(\eta\,,\eta_j\,,\lambda_k)}{\Delta(\eta_j\,,\lambda_k)}$$

$$\cdot\ \frac{d\Omega_\eta}{P_\lambda(\eta\,,\eta_j\,,\lambda_k)(\Xi + i0)^{n-m+1}}\,,$$

where

$$\Xi = \Xi_{k\ell} = x\mu_\ell\left(\eta_j\,,\lambda_k(\eta_j)\right) + x_j\eta_j + t\lambda_k(\eta\,,\eta_j) - \ell\eta(\eta_j)\ .$$

For this last step, we must assume that P is homogeneous so that all $\lambda_k(\xi\,,\xi_j)$, $\mu_k(\xi_j\,,\lambda_k)$ are homogeneous of degree one. In the contrary case, a series expansion in the style of Atiyah, Bott and Gårding (1), can be employed with first term still homogeneous as assumed here.

Each term above gives rise to a reflected wave front which is the envelope of $\Xi_{k\ell}$ with respect to the dual variables η_j . The singularity and asymptotic expansion for each of these main reflected wave fronts can be found using the method of stationary phase as in Chapter 2 above. The order of the singularity is the same as for the incident wave, and the amplitude involves the reflection coefficient as well as other integrand factors containing the $\mu_{k\ell}$, λ_k and normal surface curvature terms. For details we refer to Duff (5, pp. 205 - 207).

The geometry of these reflected wave surfaces brings in the head waves that arise from the branch points of the $\mu_{k\ell}(\xi_j\,,\lambda_k)$. Consider the case of two wave fronts, fast and slow, respectively. Reflection of the fast front creates fast and slow reflected fronts with the slow front inclined more nearly parallel to the boundary and hence able to "keep up" with the oblique motion of the intersection of the incident fast front with the boundary. The slow incident front likewise gives rise to fast and slow reflected fronts, but after some time the fast reflected front must break away ahead of the boundary intersection of the slow front. The trace of this reflected fast front on the boundary leaves a reflected slow wake called a head wave. In such a case there is a branch point of the $\mu_{k\ell}(\xi_j, \lambda_k)$ which gives rise to this term. (Figures 17,18).

By asymptotics it is found that the sharpness of a head wave front is one degree less, and the time attenuation one degree more, than the other fronts involved. (Duff, 1, p. 213). Complex branch points can also give rise to waves that are smooth except at the attachment point of a slow reflected front to the boundary, see Brekhovskikh (1, p. 290), and Deakin (2, p. 236).

Consider now the supersonic or ultrasomic or "lateral" waves that will arise from a common real zero of $\Delta(\xi_j, \lambda)$ and $P(\xi, \xi_j, \lambda)$. Let $\lambda = \lambda^b(\xi_j)$ be a real root of $\Delta(\xi_j, \lambda) = 0$; the corresponding sheet of S_B is a cylinder with generators parallel to the ξ axis and it corresponds to a wave surface on the boundary itself. We suppose that S_B has a real intersection with S, corresponding to the existence of real roots $\mu_\ell(\xi_j, \lambda^b)$ satisfying $P(\mu_\ell(\xi_j, \lambda^b), \xi_j, \lambda^b) = 0$. This $n-2$ dimensional locus (μ_ℓ, ξ_j) on S generates a corresponding ruled wave surface that joins sheets of W_R to the boundary as a supersonic wave front. (Figure 19).

In the integral for the reflected wave we choose a contour integral form for the λ variable, obtaining

$$K_2(x, x_j, t) = \frac{1}{(2\pi)^{n+1} i^{m+1}} \int_{R^m} \int_{C(\xi_j)} \sum_\ell \frac{\Delta_\ell(\xi, \xi_j, \lambda)}{\Delta(\xi_j, \lambda)} \cdot \frac{e^{i\Xi_\ell}\, d\lambda d\xi}{P(\xi, \xi_j, \lambda)} .$$

Note that the contour $C(\xi_j)$ can now be chosen independently of ξ which does not appear in $\Delta(\xi_j, \lambda)$ whose zeros are the object of study. Thus the integration over ξ can be done first, with the understanding that powers of ξ in $\Delta_\ell(\xi, \xi_j, \lambda)$ can be replaced by $i\partial/\partial\ell$ operating on $\Xi_\ell = -\ell\xi + x_j\xi_j + x\mu_\ell(\lambda, \xi_j) + t\lambda$ and on $e^{i\Xi_\ell}$. The integral over ξ then takes the form

$$\int_{-\infty}^{\infty} \frac{e^{-i\ell\xi}}{P(\xi, \xi_j, \lambda)} d\xi$$

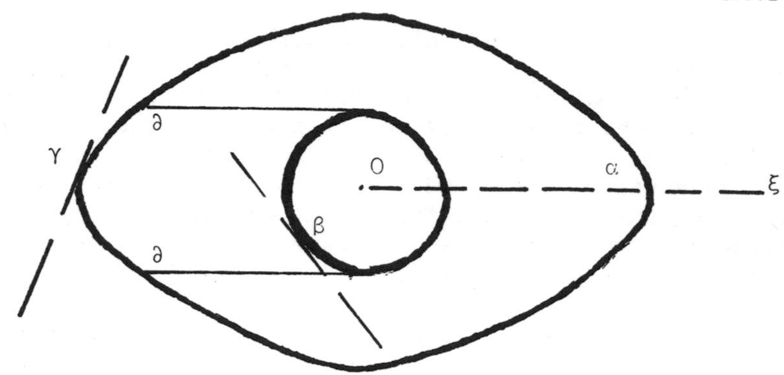

Fig. 17. Normal surface components for reflection of a slow incident front.

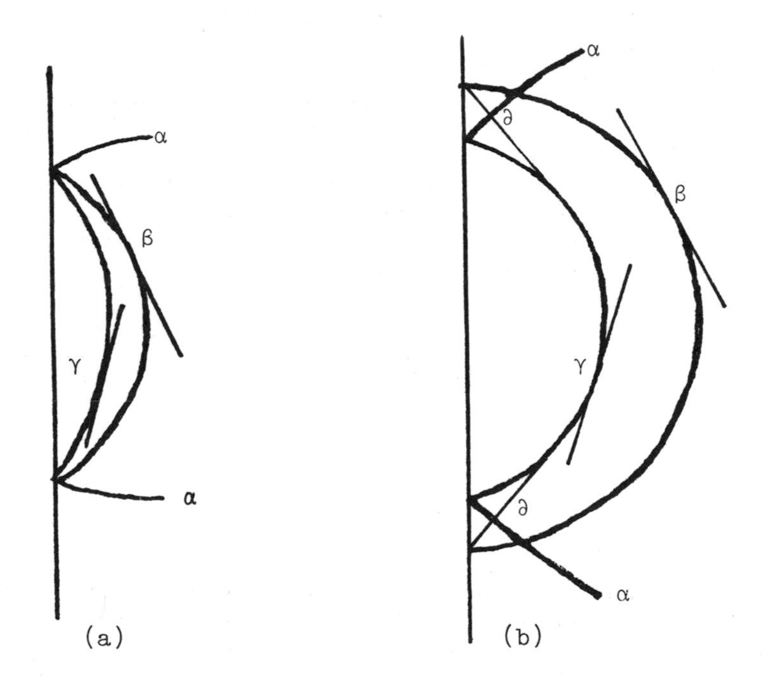

Fig. 18. Reflection of a slow incident wave front (a) early stage (b) late stage.

and may be evaluated by residues in the lower half ξ-plane. However residues exactly on the real contour will appear with factor $\frac{1}{2}$ as we interpret all these integrals as distributions, in this case with principal values. So we find

$$\pi i \sum_{\nu_m \text{ real}} \frac{e^{-i\ell\nu_m(\xi_j, \lambda)}}{P_\xi(\nu_m, \xi_j, \lambda)} + 2\pi i \sum_{\operatorname{Im}\nu_m < 0} \frac{e^{-i\ell\nu_m(\xi_j, \lambda)}}{P_\xi(\nu_m, \xi_j, \lambda)},$$

where ν_m denotes a root of $P(\nu_j, \xi_j, \lambda) = 0$.

If ν_m is a "reflected" root μ_ℓ then the determinant $\Delta_\ell(\nu_m, \xi_j, \lambda)$ will be zero if $m \neq \ell$ and equal to $\Delta(\xi_j, \lambda)$ if $m = \ell$.

Let us now reconstitute the integral for K_2:

$$K_2(x, x_j, t)$$

$$= \frac{1}{(2\pi)^n i^{m+1}} \int_{R^{n-1}} d\xi_j \int_{C(\xi_j)} d\lambda \sum_{\ell,m}{}' \frac{\Delta_\ell(\nu_m, \xi_j, \lambda)}{\Delta(\xi_j, \lambda)} \cdot \frac{e^{i\Xi_{\ell m}}}{P_\xi(\nu_m, \xi_j, \lambda)},$$

where

$$\Xi_{\ell m}(\lambda) = x_j\xi_j + x\mu_\ell(\lambda, \xi_j) + t\lambda - \ell\nu_m(\zeta_j, \lambda).$$

Here the prime on the summation sign denotes omission of the "reflected" values of m for which Δ_ℓ/Δ is zero or unity and no poles arise. With this in hand we carry out the residue evaluation of the integral over λ, obtaining

$$\frac{1}{(2\pi)^{n-1} i^m} \int_{R^{n-1}} \sum_{\ell,m}{}' \frac{\Delta_\ell(\nu_m, \xi_j, \lambda^b)}{\Delta_{(\lambda)}(\xi_j, \lambda^b)} \cdot \frac{e^{i\Xi_{\ell m}(\lambda^b)}\, d\xi_j}{P_\xi(\nu_m, \xi_j, \lambda^b)}.$$

After the radial integration which is possible as P_ξ, Δ_ℓ and Δ are homogeneous functions, we obtain

$$\frac{i^n(-1)^{m+1}}{(2\pi)^{n-1}} \int_{\Omega_{n-1}} \sum_{\ell,m}{}' \frac{\Delta_\ell(\nu_m, \eta_j, \lambda^b)}{\Delta_{(\lambda)}(\eta_j, \lambda^b)} \cdot \frac{d\Omega_n}{P_\xi(\nu_m, \eta_j, \lambda^b)\left(\Xi_{\ell m}(\lambda^b) + i0\right)^{n-m+1}}.$$

We omit details of the stationary phase evaluation of this singularity, but observe that there is one less η-integration so the order at a point of nonzero curvature is $\frac{1}{2}$ greater than that of the main sheets of W. The order is $s^{-(\frac{1}{2}n-m+2)}$ and also the leading term is homogeneous of degree $-(\frac{1}{2}n-1)$ in x, t, and ℓ jointly. For details we again refer to Duff (5, p. 217).

The geometry of the wave front $W^b_{\ell m}$ is deducible by duality from the geometry of S_B and S. As $W^b_{\ell m}$ corresponds to $S_B \cap S = \Sigma_\ell$, we see that $W^b_{\ell m}$ is tangent to W_ℓ the ordinary reflected sheet, and to W_B the "boundary wave surface". Each generator of $W^b_{\ell m}$ is a ray, or half-line, as each root μ is necessarily a reflected root only. The supersonic wave front makes its appearance at the boundary when the reflected sheet $W_{\ell m}$ first becomes tangent to it, or dually, when the expanding cone of normals of the reflected sheet first reaches Σ_ℓ. (Figure 20).

The famous example of Rayleigh waves in seismology shows that there may be zeros $\lambda^b(\eta_j)$ of the boundary discriminant $\Delta(\xi_j \lambda)$ such that μ_ℓ or ν_m in the expressions above are complex valued, not real. The wave contributions then arising are generally smooth, with certain exceptions when source and observation point lie on the boundary. There are several qualitatively different cases depending on the geometry of S_B in relation to the incident normal sheet S_m and the reflected normal sheet S_ℓ. For example, if S_B lies outside both S_ℓ and S_m so that μ_ℓ and ν_m are complex valued, then there will be a smoothly varying contribution for $\ell > 0$, or $x > 0$. It will appear inside the reflected wave sheets and will have the form of the distribution $(\Xi + i\varepsilon)^{-q}$, $q = \frac{1}{2}n + 2 - m$, where ε is related to values of x and ℓ. This contribution can be large near the boundary if $\frac{1}{2}n + 2 > m$ as in the elastic wave case where $m = 2$, $n = 3$ for this purpose.

4.5 Singularities and localization

The singularities of the reflected wave have recently been studied by Tsuji (5) and Wakabayashi (1) by the method of localization used by Atiyah, Bott and Gårding. Their methods may be somewhat more elaborate in a general case than those described above, but are also capable of great precision when carried out in full detail. The order of a singularity is defined by Tsuji using a Hilbert space which leads to numbers one half step higher than those used above which were based on orders of homogeneity. Tsuji and Shirota (1) give an example of fourth order

$$P(D) = (D_t^2 - D_x^2 - D_y^2)(a^2 D_t^2 - D_x^2 - D_y^2),$$

where $a > 1$, where a head wave appears with the boundary

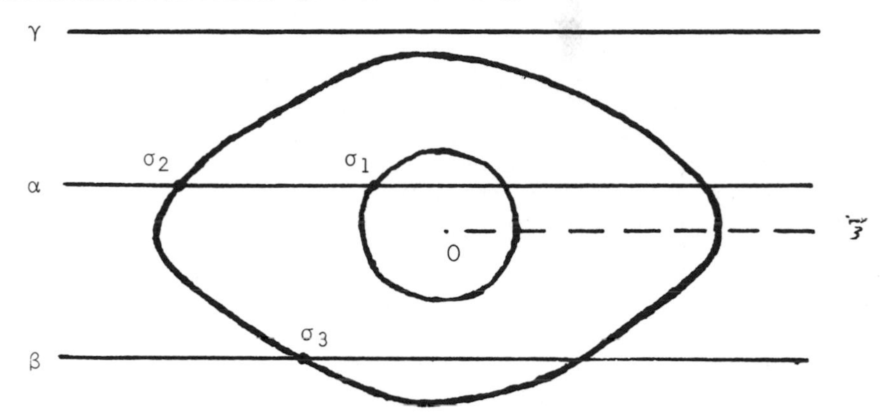

Fig. 19. Normal surface and boundary normal cylinder sheets.

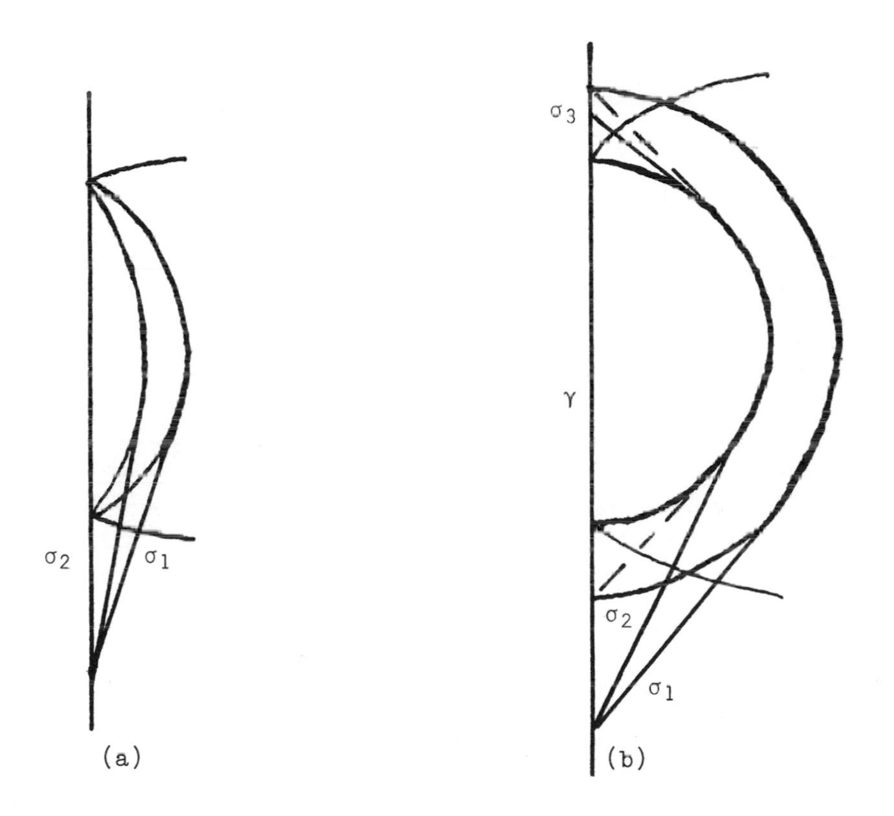

Fig. 20. Formation after reflection of supersonic boundary wave sheets. (a) early stage (b) late stage.

conditions $B_1 = 1$, $B_2 = D_x$. However when the boundary conditions are $B_1 = 1$, $B_2 = D_x^2$ the branch point disappears and there is no head wave. This could also be seen from the symmetry with respect to x as discussed for one boundary condition.

A brief account of the recent work of Wakabayashi (1) will now be given and it is noted that Tsuji (5) has also given very similar results for homogeneous operators P for the full space R^{n+1} as well as the mixed problem in $R^n \times R_+$. The notations used are those of Atiyah, Gårding and Bott (1) unless otherwise stated. For the mixed problem

$$P(D)u(x) = f(x) , \qquad x \in R^n_+ , \quad x_1 > 0 ,$$

$$D_1^k u(0, x'') = 0 , \quad 0 \le k \le m-1 , \quad x_n > 0 ,$$

$$B_j(D)u(x)\Big|_{x_n=0} = 0 , \quad 1 \le j \le \ell , \quad x_1 > 0 ,$$

the number ℓ of boundary conditions equals the number of roots λ^+ of $P(\xi' - i\gamma\phi', \lambda) = 0$ with positive imaginary parts for $\gamma > \gamma_0$. Let

$$P_+(\xi', \lambda) = \Pi_{j=1}^{\ell}\left(\lambda - \lambda_j^+(\xi')\right) ,$$

where $\xi' \in \Xi^{n-1} - i\gamma_0\phi' - i\Gamma_0$, and $\Gamma_0 = \{\xi' \in \Xi^{n-1} ; (\xi', 0) \in \Gamma\}$ with $\Gamma = \Gamma(P, \phi)$ that component of $\{\xi \in \Xi^n , P(\xi) \ne 0\}$ which contains ϕ the given timelike direction.

The Lopatinsky determinant is

$$R(\xi') = \det\left(\frac{1}{2\pi i}\oint \frac{B_j(\xi', \lambda)\lambda^{k-1}}{P_+(\xi', \lambda)}\, d\lambda\right)_{j,k=1,\ldots,\ell}$$

and it is assumed the problem is ε well posed, that is, $R(\xi' + s\phi') \ne 0$ for $\xi' \in \Xi^{n-1}$ and $\operatorname{Im} s < -\gamma$ and $R_0(\phi') \ne 0$ where $\tilde{R}_0(\xi')$ is the principal part of $R(\xi')$. The reflected wave is

$$F(x, y)$$

$$= \frac{1}{(2\pi)^{n+1} i}\int_{\Xi^{n+1} - i\tilde{\eta}} \sum_{j,k=1}^{\ell} \exp\left(i(x'-y')\cdot\xi' - y_n\xi_n + x_n\xi_{n+1}\right)$$

$$\times \frac{R_{jk}(\xi')B_k(\xi)\xi_{n+1}^{j-1}\, d\tilde{\xi}}{R(\xi')P_+(\xi', \xi_{n+1})P(\xi)} ,$$

where $\eta \in \gamma_1\Theta + \Gamma$, $\eta' \in \gamma_1\Theta' + \Gamma_0$, $\eta_{n+1} = 0$, and

$R_{jk}(\xi')$ is the k-j Lopatinsky cofactor.

The method of localization is exploited to study the singularities of this reflected wave. Let $\mathring{\Gamma} = \{\xi' \in \Xi^{n-1}, (\xi', \xi_n) \in \Gamma$ for some $\xi_n \in \Xi\}$. Then it is shown that $R(\xi')$ is holomorphic in $\Xi^{n-1} - i\gamma_1\phi' - i\mathring{\Gamma}$. The terms homogeneous of each degree in $R(\xi')$ are studied by expanding

$$R(t\xi') = t^{h_0}[\tilde{R}_0(\xi') + t^{-1}\tilde{R}_1(\xi') + t^{-2}\tilde{R}_2(\xi') + \cdots$$

h_0 being an integer. Then $\mathring{\Sigma} \subset \Xi^{n-1}$ is defined as the component of $\{\xi' \in \mathring{\Gamma}, R_0(-i\xi') \neq 0\}$ that contains ϕ'; $\mathring{\Sigma}$ is an open convex cone, $R(\xi') \neq 0$ for $\xi' \in \Xi^{n-1} - |\gamma|\phi' - i\mathring{\Sigma}$, and $\mathring{\Sigma}$ is star shaped with respect to ϕ'.

The localization of R is defined by a series

$$\nu^{h_1} R(\nu^{-1} r\xi^{0'} + \eta') = \sum_{j=0}^{N} Q_j(r\eta')\nu^{j/L} + O(r^{h_0}\nu^{N+1/L}),$$

where ν is small and h_1 rational. Then $\mathring{\Sigma}_{\xi^{0'}}$ is defined as the component of $\{\eta' \in \mathring{\Gamma}_{\zeta^{0'}} ; Q_0^0(-i\eta') \neq 0\}$ that contains ϕ',

where $Q_0^0(\eta')$ is the principal part of $Q_0(\eta')$; this set is an open convex cone. Also defined are sets

$$\tilde{\Gamma}_{(\xi^{0'}, \xi^0_{n+1})} = \bigcap_{k=1}^{r_0} \{\tilde{\xi} \in \Xi^{n+1} ; (\xi', \xi_{n+1}) \in \Gamma(P_+, \phi)\},$$

where certain simple reflected roots only are represented in P_+. Then let

$$\Gamma_{\tilde{\xi}^0} = \left(\Gamma(P_{\xi^0}, \ell) \times \Xi\right) \cap \tilde{\Gamma}(\xi^0, \xi^0_{n+1}) \cap (\mathring{\Sigma}_{\xi^0} \times \Xi^2).$$

The reflected wave has a localization expansion (we assume $y_1 = \cdots = y_{n-1} = 0$)

$$t^{p_0} \exp[-it(x'\xi^{01} - y_n\xi^0_n + x_n\xi^0_{n+1})F(x', y_n, x_n)$$
$$\sim \sum_{j=0}^{\infty} \tilde{F}_{\tilde{\xi}^0, j}(x', x_n, y_n)t^{-j/L},$$

where p_0 is rational and L an integer. Also it is shown that

$$\bigcup_{j=0}^{\infty} \operatorname{supp} \tilde{F}_{\tilde{\xi}^0, j}(x', y_n, x_n) \times [\{\xi^0, -\xi^0_n, \xi^0_{n+1}\}]$$
$$\subset WF\left(F(x', y_n, x_n)\right), \quad \text{for } \tilde{\xi}^0 \neq 0$$

and that the closed convex hull of the supports,

$$\overline{\mathrm{ch}}\ [\cup_{j=0}^{\infty} \operatorname{supp} \tilde{F}_{\tilde{\xi}^0,j}(x', y_n, x_n)] \subset \tilde{K}_{\tilde{\xi}^0},$$

where the support cone K is defined as

$$\tilde{K}_{\tilde{\xi}^0} = \{x', y_n, x_n) \in X,\ x' \cdot \eta' - y_n \eta_n + x_n \eta_{n+1} \geq 0 \ \forall \tilde{\eta} \in \tilde{\Gamma}_{\tilde{\xi}^0}\} .$$

Recall that the wave front set $WF(u)$ of a distribution u is a subset of the tangent bundle $T^*(R^{n+1})\backslash 0$, defined as the intersection of the characteristic sets $\gamma(A)$ of pseudo-differential operators of class L^0, where $Au \in C^\infty$:

$$WF(u) = \bigcap_{Au \in C^\infty} \gamma(A) ;$$

$$\gamma(A) = \{(x, \xi) \in T^*(R^{n+1})/0 ,\ \underline{\lim}_{t\to\infty} a(x, t\xi) = 0\} .$$

Since A may be taken as a C^∞ function on R^{n+1}, it follows that the spatial projection $\pi WF(n) = \{x \mid (x, \xi) \in WF(u)\}$ is a subset of SSu, and in fact $\pi WF(x) = SSu$, see Hörmander (4, p. 120).

The analytic wave front set $WF_A(u)$ is defined indirectly as the complement of that set $(x, \xi) \in T^*(R^{n+1}) \setminus 0$ of the points $((x^{0'}, y_n^0, x_n^0), \tilde{\xi}^0)$ such that for some sequence ϕ_N described below, there is a conic neighbourhood Δ of $\tilde{\xi}^0$ in $\Xi^{n+1} \setminus 0$ with

$$\hat{F}(x', y_n, x_n)[\phi_N(u)](\tilde{\xi}) \leq C(CN)^N (1 + |\hat{\xi}|)^{-N} \quad \text{for } \tilde{\xi} \in \Delta .$$

Here $\{\phi_N\} \in C_0^\infty(R^{n+1})$ satisfies $\phi_N = 1$ on a fixed neighbourhood of (x'^0, y_n^0, x_n^0) in R^{n+1} independent of N, and $|\tilde{D}^\alpha \phi_N| \leq C(CN)^{|\alpha|}$ for $|\alpha| \leq N$.

Wakabayashi defines two more sets:

$$\Gamma^0_{\tilde{\xi}^0} = (\Gamma(P_{\xi^0}, \phi) \times \Xi) \cap \tilde{\Gamma}_{(\xi^{0'}, \xi^0_{n-1})} \cap (\mathring{\Sigma}^0_{\xi^0} \times \Xi^2)$$

and

$$\tilde{K}^0_{\tilde{\xi}^0} = \{(x', y_n, x_n) \in X,\ x' \cdot \eta' - y_n \eta_n + x_n \eta_{n+1} \geq 0 \ \text{ for all } \tilde{\eta} \in \Gamma^0_{\tilde{\xi}^0}\} .$$

Here $\tilde{\xi}^0 \in \Xi^{n+1} \setminus 0$ and $\tilde{K}^0_{\tilde{\xi}^0} = \phi$ for $\xi^{0'} = 0$.

Then he shows that

$$WF(F) \subset WF_A(F) \subset \bigcup_{\tilde{\xi}\in \Xi^{n+1}\setminus 0} \tilde{K}^0_{\tilde{\xi}} \times \{\xi_1', -\xi_n, \xi_{n+1})\} .$$

This theorem gives an outer estimate of the singular supports of the reflected wave functions, and some information of the wave fronts as well. An example is given in which the Lopatinski condition is not satisfied, and a particular reflected wave front does not appear in consequence. As the complete details are intricate, we refer to the forthcoming papers of Wakabayashi (1) and Tsuji (5) for proofs and explanations.

CHAPTER 5

Well posed mixed problems for hyperbolic systems with constant coefficients.

1. Stable boundary conditions. For the initial value problem for a hyperbolic system, existence theorems have been established under very general conditions for constant coefficients and variable coefficients, Hormander (1,Ch.9), Garding (2). For the mixed initial and boundary value problem, general existence proofs and the enumeration of correctly posed boundary conditions are still quite recent even for constant coefficients. Here we shall describe work of Hersh (1,2) who analysed the algebra of boundary conditions and showed that a certain stability condition is necessary and sufficient for the mixed problem to be correctly set. We follow the treatment in Hersh (2) which applies to general hyperbolic systems with constant coefficients, and will describe the case of a single hyperbolic system of higher order. Consider

$$P(D_t, D_x, D_{y_j})U = 0 , \quad 1 \le j \le m$$

where P is an $n \times n$ matrix of polynomials and U an n-vector. Let P be correct in the sense of Petrowsky, that is, all roots τ of the characteristic equation $\det P(\tau, i\xi, i\eta) = 0$ satisfy $\mathrm{Re}\,\tau < M_0$ for some constant M_0 independent of ξ and $\eta = (\eta_1, \ldots, \eta_m)$. Equations correct in the Petrowsky sense include hyperbolic equations for which in the case where P is homogeneous all roots τ satisfy $\mathrm{Re}\,\tau = 0$ (corresponding to $\mathrm{Im}\lambda = 0$ if $\lambda = -i\tau$). Parabolic equations such as the heat flow equation, $u_t = \Delta u$, or Schrödinger's equation $u_t = i\Delta u$ or the vibrating elastic plate equation $u_{tt} + \Delta^2 u = 0$ can also be seen to satisfy the Petrowsky condition. Indeed this is the natural condition for correctness of the initial problem, because it has been shown (Shilov, 1, p.262) that the initial or Cauchy problem has a square integrable solution for all square integrable initial data if and only if the Petrowsky condition holds.

Now let $P[u] = 0$ for $x > 0$, $t > 0$ and $y \in R^m$. Assume given Cauchy data which are of integrable square for $t = 0$. Actually by subtraction of a solution of the Cauchy problem we reduce our mixed problem to the case where the Cauchy data are zero. Then there are given k boundary conditions for $x = 0$, $t > 0$ and $y \in R^m$, where k will be determined below by the character of the roots of $\det P = 0$. The boundary matrix B is $R \times n$ and

$$B[(D_t, D_x, D_y)U]_{x=0} = F(t,y)$$

where F is a k vector on the boundary.

Hersh now goes about the determination of all B such that this problem is correctly posed in L^2, and the specification of

k is an important partial step. The boundary data F are taken as a delta function set $F_0 = \delta(t)\delta(y_1) \ldots \delta(y_m) I_k$ where I_k is the $k \times k$ identity matrix. Then in $F_0 = BU$ we take U as an $n \times k$ matrix with entries consisting of k n-vectors, and we think of this U as a Green's matrix. Then a solution V for arbitrary data $F(y)$ can be written as a convolution over the boundary.

Let the boundary space N be the set of all n vectors U such that $PU = 0$ for $x > 0$ and $t > 0$, and $BU = 0$ holds for $x = 0$, $t > 0$. Consider now exponential solutions

$$e^{\tau t + i\eta y} \sum_r x^r e^{\xi_s x} C_{r,s}$$

where the sum is taken over roots ξ_s of $\det P(\tau_1\xi, i\eta) = 0$. Here $C_{r,s}$ is a constant n vector, and for a root ξ_s of multiplicity m_s, then $r = 0,1,\ldots,m_s-1$. The finite dimensional linear space $E(\tau, i\eta)$ of these exponentials divides into subspaces $E^-(\mathrm{Re}\xi < 0)$ and $E_+^-(\mathrm{Re}\xi > 0)$ respectively of dimensions d_- and d_+. Since there are no roots ξ for $\mathrm{Re}\tau > M_0$ as $P(\tau,\xi,i\eta) \neq 0$ then we have in this case $E = E^- \oplus E^+$. Notice also that when $\mathrm{Re}\tau > M_0$ we also have $d_- = $ const., because the number d_- of roots with negative real part could change only if there appeared a root with zero real part. But this has just been seen to be impossible.

The boundary conditions B and boundary space N will be called unstable if N contains exponential solutions U in $E^-(\tau, i\eta)$ for real η and τ with real part positive and arbitrarily large. Such a boundary problem cannot be well posed, as a sequence of solutions bounded initially but increasing without limit at any later time t can be selected.

If N is stable, (not unstable) but the N obtained by dropping any boundary condition is unstable, then N is called maximally stable. If N is stable, then $N \cap E^-(\tau, i\eta) = 0$ for $\mathrm{Re}\tau > M_1$, η real. Let W be an $n \times d_-$ matrix whose columns form a basis for $E^-(\tau, i\eta)$. Applying the $k \times n$ boundary operator B, we obtain a $k \times d_-$ matrix BW. Setting $x = 0$ in BW we obtain a matrix written as $e^{\tau t + i\eta y}\tilde{B}$, where $\tilde{B} = \tilde{B}(\tau, \xi_s, i\eta)$. The columns of $\tilde{B}$ span a certain vector space of functions of τ, ξ, η. If one of the columns of W lies in N, then as $B(N)|_{x=0} = 0$ the corresponding column of $\tilde{B}$ would be zero, and the linear space generated by the columns of $\tilde{B}$ would have dimension less than d_-. Likewise, if any vector of the column space of W lies in N, the column rank of $\tilde{B}$ is reduced.

Hersh uses "permitted values" of τ, η to denote real values of η, and values of τ with real part greater than M_0 and M_1. Then N is stable if $\tilde{B}$ has rank d_- for all permitted values of τ and η. Since $\tilde{B}$ must have at least d_- rows, therefore $k \geq d_-$. If B is stable, and $k = d_-$, then B

must be maximally stable. Thus $\tilde{B}$ is square and stability, or maximal rank, implies that $\tilde{B}^{-1}$ exists. Then there exists exactly one U such that $PU = 0$ and $e^{-Mt}U$ is a tempered distribution in the quarter space $x > 0$, $t > 0$. Also $BU = \delta I$ on $x = 0$, $t > 0$, and U has zero Cauchy data. This U is given by

$$U(x,t) = -i(2\pi)^{-m-1} \int_{M-i\infty}^{M+i\infty} d\tau \int_{-\infty}^{\infty} W B^{-1} \, d\eta_1 \ldots d\eta_m$$

where W is a column basis for $E^{-}(\tau,i\eta)$ and $M > \max(M_0,M_1)$.

This formula is established by taking a Laplace transform in t, and Fourier transform in $y_1,\ldots,y_m$. There results a system of ordinary differential equations in x:

$$P(\tau,D_x,i\eta)\hat{U} = \hat{P}\hat{U} = 0 \qquad , \qquad x > 0$$

with

$$B(\tau,D_x,i\eta)\hat{U} = \hat{B}\hat{U} = I \qquad\qquad x = 0$$

as boundary conditions. We look for solutions growing at most like x^r, and these form a space $\hat{E}^{-}$ of dimension d^{-}. If $\hat{W}$ is a column basis for $\hat{E}^{-}$, then $\hat{U} = \hat{W}\tilde{B}^{-1}$ since $\hat{P}\hat{U} = \hat{P}\hat{W}\tilde{B}^{-1} = 0$ and $\hat{B}\hat{U}|_{x=0} = \hat{B}\hat{W}|_{x=0}\tilde{B}^{-1} = \tilde{B}\tilde{B}^{-1} = I$.

Hersh then shows that $\hat{U}$ is independent of the choice of basis for E^{-} and hence unique. Indeed if $\hat{W}_1 = \hat{W}K$ where $K = K(\tau,\eta)$ is nonsingular, then

$$\hat{U}_1 = \hat{W}B_1^{-1} = \hat{W}_1\{\hat{B}W_1|_{x=0}\}^{-1} = \hat{W}K\{\hat{B}\hat{W}K|_{x=0}\}^{-1}$$

$$= \hat{W}K\{\hat{B}\hat{W}|_{x=0}K\}^{-1} = \hat{W}KK^{-1}\tilde{B}^{-1} = \hat{W}\tilde{B}^{-1} = U .$$

The integral formula for U is just the inverse Laplace and Fourier transformation of $\hat{U}$, which can be shown to represent a distribution of finite order. That is, $\hat{U}$ is holomorphic and of polynomial growth in (τ,η) which can be verified using the decreasing character of the exponentials in $e^{\xi x}$, and the stability of N.

Kasahara (1) has pointed out that a difficulty arises in the calculations of Hersh when multiple eigenvalues are present, since the smoothness of generalized eigenvectors ω_j as functions of τ,η cannot be assumed. This difficulty is circumvented by means of a Cauchy integral in the complex plane. Kasahara also gives a justification for the polynomial growth in τ,η of the elementary solution in the case of multiple roots, by means of the Seidenberg-Tarski elimination theorem.

We also mention here that a similar treatment of the mixed problem (at least for Dirichlet boundary conditions) has been given by Shilov (1, p.318) who makes use of a distributional formulation of the boundary conditions for well posedness in the form of analytic continuability of certain expressions in a half plane.

The Hersh criterion of stability can be applied to the equations of mathematical physics. For the wave equation $u_{tt} - c^2\Delta u = 0$ say in 3 space dimensions, there is one boundary condition, and $W = e^{\xi x}$ where

$$\xi^2 = \tau^2/c^2 + \eta^2 + \zeta^2 .$$

With $B = b_0 + b_1 D_x + b_2 D_y + b_3 D_z + b_4 D_t$ where $b_i \in R$ we set $\tau = w + \tau v$, $\xi = p + iq$ and take real and imaginary parts of $B = 0$:

$$b_0 + pb_1 + wb_4 = 0 \qquad (5.1.1)$$

$$qb_1 + \eta b_2 + \xi b_3 + \sigma b_4 = 0 . \qquad (5.1.2)$$

Then from the relation of the wave equation we find

$$p^2 - q^2 = \frac{w^2 - v^2}{c^2} + \eta^2 + \zeta^2 \qquad (5.1.3)$$

$$pq = \frac{wv}{c^2} . \qquad (5.1.4)$$

Then B is unstable if and only if these four relations have real solutions with $p < 0$ and $w \to +\infty$.

From (5.1.1) b_1 and b_4 must have the same sign, and if $b_1 = b_4 = 0$ then $b_0 = 0$. Now solve (5.1.4) for q , insert in (5.1.3), and rearrange, obtaining

$$\left(\frac{\omega^2}{c^2} - p^2\right)\left(\frac{v^2}{p^2c^2} + 1\right) + \eta^2 + \zeta^2 = 0 . \qquad (5.1.5)$$

This implies $|p| \geq \left|\frac{w}{c}\right|$. But, by (5.1.1),

$$p = \frac{(-wb_4 - b_0)}{b_1} \quad \text{so that} \quad \frac{b_4}{b_1} \geq \frac{1}{c} , \text{ while if } \frac{b_4}{b_1} = \frac{1}{c} , \text{ then}$$

$\frac{b_0}{b_1} \geq 0$. That is, the four relations can not be solved as described, and hence B is stable, unless one of the following is true:

(A) $b_0 = b_1 = b_4 = 0$

or (B) $0 < \frac{b_1}{b_4} < c$

or (C) $b_1 = cb_4$, and $\frac{b_0}{b_1} \geq 0$.

However, if any of these three conditions hold, an explicit solution of (5.1.1) to (5.1.4) is easily given, and then B is unstable. For the one space dimensional wave equation, η, ξ are not present and so $-cp = w$ whence $b_0 = 0$ and $b_1 = cb_4$. Thus all first order B are correct for the one dimensional wave equation except $D_t + cD_x$.

For the heat equation and Schrödinger's equation a similar analysis shows B is stable unless either $b_0 = b_1 = b_4 = 0$ or $b_1 b_4 > 0$.

Because of the great interest of these results, the case of the vibrating plate equation $u_{tt} + \Delta^2 u = 0$ will be mentioned. Again using Laplace transforms in (τ,ξ) and Fourier transforms in η, we find $\tau^2 + (\xi^2 - \eta^2)^2 = 0$ so $\xi = (\eta^2 \pm i\tau)^{\frac{1}{2}}$. For each choice of $\pm$ sign there is a root ξ with negative real part, so $d_- = k = 2$. Let $B = (B_1, B_2)$ so $\det B = B_1(\xi_1)B_2(\xi_2) - B_1(\xi_2)B_2(\xi_1)$ where ξ_1 and ξ_2 are the two permitted values of ξ, with $\xi_2 \neq \xi_1$ since $\operatorname{Re}\tau < M_0 < 0$.

Consider the simplest powers of D_x:

$$B_1 = D_{x_1}^{\alpha_1} \ , \quad B_2 = D_{x_2}^{\alpha_2} \ ;$$

then B is stable if and only if

$$\xi_1^{\alpha_1}\xi_2^{\alpha_2} \neq \xi_2^{\alpha_1}\xi_1^{\alpha_2}$$

for permitted values of τ, η; that is

$$\xi_1^{\alpha_1-\alpha_2} \neq \xi_2^{\alpha_1-\alpha_2} \ , \text{ or } \ (\eta^2 + \tau)^{\frac{\alpha_1-\alpha_2}{2}} \neq (\eta^2 - i\tau)^{\frac{\alpha_1-\alpha_2}{2}} \ .$$

Setting $\eta = 0$, one can get equality if $\alpha_1 - \alpha_2$ is a multiple of 4. For $\eta \neq 0$, they can be equal only if $|a_1 - a_2| > 4$ or $|a_1 - a_2| = 3$. Hence B is stable if and only if $|a_1 - a_2| = 1$ or 2. Note that for the standard problems the exponents are: 0,1 for boundary clamped, 0,2 for boundary supported, and 2,3 for a free boundary.

Hersh also considers Maxwell's equations and the isotropic elastic wave equations. For the latter he gets stability unless

(A) $b_0 = b_1 = b_4 = 0$, or

(B) $0 < b_1/b_4 \leq \max\left(\sqrt{\frac{\mu}{\rho}}\ , \sqrt{\frac{\lambda+2\mu}{\rho}}\right)$ or

(C) $b_1/b_4 = \max\left(\sqrt{\frac{\mu}{\rho}}\ , \sqrt{\frac{\lambda+2\mu}{\rho}}\right)$ and $b_0 b_1 \geq 0$.

where the linear boundary condition $B(u_i) = 0$ for each $i = 1,2,3$. More complicated boundary conditions for this elastic system are also discussed.

Observe that in any of these problems it is the zeros of the boundary forms $B(\tau,\xi,i\eta)$ that destroy stability. Hence powers of B, or other functions of B vanishing only if B vanishes, will lead to the same stability results.

The conditions for stability or well-posedness for these boundary value problems can be shown to lead to L^2 integral estimates for the solutions, Rauch (2), Sarason (1). For if a component u is L^2 on the boundary surface $x_1 = 0$, then by Parseval's theorem its Fourier transform u^2 with respect to the boundary variables is also L^2. Then the full solution is expressed as a superposition of terms of the form

$$u(\eta)e^{i(y\cdot\eta + t\tau_k + x\mu_e)}$$

where $e^{it\tau_k}$ does not increase rapidly with t, and $e^{ix\mu_s}$ does increase rapidly with x. Indeed, for $\mathrm{Im}\mu > 0$ which is the usual situation, the x-factor is integrable and square integrable over $(0,\infty)$. It is then easily seen that the solution is L^2 integrable over space with bound increasing at most like e^{At} in t. We return to the L^2 estimates in Chapter 6, for variable coefficient problems.

5.2 Propagation of Surface Waves

Here we describe further work of Hersh (3) on the condition for finite surface wave speed, uniqueness, and time reversibility of mixed problems. The notion of stable boundary conditions must be further refined to that of hyperbolic boundary conditions for these purposes.

Assume now P is corect in Petrowsky's sense and B stable, and adopt the notations of (4.1). Let $\lambda(\eta) = \mathrm{Max}\,\mathrm{Re}\tau$ for which $\hat{u}(\tau,\eta)$ is singular, so that also

$$\lambda(\eta) = \inf_{\mathrm{Re}P}\{\mathrm{Re}\tau \mid E^-(\tau,i\eta) \text{ has } \dim d^-, E^- \cap N = 0\} .$$

Since B is stable, we know $\lambda \leq M$. Now let $\Lambda(s) = \mathrm{Max}\lambda(\zeta)$ where $|\eta_j| \leq s$, $j = 1,\ldots,m$, and let $\Gamma(\eta_1)$ be the line parallel to the imaginary axis of τ such that $\mathrm{Re}\tau = \Lambda(s) + \varepsilon$, $\varepsilon > 0$.

To apply the Paley-Wiener theorem to U, we shall need to divide $\hat{U}$ by a polynomial $Q(\tau)$ that does not vanish for $\mathrm{Re}\tau \geq 0$ and which has sufficiently high degree. Thus we let $\tilde{U}$ be the inverse Laplace transform with respect to τ of $\hat{U}/Q$. Since operation by $Q(D_t)$ on $\tilde{U}$ will cancel off the polynomial $Q(\tau)$ so the support in $\{y,t\}$ of the inverse Fourier transform of $\tilde{U}$ contains the support of U.

It can be shown without difficulty that

$$\tilde{U}(\eta) = \frac{1}{2\pi i} \int_{\Gamma} \frac{e^{\tau t} \hat{U}\, d\sigma}{Q(\tau)} ,$$

so that $\hat{U}(\eta)$ is seen to be entire in η . Then we have

$$|\tilde{U}(\eta)| \le \frac{1}{2\pi} e^{t(\Lambda+\varepsilon)} \int_{\Gamma} \frac{|\hat{U}|\, d\tau}{|Q|}$$

and in this estimate $|\hat{U}|$ as shown in the preceding section is a sum of terms like $x^r e^{x\xi}$ multiplied by a ratio of polynomials R/S in (τ,ξ,η) . Note that $\mathrm{Re}\xi \le 0$ so the exponentials are bounded for $x \ge 0$. By a special argument based on the Seidenberg Tarski elimination lemma, Hersh shows that $\min|S|$ is a piecewise algebraic function of ε and so can be estimated from below by a power of ε . Hence this estimate follows for each positive ε :

$$|\tilde{U}(\eta)| \le C(\varepsilon)(1+|\eta|^{\eta})e^{t(\Lambda(s)+\varepsilon)}$$

Comparing with the Paley Wiener theorem (Ch. 1, Section 1) we see that the inverse Fourier transform $U(t,x,y)$ will have compact support only if

$$|\Lambda(s)| \le c|\eta| + K \quad , \qquad c,K \in R .$$

If this estimate holds then the velocity of propagation of waves near the boundary is at most c . If this condition holds, the boundary conditions will be called <u>hyperbolic</u>. Roughly speaking, B will be hyperbolic if in every element of $B(\tau,\xi,i\eta)$ the degree of τ or ξ equals the degree in η , and this excludes conditions such as $B = \tau - \eta^2$ of parabolic character.

A uniqueness theorem can also be deduced from the behaviour of $\tilde{U}(\eta)$, using a method similar to that of Holmgren but adapted to our special regions. If

$$P(D) = \sum_{j=0}^{\ell} D_x^i P_j(D_t, D_y)$$

then construct the bilinear form

$$H(U,V) = \sum_{j=0}^{\ell} \sum_{K=0}^{j-1} (-1)^k (D_x^k V) D_x^{j-k-1} P_j(D_t, D_y) U$$

and let (U,V) be "H-orthogonal" if $H(U,V) = 0$. Then also

$$\hat{H}(\hat{U},\hat{V}) = \sum_{j=0}^{\ell} \sum_{k=0}^{j-1} (-1)^k (D_x^k \hat{V}) D_x^{j-k-1} P_j(\tau, i\eta)\hat{U} .$$

Also let $P^*(D) = P^T(-D)$ where T denotes the transpose operation on a matrix P. Then let

$$N^* = \{V | P^*V = 0 , H(U,V) = 0 \ \forall U \in N\}$$

so that N^* is the H-orthocomplement of N. The substitution of P^* for P in the various E spaces, W matrices, and so on, may be indicated by a $*$: thus $\hat{W}^*$, λ^*, Λ^*.

The appropriate Greens formula may be written

$$\iint_{D_0} (V \cdot PU - U \cdot P^*V)\, dx\, dt = \int_{D_2} H(U,V)dt - \int_{D_1} H(U,V)dt$$

$$+ \int_{D_3} + \int_{D_4} + \lim_{y \to \pm\infty} \iint_{D_5}$$

where the last terms contain integrands that do not need to be specified. Here

$$D_0 = \{0 < x < X , 0 < t < T , y_j \in R\}$$

$$D_1 = \{x = 0 , 0 < t < T , y_j \in R\}$$

$$D_2 = \{x = X , 0 < t < T , y_j \in R\}$$

$$D_3 = \{0 < x , t = 0 , y_j \in R\}$$

$$D_4 = \{0 < x , t = T , y_j \in R\}$$

$$D_5 = \{0 < x < X , 0 < t < T\} .$$

Proof will be given that $U \equiv 0$ in the distribution sense if $PU = 0$ in D_0, $U \in N$, and U has zero Cauchy data on D_3. Assuming these data, three terms vanish from Green's formula. Now if $P^*V = h$ can be solved for a given arbitrary h in D_0 with compact support, if V has zero Cauchy data on D_4 the "final" time surface, if $V \in N^*$, and if V has compact support in y, then all other terms vanish and we have

$$\iint_{D_0} (U-h)\, dx\, dt = 0$$

so $U \equiv 0$ as stated.

The required V will exist for all h if and only if N^* is stable for P^* backwards, and if $\Lambda^*(s) = O(s)$. Thus t is replaced by $-t$, so we need N^* disjoint from $\hat{E}^{*-}(-\tau,-i\eta)$ for $\mathrm{Re}\,\tau > M_1\ \eta_j \in R$. Hersh then shows $E^-(\tau,i\eta)$ is $\hat{H}$ orthogonal to

$\hat{E}^{*-}(-\tau,-i\eta)$, but only the zero vector in $\hat{E}^{*}(-\tau,-i\eta)$ is $\hat{H}$ orthogonal to the full space $\hat{E}(\tau,i\eta)$. For if $\hat{U} \in \hat{E}^{-}(\tau,i\eta)$, $\hat{V} \in \hat{E}^{*-}(-\tau,-i\eta)$ then $\hat{H}(\hat{U},\hat{V})$ is a well defined function $\hat{H}(x)$ of x, and an integration by parts gives

$$\int_a^b \{\hat{V} \cdot \hat{P}\hat{U} - \hat{U} \cdot \hat{P}^{*}V\}\, dx = \hat{H}(b) - \hat{H}(a) .$$

By differentiation we obtain

$$\frac{d\hat{H}}{dx} = \hat{V} \cdot \hat{P}\hat{U} - \hat{U} \cdot \hat{P}^{*}\hat{V} = 0$$

so H is constant. However H contains only exponential terms with negative exponents for $x > 0$, so the only possible constant value is $\hat{H} = 0$. To show that only zero is $\hat{H}$ orthogonal to $E^{-}(\tau,i\eta)$, suppose $\hat{U} \in \hat{E}^{*}(-\tau,-i\eta)$ is not, and consider $\hat{H}(\hat{U},\hat{V}) = 0$ as a differential equation for $\hat{U}$ with $\hat{V}$ fixed. If $n = 1$, this equation has order less than d and hence fewer than d independent integrals. But $\hat{E}$ has dimension d so a contradiction is reached for every $\hat{H}$ orthogonal vector of $\hat{E}$ yields an integral. For a system with $n > 1$, regard all components of $\hat{U}$ but one as free parameters, and the same argument will succeed. Hence $N^{*} \cap E^{*-}(-\tau,-i\eta) = 0$ for $\mathrm{Re}\,\tau > M$, $\eta_j \in R$. In fact, the result holds even for complex η. It follows that $\lambda^{*}(\eta) = \lambda(-\eta)$ and so $\Lambda^{*}(s) = \Lambda(s)$, as can be seen since N intersects $E^{-}(\tau,i\eta)$ if and only if N^{*} intersects $E^{*-}(-\tau,-i\eta)$. Also the dimensions of $E^{-}(\tau,i\eta)$ and $E^{*-}(-\tau,-i\eta)$ are equal.

Since now $\Lambda^{*}(s) = O(s)$ can be assumed, it follows that V has compact support in y. This finally shows that for $\Lambda(\eta) = O(\eta)$, the solution U is unique in the space of distributions, with no limitation of behaviour at infinity.

From the Gelfand Shilov theory of Fourier transforms of entire functions, it can be shown that if $\Lambda(\eta) = O(|\eta|^p)$, $p > 1$, then uniqueness holds among functions of growth not exceeding

$$\exp(|\eta|^{p'}), p' = \frac{p}{p-1} .$$

The hyperbolic boundary conditions are actually the only ones for which such a uniqueness property holds, as will now be shown. Recall that B is hyperbolic if the speed of surface wave propagation is finite and $\Lambda(|\eta|) \leq c|\eta| + K$. Thus if B is stable but non-hyperbolic it will follow that some root $\tau(\eta)$ of

$$\det \tilde{B}(\tau,\xi(\tau,\eta),\eta) = 0$$

satisfies

$$\mathrm{Re}\,\tau \geq \text{const.}|\eta|^p , \qquad p > 1$$

as $|\eta| \to \infty$. Then the problem

$$Pu = 0 \quad \text{for} \quad x > 0 \quad \text{and} \quad Bu = 0 \quad \text{at} \quad x = 0$$

has a non-vanishing solution $U \in C^\infty$ for all t, and identically zero for $t < 0$.

To establish this theorem recall that a function $\tau(\eta)$ is a root of $\det\tilde{B}$ when $\tilde{B}(\tau,\xi(\tau,\eta),\eta)$ is singular. That is some vector $\hat{W}$ in the column space of E is annihilated by applying $B(\tau(\eta),D_x,\eta)$ and then setting $x = 0$. Hence $\hat{W}$ satisfies

$$P(\tau,D_x,\eta)\hat{W} = 0 \quad \text{for} \quad x > 0$$

$$B(\tau,D_x,\eta)\hat{W} = 0 \quad \text{for} \quad x = 0$$

and $\hat{W} = O(|x|^r)$ for some r as $x \to \infty$. But then $W = e^{\tau t+\eta y}\hat{W}$ satisfies

$$P(D_t,D_x,D_y)W = 0 \quad \text{for} \quad x > 0$$

$$B(D_t,D_x,D_y)W = 0 \quad \text{on} \quad x = 0 .$$

Now we construct U as a superposition of such solutions W, which vanishes for $t < 0$. The essential hypothesis $\text{Re}\tau \geq c|\eta|^p$, $p > 1$ shows that for $\text{Re}\tau \to +\infty$, there exist values $\eta(\tau)$ with

$$\det\tilde{B}(\tau,\eta(\tau)) = 0 , \qquad |\eta(\tau)| \leq \frac{1}{c}|\text{Re}\tau|^{1/p}$$

and therefore with

$$|\eta(\tau)| \leq \frac{1}{c}|\tau|^{1/p} .$$

We therefore define η_1 as that solution of $\det\tilde{B}(\tau,\eta(\tau)) = 0$ which for each $\text{Re}\tau$ (and for all $\text{Im}\tau$ values) has the smallest modulus $|\eta_1|$. This quantity is piecewise algebraic and hence algebraic for $\text{Re}\tau$ large enough, and because of its minimal property

$$|\eta_1(\tau)| \leq \frac{1}{c}|\tau|^{1/p}$$

holds. At this stage we can drop the subscript and let $\eta(\tau)$ be that algebraic function root equal to $\eta_1(\tau)$ for large τ. As it has only a finite number of singularities it will be holomorphic for $\text{Re}\tau > M$, and so will be $W(\tau,\eta(\tau))$. As $\hat{W}$ is a sum of terms

$$x^r e^{\xi_s x} C(\tau,\eta) \quad \text{where} \quad x > 0 , \ \text{Re}\xi < 0 ,$$

and C is algebraic, we have

$$|W| = |\hat{W} e^{\tau t+\eta y}| \leq K \exp(Mt + y|\tau|^{1/p})$$

for $\tau = M + bi$ and M large enough. Also $K = K(\tau)$ can be majorized by a polynomial.

Now construct

$$U = \int_{M-i\infty}^{M+i\infty} W(\tau)\exp(-\tau^q)d\tau$$

where $\frac{1}{p} < q < 1$, choosing the branch of τ^q that is real and positive when τ is. On the line of integration $\mathrm{Re}\tau = M$, we find

$$\mathrm{Re}\tau^q = |\tau|^q\cos(q\arg\tau) \geq |\tau|^q\cos\frac{\pi q}{2} .$$

This immediately shows the integral for U converges uniformly for $M > M_1$, and so by Cauchy's theorem gives values independent of M. Also differentiation under the integral sign with respect to t, x or y, gives integrals with the same property. This shows $U \in C^\infty$.

Furthermore,

$$|U| \leq e^{Mt}\int |\hat{W}|\exp\left(y\eta(\tau) - \tau^q\right) d\tau$$

where the integral can be estimated independently of M. Therefore, if $t < 0$ we let $M \to \infty$ and conclude that $U = 0$, for $t < 0$. Since $U \not\equiv 0$ is easily shown, it follows that the desired "nonunique" solution has been constructed.

A detailed discussion of time reversibility for well posed mixed boundary problems has also been given by Hersh (4). For two variable problems, it is necessary and sufficient if $k = \frac{1}{2}n$ and the given boundary conditions are independent in the two senses required. For more than two variables (and then with constant coefficients), the boundary operator B should evidently be stable both ways, implying $\det\tilde{B} \neq 0$ for $|\mathrm{Re}\tau| > M_0$ not just for $\mathrm{Re}\tau > M_0$. This would assume that the column space of E has dimension k both for $\mathrm{Re}\tau > M_0$ and $\mathrm{Re}\tau < -M_0$. However it can be shown that $\dim E(-\tau) = \dim F(\tau)$, where F is the solution space of $P(\tau, D_x, \eta)U = 0$ which grow no faster than a power of $|x|$ as $x \to -\infty$. Hence the boundary operator can be stable both ways only if the degree of D_x in P is equal to $2k$.

For some "parabolic" boundary conditions such as $u_t = u_{yy}$ for $x = 0$, no reversibility can be expected. However if the operators P and B are homogeneous in D_t, D_x and D_y, then the "mirror" problem for $t < 0$, $x < 0$ should be correct if the given problem is correct. By a detailed argument involving Puiseux series for $\tau(\xi,\eta)$, Hersh shows that if P is hyperbolic and B stable and hyperbolic for P in $t > 0$, $x > 0$, then also B is stable and hyperbolic for P in $x < 0$, $t < 0$.

5.3 Singularities of the reflected Riemann matrix

For a hyperbolic system of first order

$$P(D_t,D_x,D_y)u(x,y,t) \equiv \left(Im\frac{\partial}{\partial t} - A\frac{\partial}{\partial x} - \sum_{j=1}^{n-1} A_j \frac{\partial}{\partial y_j} \right) u = 0$$

the asymptotic expansions about the wave fronts of the reflected Riemann matrix have been calculated by Deakin (1). Here the coefficients A, A_j are $m \times m$ constant matrices such that the system is hyperbolic, Im is the identity matrix, and u an m-vector. If $f(t)$ is an m-vector point source, located at $(\ell,0,0..)$ then

$$Pu = \delta(x-\ell)\,\delta(y)\,f(t) ,$$

and u is expressed as $u = u_1 + u_2$ where

$$Pu_1 = \delta(x-\ell)\,\delta(y)\,f(t) , \quad u_1 \equiv 0 , \; t < t_0 , \; -\infty < x < \infty .$$

$$Pu_2 = 0 , \; Bu_2 = -Bu_1, \text{ on } x = 0 ; \quad u_2 \equiv 0 , \; t < 0 .$$

Then the reflected solution u_2 is expressed as

$$u_2 = -U * \left[Bu_1\right]_{x=0} \quad \text{where}$$

$$U = \frac{-i}{(2\pi)^n} \int_{t-i\infty}^{t+i\infty} d\tau \int_{-\infty}^{\infty} W \tilde{B}^{-1}\, dz_i , \ldots , dz_{n-1}$$

and W, $\tilde{B}$ are as defined by Hersh. The boundary conditions are assumed stable and hyperbolic, giving finite propagation speeds. If R is the Riemann matrix, then

$$u = R(x-\ell , y , t) * f(t) - U * \left[BR\right]_{x=0} * f(t) .$$

Considering now only the reflected terms, it is possible to carry out certain integrations in the convolution and to obtain

$$U * \left[BR\right]_{x=0} = \sum_{\ell,j} \frac{-1}{(2\pi)^n} \int_{-\infty}^{\infty} d\eta \int_{\varepsilon-i\infty}^{\varepsilon+i\infty} d\tau\, W\, \tilde{B}^{-1}(\tau,i\lambda^\ell,i\eta)$$

$$\times\, B(\tau,i\lambda^\ell,i\eta)\, t_j \exp(-i\,\ell\,\lambda^j)$$

where $\lambda^\ell = \lambda^\ell(\tau/i,\eta)$ is a root of $\det P(\tau/i,\lambda^\ell,\eta)$, and $t_j = t_j(x-\ell,\tau/i,\eta,\lambda^\ell)$ is a column null vector or generalized null vector of P_j . This expression can be rewritten as

$$\sum_{\ell_j} \frac{1}{(2\pi)^n} \int_{-\infty}^{\infty} d\eta \int_{\varepsilon-i\infty}^{\varepsilon+i\infty} d\tau\, S_{\ell j}(\tau,\eta) \exp\left(i(x\lambda^{\ell}-\ell\lambda^{j}) + iy\eta + \tau t\right)$$

and its asymptotic expansion has the form

$$\sum_{\ell_j} \sum_{\alpha=0}^{\infty} \left(I^{\alpha}_{\ell,j} + (-1)^{n} I^{*\alpha}_{\ell,j}\right) * \delta(t)\, I_m \ ,$$

where

$$I^{\alpha}_{\ell,j} = \frac{+1}{(2\pi)^n} \int_{-\infty}^{\infty} dp \frac{t^{\alpha}_{\ell,j}(1-i0,p,x)}{\Delta^0(1-i0,p)} (\Xi_{\ell,j} + i0)^{\alpha-n-\beta_j}$$

and $I^{*\alpha}_{\ell,j}$ is defined similarly but with $i0$ replaced by $-i0$. The phase is

$$\Xi_{\ell,j} = x\,p^{\ell}(1-i0,y) + y\cdot p - \ell p^{j}(1-i0,p) + t \ ,$$

and $P_\ell = 0$, $\operatorname{Im} p\ell \geq 0$. If characteristic roots are simple, $\beta_j = 0$, and in general β_j is less than the multiplicity of p^j. Also, if B is row homogeneous in its derivatives, then $T^{\alpha}_{\ell,j} = 0$ for $\alpha > \beta_j$.

At a point distant s from an ordinary point of a wave surface, the asymptotic expansion of the reflected wave is

$$\sum_{\alpha=-\beta_j}^{\infty} (Z^{\alpha}_{\ell j} + Z^{*\alpha}_{\ell j}) F_{\alpha}(s) + i(Z^{\alpha}_{\ell j} - Z^{*\alpha}_{\ell j}) \overline{F}_{\alpha}(s)$$

where, for example.

$$F_0(s) = \begin{cases} \pi^{-\frac{1}{2}} \left(\frac{\partial}{\partial s}\right)^{\frac{n}{2}} s_+^{-\frac{1}{2}} I_m \\ \delta^{(n-1)/2}(s)\, I_m \end{cases} , \quad \overline{F}_0(s) = \begin{cases} \pi^{-\frac{1}{2}} \left(\frac{\partial}{\partial s}\right)^{\frac{n}{2}} s^{-\frac{1}{2}} I_m & n \text{ eve} \\ -\pi^{-1} \left(\frac{\partial}{\partial s}\right)^{(n-1)/2} I_m & n \text{ odd} \end{cases}$$

Near a lateral or branch wave surface the asymptotic expansion contains two groups of terms. The first is as above for the ordinary points but with s replaced by

$$s - \frac{1}{2} w_1^2$$

where w_1 is a lateral coordinate measured from the point of tangency. The second group, after reduction, becomes

$$w_1^{-3/2} \sum_{\alpha=-\beta_j+1}^{\infty} (C^{\alpha}_{\ell j} + C^{*\alpha}_{\ell j}) F_{\alpha}(s) + i(C^{\alpha}_{\ell j} - C^{*\alpha}_{\ell j}) \overline{F}_{\alpha}(s)$$

where $C^{\alpha}_{\ell j}$ is a matrix with $C^{\alpha}_{\ell j} = C^{*\alpha}_{\ell j}$ for the outermost lateral wave for each reflected sheet. These terms are one order less singular than for an ordinary point. However at the point of tangency with the main reflected front, Deakin shows that this second group of terms is only ¼ order less singular.

5.4 Interface problems

Suppose that in a vibrating medium there is a change in the speed of wave propagation across a certain plane, as when light passes from air to water. Laws of reflection and refraction, such as Snell's law, have been established for the behaviour of rays, or for plane wave solutions. However the general radiation problem can be formulated first as an interface problem for transmission from one medium to the other; and then as a mixed problem for a system in a single "composite" region with plane boundary. Here we shall describe an interface problem discussed by Hersh (4) and then show how the reduction to a mixed problem is achieved.

Let two media occupy domains $D_\pm$ where $x > 0$ in D_+ and $x < 0$ in D_-. Let the wave speeds be $c_\pm$, and solutions $u_\pm$ respectively. Let

$$u^{\pm}_{tt} = c^2_{\pm} \Delta u^{\pm} \qquad \text{for } x \gtrless 0 , \ t > 0 ,$$

with $u^{\pm}(0,x,y) = u_t(0,x,y) = 0$ and let

$$\left(B + C\frac{\partial}{\partial x}\right)\begin{pmatrix}u_+\\u_-\end{pmatrix} = \delta I \qquad \text{on } x = 0 .$$

Here $\delta = \delta(t)\delta(y)$ where for simplicity we assume one lateral variable y only. Also

$$B = \begin{pmatrix}b_{11} & b_{12}\\ b_{21} & b_{22}\end{pmatrix} , \quad C = \begin{pmatrix}c_{11} & c_{12}\\ c_{21} & c_{22}\end{pmatrix}$$

where $R = b_{12}c_{21} - b_{22}c_{11} \geq 0$ and $S = b_{11}c_{22} - b_{21}c_{12}$. Then Hersh shows by a detailed study of the transform algebra that a unique solution u exists, with $e^{-Mt}u$ a tempered distribution for some $M > 0$, unless $\det C = 0$ and one of the following four conditions holds:

$$\text{I.} \quad \|B\| = \frac{R}{c_+} + \frac{S}{c_-} = 0$$

$$\text{II.} \quad S = -R < 0 , \ \operatorname{sgn}\det B = \operatorname{sgn}(c_+ - c_-)$$

$$\text{III.} \quad \operatorname{sgn}(R + S) = -\operatorname{sgn}\left(\frac{R}{c_+} + \frac{S}{c_-}\right) \neq 0$$

IV. $\frac{R}{c_+} + \frac{S}{c_-} = 0$, $\mathrm{sgn}(R+S) = -\mathrm{sgn}\,\det B$.

In the one dimensional problem $u^{\pm}_{tt} = c^2_{\pm} u^{\pm}_{xx}$ the conditions simplify and a solution exists unless $\det B = \det C = Rc^- + Sc^+ = 0$.

For the detailed proof we refer to Hersh (4); however we remark that essentially the same result is shown to hold for heat diffusion with the wave equation replaced by the parabolic heat flow equation $u_t = k\,\Delta u$, and the differing wave velocities $c_\pm$ by differing heat diffusivities $k^\pm$, with $\sqrt{k^\pm}$ in place of $c_\pm$ in the algebra. Hersh also considers the case of a vibrating medium (wave equation $u^+_{tt} = c^2_+ \Delta u_+$) coupled to a diffusing medium (heat equation $u^-_t = k_- \Delta u^-$) , for which existence is shown unless $\det C = 0$ and either $\{\det B = R = S = 0\}$ or $\{R = -S > 0$, $\det B \le 0\}$ or $\{R + S\} < 0$ in which cases there is no solution. He also treats the case of infinite wave speed $c \to \infty$ in which the problem becomes "stationary" on one or both sides of the interface. Thus a complete treatment of all the combinations is possible.

These classes of problems can all be reduced to one-sided mixed problems by the artifice of reflecting the far side $x < 0$ in the interface itself and working with the reflected functions and equations. Thus a new system, say

$$P_+(D_t,D_x,D_y)\,u_+ = f_+ \quad , \quad P_-(D_t,-D_x,D_y)\,u_- = f_-$$

is obtained, with boundary conditions involving both u_+ and u_- . These form a system of the type discussed in §4.1, and the correctness and stability conditions also carry over to the new system. For equation $P_- u_- = 0$ in $x < 0$ there should be one boundary condition for each root ξ_- of $P_-(\tau,\xi,i\eta) = 0$ with $\mathrm{Re}\xi_- > 0$ and $\mathrm{Re}\tau > M$; each such root goes over into a root ξ_+ of $P_-(\tau,-\xi,i\eta) = 0$ such that $\mathrm{Re}\xi_+ < 0$ for $\mathrm{Re}\tau > M$. We thus obtain the appropriate number of roots ξ_+ of $\det P(\tau,\xi,i\eta) = 0$ for the new one-sided problem, and this is sufficient for the proof described earlier. Existence having been shown for the mixed problem, a suitable reflection back to the two-sided domain completes the demonstration.

Problems involving one or more parallel layers are also physically relevant, and solutions can sometimes be constructed by multiple reflections. Hersh gives in (5) a necessary algebraic condition for P in the problem of a layer of finite thickness: For all $M > 0$ there exists $K(M)$ such that if $\det P(\tau,\xi,i\eta) = 0$ and $\mathrm{Re}\tau > K$ then $|\mathrm{Re}\xi| > M$. This holds for hyperbolic or parabolic operators.

CHAPTER 6.

Mixed problems for equations and systems with variable coefficients.

6.1. Historical survey. For the Cauchy problem existence theorems were known in the analytic case through the Cauchy-Kowalewska theorem, in the nineteenth century. The use of integral estimates for existence proofs dates from the 1920's and 1930's when such methods were studied by Courant, Friedrichs and Lewy (1) and Sobolev (1). These were first applied to mixed initial and boundary value problems by Schauder and Kryzanski (1) in the 1930's, using estimates as well as analytic approximation methods in the second order case and treating the Dirichlet and Neumann boundary conditions.

Mixed problems of higher order were first treated in the case of two variables, when the number of boundary conditions is equal to the number of characteristic lines entering the space-time domain from the corner. Campbell and Robinson (1) treated the case of semi-linear systems of first order, and higher order equations. A Lax (1) showed that in the case of multiple characteristics, differentiability is lost on reflection at the boundary. Thomée (1) gave integral estimates for the two variable problems.

Attention then returned to the higher dimensional situations. Duff (1) extended the Kryzanski Schauder results to wider classes of boundary conditions, and gave some isolated results for specific symmetric boundary conditions for the higher order equation. Duff (2) and Eisen (1) also treated the analytic case for general non-linear systems. Then Agmon (1) established an integral estimate for higher order equations with constant coefficients in the leading terms. Interest then swung to the most general boundary conditions in the constant coefficient problems, which were treated in detail by Hersh as described in the preceding chapter. This work brought out the roles of reflection coefficients and the possibility of ultrasonic boundary waves. With a view to subsequent generalization, Sarason (1) obtained L^2 estimates for the solution in the constant coefficient case.

The securing of L^2 estimates in the most general cases of first order systems with variable coefficients then became central, and this was established by Kreiss (1). The analogous problem for hyperbolic equations of higher order was solved by Sakamoto (1) at almost the same time, using the algebra of differential bilinear forms created by Hörmander (1). A general existence proof for the higher order equation was then given by Balaban (1) who brought together all of the intricate parts of the necessary proof, including an L^2 estimate for the equation and for the dual problem. In the succeeding five years, numerous extensions and additional contributions have been made to both second order and higher order mixed problems by Ikawa (1,2,3,4), Mizohata (1), Kajitani (1,2,3,4), Kubota (1), Miyatake (1,2), Peyser (1) and by Agemi (1,2,3,4), Shirota (1,2), Okhubo (1), Asano, Tsuji

(1,2,3,4,5) and Sato (1), as well as by Chazarain (1) and Piriou.

Here we can give only a brief description of some selected aspects of this extensive volume of results. In the next section estimates for second order equations are considered and very general boundary conditions established. Then the algebra of higher order estimates will be described, and an account of an existence proof given. As more general results and wider definitions of well posedness, such as the ε-well posed definition used recently by Ikawa and others, are currently being established, this subject may yet develop considerably before it reaches a definitive stage.

6.2. Boundary conditions for second order equations.

For constant coefficients the work of Hersh (1,2,3,4,5,6) has shown that a wide variety of boundary conditions lead to a well posed initial boundary problem. To extend these results to problems with variable coefficients calls for the combining of Fourier transform techniques with the type of global extension used in the theory of elliptic boundary value problems (Hormander 1, Ch. X). Here we give a treatment of the wave equation with a variable coefficients boundary condition

$$Bu = D_x u + b_0(t,y)D_t^{u} + b_1(t,y)D_y^{u} + c(t,y)u = g(t,y)$$

which is due to Tsuji (3) and in which the technique of pseudo-differential operators is employed to solve the boundary condition in conjunction with the wave equation.

The mixed problem to be studied is then

$$\left(D_t^2 - D_x^2 - \sum_{i=1}^{n-1} D_y^2\right)u = f(t,x,y) , \qquad D = -i\frac{\partial}{\partial x}$$

where $y \in R^{n-1}$ and the boundary is $x = 0$. Taking zero Cauchy data for simplicity we set $u \equiv 0$ for $t < 0$ with f and g also vanishing then. This problem is said to be H-well-posed at $t = t_0$ if for any integer $k > 0$ there exists an integer m and a positive number γ_k such that for any square integrable data which satisfy the corner compatibility conditions of order m-3, there exists a unique solution $u(t,x,y)$ in H_γ^k $(t > t_0 , x > 0 , y \in R^{n-1})$. The Hilbert space H_0^k is the space of square integrable k'th derivatives over the given domain, and $u \in H_\gamma^k$ if $u e^{-\gamma t} \in H_0^k$. Tsuyi shows that this problem is uniformly H-well-posed, if $\sup b_0(t,y) \leq 1 - \varepsilon$, where $\varepsilon > 0$. This condition may be compared with the condition $p > 0$, $q < 0$, $|r| < p$ of Chapter 1, which was recently shown by Miyataki (2) to be the necessary and sufficient condition for L^2 well posedness. Also for $n = 2$, Ikawa (3) has shown that if $b_0 = 0$, $b_1 \neq 0$, the problem is ε well posed (i.e. is H-well-posed and has a finite propagation speed).

For the following sketch of the proof we depend on Tsuji (3) and refer the reader to a detailed forthcoming account. With $g(t,y) \equiv 0$ for $t < 0$, the compatibility conditions of order m-3 imply $g(t,y) \in H^{m-2}_{\gamma}(R^n)$ for $\gamma \geq c$.

If we first assume that the problem is H-well-posed, and take m sufficiently large, then there is a solution $v(t,x,y) \in H^k_\gamma$; for $\gamma \geq \gamma_k$. Taking boundary values on $x = 0$ and using the trace imbedding theorems, it follows that

$$v_0(x,t) = \lim_{x\to 0} u(t,x,y)$$

and

$$v_1(x,t) = \lim_{x\to 0} D_x\, u(t,x,y)$$

exist and belong to $H_\gamma^{k-1/2}$ and $H_\gamma^{k-3/2}$ respectively. Now

$$v_1(t,y) = (2\pi)^{-n} \int_{R^n} e^{i(t\tau+y\eta)} \sqrt[+]{\tau^2-\eta^2}\, \hat{u}_0(\tau,\eta)\, d\sigma\, d\eta$$

which is the expression of the pseudodifferential operator defined by $\sqrt[+]{D_t^2 - D_y^2}$. The square root chosen has positive imaginary part, marked by the + sign. This formula is an easy consequence of the relation

$$u(t,x,y) = (2\pi)^{-n} \int_{R^n} e^{i(t\tau+x\mu^+ +y\eta)} \hat{u}_0(\tau,\eta)\, d\sigma\, d\eta$$

where $\mu^+ = \sqrt[+]{\tau^2-\eta^2}$ is the characteristic root with positive imaginary part that yields bounded solutions for γ large and positive. Comparing v_1 and v_0, we see that

$$v_1(t,y) = \sqrt[+]{D_t^2 - D_y^2}\, v_0 .$$

The boundary condition $Bu = g$ can now be written as a pseudo-differential equation

$$\left(\sqrt[+]{D_t^2 - D_y^2} + b_0(t,y)D_t + b_1(t,y)D_{y_1} + c(t,y)\right)v_0 = g(t,y) .$$

Here $\tau = \sigma - i\gamma$ $(\gamma > 0)$, and $(\sigma,\eta) \in R^n$.

Conversely, and without assuming existence of an overall solution, let us suppose $v_0(t,y)$ and $v_1(t,y)$ satisfy the pseudodifferential equation of the boundary condition, and the defining pseudodifferential relation. Then the given mixed boundary value problem can easily be solved by taking $v_0(t,y)$ as a Dirichlet boundary value. We note also that if $v_0(t,y) = 0$ for $t < 0$, then also $u(t,x,y)$ satisfying the homogeneous wave equation for $t < 0$ and the boundary condition $u(t.0,y) = v(t,y)$, will also vanish for $t < 0$.

To solve the above pseudodifferential equation we set

$$r(t,y;\tau,\eta) = \sqrt[+]{\tau^2-\eta^2} + b_0(t,y)\tau + b_1(t,y)\eta + c(t,y)$$

where $\tau = \sigma - i\gamma$, $\gamma > 0$, $(\sigma,\eta) \in R^n$ and $(t,y) \in R^n$.

Let the pseudodifferential operator be denoted by $R(t,y,D_t,D_y)$ where

$$R(t,y,D_t,D_y)u = (2\pi)^{-n} \int_{R^n} e^{i(t\tau+y\eta)} r(t,y;\tau,\eta)\hat{u}(\tau,\eta)\, d\tau\, d\eta \ .$$

Then the adjoint operator R^* is defined by

$$(Ru,v) = (u,R^*v) \qquad u,v \in C_0^\infty(R^n)$$

and it is easily shown that

$$R^*(t,y,D_t,D_y)v = (2\pi)^{-n} \int_{R^n} e^{itt\tau+y\eta)} r^*(t,y,\tau,\eta)\hat{v}(\tau,\eta) d\tau\, d\eta$$

where

$$r^*(t,y,\bar{\tau},\eta) = \sqrt[-]{\bar{\tau}^2-\eta^2} + b_0(t,y)\bar{\tau} + b_1(t,y)\eta + \bar{c}(t,y)$$
$$+ D_t b_0(t,y) + D_{y'} b_1(t,y)$$

and $\bar{\tau} = \sigma + i\gamma$ while $\sqrt[-]{\bar{\tau}^2-\eta^2}$ is the root of $\bar{\tau}^2-\eta^2$ with negative imaginary part.

Now assume $\sup b_0(t,y) \le 1 - \varepsilon$, where $\varepsilon > 0$. Then there exist positive constants c_k, γ_k , $k = 0, \pm 1, \pm 2, \ldots$ such that

1) $\langle Ru\rangle_{k,\gamma,R^n} \ge c_k \gamma \langle u\rangle_{k,\lambda,R^n}$, $u \in H^k_\gamma$, $\gamma \ge \gamma_k$

and

2) $\langle R^* v\rangle_{k,-\gamma,R^n} \ge c_k \gamma \langle v\rangle_{k,-\gamma,R^n}$, $v \in H^k_{-\gamma}$, $\gamma \ge \gamma_k$.

Here the Hilbert space norm of order k for H^k_γ is given by

$$\langle v\rangle^2_{k,\gamma,R^n} = \sum_{i+|\alpha|=k} \int_{R^n} |e^{-\gamma t} \gamma^i (D_t,D_y)^\alpha v|^2\, dt\, dy \ ,$$

where $\gamma^i(D_t,D_y)^\alpha$ denotes a homogeneous positive definite polynomial of degree $|\alpha|$ in the derivations D_t, D_y .

The proof of the first inequality will be given for $k = 0$, the other cases being similar. Let $\langle u,v\rangle_\alpha = \langle u,v\rangle_{0,\gamma,R^n}$ denote the scalar product, and form

$$\langle Ru,u\rangle_\gamma - \langle u,Ru\rangle_\gamma = \Big((\overset{+}{\sqrt{\tau^2-\eta^2}} - \overline{\overset{+}{\sqrt{\tau^2-\eta^2}}}\ \hat{u}(\tau,\eta)\,,\,\hat{u}(\tau,\eta)\Big)$$

$$-2i\gamma \langle u\, b_0\, u\rangle_\gamma + \langle u,D_t b_0\, u\rangle_\gamma$$

$$+ \langle u,D_y b,u\rangle_\gamma + \langle u,(\bar{c}-c)u\rangle_\gamma \ .$$

If the imaginary part of $\overset{+}{\sqrt{\tau^2-\eta^2}}$ is denoted by $f(\tau,\eta)$, then

$$f(\tau,\eta) = \frac{1}{\sqrt{2}} [\{(\gamma^2+\eta^2-\sigma^2)^2 + 4\sigma^2\gamma^2\}^{\frac{1}{2}} + \gamma^2+\eta^2-\sigma^2]^{\frac{1}{2}} \geq \gamma$$

as may be verified for all real γ,η,σ .

Thus we obtain

$$\mathrm{Im}\langle Ru,u\rangle_\gamma \geq \big(f(\tau,\eta)\hat{u}(\tau,\eta)\big) - \gamma\langle u,b_0 u\rangle_\gamma$$

$$+ \frac{1}{2i} \langle u,D_t b_0 u\rangle_\gamma + \frac{1}{2i} \langle u,D_y b_1 u\rangle$$

$$+ \langle \mathrm{Im}c.u \cdot u\rangle_\gamma$$

$$\geq (\varepsilon_\gamma - M_0)\langle u,u\rangle_\gamma$$

where the ε is the positive number in the bound for b_0 , and M_0 is a positive constant. Now if γ_0 is chosen as $\gamma_0 = 2M_0/\varepsilon$, it follows that for any $\gamma \geq \gamma_0$,

$$\mathrm{Im}\langle Ru,u\rangle_\gamma \geq \frac{\varepsilon}{2}\gamma\langle u\rangle_\gamma^{\ 2} \ .$$

Thus

$$\langle Ru\rangle_\gamma \langle u_\gamma\rangle \geq |\mathrm{Im}\langle Ru,u\rangle_\gamma| > \frac{c}{2}\gamma \langle u\rangle_\gamma^{\ 2}$$

whence

$$\langle Ru\rangle_\gamma \geq \frac{\varepsilon}{2}\gamma\langle u\rangle_\gamma \ , \qquad \gamma \geq \gamma_0 \ .$$

The higher inequalities for R and R^* are established similarly.

These inequalities show that R is an invertible operator on a dense subset of H^k_γ . To show that the range of R is H^k_γ we need to know (Friedrichs and Lax, 1) that the adjoint operator R^* is also invertible - but this is precisely the significance of the inequalities for R^* on $H^k_{-\gamma}$. Hence $Ru = g \in H^k_\gamma$ has a unique solution $u \in H^k_\gamma$. Combining these results we conclude that the wave equation has a solution with Dirichlet boundary values v_0 which then satisfies the given first order boundary condition $Bu = g$.

We also mention that Ikawa (2,3) has given a similar account for wave equations with variable coefficients which in the above case is equivalent to $b_0 = 0$, $b_1 = 0$.

For an equation with variable coefficients, constant coefficient approximations in small patches can be used to find approximations to solutions and to prove existence theorems. This leads to the concept of "freezing" the coefficients at a point, that is, taking these constant coefficient values over a neighbourhood or region. Agemi (3) has shown that a second order hyperbolic equation with variable coefficients and a first order boundary condition form a well posed problem in L^2 if and only if every constant coefficient problem obtained by freezing the coefficients at a boundary is L^2 well posed.

6.3 Estimates for hyperbolic systems and equations of higher order.

Solution of the mixed problem for systems or higher order equations involves L^2 estimates of solution functions and their derivatives, which in turn depend on certain positivity properties of the boundary conditions. The necessary algebra has been performed by Kreiss (1) for first order systems, and by Sakamoto (1) and Balaban (1) for higher order equations. Here we shall follow the work of Kreiss, Ralston (1,2) and Rauch (1,2) which leads to a general existence theorem for the mixed problem for a first order hyperbolic system. Related estimates for first order systems were also given by Sarason (1).

In this section we describe Kreiss estimates for the constant coefficient problem assuming it is stable hyperbolic in the sense of Hersh. In the following section estimates for the variable coefficient problem are described and the existence proof is outlined.

Consider the system

$$Lu = \frac{\partial u}{\partial t} - A\frac{\partial u}{\partial x_1} - \sum_{j=2}^{n} B_j \frac{\partial u}{\partial x_j} = f$$

$$u = u_0 \quad \text{for} \quad t = 0 \ , \ x > 0$$

$$B_j u = g_j \ , \ j = 1,\dots,k \quad \text{for} \quad x = 0 \ , \ t > 0 \ .$$

where u is an m-vector, A, B_j are $m \times m$ matrices, and f, u_0 are m vectors, g_j data functions given on the boundary $x = 0$, $t > 0$. We assume the system is strictly hyperbolic, that is $A(\xi) = A(\xi_1) + \Sigma B_j \xi_j$ has distinct real eigenvalues for all $\{\xi\} \in R^n$. Also we assume the boundary $x = 0$ is non-characteristic, that is $\det A \neq 0$. The boundary conditions are assumed to be well posed in the sense of Hersh (2) so that the homogeneous system does not have any exponential eigensolution with unlimited growth with respect to t.

Let N denote the boundary space defined by the given boundary conditions and let $E_-(\tau,\xi')$ be the linear space of boundary values of growing exponential solutions in t. Then we require $N \cap E_- = 0$, and assume N is maximal with this property. Following

Kreiss, we may write the boundary conditions in the form

$$u^{I}(0,x',t) = S\, u^{II}(0,x',t) + g(x',t)$$

where S is a $k \times k$ matrix and g a k vector. The estimate to be derived is

$$\int_0^T \|u(0,x',t)\|^2_{R^{n-1}}\,dt + \int_0^T \|u(x,t)\|^2_{R^n_+}\,dt \leq K_T\left(\int_0^T \|g(x',t)\|^2_{R^{n-1}}\,dt + \int_0^T \|f(x,t)\|^2_{R^n_+}\,dt\right)$$

where K_T may depend on T but not on f and g.

By a Fourier transform with respect to x' and Laplace transform with respect to t we obtain the system

$$(i\xi+\eta)\hat{v} = s\hat{v} = A\frac{d\hat{v}}{dx_1} + i\,B(\omega)\,\hat{v} + \hat{f}$$

$$\hat{v}^{I} = S\,\hat{v}^{II} + \hat{g}$$

where $u(x,t) = e^{\eta t}v(x,t)$. Then s is an eigenvalue if there is a nontrivial solution of these equations when $\hat{f}$ and $\hat{g}$ are zero. The assumption made on the boundary conditions is that there is no eigenvalue s with $\operatorname{Re} s \geq 0$.

Then Kreiss constructs a symbol or multiplier $\hat{R} = \hat{R}(\omega,\zeta,\eta)$ which for $\eta_0 < \eta < \infty$ has the following properties:

(1) $\hat{R}A$ is Hermitian

(2) $\hat{R}$ is uniformly bounded and is a smooth function of ω, η, ξ, and of the matrices A, B, S

(3) With $\zeta = (i\xi,\omega)$, $|\zeta|^2 = |\xi|^2 + |\omega|^2$,

$$\zeta' = \frac{\zeta}{|\zeta|}, \quad \eta' = \frac{\eta}{|\zeta|}$$, the symbol $\hat{R}$ is a function of ζ', η' for $|\zeta| > 1$

(4) $y^*\hat{R}Ay \geq \delta_1|y|^2 - C|g|^2$ for all vectors y that satisfy the boundary conditions; here $\delta_1, C > 0$

(5) $\operatorname{Re}\hat{R}\big(sI - iB(\omega)\big) \geq \delta_2\,\eta\, I$, where $\delta_2 > 0$.

Assuming for the moment the existence of $\hat{R}$, observe that

$$\operatorname{Re}(\hat{v},Rf) = \operatorname{Re}\left[-(\hat{v},\hat{R}A\frac{dv}{dx_1}\right]_0 + \left(\hat{v},\hat{R}(sI-iB(\omega))\hat{\sigma}\right)$$

$$= \operatorname{Re}\left(-\hat{v}^* R\,Av\Big|_{x_1=0}^{\infty}\right) + \left(\hat{v},\hat{R}(sI-iB(\omega)),\hat{v}\right)_0$$

$$\geq \delta_1|v(0,\omega,s)|^2 + \delta_2\eta\|v(x_1,\omega,s)\|^2 - C|g|^2 .$$

Estimating the left hand side by Schwartz inequality we easily find the estimate

$$\|\hat{v}(x_1,\omega,s)\|_0^2 + |\hat{v}(0,\omega,s)|^2 \leq \text{const.}(\|\hat{f}\|^2 + \|\hat{g}\|^2) .$$

The above estimate in the space variables now follows by means of Parseval's formula. We next discuss the construction of the multiplier, leaving to the following section the description of estimates for the case of variable coefficients.

Consider first the "resolvent" equation

$$\left(sI - P(\partial/\partial x)\right)v = f \quad , \qquad P \equiv A\frac{\partial}{\partial x_1} + \sum B_j\frac{\partial}{\partial x_j} \quad ,$$

and take its Fourier transform with respect to x' . Then $\hat{v}(x_1,\xi)$ satisfies

$$\frac{d\hat{v}}{dx_1} = A^{-1}\left(sI - iB(\omega)\right)\hat{v} + A^{-1}\hat{f} = M\hat{v} + A^{-1}\hat{f} ,$$

where the matrix $M = A^{-1}\left(sI - iB(\omega)\right)$ has for $\operatorname{Re}s \geq 0$ precisely ℓ eigenvalues κ with $\operatorname{Re}\kappa < 0$ and $m-\ell$ eigenvalues κ with $\operatorname{Re}\kappa > 0$ (Hersh (2)). It is easily shown that for every $f \in L^2[0,\infty)$, for $\operatorname{Re}s > 0$, the resolvent equation has a unique solution analytic in s , and such that

$$\|\hat{v}\| \leq \kappa(\operatorname{Re}s)^{-1}\|\hat{f}\|_0 .$$

Construct a unitary transformation U satisfying

$$UMU^{-1} = \begin{pmatrix} M_{11} & M_{12} \\ 0 & M_{22} \end{pmatrix}$$

where M_{11} is $\ell\times\ell$ with eigenvalues κ such that $\operatorname{Re}\kappa < 0$, and M_{22} is $(m-\ell)\times(m-\ell)$ with $\operatorname{Re}\kappa > 0$. Then with $h = (h^I,h^{II})$, $w = (w^I,w^{II}) = U^{-1}\hat{v}$, the solution can be written

$$w^{II} = -\int_{x_1}^{\infty} e^{M_{22}(x_1-\tau)}\, h^{II}(\tau,\omega)\,d\tau$$

$$w^{I} = \int_{-\infty}^{x_1} e^{M_{11}(x_1-\tau)}\, r^{I}(\tau,\omega)\,d\tau$$

where $r^{I} = h^{I} + M_{12}w^{II}$. Also, by Fourier transformation it follows that

$$|(M - i\beta I)^{-1}| \le \kappa |A^{-1}| \eta^{-1} .$$

Now let $\zeta = i\xi + \eta$, $\eta \ge 0$, $\zeta = (i\xi,\omega)$ and $M = M(\zeta,\eta) = A^{-1}((i\xi+\eta)I - iB(\omega)) = M(\zeta,0) + \eta A^{-1}$. The following lemmas will be stated, and for the proofs we refer to Kreiss' paper.

1. There exists a transformation T_0 such that for $\zeta_0 = (i\xi_0,\omega_0) \ne 0$, $\eta = 0$, we have

$$T_0 M T_0^{-1} = \mathrm{diag}(M_1,M_2,\dots,M_r) ,$$

where (a) the eigenvalues κ of M , have $\mathrm{Re}\kappa \ne 0$, (b) the eigenvalues κ of M_j for $j \ge 2$ have $\mathrm{Re}\kappa = 0$, and

$$M_j(\zeta,0) = \begin{pmatrix} \kappa_j & i & 0 & 0 & 0 \\ 0 & \kappa_j & i & 0 & 0 \\ 0 & 0 & \kappa_j & i & \\ & & & \ddots & \\ & & & & \kappa_j \end{pmatrix}$$

while the κ_j are distinct.

2. For every $\zeta_0 = (i\xi_0,\omega_0)$, $\eta = 0$. there is a T ,

$$T(\zeta,\eta) = T_0 + \eta T_1 + \eta^2 T_2 + \dots$$

analytic in ζ and η in a neighbourhood of ζ_0 , $\eta = 0$, such that

$$T M T^{-1} = \mathrm{diag}(M_1,M_2,\dots,M_r) ,$$

which reduces to that above for $\zeta = \zeta_0$, $\eta = 0$. Also T_0 is real, T_1 pure imaginary.

3. In a neighbourhood of ζ_0 , $\eta = 0$, there is a T_1 , with

$$T_{11} M_1 T_{11}^{-1} = \mathrm{diag}(N_{11},N_{12})$$

where

$$N_{11} + N_{11}^{*} \le -\delta I , \quad N_{12} + N_{12}^{*} \ge \delta I , \quad \delta > 0 .$$

4. If M_j has order $s \times s$, for $j \ge 2$, there is a continuous transformation $U_j(\zeta,\eta)$ with uniformly bounded inverse such that $U_j(\zeta_0,0) = I$, and

$$U_j M_j U_j^{-1} = \begin{pmatrix} \kappa_{j1} & \gamma_{j1} & 0 & \cdots & 0 \\ 0 & \kappa_{j2} & \gamma_{j2} & \cdots & 0 \\ 0 & 0 & \kappa_{j3} & \cdots & \\ 0 & 0 & & & \kappa_{js} \end{pmatrix}$$

while for $\eta > 0$ the eigenvalues κ with $\mathrm{Re}\kappa < 0$ stand in the first rows.

By the preceding each block $M_j(\zeta,\eta)$ can be written as

$$M_j(\zeta,\eta) = |\zeta|\left(M_j(\zeta_0') + \eta' N_j(\zeta_0) + O(\zeta'^2)\right)$$

where $N_j = (n_{rs})$ is real. Thus

5. There is a $c > 0$ such that $|ns_1| \geq c > 0$. Then the eigenvalues $\kappa_{j\nu}$ of M_j $(j \geq 2)$ and the corresponding eigenvectors $\phi_{j\nu}$ have the form

$$\kappa_{j\nu} = \kappa_j + |\zeta|(i^{s-1} n_{s1}\eta')^{1/s} + \ldots$$

$$= \kappa_j + |\zeta|^{(s-1)/s}(i^{s-1} n_{s1}\eta)^{1/s} + O(|\zeta|^{(s-2)/s}$$

$$\phi_{j\nu} = (1,\alpha,\alpha^2, \ldots, \alpha^{s-1})' + O(\eta') , \quad \alpha = i(i^{s-1} n_{s1}\eta')^{1/s} .$$

Also there are exactly

$$\rho_j = \begin{cases} \frac{1}{2}s & \text{eigenvalues} \\ \frac{1}{2}(s-1) & \kappa \text{ with} \\ \frac{1}{2}(s+1) & \mathrm{Re}\kappa < 0 \text{ if} \end{cases} \quad \begin{cases} s \equiv 0 \pmod 2 \\ s \equiv 1 \pmod 2, \; n_{s1} > 0 \\ s \equiv 1 \pmod 2, \; n_{s1} < 0 . \end{cases}$$

To derive necessary algebraic conditions, Fourier and Laplace transform the homogeneous equations obtaining

$$s\hat{v} = A\frac{d\hat{v}}{dx} + i\,B(\omega)\,\hat{v} \qquad x_1 \geq 0$$

$$\hat{v}^{I} = S\,\hat{v}^{II} .$$

Agmon (1) showed that if the problem is well posed, there is no eigenvalue s with $\mathrm{Re}s > 0$, for the solution $e^{\alpha st}\psi(\alpha x_1,\alpha\omega)$ would overpower any estimate as $\alpha \to \infty$. Let $\phi = U\psi$ so that

$$\frac{d}{dx_1}\begin{pmatrix}\psi_I \\ \psi_{II}\end{pmatrix} = \begin{pmatrix} M_{11} & M_{12} \\ 0 & M_{22}\end{pmatrix}\begin{pmatrix}\psi_I \\ \psi_{II}\end{pmatrix}$$

and

$$\begin{pmatrix} 1 & -S \\ 0 & 0 \end{pmatrix} U\psi = 0 \quad \text{or} \quad S_I\,\psi_I + S_{II}\,\psi_{II} = 0 .$$

Now s is an eigenvalue if and only if $\det S_I = 0$ and by homogeneity, if there is no zero on the unit sphere $|s|^2 + |\omega|^2$, we have $\det S_1 \geq \delta > 0$ there if and only if there is no eigenvalue s with $\operatorname{Re} s \geq 0$. It follows that the set of parameter points of the boundary condition space for which an L^2 estimate holds is the interior of the set M for which the problem is correctly set in the sense of Hersh. This is shown by perturbing the boundary conditions, in the case when an eigenvalue s with zero real part is present.

To construct the symmetrizing symbol $\hat{R}$ we write $\hat{R} = \hat{R}A$ and note that the earlier conditions for $\hat{R}$ amount to

1) $\hat{R}$ is Hermitian with properties 2) and 3)

4) $y^*\hat{R}y \geq \delta_1|y|^2 - c_1|g|^2$ for all y with $y^I = S\,y^{II} + g$.

5) $\operatorname{Re}\hat{R}\,A^{-1}\big(sI - iB(\omega)\big) = \operatorname{Re}\hat{R}\,M(\zeta,\eta) \geq \delta_2\eta\,I$.

Consider first the region $|\zeta| \leq 1$ wherein the eigenvalues κ of M divide into those with $\operatorname{Re}\kappa < -2\delta$ and those with $\operatorname{Re}\kappa > 2\delta$, where $\delta > 0$. Choose U as analytic in ζ,η so that $M_{12} = 0$. It is well known that there exist matrices

$$D_j \geq 0\,,\; D_j = D_j^* = D_j(\zeta,\eta) \in C^\infty(|\zeta| \leq 1\,,\, \eta_0 \leq \eta)\,,\; j = 1,2$$

such that

$$(-1)^j\,(D_jM_{jj} + M_{jj}^*D_j) \geq \delta I .$$

With $c > 0$ fixed later, set

$$\hat{R} = U^*\begin{pmatrix} -cD_1 & 0 \\ 0 & D_2 \end{pmatrix} U .$$

Then 1) holds. Also if we set $Uy = w$ then

$$y^*\hat{R}y = (Uy)^*\begin{pmatrix} -cD_1 & 0 \\ 0 & D_1 \end{pmatrix}(Uy) = -cw^{I*}D_1w^I + w^{II*}D_2w^{II}$$

and since $S_Iw^I + S_{II}w^{II} = g$ we obtain 4) if c is chosen small enough. Finally

$$2\,\mathrm{Re}\,\hat{R}M = \hat{R}M + M^*\tilde{R} = U^*\begin{pmatrix} -c(D_1M_{11} + M_{11}^*D_1) & 0 \\ 0 & D_2M_{22} + M_{22}^*D_2 \end{pmatrix}$$

so that 5) follows at once from the relations satisfied by D_1 and D_2.

Now consider $|\zeta| \geq \frac{1}{2}$. Relation 5) above can be written

$$\mathrm{Re}\,\tilde{R}M(\zeta,\eta) = |\zeta|\,\mathrm{Re}\,\tilde{R}M(\zeta',\eta') ,$$

and we construct R as a function of $\zeta' = \zeta/|\zeta|$, $\eta' = \eta/|\zeta|$. The argument used above works if $\eta \geq \eta_0' > 0$ for any positive η_0', so we need only consider neighbourhoods of ζ_0 and $\eta' = 0$. By the second and third lemmas there is a transformation $U = \mathrm{diag}(T_{II},T)T \in C^\infty$ with UMU^{-1} having block diagonal form $\mathrm{diag}(M_1,\ldots,M_r)$ with $M_1 = \mathrm{diag}(N_{11},\ldots,N_{12})$ while for $j \geq 2$,

$$M_j(\zeta',\eta') = M_j(\zeta',0) + \eta' N_j(\zeta') + O(\eta'^2) .$$

Here elements $n_{\ell k}$ of N_j $(\ell,k = 1,\ldots,s(j))$ are real while elements of $M_j(\zeta',0) = M_j(\zeta_0',0) + M_j(\zeta' - \zeta_0')$ are pure imaginary and $M_j(\zeta_0,0)$ has the form given in the first lemma. Let ρ_j be the number of eigenvalues κ of M_j with $\mathrm{Re}\,\kappa < 0$ for $\zeta' > 0$, and write

$$w^{(j)} = \left(w_I^{(j)'}, w_{II}^{(j)'}\right)$$

where $w_I^{(j)'}$ consists of the first ρ_j components of $w^{(j)}$. Then the boundary conditions become

$$S_I w^I + S_{II} w^{II} = g$$

where

$$w^I = \left(w_I^{(1)'}, w_I^{(2)'}, \ldots, w_I^{(r)'}\right)$$

and

$$w^{II} = \left(w_{II}^{(1)'}, w_{II}^{(2)'}, \ldots, w_{II}^{(r)'}\right) .$$

If for each block M_j we can find an $\hat{R}$ which is Hermitean, has the same smoothness as $\tilde{R}$, satisfies

$$w^{(j)*} R_j w^{(j)} \geq 2\delta_1\left(-c|w_I^{(j)}|^2 + |w_{II}^{(j)}|^2\right)$$

where c is a (small) positive constant, and $\mathrm{Re}\,\tilde{R}_j M_j(\zeta',\eta') \geq \delta_2 \eta' I$. Then $\tilde{R} = U^*\mathrm{diag}(\tilde{R}_1,\ldots,\tilde{R}_n)U$ has the necessary properties.

For $j = 1$, we can choose $\tilde{R}_1 = \mathrm{diag}(-cI, I)$ by the fourth lemma. For $j \geq 2$, a number of further lemmas are needed, which we again quote without proof.

(6) Let

$$C = \begin{pmatrix} 0 & 1 & & & & \\ & 0 & 1 & 0 & & \\ & & 0 & 1 & & \\ & 0 & & 0 & 1 & \\ & & & & & 0 \end{pmatrix}, \quad D = \begin{pmatrix} 0 & 0 & 0 & 0 & d_{1s} \\ 0 & & & d_{1s} & d_{2s} \\ 0 & & & & \\ 0 & & & & \\ d_{1s} & d_{2s} & & & d_{ss} \end{pmatrix} .$$

Then $DC = C^*D$, that is, D symmetrizes C.

(7) Let B be $s \times s$ symmetric of the form

$$B = \begin{pmatrix} b_{11} & b_{1s-1} & 0 \\ & & \\ b_{s-11} & b_{s-1\,s-1} & 0 \\ & & \\ 0 & 0 & 0 \end{pmatrix} .$$

Then $BC = C^*B$ implies $B = 0$.

(8) Consider $C + \varepsilon E$ where E is $s \times s$ symmetric. For $0 < \varepsilon < \varepsilon_0$ there is a matrix B such that

$$(D + \varepsilon B)(C + \varepsilon E) = S = S^*$$

is symmetric. The elements of B are rational functions of ε and of the elements of E.

(9) For $d \geq 2$, there exists a real antisymmetric matrix

$$F = \begin{pmatrix} 0 & -f_{12} & & 0 \\ f_{12} & 0 & -f_{23} \cdots & -f_{s-1,s} \\ 0 & & f_{s-1,s} & 0 \end{pmatrix}$$

such that $FC + C^*F^* > \mathrm{diag}(-1, \frac{1}{2}d, \ldots, \frac{1}{2}d)$. The choice $f_{i,i+1} = di^2$ suffices.

Now we can choose $\tilde{R}_j$ in the form

$$\tilde{R}_j = (D + \varepsilon B) - i\eta' F$$

with $d_{1s}n_{1s} \geq 2$. Since $M_j(\zeta',\eta')$ has the form

$$M_j(\zeta',\eta') = \kappa_j I + i(C + \varepsilon E) + \eta' N_i(\zeta') + O(\eta^{12})$$

where $\varepsilon = |\zeta' - \zeta_0'|$ and the elements of $\kappa_j I + iC = M_j(\zeta',0)$; $i\varepsilon E = M_j(\zeta' - \zeta_0',0)$ are pure imaginary and the elements $(n_{\ell k})$ of $N_j(\zeta')$ are real. By Lemma 5) we know that $|n_{1s}| \geq c > 0$. By lemmas 6), 7) and 8) we see that

$$\tilde{R}_j M_j + M_j^* \tilde{R}_j = \eta'(DN_j + N_j^* D + FC + C^*F) + O(|\varepsilon\eta'| + |\eta'|^2) .$$

But the upper left hand corner element of $DN + N^*D$ is $d_{1s}n_{1s} \geq 2$ so there is a constant $K = K(|N_j|)$ such that

$$DN_j + N_j^* D \geq \begin{pmatrix} \frac{3}{2} & 0 \\ 0 & 0 \end{pmatrix} - K|D| \begin{pmatrix} 0 & 0 \\ 0 & I \end{pmatrix} .$$

By lemma 9), we can choose F so that

$$DN_j + N_j^* D + FC + C^*F > 2\delta_2 I > 0$$

and thus the last of the required properties for the block M_j is satisfied, provided ε and η' are small enough. To fulfil the second last property, we choose the d_{js} as a sufficiently rapidly increasing sequence, for

$$j = 1,2,\ldots,s \quad \text{if} \quad s \equiv 1(2) \quad \text{and} \quad n_{s1} > 0 ,$$

and

$$j = 2,3,\ldots,s \quad \text{if} \quad s \equiv 0(2) \quad \text{or} \quad n_{s1} < 0 .$$

The Hermitean and smoothness properties of $\tilde{R}_j$ are obvious. This concludes the construction of the symmetrizing symbol $\hat{R}$, where $\tilde{R} = \hat{R}A$, and thus yields the existence of the L^2 estimates, when there is no eigenvalue s with $\mathrm{Re}\, s \geq 0$. Ralston (2) has shown that the same estimates hold if the boundary conditions are complex.

For a problem with variable coefficients, we may consider at each boundary point, the corresponding "frozen" problem with constant coefficients. The next step of the existence theory aims to use these estimates to show that suitable L^2 estimates also hold for the variable coefficients problem, a step first taken by Gårding (7) for elliptic estimates.

6.4. The existence theorem for the mixed problem with variable coefficients.

We first describe briefly the method originated by Gårding by which estimates, such as those of the preceding section, can be established for equations with variable coefficients. Supposing the coefficients vary continuously, we select a partition of unity $1 = \Sigma\psi_n$, where the ψ_n are C^∞, $\psi_n \geq 0$, and each ψ_n vanishes outside a neighbourhood of radius ε. We then form the constant coefficient estimates for every $\psi_n u$, using "frozen" or "local" constant coefficients. Comparing these with the variable coefficients, the difference can be shown to be small for small ε, so that by an adjustment of the constant, the estimates still hold with variable coefficients for $\psi_n u$. Summing over n, we find terms that add precisely to u and other terms with derivatives of ψ_n that contain lower derivatives or values of u which can be estimated in terms of the L^2 norms of the leading (first) order of derivatives of u. From this calculation there emerges a similar estimate for u in the L^2 norm in the variable coefficients case, the constant depending now on the variability of the coefficients.

However it is necessary to assume that the multipliers $\hat{R}$ for the various constant coefficient problems are uniformly bounded when the constants δ_1, δ_2 and C of the preceding work are fixed. This amounts to an assumption that the eigenvalue condition of Kreiss is uniformly satisfied and that therefore $\operatorname{Re}s < -\delta < 0$ holds for any eigenvalue of the homogeneous frozen problem. In this case the frozen problem may be called stable hyperbolic (Rauch (2,3)). Another way to describe this condition is to require that the angle between the subspaces $E_+(\tau,\zeta)$ and $E_-(\tau,\zeta)$ of the frozen problem should be bounded away from zero. That this angle can approach zero within the set of hyperbolic boundary conditions is due to the non-open character in the parameter space of the set of well posed hyperbolic problems. In this respect the hyperbolic mixed problems are deeper than elliptic boundary value problems.

We now describe briefly work of Rauch (2) who showed that L^2 is a continuable initial condition for the variable coefficient mixed problems for first order hyperbolic systems:

$$L_u = \partial_t u - \sum_{j=1}^{m} A_j(t,x)\,\partial_j u - B(t,x)\,u = F$$

in $[0,T] \times R^m_+$, with initial conditions $u(0,x) = f(x)$ and boundary conditions $Mu = u^I - Su^{II} = g$. Here all data are assumed to lie in a suitable Sobolev-Hilbert space $H_{s,\alpha}(V)$ with norm

$$||\psi||^2_{s,V,\alpha} = \sum_{|\nu|\leq s} \int_V ||D^\nu\psi||^2 \; e^{-2\alpha t}\, dV\, dt$$

and where V is R_+^m or $R^{m-1} = \partial R_+^m$, or R_+^{m-1} the initial manifold.

Rauch first shows there is a strong solution, which is a function $u \in L^2([0,T] \times R_+^m)$ for which there exist approximations $u_n \in C_0^\infty([0,T] \times R_+^m)$ and $u_0 \in C_0^\infty([0,T] \times R^{m-1})$ with

$$||u_n - u||_{[0,T] \times R_+^m} \to 0$$

$$||u_n - u_0||_{[0,T] \times R^{m-1}} \to 0$$

$$||Lu_n - F||_{[0,T] \times R_+^m} \to 0$$

and

$$||u_n(0,x) - f||_{R^{m-1}} \to 0$$

as $n \to \infty$, as well as $Mu_0 = g$.

The chief tool used for this purpose is a form of Kreiss' estimate which contains a time factor $e^{-2\alpha t}$ and is suitable for deriving similar estimates of derivatives of the solutions, namely,

$$\alpha \int_{-\infty}^{\infty} ||u(t)||^2_{R_+^m} e^{-2\alpha t}\, dt + \int_{-\infty}^{\infty} ||u(t)||^2_{R^{m-1}} e^{-2\alpha t}\, dt$$
$$\le c\left(\frac{1}{\alpha}\int_{-\infty}^{\infty} ||Lu(t)||^2_{R_+^m} e^{-2\alpha t}\, dt + \int_{-\infty}^{\infty} ||Mu(t)||^2 e^{-2\alpha t}\, dt\right) .$$

This holds for α sufficiently large and c independent of α and u as can be shown from Kreiss' estimate by elementary means. By successive differentiation, and use of the differential equation and boundary conditions, similar estimates can be established for derivatives of u, that is, estimates in the higher Sobolev norms. These can be written

$$\sqrt{\alpha}||u||_{s,R_+^{m+1},\alpha} + ||u||_{s,R^m,\alpha} \le c_s\left(\frac{1}{\sqrt{\alpha}} ||Lu||_{s,R_+^{m+1},\alpha} + ||Mu||_{s,R^m,\alpha}\right)$$

where R_+^{m+1} is $\{t,x_1,\ldots,x_m ; x_1 \ge 0\}$. Then by applying a differentiability theorem of Tartakoff (1, Theorem 3), which requires α large for s large, it can be shown that u has derivatives of all orders provided the corner or compatibility conditions are satisfied relative to the "edge" $t = 0$, $x_1 = 0$.

To establish the existence of u, however, similar estimates (with $e^{2\alpha t}$ factors) are obtained for solutions of the adjoint problem. A result of Friedrichs and Lax (1) then shows the existence of a unique strong solution, for suitable square integrable data. The main estimate used for a solution u of the full non-homogeneous problem is

$$\|u(t)\|_{s,R^m_+,\alpha} + \sqrt{\alpha}\,\|u\|_{s,R^{m+1}_+,\alpha} + \|u\|_{s,R^{m-1}_+\times[0,\tau],\alpha}$$

$$\leq c_s\left(\|f\|_{s,R^m_+,\alpha} + \frac{1}{\sqrt{\alpha}}\|F\|_{s,R^{m+1}_+,\alpha} + \|g\|_{s,R^{m-1}_+\times[0,\tau],\alpha}\right),$$

where c_s is independent of t, F, g, f and α.

In the most general case when the compatibility conditions are not satisfied, the main existence theorem obtained is the existence of a strong solution. If the compatibility conditions hold up to a certain order k of derivatives, then smoothness of the solution to that order can also be shown.

We omit the extensive calculations and refer instead to the paper of Rauch (1, 2, 3) who also gives similar estimates for hyperbolic polynomials of higher order. Similar estimates for higher order hyperbolic polynomials, and an existence theorem, have been given by Sakamoto (1).

Recently Majda and Osher (2) have extended the results of Kreiss, Ralston and Rauch for first order systems to the case when the boundary is characteristic, that is when one or more roots of the matrix $\Sigma A^j n_j = A_1$ are zero. Geometrically, a zero root corresponds to tangency of a characteristic surface to the boundary in space-time, and it is assumed that this holds throughout the space-time region considered. Several physically important systems including Maxwell's equations and the linearized shallow water equations have this property.

For the constant coefficient problem, Majda and Osher extend Kreiss' calculations based on the Laplace - Fourier transform, and develop a new symmetrizing construction for the matrix $M(s, i\omega)$ in the conical neighbourhood $|s| < \varepsilon|\omega|$ to cover the singularities that can now arise near $s = 0$. Here $s = \eta + i\xi$ is the Laplace transform variable with respect to t, and ω denotes the Fourier transform variable with respect to $x' = (x_2, x_3, \ldots, x_n)$, with x_1 the variable normal to the boundary $x_1 = 0$. In the transformed equations

$$(Es - iB(\omega))\hat{u} - \begin{bmatrix} 0 \\ A\hat{u}_x \end{bmatrix} = \hat{F},$$

the first ℓ_0 equations, corresponding to the zero roots of A_1 contain no term differentiated with respect to x_1. These equations are solved algebraically for the "$\hat{z}$" variables that do not appear differentiated with respect to x_1, and are then used to eliminate these "z" variables from the remaining equations. This yields a system of reduced rank and of the same form

$$\frac{d\hat{u}}{dx_1} = M(s, i\omega)\hat{u}$$

as that studied by Kreiss, but with possible poles as $s \to i\lambda_j(\omega)$ for any root $\lambda_j(\omega)$, $j = 1, 2, \ldots, \ell_0$ of the characteristic part $B_{11}(\omega)$ of $B(\omega) = i\Sigma_{j=2}^m A^j\omega_j$. For the extensive calculations needed for these constructions, we refer to the paper of Majda and Osher.

For the extension to variable coefficients L^2 estimates of higher derivatives in suitable Sobolov norms are required, and here a new phenomenon appears because the usual calculation for derivatives taken normal to the boundary requires use of the differential equation and thus fails for the characteristic variables that do not appear because the boundary is characteristic. It turns out in consequence that higher derivative estimates for this characteristic problem entail a "loss of derivatives" so that less complete differentiability results must be expected. In their main theorem on higher derivative estimates, Majda and Osher give five distinct cases for which different orders of estimates hold. They show that for the curl operator and the linearized shallow water equations the well posed conditions are those which are maximal dissipative (Phillips, 1), and for these estimates involving no loss of derivatives are found. For Maxwell's equations the well posed boundary conditions include the maximal dissipative conditions and are characterized by $\rho(S) < 1$ where $\rho = \lim\|S^n\|^{1/n}$ is the spectral radius and S the coefficient matrix in the boundary condition $u^I = Su^{II} + g$ in Kreiss' notation. Only for these boundary conditions does the usual estimate hold. A special estimate is given for energy-conserving boundary conditions for which $\|S\| = 1$. In general, it is also shown that non-symmetric coefficient matrices and boundary conditions involving the characteristic 'z' variables may lead to loss of derivatives. An example is given in which k^{th} derivatives of the solution behave like $2k^{th}$ derivatives of the initial data. For further details reference is again made to the paper of Majda and Osher.

To summarize, the chief stages of the existence theory are as follows.

1. Reduction of the variable coefficient problem to an assemblage of constant coefficient problems by "freezing".
2. Derivation of an estimate or inequality for the constant coefficient problem by means of a multiplier.
3. Extension of the estimate to the variable coefficient problem and to higher derivatives.
4. Construction of adjoint estimates and derivation of existence theorem in L^2 with appropriate smoothness for the solution.

A complete existence proof along these lines for the higher order hyperbolic equation $P(t, x, D_t, D_x)u = f$ with boundary conditions $Q_j(t, x, D_t, D_x)u = g_j$, where $Q_j(t, x, \tau, \xi_j)$ are linearly independent in ξ_1 modulo $P^+(t, x, \tau, \xi_j)$, has been given by Balaban (1) in a work of 117 pages in which extensive use is made of pseudo-differential operators. More recently, Okhubo and Shirota (1) have given a self-contained proof for first order hyperbolic systems with boundary conditions well posed in L^2 and with a multiplicity condition that roots of the determinantal characteristic equation are at most double. They further assume a condition in the complex plane if the Lopatinski determinant or boundary determinant vanishes at a double characteristic root. Also Ikawa (4) has treated the higher order hyperbolic equation with mixed conditions when the problem is well posed, not in the L^2 sense, but in a "sense of $\mathcal{E}$" which is determined by properties of the highest order or principal parts of equation and boundary conditions. Here $\mathcal{E}$ denotes a Fréchet space with seminorms $\sum_{|\nu|\leq \ell} \sup |D^\nu u(x)|$ taken over compact subsets K of the given domain E. Again extensive use is made of pseudo-differential operators and L^2 estimates of successive derivatives. The extensive character of each of these papers is an indication of the technical complexity of the necessary theory, and simplifications while desirable may not easily be found.

The essential hypotheses for the existence of a solution to a general mixed problem seem to include the following:

1. Condition of hyperbolicity of the differential equation or system, with restriction such as single or double on the multiplicity of roots.

2. Number of boundary conditions fixed by number of characteristic roots in, say, the upper half plane for $\mathrm{Im}\ \tau > 0$, (or, the number of inward oriented characteristic surfaces).

3. Roots condition of algebraic independence in ξ of boundary conditions, or else a condition on the Lopatinsky determinant in the vicinity of single or double zeros, together with a reflection coefficient restriction.

4. Any "frozen" constant coefficients problem at a boundary point is well posed.

5. Smoothness of coefficients and satisfaction to given order of corner compatibility conditions.

While some more general cases of higher multiplicities remain to be studied, little improvement in the existing results can be expected as many nearby counterexamples are known. Extensions to other types of partial differential equations, and to pseudodifferential equations, may still offer challenging future problems.

Bibliography.

AGEMI, R. (1) On energy inequalities of mixed problems for hyperbolic equations of second order. Jour. Fac. Sci. Hokkaido Univ. Ser. I. 21 (1971) 221-236.

(2) Remarks on L^2 well posed mixed problems for hyperbolic equations of second order. Hokkaido Math. Journal 2 (1973) 214-230.

(3) On a characterization of L^2 well-posed mixed problems for hyperbolic equations of second order. Proc. Jap. Acad. 51, (1975), 247-251.

(4) Iterated mixed problems for d'Alembertians, Hokkaido Math. Jour., I,3 (1974) 104-128, II,4 (1975), 281-294.

AGEMI, R. and SHIROTA, T. (1) On necessary and sufficient conditions for L^2 well posedness of mixed problems for hyperbolic equations, I,II, J. Fac. Sci. Hokkaido Univ. 21 (1970), 133-151, 22(1972) 137-149.

AGMON, S. (1) Problèmes mixtes pour les equations hyperboliques d'ordre superieure, Colloques Internationaux du C.R.N.S., 1963, 13-18.

ATIYAH, M.F., BOTT, R. and GÅRDING, L. (1) Lacunas for hyperbolic differential equations with constant coefficients, I Acta Math 124 (1970), 109-189, II Acta Math 131 (1973), 145-206.

BALABAN, T. (1) On the mixed problem for a hyperbolic equation, Mem Amer Math Soc., 112 (1971), 117 pp.

BARDOS, C. (1) Comportement asymptotique de la solution d'un system hyperbolique, Seminaire, College de France, 1973-74.

BAZER, J. and YEN, D.H.Y. (1) Lacunas of the Riemann matrix of Symmetric-Hyperbolic Systems in Two Space Variables, Comm. Pure and Appl. Math. 22 (1969), 279-333.

BLEISTEIN, N. (1) Uniform asymptotic expansions of Integrals with Stationary point near Algebraic Singularity, Comm. Pure and Appl. Math. 19 (1966), 353-370.

(2) Uniform asymptotic expansions of integrals with many nearby stationary points and algebraic singularities, J. Math. Mech. 17 (1967), 533-559.

BONDI, H. (1) Waves on the surface of a compressible liquid, Proc. Camb. Phil. Soc. 43 (1947), 75-95.

BRECKOVSKIKH, L.M. (1) Waves in Layered Media, Academic Press, N.Y. 1960.

BUREAU, Florent J. (1) Divergent integrals and partial differential equations, CPAM 8 (1955), 143-202.

BURRIDGE, R. (1) Lacunas in two dimensional wave propagation, Proc. Camb. Phil. Soc. 63 (1967), 819-825.

CAGNIARD, L. (1) Reflection and refraction of progressive seismic waves, McGraw-Hill 1962, xx+282pp. trans. Flinn and Dix.

CAMPBELL, L.L. and ROBINSON, A., (1) Mixed problems for hyperbolic partial differential equations, Proc. Lond. Math. Soc. 5 (1955), 129-147.

CEHLOV, V.I. (1) A mixed problem with a discontinuous boundary operator for a hyperbolic equation, in Boundary Value Problems for differential equations III Trudy Mat. Inst. Steblov, 126 (1973) 171-223.

CHARAZAIN, J. and PIRIOU, A. (1) Characterization des problemes mixtes hyperboliques bien posés, Ann. Inst. Fourier Grenoble, 22, 4 (1972) 193-237.

CHARAZAIN, J. (1) Sur quelques problemes mixte, Paris C.R. Ser A,B (268) (1969) 1197-1199.

COPSON, E.T. (1) Asymptotic Expansions, Cambridge, 1965, 120p.

COURANT, R. (1) Methods of Mathematical Physics, Vol. II, Partial Differential Equations, Interscience, New York, 1962, 830p.

COURANT, R., FRIEDRICHS, K.O. and LEWY, H. (1) Uber die partiellen Differenzengleichungen Physik, Math. Annalen, 100 (1928), 32-74.

DEAKIN, A.S. (1) Singularities of the reflected Riemann matrix, Jour. of Math. and Mechanics 17 (1967), 279-298.
(2) Asymptotic expansions for a hyperbolic boundary problem, Comm. Pure and Applied Math 24 (1971), 227-252.
(3) Asymptotic solution of the wave equation with variable velocity and boundary conditions, Siam J. Appl. Math 23 (1972), 87-98.

DUFF, G.F.D. (1) A mixed problem for normal hyperbolic linear partial differential equations of second order, Canad. J. Math. (1957) 141-160.
(2) Mixed problems for linear systems of first order equations, Canad. J. Math. 10 (1958), 127-160.
(3) The Cauchy Problem for elastic waves in an anisotropic medium, Phil. Trans. Roy. Soc. A 252 (1960), 249-273.
(4) On the Riemann matrix of a hyperbolic system, M.R.C. Technical Report 246 (1961), 58pp., Madison, Wisconsin.
(5) On wave fronts and boundary waves, Comm. Pure and Applied Math. 17 (1964), 189-225.

DUFF, G.F.D. and ROSS, R.A. (1) Indefinite Green's functions and elementary solutions, Canad. Math. Bull. 6 (1963), 71-103.

DUFF, G.F.D. and TSUTSUMI, A. (1) On domains of dependence and partial lacunas for Symmetric Hyperbolic Systems, Jour. of Math. and Mechanics 19 (1969), 219-238.

DUISTERMAAT, J.J. and HÖRMANDER, L. (1) Fourier integral operators II, Acta. Math. 128 (1972), 183-269.

EISEN, M. (1) Piecewise analytic solutions of mixed boundary value problems, Can. J. Math. 18 (1966), 1121-1147.

ERDELYI, A. (1) Asymptotic expansions, New York, 1956.

EWING, W.M., JARDETSKY, W.S., and PRESS, F. (1) Elastic waves in layered media, McGraw-Hill, New York (1957), xi + 380p.

FRIEDLANDER, F.G. (1) The wave front set of the solution of a simple initial boundary value problem with glancing rays, Math. Proc. Camb. Phil. Soc. 79, (1976), 145-159.

FRIEDRICHS, K.O. (1) Symmetric hyperbolic linear differential equations, Comm. Pure and App. Math. 7 (1954), 345-392.
(2) Symmetric Positive linear differential equations, Comm. Pure and App. Math. 11 (1958), 333-418.

FRIEDRICHS, K.O. and LAX, P.D. (1) Boundary value problems for first order operators, Comm. Pure and App. Math. 18 (1965), 355-388.
(2) On symmetrizable differential operators, AMS Proc. Symposia in Pure Math. Vol. 10, Providence, R.I., 1967

GÅRDING, L. (1) Linear hyperbolic partial differential equations with constant coefficients, Acta Math. 85 (1951), 1-62.
(2) Solution directe du problème de Cauchy pour les equations hyperboliques, Proc. Coll. Int. du CNRS, LXXI (1956), 71-90.
(3) Cauchy's problem for hyperbolic equations, Univ. of Chicago lecture notes, 151 pp. (1958).
(4) An inequality for hyperbolic polynomials, J. Math. Mech 8 (1959), 957-966.
(5) Transformation de Fourier des distributions homogènes, Bull. Soc. Math. de France 89 (1961), 381-428.
(6) The theory of lacunas, Battelle Seattle 1968 Rencontres, Springer, 1970, 13-21.
(7) Dirichlet's problem for linear elliptic partial differential equations, Math. Scand. 1 (1953), 55-72.

GELFAND, I.M. and SHILOV, G.E. (1) Generalized functions, I (Moscow 1958) Vol. I Properties and operations, 423 p. trans. 1964 Academic Press; Vol. III Theory of Differential Equations (trans. 1967, Academic Press), 222 p.

GARDNER, C.S., GREENE, J.M., KRUSKAL, M.D. and MIURA, R.M. (1) The Korteweg-de Vries equation and generalizations, VI Methods for exact solution, CPAM 27 (1974), 97-133.

HADAMARD, J. (1) Le problème de Cauchy et les equations aux derivées partielles lineaires hyperboliques, Paris, 1932.

HAYASHIDA, K. (1) On a mixed problem for hyperbolic equations with discontinuous boundary conditions, Publ. RIMS, Kyoto Univ. 7, (1971), 57-67.

HERGLOTZ, G. (1) Uber die Integration Linearer Partieller Differentialglaichungen mit Konstanten Koeffizienten I (Anwendung Abelscher Integrale); II and III (Anwendung Fouriersche Integrale) Leipzig, Ber. Süch. Akad. Wiss, Math-Phys. Kl, 78 (1926), 93-106; 80 (1928), 6-114.

HERSH, R. (1) Mixed Problems in Several Variables, J. Math. Mech. 12 (1963), 317-334.
(2) Boundary conditions for equations of evolution, Arch. Rat. Mech. Anal. 16 (1964) 243-263.
(3) On surface waves with finite and infinite speed of propagation, Arch. Rat. Mech. Anal. 19 (1965), 309-316.
(4) On vibration, diffusion, or equilibrium across a plane interface, Arch. Rat. Mech. Anal. 21 (1966), 368-390.
(5) On the general theory of mixed problems, Battelle Seattle 1968 Rencontres, Springer 1970, 85-95 (Hyperbolic equations and waves).
(6) How to classify differential polynomials, Amer. Math. Monthly 80 (1973), 641-654.

HÖRMANDER, L. (1) Linear partial differential operators, Berlin (1963), 284 pp.

(2) On the singularities of solutions of partial differential equations, CPA, 23 (1970), 329-358.
(3) Uniqueness theorems and wave front sets for solutions of linear differential equations, CPAM 24, (1971), 671-703.
(4) Fourier Integral Operators I, Acta Math. 127 (1971), 79-183.

IKAWA, M. (1) A mixed problem for hyperbolic equations of second order with nonhomogeneous Neumann type boundary condition, Osaka J. Math. 6 (1969), 339-374.
(2) On the mixed problem for hyperbolic equations of second order with the Neumann boundary condition, Osaka J. Math 7, (1970), 203-223.
(3) Mixed problem for the wave equation with an oblique derivative boundary condition, Osaka J. Math. 7 (1970), 495-525.
(4) Problemès mixtes mais pas necessairement bien posés pour les equations strictement hyperboliques, Osaka J. of Math. 12 (1975), 69-116.
(5) A mixed problem for hyperbolic equations of second order with a first order derivative boundary condition, Publ. RIMS, Kyoto Univ. 5 (1969), 119-147.
(6) Mixed problem for a hyperbolic system of first order, Publ. RIMS, Kyoto Univ. 7 (1971), 427-454.
(7) Sur les problèmes mixtes pour l'equation des ondes, Publ. RIMS, Kyoto Univ. 10 (1975), 669-690.

INOUE, A. (1) On the mixed problem for the wave equation with an oblique boundary condition, J. Fac. Sci. Univ. Tokyo, Sec. I 16 (1970), 313-329.
(2) On a mixed problem for d'Alembertian with a mixed boundary condition - an example of a moving boundary, Publ. RIMS, Kyoto Univ. 11 (1976), 339-401.

JOHN, F. (1) Partial differential equations, A.M.S. Lectures in Applied Math. Vol. 3 (1964), Part I Hyperbolic and Parabolic Equations by F. John.

KASHIWARA, M. and KAWAI, T. (1) Micro hyperbolic pseudodifferential operators I, Jour. Math. Soc. Japan 27 (1975), 359-404.

KAJITANI, K. (1) First order hyperbolic mixed problems, J. Math. Kyoto Univ. 11-3 (1971), 449-484.
(2) Initial boundary value problems for first order hyperbolic systems, R.I.M.S. Kyoto Univ. 7 (1971/2), 181-204.
(3) A necessary condition for the L^2 well posed Cauchy problem with variable coefficients, J. Math. Kyoto Univ. 13-2 (1973), 391-402.
(4) A necessary condition for the well posed hyperbolic mixed problem with variable coefficients, J. Math. Kyoto Univ. 14-2, (1974), 231-242.
(5) Sur la condition necessaire du problème mixte bien posé pour les systemes hyperboliques à coefficients variables, Publ. RIMS, Kyoto Univ. 9 (1974), 261-284.

KASAHARA, K. (1) On weak well posedness of mixed problems for hyperbolic systems, Publ. RIMS, Kyoto Univ. 6 (1970), 503-514.

KOLAKOWSKI, H. (1) Non-coercive mixed problems, Commentationes

Mathematicae 18 (1975), 189-192.

KREISS, H.O. (1) On difference approximations of the dissipative type for hyperbolic differential equations, CPAM 17 (1964), 335-353.

(2) Initial boundary value problems for hyperbolic systems, Comm. Pure and Applied Math. 23 (1970), 277-298.

(3) Generalized eigenvalues for mixed boundary value problems, to appear.

KRYZANSKI, M. and SCHAUDER, J. (1) Quasilineare Differentialgleichungen zweiter Ordnung vom hyperbolischen Typus Gemischte Randwertaufgaben, Studia Math. 6 (1936), 162-189.

KOHN, J.J. and NIRENBERG, L. (1) An algebra of pseudo differential operators, Comm. Pure and App. Math. 18 (1965), 269-305.

KUBOTA, K. (1) Remarks on boundary value problems for hyperbolic equations, Hobkaido Math. Journal 2 (1973), 202-213.

KUPKA, I.A.K. and OSHER, S. (1) On the wave equation in a multi dimensional corner, Comm. Pure and App. Math. 24 (1971), 381-394.

LADYZHENSKAYA, O.A. (1) A mixed problem for a hyperbolic equation (in Russian) GITTL, 1953.

LAGERSTROM, P.A., COLE, J.D. and TRILLING, L. (1) Problems in the theory of viscous compressible fluids, California Institute of Technology, 1949.

LAMB, H. (1) On the propagation of tremors over the surface of an elastic solid, Phil. Trans. Roy. Soc. A 203 (1904), 1-42.

LAPWOOD, E.R. (1) The disturbance due to a line source in a semi-infinite elastic medium, Phil. Trans. Roy. Soc. A 242 (1949), 63-100.

LAX, A. (1) On Cauchy's problem for partial differential equations with multiple characteristics, Comm. Pure Appl. Math. 9 (1956), 135-169.

LAX, P.D. (1) On Cauchy's problem for hyperbolic equations and the differentiability of solutions of elliptic equations, Comm. Pure App. Math. 8 (1955), 615-633.

LAX, P.D. and NIRENBERG, L. (1) On stability for difference schemes, a sharp form of Gårding's inequality, C.P.A.M. 19 (1966), 473-492.

LERAY, J. (1) Hyperbolic differential equations, Institute for Advanced Study, Lecture Notes, 1953, 238 pp.

(2) Unprolongement de la transformation de Laplace ... (Problème de Cauchy IV), Bull. Soc. Math. France, 90 (1962), 39-156.

(3) Equations hyperboliques non-stricts ... Battelle Seattle 1968 Réncontres, Springer 1970, 274-282.

(4) Solutions asymptotiques des equations aux-derivées partielles, Seminaire, College de France, 1972-73.

LERAY, J. and OHYA, Y. (1) Systèmes lineaires, hyperboliques non-stricts, Battelle Seattle 1968 Rencontres, 283-322.

(2) Equations et Systemes Non-lineaires, hyperboliques non-stricts, ibid, 331-369.

LERAY, J. and WAELBROECK, L. (1) Norme formelle d'une fonction Composée, ibid, 323-330.

LIGHTHILL, M.J. (1) On waves generated in dispersive systems ..., Battelle Seattle 1968 Rencontres, Springer, 1970, 124-152.
(2) Contributions to the Theory of Waves in Nonlinear dispersive systems, ibid, 173-210.
(3) Some Special Cases treated by the Whitham Theory, ibid, 237-262.

LUDWIG, D. (1) Exact and asymptotic solutions of the Cauchy problem, Comm. Pure and App. Math. 13 (1960), 473-508.
(2) Singularities of superpositions of distributions, Pacific J. Math. 15 (1965), 215-239.
(3) The Radon transform on Euclidean space, Com. Pure and App. Math. 19 (1966), 49-81.
(4) Uniform asymptotic expansions at a caustic, Com. Pure and Appl. Math. 19 (1966), 215-250.

LUDWIG, D. and GRANOFF, B. (1) Propagation of singularities along characteristics with non uniform multiplicity, Jour. Math. Anal. Appl. 21 (1968), 556-574.

MAJDA, A.J. (1) Coercive inequalities for nonelliptic symmetric systems, Comm. Pure and App. Math. 28 (1975), 49-89.

MAJDA, A. and OSHER, S. (1) Reflection of singularities at the boundary, CPAM 28 (1975), 479-499.
(2) Initial-boundary value problems for hyperbolic equations with uniformly characteristic boundary, Comm. Pure App. Math. 28 (1975), 607-676.

MATSUMOTO, W. (1) Uniqueness in the Cauchy problem for partial differential equations with multiple characteristic roots, J. Math. Kyoto Univ. 15-3 (1975), 479-525.

MATSUMURA, M. (1) Comportement des solutions de quelques problèmes mixtes pour certains systèmes hyperboliques symmetriques á coefficients constants, Publ. RIMS, Kyoto Univ. 4 (1968), 309-359.
(2) Comportement asymptotique de Solutions de certains problèmes mixtes pour des systèmes hyperboliques symmetriques á coefficients constants, Publ. RIMS, Kyoto Univ. 5 (1969), 301-360.

MIYAKE, M. (1) On the initial value problems with data on a characteristic surface for linear systems of first order equations, Publ. RIMS, Kyoto Univ. 8 (1972), 231-264.

MIYATAKE, S. (1) An approach to hyperbolic mixed problems by singular integral operators, J. Math. Kyoto Univ. 10-3 (1970), 439-474.
(2) Mixed problem for hyperbolic equation of second order, J. Math. Kyoto Univ. 13-3 (1973), 435-487.

MIZOHATA, S. (1) Quelques problèmes au bord, du type mixte, pour des equations hyperboliques, Seminaire sur les equations aux derivées partielles, College de France, (1966-67), 23-60.

(2) The theory of partial differential equations, Cambridge U.P. (1973), xii + 490 p.
MIZOHATA, S. and OHYA, Y. (1) Sur la condition d'hyperbolicité pour les equations à characteristiques multiples, Japanese Jour. of Math. XL (1971), 63-104.
MUSGRAVE, M.J.P. (1) On the propagation of elastic waves in aelotropic media, Proc. Roy. Soc. London Ser. A, 226 (1954),
I General principles, 339-355;
II Media of hexagonal symmetry, 356-366;
III Media of cubic symmetry, 236 (1956), 352-383 (with G.F. Miller).
(2) On whether elastic wave surfaces possess cuspidal edges, Proc. Camb. Phil. Soc. 53 (1957), 897-906.
(3) Crystal Accoustics, Holden Day, San Francisco, 1970.
NAKANO, H. (1) On Rayleigh waves, Japan J. Astron. Geophys. 2 (1925), 233-326.
(2) Some problems concerning the propagation of the disturbances in and on semi-infinite elastic solid, Geophys. Mag. (Tokyo) 2 (1930), 189-348.
NIRENBERG, L. (1) Lectures on linear partial differential equations, CBMS Regional Conference Series, 17 (1973), 58 p.
OKHUBO, T. and SHIROTA, T. (1) On structures of certain L^2 well posed mixed problems for hyperbolic systems of first order. Hokkaido Math. Jour. 4 (1975), 82-158.
OSHER, S. (1) Initial boundary value problems for hyperbolic systems in regions with corners, I. Trans. A.M.S. 176 (1973), 141-164.
(2) An ill posed problem for a hyperbolic equation near a corner, Bull. Amer. Math. Soc. 19 (1973), 1043-1044.
PAPADOPOULOS, M. (1) Diffraction of singular fields by a half-plane, Archiv for Rational Mechanics and Analysis 13 (1963), 279-295.
(2) The use of singular integrals in wave propagation problems, Proc. Roy. Soc. A 276 (1963), 204-237.
PEKERIS, C.L. (1) The seismic surface pulse, Proc. Nat. Acad. Sci. 41 (1955), 469-480.
PETROWSKI, I.G. (1) On the diffusion of waves and the lacunas for hyperbolic equations, Mat. Sbornik, 8617 (59)(1945), 289-370.
(2) Partial Differential Equations, Interscience, New York 1954, 245 p.
PETUKHOV, L.V. and TROITSKII, V.A. (1) Variational problems of optimization for equations of the hyperbolic type in the presence of boundary controls, PMM (App. Mathematics and Mechanics) 39 (1975), 245-253.
PEYSER, G. (1) Energy integrals for mixed problems in hyperbolic differential equations of higher order, J. of Math. and Mech. 6 (1957), 641-653.
PHILLIPS, R.S. (1) Dissipative operators and hyperbolic systems of partial differential equations, Trans. Amer. Math. Soc.

90 (1959), 193-254.
PRATO, DA, G. (1) Problèmes au bord de type mixte pour des equations paraboloques ou hyperboliques, Seminaire, College de France, 1967-8.
RALSTON, J.V. (1) Solutions of the wave equation with localized energy, Comm. Pure and App. Math. 22 (1969), 807-823.
(2) Note on a paper of Kreiss, Comm. Pure and App. Math. 24 (1971), 759-762.
RAUCH, J. (1) Energy and resolvent inequalities for hyperbolic mixed problems, Jour. Diff. Equations 11 (1972), 528-540.
(2) L_2 is a continuable initial condition for Kreiss' mixed problems, CPAM 15 (1972), 265-285.
(3) General Theory of Hyperbolic Mixed Problems, Proc. 1971 Berkeley Symposium in Pure Mathematics, A.M.S. (1973), vol. 23, 161-166.
RAUCH, J. and MASSEY, F.J. (1) Differentiability of solutions to hyperbolic initial boundary value problems, Trans. Amer. Math. Soc. 189 (1974), 303-318.
RAUCH, J. and TAYLOR, M. (1) Penetrations into shadow regions and unique continuation properties in hyperbolic mixed problems. Indiana Univ. Math. Journal 22 (1972-3), 277-285.
RAYLEIGH, LORD (1) On waves propagated along the plane surface of an elastic solid, Proc. Lond. Math. Soc. 17 (1885), 4-11.
RIESZ, M. (1) L'integrale de Riemann-Liouville et le problème de Cauchy, Acta Math. 81 (1949), 1-223.
ROGAK, E.D. (1) Mixed problem for the wave equation in a time dependent domain, Arch. Rat. Mech. Anal. 22 (1966), 24-36.
RUSSELL, D.L. (1) On boundary value controllability of linear symmetric hyperbolic systems. In Mathematical Theory of Control, ed. A.V. Balakrichnan and L.W. Neustadt, New York, 1967.
(2) Boundary value control of the higher dimensional wave equation, SIAM Journal on Control 9 (1971), 29-42.
(3) Control Theory of hyperbolic equations related to certain questions in harmonic analysis and spectral theory, Jour. of Math. Anal. and App. 40 (1972), 336-368.
(4) Controllability theory for hyperbolic and parabolic partial differential equations, Studies in Applied Mathematics 52 (1973), 189-211.
SADAMATSU, T. (1) On mixed problems for hyperbolic systems of first order with constant coefficients, J. Math. Kyoto Univ. 9 (1969), 339-361.
SAKAMOTO, R. (1) Mixed problems for hyperbolic equations, I, II, J. Math. Kyoto Univ. 10 (1970), 349-373, 403-417.
(2) Iterated hyperbolic mixed problems, Publ. RIMS, Kyoto Univ. 6 (1970), 1-42.
(3) L^2 well posedness for hyperbolic mixed problems, Publ. RIMS Kyoto Univ. 8 (1972), 265-293.

(4) Σ -well posedness for hyperbolic mixed problems with constant coefficients, J. Math. Kyoto Univ. 14-1 (1974), 93-118.

SARASON, L. (1) On hyperbolic mixed problems, Arch. National Mech. Anal. 18 (1965), 310-334.
(2) Elliptic regularization for symmetric positive systems, Jour. Math. Mech. 16 (1967), 807-827.
(3) Symmetrizable systems in regions with corners and edges, J. Math. Mech. 19 (1970), 601-607.

SARASON, L. and SMOLLER, J. (1) Geometrical Optics and the Corner Problem, Arch. Rat. Mech. Anal. 56 (1974), 35-69.

SATO, S. and SHIROTA, T. (1) Remarks on modified symmetrizers for 2×2 hyperbolic mixed problems, Hokkaido Math. Jour. 5 (1976), 120-138.

SHILOV, G.E. (1) Generalized Functions and Partial Differential Equations, New York, (1968), xii + 345 p.

SHIROTA, T. (1) On the propagation speed of hyperbolic operators with mixed boundary conditions, J. Fac. Sci. Hokkaido Univ. 22 (1972), 25-31.
(2) On certain L^2 well posed mixed problems for hyperbolic systems of first order, Proc. Jap. Acad. 50 (1974), 143-147.

SHIROTA, T. and ASANO, K. (1) On mixed problems for regularly hyperbolic systems, J. Fac. Sci. Hokkaido Univ. 121 (1970), 1-45.

SMOLLER, J.A. (1) A survey of hyperbolic systems of conservation laws in two independent variables, Battelle Seattle 1968 Rencontres, Springer, 1970, 51-60.

SOBOLEV, S.L. (1) Methode nouvelle à resoudre le problème de Cauchy pour les equations lineaires hyperboliques, Mat. S61, 43 (1936), 39-71.
(2) Application de la theorie des ondes planes a la solution du probleme de H. Lamb, Publ. Inst. Seism. Acad. Sci. URSS, 18 (1932), 1-41.
(3) Sur les vibrations d'un demiplan d'un couche a conditions initiales arbitraires, Mat. Sbornik (Moscow) 40 (1933), 236-266.

STOKES, G.G. (1) On the dynamical theory of diffraction, Trans. Camb. Phil. Soc. 9 (1849), reprinted in Mathematical and Physical Papers, vol. 2 (1966), 242-328.

STRANG, G. (1) On strong hyperbolicity, J. Math. Kyoto Univ. 6 (1967), 397-417.
(2) Hyperbolic initial boundary value problems in two unknowns, J. Diff. Equations 6 (1969), 161-171.

STRAUSS, W.A. (1) Dispersion of low energy waves for two conservative equations, Arch. Rat. Mech. Anal. 55 (1974), 86-92.

TANIGUCHI, M. (1) Mixed problem for hyperbolic systems of first order, Kyoto Univ., Publ. RIMS I, 8(1972), 471-481, II, 10 (1974), 91-100.

TARTAKOFF, D.S. (1) Regularity of solutions of Boundary Value Problems for first order systems, Indiana University Math. Journal 21 (1972), 1113-1129.

THOMÉE, V. (1) Estimates of the Friedrichs -Lewy type for mixed problems in the theory of linear partial differential equations in two independent variables, Math. Scand. 5 (1957), 93-113.

TSUJI, M. (1) Regularity of solutions of hyperbolic mixed problems with characteristic boundary, Proc. Jap. Acad. 48 (1972), 719-724.

(2) Analyticity of solutions of hyperbolic mixed problems, J. Math. Kyoto University 13-2 (1973), 323-371.

(3) Characterization of the well posed mixed problem for wave equation in a quarter space, Proc. Japan. Acad. 50 (1974), 138-142.

(4) Mixed problems for hyperbolic equations with constant coefficients, Proc. Japan. Acad. 51 (1975), 369-373.

(5) Propagation of the singularities for hyperbolic equations with constant coefficients, to appear.

VAILLANT, J. (1) Solutions asymptotiques d'un système à caractéristiques de multiplicité variable, Seminaire, College de France, 1972-73.

WAKABAYASHI, S. (1) Singularities of the Riemann functions of hyperbolic mixed problems in a quarter space, Publ. RIMS, Kyoto Univ. 11 (1976), 417-440.

WEITZNER, H. (1) Green's function for two-dimensional magneto hydrodynamic waves, I, II, Phys. Fluids 4 (1961), 1250-1258.

WHITHAM, G.B. (1) Linear and Nonlinear Waves, Wiley, New York, 1974, xvi + 636 pp.

ULTRADISTRIBUTIONS AND HYPERBOLICITY

Hikosaburo Komatsu

Department of Mathematics, University of Tokyo,
Hongo, Tokyo, Japan

ABSTRACT. There are infinitely many classes of generalized functions, called ultradistributions, between the distributions of L. Schwartz [34] and the hyperfunctions of M. Sato [32]. Each class of ultradistributions have similar properties as the distributions or the hyperfunctions. They form a sheaf on which linear differential operators act as sheaf homomorphisms. We have, among others, two structure theorems of ultradistributions, the structure theorem of ultradistributions with support in submanifold (which implies a Whitney type extension theorem for ultradifferentiable functions), and the kernel theorem for ultradistributions.

Ultradistributions are not only interesting for their own sake but also important for their applications to other branches of analysis and especially to the theory of linear differential equations. As an example we have formerly discussed the regularity of solutions of linear ordinary differential equations ([12], [14]).

Here we consider the hyperbolicity of partial differential equations and show that the Gevrey classes of ultradistributions come in naturally in the problem.

1. ULTRADISTRIBUTIONS

Let M_p, $p = 0, 1, 2, \ldots$, be a sequence of positive numbers. Usually we assume the following:

(M.0) $M_0 = 1$;

(M.1) $m_p = M_p/M_{p-1}$ is increasing ;

Garnir (ed.), Boundary Value Problems for Linear Evolution Partial Equations. 157-173.

(M.2) $M_p \leq AH^p \min_{0\leq q\leq p} M_q M_{p-q}$,

(M.3) $\sum_{q=p+1}^{\infty} m_q^{-1} \leq Apm_p^{-1}$,

where A and H are constants independent of p. In some cases we replace the last three conditions by the following:

(M.1)' m_p/p is increasing ;

(M.2)' $M_{p+1} \leq AH^p M_p$;

(M.3)' $\sum_{p=1}^{\infty} m_p^{-1} < \infty$.

(M.1)' is stronger than (M.1) and (M.2)' and (M.3)' are weaker than (M.2) and (M.3) respectively. (M.3)' is the non-quasi-analyticity condition of Denjoy-Carleman.

An infinitely differentiable function φ on an open set Ω in $\mathbb{R}^n$ is said to be an ultradifferentiable function of class (M_p) (resp. of class $\{M_p\}$) if for each compact set K in Ω and $h > 0$ there is a constant C (resp. there are constants h and C) such that

(1.1) $\sup_{x\in K} |\partial^{\alpha} \varphi(x)| \leq Ch^{|\alpha|} M_{|\alpha|}, \quad |\alpha| = 0, 1, 2, \ldots$

where

(1.2) $\partial^{\alpha} = \partial_1^{\alpha_1} \cdots \partial_n^{\alpha_n} = (\partial/\partial x_1)^{\alpha_1} \cdots (\partial/\partial x_n)^{\alpha_n}$

and

(1.3) $|\alpha| = \alpha_1 + \cdots + \alpha_n$.

If $s > 1$, the Gevrey sequence

(1.4) $M_p = (p!)^s$ or p^{ps} or $\Gamma(1+ps)$

satisfies the conditions (M.0), (M.1)', (M.2) and (M.3). In this case we write (s) and $\{s\}$ for (M_p) and $\{M_p\}$.

In general the asterisk $*$ stands for either (M_p) or $\{M_p\}$. The space $\mathcal{E}^*(\Omega)$ (resp. $\mathcal{D}^*(\Omega)$) of all ultradifferentiable functions φ of class $*$ on Ω (resp. with compact support) has the following expression:

(1.5) $\mathcal{E}^{(M_p)}(\Omega) = \varprojlim_{K\subset\subset\Omega} \varprojlim_{h\to 0} \mathcal{E}^{\{M_p\},h}(K)$,

(1.6) $\mathcal{E}^{\{M_p\}}(\Omega) = \varprojlim_{K\subset\subset\Omega} \varinjlim_{h\to\infty} \mathcal{E}^{\{M_p\},h}(K)$,

(1.7) $\mathcal{D}^{(M_p)}(\Omega) = \varinjlim_{K\subset\subset\Omega} \varprojlim_{h\to 0} \mathcal{D}^{\{M_p\},h}_K$,

$$(1.8) \quad \mathcal{D}^{\{M_p\}}(\Omega) = \varinjlim_{K \subset\subset \Omega} \varinjlim_{h\to\infty} \mathcal{D}^{\{M_p\},h}_K ,$$

where $\mathcal{E}^{\{M_p\},h}(K)$ is the Banach space of all infinitely differentiable functions φ in the sense of Whitney on the regular compact set K which satisfies condition (1.1) and $\mathcal{D}^{\{M_p\},h}_K$ is its closed linear subspace composed of all functions φ on $\mathbb{R}^n$ with support in K.

We introduce in the spaces $\mathcal{E}^*(\Omega)$ and $\mathcal{D}^*(\Omega)$ the locally convex topologies defined by these expressions. The inductive and the projective limits appearing here are good ones and we can prove the following.

Theorem 1.1 ([13], [17]). $\mathcal{E}^{(M_p)}(\Omega)$ is an (FG)-space, $\mathcal{E}^{\{M_p\}}(\Omega)$ is a (DLFG)-space, $\mathcal{D}^{(M_p)}(\Omega)$ is an (LFG)-space and $\mathcal{D}^{\{M_p\}}(\Omega)$ is a (DFG)-space.

Here we employed the terminology of [15]. G stands for Grothendieck (= nuclear), F Fréchet, L strict inductive limit and D strong dual. Thus a (DLFG)-space is the strong dual of the strict inductive limit of a sequence of Fréchet-Grothendieck spaces. In particular, all the above spaces as well as their strong duals are complete bornologic Grothendieck spaces.

Under the assumption of (M.1) and (M.3)' Petzsche [29] has proved that (M.2)' is a necessary and sufficient condition in order that any one of the above spaces is a Grothendieck space.

We denote by $\mathcal{D}^{*\prime}(\Omega)$ the strong dual of $\mathcal{D}^*(\Omega)$ and call its elements ultradistributions of class $*$. $\mathcal{D}^{\{M_p\}\prime}(\Omega)$ is the space of ultradistribution defined by Roumieu [30] and, if M_p satisfies certain conditions which Gevrey sequences satisfy (see Petzsche [29]), $\mathcal{D}^{(M_p)\prime}(\Omega)$ is the space of generalized distributions due to Beurling and Björck [1].

We can prove the existence of a partition of unity of class $*$ subordinate to any open covering. Moreover, the multiplication by an ultradifferentiable function of class $*$ and the differentiation ∂^α are continuous on the space $\mathcal{D}^*(\Omega)$. Hence we can prove that $\mathcal{D}^{*\prime}(\Omega)$, $\Omega \subset \mathbb{R}^n$, form a soft sheaf on which the differential operators

$$(1.9) \quad P(x, \partial) = \sum_{|\alpha|\leq m} a_\alpha(x)\,\partial^\alpha , \quad a_\alpha \in \mathcal{E}^*(\Omega) ,$$

act as sheaf homomorphisms and that the dual $\mathcal{E}^{*\prime}(\Omega)$ of $\mathcal{E}^*(\Omega)$ is identified with the linear subspace of $\mathcal{D}^{*\prime}(\Omega)$ composed of all ultradistributions with compact support.

We write $M_p \prec N_p$ if for each $L > 0$ there is a constant C such that

$$(1.10) \quad M_p \leq CL^p N_p , \quad p = 0, 1, 2, \ldots .$$

If $M_p \prec N_p$ and N_p satisfies conditions (M.0)-(M.3), then we have the inclusion relations:

(1.11) $\mathcal{A}(\Omega) \subset \mathcal{E}^{(M_p)}(\Omega) \subset \mathcal{E}^{\{M_p\}}(\Omega) \subset \mathcal{E}^{(N_p)}(\Omega) \subset \mathcal{E}(\Omega)$,

where $\mathcal{A}(\Omega)$ is the space of all real analytic functions on Ω equipped with the locally convex topology as $\mathcal{E}^{\{1\}}(\Omega)$. Since the imbedding mappings are continuous and of dense range, the duals give rise to the inclusion relations:

(1.12) $\mathcal{E}'(\Omega) \subset \mathcal{E}^{(N_p)'}(\Omega) \subset \mathcal{E}^{\{M_p\}'}(\Omega) \subset \mathcal{E}^{(M_p)'}(\Omega) \subset \mathcal{A}'(\Omega)$.

These imbedding mappings keep the support (Harvey [8], Komatsu [13] and hence can be extended to the sheaf isomorphisms:

(1.13) $\mathcal{D}' \subset \mathcal{D}^{(N_p)'} \subset \mathcal{D}^{\{M_p\}'} \subset \mathcal{D}^{(M_p)'} \subset \mathcal{B}$,

where $\mathcal{B}$ is the sheaf of hyperfunctions. The action of differential operators (1.9) is compatible with these isomorphisms.

Convolutions are discussed in the same way as for distributions under the condition (M.2) ([13]).

Moreover, a differential operator

(1.14) $P(x, \partial) = \sum_{|\alpha|=0}^{\infty} a_\alpha(x)\partial^\alpha$

of infinite order acts locally and continuously on the spaces of ultradifferentiable functions and ultradistributions of class (M_p) (resp. $\{M_p\}$) on Ω if for each compact set K in Ω and each $h > 0$ there are constants L and B (resp. there is a constant h such that for each $L > 0$ there is a constant B) such that the coefficients satisfy

(1.15) $\sup_{x \in K} |\partial^\beta a_\alpha(x)| \leq B h^{|\beta|} L^{|\alpha|} M_{|\beta|}/M_{|\alpha|}$.

Such an operator will be called an _ultradifferential operator of class_ (M_p) (resp. $\{M_p\}$).

Theorem 1.2 (First structure theorem). _An_ f _is an ultradistribution in_ $\mathcal{D}^{(M_p)'}(\Omega)$ _(resp._ $\mathcal{D}^{\{M_p\}'}(\Omega)$) _if and only if on each relatively compact open set_ G _(resp. on_ Ω) _it is represented as_

(1.16) $f = \sum_{|\alpha|=0}^{\infty} \partial^\alpha f_\alpha$

with measures f_α _on_ G _(resp. on_ Ω) _such that_

(1.17) $\|f_\alpha\|_{C'(\bar{G})} \leq C L^{|\alpha|}/M_{|\alpha|}$, $|\alpha| = 0, 1, 2, \ldots$

for some L _and_ C _(resp. for every relatively compact open set_ G, _every_ $L > 0$ _and some_ C).

We can prove the theorem under the assumptions (M.1), (M.2)' and (M.3)' (see [13]). For the class $\{M_p\}$ the theorem is due to

Roumieu [30], [31]. He claims that it holds under (M.1) and (M.3)' but there seems to be a flaw in his proof.

Theorem 1.3 (Second structure theorem [13]). Every ultradistribution $f \in \mathcal{D}^{*\prime}(\Omega)$ may be represented on each relatively compact open set G in Ω as

(1.18) $f = P(\partial)g$

with an ultradifferential operator $P(\partial)$ with constant coefficients and a measure g.

To formulate the structure theorem of ultradistributions with support in a submanifold we introduce some notations. Let F be a linear submanifold of $\mathbb{R}^n$. Under a suitable coordinate system it is written

(1.19) $F = \{(x, 0);\ x \in \mathbb{R}^{n'},\ 0 \in \mathbb{R}^{n''}\}$.

A point in $\mathbb{R}^n$ is denoted as (x, y) with $x \in \mathbb{R}^{n'}$ and $y \in \mathbb{R}^{n''}$. If Ω is an open set in $\mathbb{R}^n$, we write

(1.20) $\Omega' = F \cap \Omega$

and $\mathcal{D}^*(\Omega')$ stands for the space of functions on Ω' of n' variables.

Theorem 1.4. Every ultradistribution $f(x, y) \in \mathcal{D}^{(M_p)\prime}(\Omega)$ (resp. $\mathcal{D}^{\{M_p\}\prime}(\Omega)$) with support in F is uniquely represented as

(1.21) $f(x, y) = \sum f_\beta(x) \otimes \partial^\beta \delta(y)$

with

$$f_\beta(x) \in \mathcal{D}^{(M_p)\prime}(\Omega') \quad (\text{resp. } \mathcal{D}^{\{M_p\}\prime}(\Omega'))$$

satisfying the following conditions:

For every compact set K' in Ω' there are constants L, h and C (resp. and for every $L > 0$ and $h > 0$ there is a constant C) such that

(1.22) $\|f_\beta\|_{(\mathcal{D}^{\{M_p\}\prime, h}_{K'})'} \leqq CL^{|\beta|}/M_{|\beta|}$.

Conversely if a family of ultradistributions $f_\beta \in \mathcal{D}^{*\prime}(\Omega')$ satisfies estimates (1.22), then (1.21) converges in $\mathcal{D}^{*\prime}(\Omega)$ and represents an $f \in \mathcal{D}^{*\prime}(\Omega)$ with support in F. We have moreover

(1.23) $\operatorname{supp} f = \bigcup \operatorname{supp} f_\beta$.

Theorems 1.3 and 1.4 are proved by Fourier Analysis. We employ, in particular, the Paley-Wiener theorems for ultradifferentiable functions and ultradistributions. To formulate them we need the *associated function*

$$(1.24)\quad M(\rho) = \sup_p \log(\rho^p/M_p), \quad 0 \leqq \rho < \infty$$

and the *support function*

$$(1.25)\quad H_K(\zeta) = \sup_{x\in K} \operatorname{Im}\langle x, \zeta\rangle, \quad \zeta \in \mathbb{C}^n,$$

where K is a compact convex set in $\mathbb{R}^n$. If $\zeta \in \mathbb{C}^n$, we write $M(\zeta) = M(|\zeta|)$. We note that for the Gevrey sequence $p!^s$ $M(\rho)$ is equivalent to $\rho^{1/s}$.

Theorem 1.5 (Paley-Wiener theorem for ultradifferentiable functions [13]). *An entire function $\tilde{\varphi}(\zeta)$ on $\mathbb{C}^n$ is the Fourier-Laplace transform of an ultradifferentiable function $\varphi(x)$ of class (M_p) (resp. $\{M_p\}$) with support in a compact convex set K in $\mathbb{R}^n$ if and only if for each $h > 0$ there is a constant C (resp. there are constants h and C) such that*

$$(1.26)\quad |\tilde{\varphi}(\zeta)| \leqq C \exp\{-M(\zeta/h) + H_K(\zeta)\}.$$

A sequence φ_j of ultradifferentiable functions with support in K converges in $\mathcal{D}^{(M_p)}(\mathbb{R}^n)$ (resp. $\mathcal{D}^{\{M_p\}}(\mathbb{R}^n)$) if and only if for each $h > 0$ (for some $h > 0$) one of the following equivalent conditions holds:

(a) $\exp\{M(\zeta/h) - H_K(\zeta)\}\tilde{\varphi}_j(\zeta)$

converges uniformly on $\mathbb{C}^n$;

(b) $\exp\{M(\zeta/h)\}\varphi_j(\zeta)$

converges uniformly on a strip $|\operatorname{Im}\zeta| \leqq a < \infty$;

(c) $\exp\{M(\xi/h)\}\tilde{\varphi}_j(\xi)$ *converges uniformly on* $\mathbb{R}^n$.

Theorem 1.6 (Paley-Wiener theorem for ultradistributions [17]). *The following conditions are equivalent for an entire function $\tilde{f}(\zeta)$ on $\mathbb{C}^n$.*

(i) $\tilde{f}(\zeta)$ *is the Fourier-Laplace transform of an ultradistribution of class (M_p) (resp. $\{M_p\}$) with support in the compact convex set K in $\mathbb{R}^n$*;

(ii) *There are constants L and C (resp. for each $L > 0$ there is a constant C) such that*

$$(1.27)\quad |\tilde{f}(\xi)| \leqq C \exp\{M(L\xi)\}, \quad \xi \in \mathbb{R}^n,$$

and for each $\varepsilon > 0$ there is a constant C_ε such that

$$(1.28)\quad |\tilde{f}(\zeta)| \leqq C_\varepsilon\{H_K(\zeta) + \varepsilon|\zeta|\}, \quad \zeta \in \mathbb{C}^n;$$

(iii) There are constants L and C (resp. for each $L > 0$ there is a constant C) such that

$$(1.29) \quad |\tilde{f}(\zeta)| \leq C \exp\{M(L\zeta) + H_K(\zeta)\}, \quad \zeta \in \mathbb{C}^n.$$

A sequence f_j of ultradistributions with support in K converges in $\mathcal{D}^{(M_p)'}(\mathbb{R}^n)$ (resp. $\mathcal{D}^{\{M_p\}'}(\mathbb{R}^n)$) if and only if for some L (resp. for each $L > 0$) one of the following equivalent conditions holds:

(a) $\exp\{-M(L\zeta) - H_K(\zeta)\}\tilde{f}_j(\zeta)$

converges uniformly on $\mathbb{C}^n$,

(b) $\exp\{-M(L\zeta)\}\tilde{f}_j(\zeta)$ converges uniformly on a strip $|\mathrm{Im}\,\zeta| \leq a < \infty$;

(c) $\exp\{-M(L\xi)\}\tilde{f}_j(\xi)$ converges uniformly on $\mathbb{R}^n$.

The equivalence of conditions (i) and (ii) of Theorem 1.6 has been proved by Roumieu [31] and Neymark [27] without condition (M.3). The implication (ii) $\Rightarrow$ (iii) is a type of the Phragmén-Lindelöf theorem and we need conditions (M.2) and (M.3) to prove it. To be more precise, we have the following generalization of the Phragmén-Lindelöf theorem.

Lemma 1.7. Let $F(z)$ be a holomorphic function defined on the upper half plane $\mathrm{Im}\, z > 0$. If for each $\varepsilon > 0$ there is a constant C such that

$$(1.30) \quad |F(z)| \leq Ce^{\varepsilon|z|}, \quad \mathrm{Im}\, z > 0,$$

then the non-tangential boundary value

$$(1.31) \quad F(x) = \lim_{z \to x} F(z)$$

exists for almost every $x \in \mathbb{R}$ and

$$(1.32) \quad \log|F(x+iy)| \leq \frac{1}{\pi}\int_{-\infty}^{\infty} \frac{y \log|F(t)|}{(x-t)^2+y^2}\,dt, \quad y > 0$$

Then if M_p satisfies (M.2) and (M.3), the Poisson integral of $M(Lx)$ is bounded by $M(L'z) + C'$ with constants C' and L'.

We also obtain a Whitney type extension theorem for ultradifferentiable functions as the dual of Theorem 1.4. We denote by $\mathcal{E}^*(\Omega)^F$ (resp. $\mathcal{D}^*(\Omega)^F$) the closed linear subspace of $\mathcal{E}^*(\Omega)$ (resp. $\mathcal{D}^*(\Omega)$) of all elements $\varphi(x, y)$ such that $\partial_y^\beta \varphi(x,0) = 0$ for all β. Next we consider the space $\mathcal{E}^*_\Omega(\Omega')$ (resp. $\mathcal{D}^*_\Omega(\Omega')$) of all ultradifferentiable functions of class $*$ in the sense of Whitney. Namely, $\mathcal{E}^{(M_p)}_\Omega(\Omega')$ (resp. $\mathcal{E}^{\{M_p\}}_\Omega(\Omega')$) is the space of all arrays $(\varphi_\beta(x);\ \beta \in \mathbb{N}^{n''})$ of functions

$\varphi_\beta \in \mathcal{E}^{(M_p)}(\Omega')$ (resp. $\mathcal{E}^{\{M_p\}}(\Omega')$) such that for each compact set K' in Ω' and $h>0$ there is a constant C (resp. there are constants h and C) satisfying

$$(1.33) \quad \sup_{x\in K'} |\partial^\alpha \varphi_\beta(x)| \leq Ch^{|\alpha+\beta|}M_{|\alpha+\beta|}.$$

$\mathcal{D}^*(\Omega')$ is defined similarly. Since $\mathcal{E}^*_\Omega(\Omega')$ and $\mathcal{D}^*_\Omega(\Omega')$ have similar expressions to (1.5)-(1.8), we can introduce in them natural locally convex topologies.

Let $\iota: \mathcal{E}^*(\Omega)^F \to \mathcal{E}^*(\Omega)$ be the canonical injection and let $\rho: \mathcal{E}^*(\Omega) \to \mathcal{E}^*_\Omega(\Omega')$ be the mapping defined by $\rho(\varphi(x, y)) = (\partial_y^\beta \varphi(x, 0))$. Then we have

<u>Theorem 1.8</u> ([17]).

$$(1.34) \quad 0 \to \mathcal{E}^*(\Omega)^F \xrightarrow{\iota} \mathcal{E}^*(\Omega) \xrightarrow{\rho} \mathcal{E}^*_\Omega(\Omega') \to 0$$

<u>and</u>

$$(1.35) \quad 0 \to \mathcal{D}^*(\Omega)^F \xrightarrow{\iota} \mathcal{D}^*(\Omega) \xrightarrow{\rho} \mathcal{D}^*_\Omega(\Omega') \to 0$$

<u>are topologically exact sequences of locally convex spaces</u>.

In case $F = \{0\}$, the theorem says that if a sequence c_α, $\alpha \in \mathbb{N}^n$, of complex numbers satisfies

$$(1.36) \quad |c_\alpha| \leq Ch^{|\alpha|}M_{|\alpha|}$$

for each $h>0$ and some C (resp. some h and C), then there is an ultradifferentiable function $\varphi(x)$ of class (M_p) (resp. $\{M_p\}$) such that

$$(1.37) \quad \partial^\alpha \varphi(0) = c_\alpha .$$

When $n = 1$, this has been proved by L. Carleson [3].

We say that a subset K of $\mathbb{R}^n$ has the <u>cone property</u> if for each $x \in K$ there are a neighborhood $U \cap K$ of x, a unit vector e in $\mathbb{R}^n$ and a positive number ε_0 such that $(U \cap K) + \varepsilon e$ is in the interior of K for any $0 < \varepsilon < \varepsilon_0$.

We assume that Ω' and Ω'' are open sets in $\mathbb{R}^{n'}$ and $\mathbb{R}^{n''}$ respectively and that K' and K'' are compact sets with the cone property in $\mathbb{R}^{n'}$ and $\mathbb{R}^{n''}$ respectively.

<u>Theorem 1.9</u>. <u>The bilinear mapping which assigns to each pair of functions $\varphi(x)$ and $\psi(y)$ the product $\varphi(x)\psi(y)$ induces the following isomorphisms of locally convex spaces</u>:

$$(1.38) \quad \mathcal{E}^*(\Omega') \hat{\otimes} \mathcal{E}^*(\Omega'') \cong \mathcal{E}^*(\Omega' \times \Omega'');$$

$$(1.39) \quad \mathcal{D}^*_{K'} \hat{\otimes} \mathcal{D}^*_{K''} \cong \mathcal{D}^*_{K' \times K''};$$

$$(1.40) \quad \mathcal{D}^{\{M_p\}}(\Omega') \hat{\otimes} \mathcal{D}^{\{M_p\}}(\Omega'') \cong \mathcal{D}^{\{M_p\}}(\Omega' \times \Omega'').$$

Hence we obtain the kernel theorem similar to the case of distributions [35].

Theorem 1.10. We have the canonical isomorphisms of locally convex spaces:

$$(1.41)\quad B^s_\beta(\mathcal{D}^*(\Omega'), \mathcal{D}^*(\Omega'')) = L_\beta(\mathcal{D}^*(\Omega'), \mathcal{D}^{*\prime}(\Omega''))$$
$$= L_\beta(\mathcal{D}^*(\Omega''), \mathcal{D}^{*\prime}(\Omega')) = \mathcal{D}^{*\prime}(\Omega') \hat{\otimes} \mathcal{D}^{*\prime}(\Omega'') = \mathcal{D}^{*\prime}(\Omega' \times \Omega'').$$

Here $B^s(X, Y)$ and $L_\beta(X, Y)$ denote the space of all separately continuous bilinear functionals equipped with the bi-bounded convergence topology and that of all continuous linear mappings equipped with the bounded convergence topology.

Theorems 1.9 and 1.10 have been proved by Komatsu [17] and Petzsche [29] independently by different methods. Petzsche proves (1.39) for compact sets K' and K'' regular in the sense of Whitney.

Combining Theorems 1.4 and 1.10, we can characterize continuous linear mappings $T: \mathcal{E}^*(\Omega) \to \mathcal{D}^{*\prime}(\Omega)$ which is local in the sense that $\operatorname{supp} Tf \subset \operatorname{supp} f$ as in Schwartz [35].

Since the ultradistributions are imbedded in the hyperfunctions, every ultradistribution may be represented as a sum of boundary values of holomorphic functions. We define

$$(1.42)\quad M^*(\rho) = \sup_p \log \frac{\rho^p p!}{M_p}.$$

If $M_p = p!^s$, $M^*(\rho)$ is equivalent to $\rho^{1/(s-1)}$.

Let V be an open set in $\mathbb{C}^n$ which contains the open set Ω in $\mathbb{R}^n$ as a relatively closed set and let Γ be a proper convex open cone in $\mathbb{R}^n$. We denote by V_Γ the wedge domain $V \cap (\mathbb{R}^n + i\Gamma)$.

Theorem 1.11. We assume (M.1)'. Then the following conditions are equivalent for a holomorphic function $F(x+iy)$ on V_Γ:

(a) For each compact set K in Ω and closed subcone Γ' of Γ there are constants L and C (resp. for each $L > 0$ there is a constant C) such that

$$(1.43)\quad \sup_{x\in K} |F(x+iy)| \le C \exp\{M^*(L/|y|)\} \quad \text{for } y \in \Gamma';$$

(a)' (1.43) holds for a ray Γ' in Γ;

(b) $F(x+iy)$ tends to an ultradistribution $F(x+i\Gamma 0)$ as y tends to 0 in a subcone Γ' of Γ in the topology of $\mathcal{D}^{(M_p)\prime}(\Omega)$ (resp. $\mathcal{D}^{\{M_p\}\prime}(\Omega)$);

(b)' $F(x+iy)$ tends to $F(x+i\Gamma 0)$ as y tends to 0 in a ray Γ' in Γ in the topology of $\mathcal{D}^{*\prime}(\Omega)$;

(c) The boundary value $F(x+i\Gamma 0)$ in the sense of hyperfunction is in $\mathcal{D}^{(M_p)\prime}(\Omega)$ (resp. $\mathcal{D}^{\{M_p\}\prime}(\Omega)$).

Then the topological boundary value in (b) and the cohomological boundary value in (c) coincide.

The implications (a) $\Leftrightarrow$ (a)' $\Rightarrow$ (b) $\Rightarrow$ (c) are proved in [13].

The other are also proved in [13] in the one-dimensional case. When the dimension is greater than one, they are derived from the edge of the wedge theorem and Kataoka's theorem [11] saying that the singularity spectrum in the sense of Sato-Kawai-Kashiwara [33] coincides with the analytic wave front set in the sense of Hörmander for distributions and hence for ultradistributions.

Suppose that $\Gamma_1, \ldots, \Gamma_m$ are convex open cones in $\mathbb{R}^n$ such that the dual cones

$$(1.44) \quad \Gamma_j^0 = \{ \xi \in \mathbb{R}^n; \langle y, \xi \rangle \geqq 0, \quad y \in \Gamma_j \}$$

cover the dual space of $\mathbb{R}^n$. An example is the 2^n connected components of $(\mathbb{R} \setminus \{0\})^n$.

Theorem 1.12 ([13]). _Let_ $f \in \mathcal{D}^{*\prime}(\Omega)$, _let_ G _be a relatively compact open set in_ Ω _and let_ V _be a complex open neighborhood of_ G. _Then there are holomorphic functions_ F_j _on_ V_{Γ_j} _satisfying the estimate_ (1.43) _such that_

$$(1.45) \quad f(x) = F_1(x+i\Gamma_1 0) + \cdots + F_m(x+i\Gamma_m 0).$$

When $\Omega = \mathbb{R}^n$, Körner [19] and Petzsche [29] prove more strongly that (1.45) holds on $\mathbb{R}^n$. Körner proves it without condition (M.3).

2. HYPERBOLICITY

Let

$$(2.1) \quad P(x, \partial) = \sum_{|\alpha| \leqq m} a_\alpha(x) \partial^\alpha$$

be a linear partial differential operator with real analytic coefficients $a_\alpha(x)$ defined on an open set Ω in $\mathbb{R}^n$. We assume that the principal part

$$(2.2) \quad p(x, \partial) = \sum_{|\alpha| \leqq m} a_\alpha(x) \partial^\alpha$$

is non-degenerate or that $p(x, \partial) \neq 0$ for any fixed $x \in \Omega$.

The operator $P(x, \partial)$ is said to be _hyperbolic_ with respect to the hypersurface $S = \{x_1 = 0\}$ in the function spaces $\mathcal{F}$ of n variables and $\mathcal{G}$ of $n-1$ variables if the Cauchy problem

$$(2.3) \quad \begin{cases} P(x, \partial)u(x) = 0 \\ \dfrac{\partial^{k-1} u(0, x')}{\partial x_1^{k-1}} = w_k(x'), \quad k = 1, \ldots, m \end{cases}$$

has a unique solution $u \in \mathcal{F}$ for any data $(w_k) \in \mathcal{G}^m$.

We restrict ourselves to the local solvability problem in the case where $\mathcal{F}$ and $\mathcal{G}$ are the same Gevrey class of ultradifferentiable functions or ultradistributions and give an almost necessary and sufficient condition for hyperbolicity.

As usual the condition consists of three parts.

Condition A. The initial surface is non-characteristic;

Condition B. The characteristic roots are real, i.e. the roots ζ_1 of the algebraic equation

$$(2.4) \quad p(x; \zeta_1, \xi') = 0$$

are real for any $x \in \Omega$ and $\xi' \in \mathbb{R}^{n-1}$.

To formulate the third condition concerning the lower order terms, we define the notion of irregularity of characteristic elements.

We assume that the coefficients $a_\alpha(x)$ are continued analytically to the complex open neighborhood V of Ω. If an element $(x^0, \xi^0\infty) \in P^*V = (T^*V \setminus V)/\mathbb{C}^\times$ ($\xi^0\infty$ stands for the class of $\xi^0 \in T^*_{x_0}V$) is on the non-singular part of the characteristic variety

$$(2.5) \quad N(p) = \{(x, \xi\infty) \in P^*V, \ p(x, \xi) = 0\},$$

it is called a non-singular characteristic element of $P(x, \partial)$. Let d be the multiplicity. Then we can find an irreducible homogeneous polynomial $K(x, \xi)$ in ξ with coefficients in the ring of germs of holomorphic functions at x^0 and a homogeneous polynomial $Q(x, \xi)$ with $Q(x^0, \xi^0) \neq 0$ such that

$$(2.6) \quad p(x, \xi) = Q(x, \xi)K(x, \xi)^d .$$

We can further find homogeneous polynomials $Q_i(x, \xi)$ which is either identically zero or not identically zero on a neighborhood of (x^0, ξ^0) in the characteristic variety so that

$$(2.7) \quad P(x, \partial) = Q(x, \partial)K(x, \partial)^d + Q_{m-1}(x, \partial)K(x, \partial)^{d_{m-1}} + \cdots + Q_0(x, \partial)K(x, \partial)^{d_0} ,$$

and $Q_i(x, \partial)K(x, \partial)^{d_i}$ is of order i (see [16]). Then we define

$$(2.8) \quad \sigma = \max\left\{1, \frac{d - d_i}{m - i} \quad (i = 0, 1, \ldots, m-1)\right\}$$

to be the irregularity of the characteristic element $(x^0, \xi^0\infty)$. We note that $1 \leq \sigma \leq d$. When $\sigma = 1$, $P(x, \partial)$ is said to satisfy Levi's condition at $(x^0, \xi^0\infty)$.

Now our third condition is the following.

Condition C. In case $\mathcal{F} = \mathcal{G} = \mathcal{E}^{(s)}$ or $\mathcal{D}^{(s)\prime}$ (resp. $\mathcal{E}^{\{s\}}$ or $\mathcal{D}^{\{s\}\prime}$), the irregularity $\sigma \leq s/(s-1)$ (resp. $\sigma < s/(s-1)$) at every non-singular characteristic element.

We note that when Conditions A and B are satisfied and every characteristic element is non-singular, Levi's condition $\sigma = 1$

is known to be necessary (Mizohata-Ohya [26], Flaska-Strang [5]) and sufficient (Mizohata-Ohya [25], Chazarain [4]) in order that $P(x, \partial)$ is hyperbolic in $\mathcal{E}$ or $\mathcal{D}'$.

Our proof of both necessity and sufficiency depends on the convergence of formal solutions associated with the irreducible factor $K(x, \xi)$ of $p(x, \xi)$. A holomorphic function $\varphi(x)$ on a complex neighborhood of x^0 is said to be a <u>characteristic function</u> of $P(x, \partial)$ if it satisfies

$$(2.9) \quad p(x, \operatorname{grad} \varphi(x)) = 0.$$

We assume that $\operatorname{grad} \varphi(x^0)\infty = \xi^0\infty$. Then $\varphi(x)$ is actually a solution of

$$(2.10) \quad K(x, \operatorname{grad} \varphi(x)) = 0.$$

Let $\Phi_j(t)$, $j \in \mathbf{Z}$, be a sequence of functions of one variable t satisfying

$$(2.11) \quad \frac{d}{dt}\Phi_j(t) = \Phi_{j-1}(t), \quad j \in \mathbf{Z}.$$

Then we can construct a unique formal solution

$$(2.12) \quad u(x) = \sum_{j=-\infty}^{\infty} u_j(x)\, \Phi_j(\varphi(x))$$

of

$$(2.13) \quad P(x, \partial)u(x) = 0$$

under the initial conditions

$$(2.14) \quad \left.\frac{\partial^{k-1}u_j}{\partial x_1^{k-1}}\right|_{x_1=x_1^0} = \delta_{j,0}\, f_k(x'), \quad k = 1,\dots,d, \quad j \in \mathbf{Z},$$

where $f_k(x')$ are arbitrary holomorphic functions defined on a neighborhood of $x^{0\prime}$. $u_j(x)$ do not depend on the sequence $\Phi_j(t)$. Employing Hamada's method in [6], we can prove that $u_j(x)$ are holomorphic on a fixed complex neighborhood V_0 of x^0 and that there is a constant C depending only on $f_k(x')$ such that

$$(2.15) \quad |u_j(x)| \leqq C^{j+1} j!, \quad j \geqq 0;$$

$$(2.16) \quad |u_j(x)| \leqq \begin{cases} C^{-j+1}\left(\dfrac{|x_1-x_1^0|^{-j}}{(-j)!}\right)^{\frac{\sigma}{\sigma-1}}, & \sigma > 1, \quad j<0 \\ 0, & \sigma = 1, \quad j<0 \end{cases}$$

(see [16]). This is the best estimates of this form in general. Namely if $\sigma > 1$ and $Q_i(x^0, \operatorname{grad}\varphi(x^0)) \neq 0$ for some $i < m$ with $d - d_i = \sigma(m - i)$, then for a suitable choice of initial values $f_k(x')$ we have

(2.17) $|u_j(x)| \geq c^{-j+1}(|x_1 - x_1^0|^{-j}/(-j)!)^{\sigma/(\sigma-1)}$

on a neighborhood of x^0 with a constant $c > 0$ for infinitely many $j < 0$.

The necessity of Condition A is proved by constructing null-solutions. (Cf. Mizohata [23] and Persson [28].)

Theorem 2.1 ([16]). *Let* $S : \psi(x) = 0$ *be a real analytic hypersurface such that* $(x, \text{grad}\,\psi(x)\infty)$ *is a non-singular characteristic element for every* $x \in S$. *If* σ *is the irregularity, then for each* $1 < s \leq \sigma/(\sigma-1)$ *and* $x^0 \in S$ *there are null-solutions exactly in* $\mathcal{E}^{\{s\}}$ *and exactly in* $\mathcal{D}^{(s)'}$ *on a neighborhood of* x^0.

Since the local existence implies a semi-global existence (cf. Lax [20]), the following is enough to prove the necessity of Condition B. (See Kataoka [11] for a more direct proof.)

Theorem 2.2. *Suppose that* $P(x, \partial)$ *has a non-singular characteristic element* $(0, (\zeta_1, \xi')\infty)$ *such that* $\text{Im}\,\zeta_1 > 0$ *and* $\xi' \in \mathbb{R}^{n-1} \setminus 0$. *Then for any sufficiently small neighborhood* Ω_0 *of* 0 *and any* $\varepsilon > 0$ *there is a real analytic solution* $u(x)$ *of* (2.13) *defined on* $\{x \in \Omega_0;\ x_1 > -\varepsilon\}$ *which cannot be extended to any hyperfunction solution across the hyperplane* $x_1 = -2\varepsilon$.

The necessity of Condition B in $\mathcal{E}$ and $\mathcal{D}'$ have been proved by Lax [20] and Mizohata [22]. Bony-Schapira [2] show that Conditions A and B are sufficient for hyperbolicity in $\mathcal{B}$.

Theorem 2.3. *Suppose that Conditions* A *and* B *are satisfied and that* $P(x, \partial)$ *has a non-singular characteristic element* $(0, \xi\infty)$ *of irregularity* $\sigma > 1$.

Then for each $s > \sigma/(\sigma-1)$ *there are Cauchy data* $w_k(x') \in \mathcal{E}^{(s)}(\Omega')$ *on a neighborhood* Ω' *of* $0'$ *in* $\mathbb{R}^{n-1}$ *such that the solution* $u(x)$ *of* (2.3) *does not belong to* $\mathcal{D}^{(\sigma)'}(\Omega_0)$ *on any neighborhood* Ω_0 *of* 0 *in* $\mathbb{R}^n$.

On the other hand, for each neighborhood Ω *of* 0 *in* $\mathbb{R}^n$ *there are Cauchy data* $w_k(x') \in \mathcal{E}^{\{s\}}(\Omega')$, *where* $s = \sigma/(\sigma-1)$ *and* $\Omega' \supset \Omega \cap \mathbb{R}^{n-1}$, *such that the solution* $u(x)$ *of* (2.3) *does not belong to* $\mathcal{D}^{\{s\}'}(\Omega)$.

Theorems 2.1, 2.2 and 2.3 are proved by constructing a solution $u(x)$ of (2.13) as the series (2.12) for a suitable choice of $\Phi_j(t)$. If we take the characteristic function $\varphi(x)$ as a local coordinate function x_n, then (2.12) is written

$$(2.12)' \quad u(x) = \int^{x_n}\Big\{\sum_{j=1}^{\infty} u_j(x)\frac{(x_n-t)^{j-1}}{(j-1)!}\Big\}\Phi_0(t)dt + \sum_{k=0}^{\infty} u_k(x)\frac{d^k}{dx_n{}^k}\Phi_0(x_n).$$

It follows from estimates (2.15) that the integral operator of the first term behaves well. On the other hand, estimates (2.16) show that the ultradifferential operator of the second term is continuous on $\mathcal{E}^{(s)}$ and $\mathcal{D}^{(s)'}$ (resp. $\mathcal{E}^{\{s\}}$ and $\mathcal{D}^{\{s\}'}$) as far as $s \leq \sigma/(\sigma-1)$ (resp. $s < \sigma/(\sigma-1)$). Hence the solution $u(x)$ has the desired regularity for the proof of Theorem 2.1 if $\Phi_0(x_n)$ has.

Theorem 2.2 is then proved by employing Sato's fundamental theorem of regularity ([33], p.356).

If Theorem 2.3 were not true, the ultradifferential operator would map every element $\Phi_0(x_n)$ in $\mathcal{E}^{(s)}(\mathbb{R})$ to an element in $\mathcal{D}^{(s)'}(\Omega_0)$. By the closed graph theorem the mapping is continuous. Hence it follows from the kernel theorem that it is represented as

$$(2.18)\quad \sum_{k=0}^{\infty} u_{-k}(x) \frac{d^k}{dx_n^k} \Phi_0(x_n) = \int K(x, t)\Phi_0(t)dt$$

with a kernel $K(x, t) \in \mathcal{D}^{(s)'}(\Omega_0 \times \mathbb{R})$. Since the left hand side is a local operator in x_n, we have $\operatorname{supp} K \subset \{x_n = t\}$. Hence we have by the structure theorem of ultradistributions with support in a submanifold

$$(2.19)\quad K(x, t) = \sum_{k=0}^{\infty} v_k(x)\, \delta^{(k)}(x_n - t)$$

with $v_k(x) \in \mathcal{D}^{(s)'}(\Omega_0)$ such that

$$(2.20)\quad \|v_k(x)\|_{(\mathcal{D}^{\{s\},h}_K)'} \leq CL^k/(k!)^s .$$

From the uniqueness of decomposition (2.19) we have $v_k(x) = u_{-k}(x)$. Then (2.20) contradicts (2.17).

The proof is similar for the case $\{s\}$. (Cf. Ivrii [9], [10].)

Conversely we have

Theorem 2.4. Suppose that Conditions A, B and C are satisfied and that every characteristic element is non-singular. Then for each sufficiently small open neighborhood Ω' of 0 in $\mathbb{R}^{n-1}$ there is an open neighborhood Ω_0 of 0 in $\mathbb{R}^n$ such that the Cauchy problem has a unique solution $u(x) \in \mathcal{E}^*(\Omega_0)$ (resp. $\mathcal{D}^{*'}(\Omega_0)$) for any data $(w_k(x')) \in \mathcal{E}^*(\Omega')^m$ (resp. $\mathcal{D}^{*'}(\Omega')^m$).

A little weaker results have been obtained by Leray-Ohya [21] and Hamada-Leray-Wagschal [7]. We can prove the theorem by their methods. Another proof is obtained by constructing a fundamental solution $E_k(x, y')$ of the Cauchy problem as the integral

$$(2.21)\quad E_k(x, y') = \int_{S^{n-2}} E_k(x, y', \xi')\,\omega(\xi')$$

of the solution $u(x) = E_k(x, y', \xi')$ of

$$(2.22)\quad \begin{cases} P(x, \partial)u(x) = 0 \end{cases}$$

$$\left\{ \left. \frac{\partial^{\ell-1} u}{\partial x_1^{\ell-1}} \right|_{x_1=0} = \delta_{k,\ell} \frac{(n-2)!}{(-2\pi i)^{n-1}} \frac{1}{(\langle x'-y', \xi' \rangle + i0)^{n-1}} \right.$$

or

$$\delta_{k,\ell} \frac{(n-2)!}{(-2\pi i)^{n-1}} \frac{(1-\alpha\langle x'-y',\xi'\rangle)^{n-2} + \alpha^2 (1-\alpha\langle x'-y',\xi'\rangle)^{n-3} (|x'-y'|^2 - \langle x'-y',\xi'\rangle^2)}{(\langle x'-y',\xi'\rangle + \alpha(|x'-y'|^2 - \langle x'-y',\xi'\rangle^2) + i0)^{n-1}}$$

with $\mathrm{Re}\,\alpha > 0$. According to Hamada [6] the solution $u(x)$ is obtained as series (2.12) for a suitable choice of $\Phi_j(t)$. Then it is easy to locate the singularity spectrum of the fundamental solution. Thus we obtain a result on the propagation of singularity of solution similar to Mizohata [24] and Chazarain [4].

REFERENCES

1. G. Björck, Linear partial differential operators and generalized distributions, Ark. Mat., 6 (1966), 351-407.
2. J.-M. Bony et P. Schapira, Solutions hyperfonctions du problème de Cauchy, Hyperfunctions and Pseudo-Differential Equations, Lecture Notes in Math. No.287, 1973, pp.82-98.
3. L. Carleson, On universal moment problems, Math. Scand., 9 (1961), 197-206.
4. J. Chazarain, Opérateurs hyperboliques à caractéristiques de multiplicité constante, Ann. Inst. Fourier, Grenoble, 24 (1974), 173-202.
5. H. Flaschka and G. Strang, The correctness of the Cauchy problem, Advances in Math., 6 (1971), 347-379.
6. Y. Hamada, Problème analytique de Cauchy à caractéristiques multiples dont les données de Cauchy ont des singularités polaires, C. R. Acad. Sci. Paris, Sér. A, 276 (1973), 1681-1684.
7. Y. Hamada, J. Leray et C. Wagschal, Systèmes d'équations aux derivées partielles à caractéristiques multiples — Problème de Cauchy ramifié — Hyperbolicité partielle, à paraître.
8. R. Harvey, Hyperfunctions and linear partial differential equations, Thesis, Stanford Univ., 1966.
9. V. Ja. Ivrii, Conditions for the correctness in Gevrey classes of the Cauchy problem for nonstrictly hyperbolic operators, Soviet Math. Dokl., 16 (1975), 415-417.
10. V. Ja. Ivrii, Well-posedness in projective Gevrey classes of the Cauchy problem for nonstrictly hyperbolic operators, Soviet Math. Dokl., 16 (1975), 570-573.
11. K. Kataoka, Radon transforms of hyperfunctions and their applications, Master's thesis, Univ. Tokyo, 1976 (in Japanese).
12. H. Komatsu, Ultradistributions, hyperfunctions and linear differential equations, Colloque Internat. C. N. R. S. sur les Équations aux Dérivées Partielles Linéaires, Astérisque 2

et 3 (1973), 252-271.

13. H. Komatsu, Ultradistributions, I, Structure theorems and a characterization, J. Fac. Sci. Univ. Tokyo, Sec. IA, 20 (1973), 25-105.
14. H. Komatsu, On the regularity of hyperfunction solutions of linear ordinary differential equations with real analytic coefficients, J. Fac. Sci. Univ. Tokyo, Sec. IA, 20 (1973), 107-119.
15. H. Komatsu, Theory of Locally Convex Spaces, Dept. of Math., Univ. of Tokyo, 1974.
16. H. Komatsu, Irregularity of characteristic elements and construction of null-solutions, J. Fac. Sci. Univ. Tokyo, Sec. IA, 23 (1976), 297-342.
17. H. Komatsu, Ultradistributions, II, The kernel theorem and ultradistributions with support in a submanifold, to appear in J. Fac. Sci. Univ. Tokyo, Sec. IA.
18. H. Komatsu, Irregularity of characteristic elements and hyperbolicity, Symp. on Algebraic Analysis, to appear in Publ. RIMS, Kyoto Univ.
19. J. Körner, Roumieu'sche Ultradistributionen als Randverteilung holomorpher Funktionen, Dissertation, Kiel, 1975, 55SS.
20. P. D. Lax, Asymptotic solutions of oscillatory initial value problems, Duke Math. J., 24 (1957), 627-646.
21. J. Leray et Y. Ohya, Systèmes linéaires, hyperboliques non stricts, Colloque CBRM, 1964, pp.105-144.
22. S. Mizohata, Some remarks on the Cauchy problem, J. Math. Kyoto Univ., 1 (1961), 109-127.
23. S. Mizohata, Solutions nulles et solutions non analytiques, J. Math. Kyoto Univ., 1 (1962), 271-302.
24. S. Mizohata, Analyticity of the fundamental solutions of hyperbolic systems, J. Math. Kyoto Univ., 1 (1962), 327-355.
25. S. Mizohata et Y. Ohya, Sur la condition de E. E. Levi concernant des équations hyperboliques, Publ. RIMS, Kyoto Univ. Ser. A, 4 (1968), 511-526.
26. S. Mizohata et Y. Ohya, Sur la condition d'hyperbolicité pour les équations à caractéristiques multiples, II, Japan. J. Math., 40 (1971), 63-104.
27. M. Neymark, On the Laplace transform of functionals on classes of infinitely differentiable functions, Ark. Mat., 7 (1968), 577-594.
28. J. Persson, Non-uniqueness in the characteristic Cauchy problem when the coefficients are analytic, Matematiche, 27 (1972), 1-8.
29. H.-J. Petzsche, Darstellung der Ultradistributionen vom Beurlingschen und Roumieuschen Typ durch Randwerte holomorpher Funktionen, Dissertation, Düsseldorf, 1976, 114SS.
30. C. Roumieu, Sur quelques extensions de la notion de distribution, Ann. Ecole Norm. Sup., 77 (1960), 41-121.
31. C. Roumieu, Ultra-distributions définies sur R^n et sur certaines classes de variétés différentiables, J. Anal. Math.,

10 (1962-63), 153-192.
32. M. Sato, Theory of hyperfunctions, J. Fac. Sci. Univ. Tokyo, Sec. I, 8 (1959-60), 139-193 and 398-437.
33. M. Sato, T. Kawai and M. Kashiwara, Microfunctions and pseudo-differential equations, Hyperfunctions and Pseudo-Differential Equations, Lecture Notes in Math. No.287, Springer, Berlin-Heidelberg-New York, 1973, pp.265-529.
34. L. Schwartz, Théorie des Distributions, Hermann, Paris, 1950-51.
35. L. Schwartz, Théorie des distributions à valeurs vectorielles, Ann. Inst. Fourier, Grenoble, 7 (1957), 1-141 et 8 (1958), 1-209.

SOME ASPECTS OF THE THEORY OF LINEAR EVOLUTION EQUATIONS

J.L. Lions

Collège de France and Laboria

INTRODUCTION

The goal of these lectures is to give (Chapter 1) a short survey of some of the methods available for proving existence and uniqueness in linear evolution equations, and in the following chapters, to indicate some trends and problems in this theory.

Chapter 2 gives an introduction to the theory of homogenization of evolution operators with highly oscillating coefficients ; this chapter can be used as an introduction to the forthcoming book by Bensoussan-Papanicolaou and the A. on these topics.

Chapter 3 shows how some questions of optimal control lead to the necessary introduction of "generalized" solutions, and we shortly present the transposition method, for which we refer to the book of Magenes and the A. on non homogeneous boundary value problems. We also briefly give an extension of the Hadamard's formula (which expresses the 1st variation of the Green's kernel with respect to variations of the domain) to problems of evolution.

In Chapter 4 we recall a result of Baouendi and Grisvard relative to an equation of mixed type and we show how this equation can be "transformed" into an apparently new singular evolution equation (whose direct study is open).

Each Chapter ends with a short Bibliography.

Garnir (ed.), Boundary Value Problems for Linear Evolution Partial Equations. 175-238.

The detailed plan is as follows :

CHAPTER I

METHODS FOR PROVING EXISTENCE AND UNIQUENESS

1. PARABOLIC EVOLUTION EQUATIONS. ABSTRACT SETTING.

1.1 Notations.

Let V and H be two Hilbert spaces ; we shall assume these spaces to be real Hilbert spaces, changes for the complex case being straightforward. We shall assume

$$V \subset H \text{ , } V \text{ dense in } H,\ V \to H \text{ continuous ;} \tag{1.1}$$

the norm in V (resp. H) is denoted by $\| \ \|$ (resp. $| \ |$), the scalar product in H is denoted by (,). We identify H to its dual, and if we identify the dual of V in a compatible way, we have

$$\left.\begin{array}{l} V \subset H \subset V' \text{ ,} \\ (\ ,\ ) \text{ denotes the scalar product in H or the scalar} \\ \text{product between V and V' (or V' and V).} \end{array}\right\} \tag{1.2}$$

Let t be the time variable. We shall assume

$$t \in [0,T], \quad 0 < T < \infty. \tag{1.3}$$

We consider a family of continuous bilinear forms on $V \times V$:

$$\left.\begin{array}{l} |a(t;u,v)| \le M \|u\| \|v\| \quad \forall t \in (0,T) \\ t \to a(t;u,v) \text{ is measurable in } (0,T) \quad \forall u,v \in V, \end{array}\right\} \tag{1.4}$$

and we assume that (<u>ellipticity hypothesis</u>)

$$\left.\begin{array}{l} \exists\lambda \text{ such that} \\ a(t;v,v) + \lambda|v|^2 \geq \alpha\|v\|^2 \quad \forall v \in V, \quad \alpha > 0 , \\ \text{a.e. in } t \in (0,T). \end{array}\right\} \tag{1.5}$$

We define $A(t)$ by

$$\left.\begin{array}{l} (A(t)u,v) = a(t;u,v) , \\ A(t) \in \mathcal{L}(V;V'). \end{array}\right\} \tag{1.6}$$

We want to consider the following problem (in a loose form first) find $u = u(t)$ such that

$$\frac{\partial u}{\partial t} + A(t)u = f \text{ in } (0,T), \tag{1.7}$$

f given with values in V', u being subject to

$$u(o) = u_o \text{ , } u_o \text{ given.} \tag{1.8}$$

One has now to make the above problem precise, by defining the class of functions where we look for u.

1.2. Functional spaces.

In order to avoid once for all any difficulty related to <u>measurability</u> in t, we assume that

$$V \text{ is separable.} \tag{1.9}$$

We define

$$W(0,T) = \{v| \quad v \in L^2(0,T;V), \frac{\partial v}{\partial t} \in L^2(0,T;V')\} . \tag{1.10}$$

Here $L^2(0,T;V)$ denotes the classical space of (classes of) functions v which are measurable with values in V and are such that

$$\int_o^T \|v(t)\|^2 dt < \infty.$$

In (1.10) $\frac{\partial v}{\partial t}$ is taken in the weak sense of distributions with values in V, i.e.

$$\frac{\partial v}{\partial t}(\phi) = -\int_o^T v(t)\frac{d\phi(t)}{dt} dt \; \forall\phi \text{ smooth with real values and}$$

compact support in $]0,T[$.

We provide (1.10) with the norm given by

$$\|v\|^2_{W(0,T)} = \int_o^T \left[\|v(t)\|^2 + \| \frac{dv(t)}{dt} \|^2_* \right] dt \tag{1.11}$$

where

$$\|\lambda\|_* = \sup. \frac{|(\lambda,v)|}{\|v\|} \quad , \lambda \in V'.$$

Equipped with this norm, W(0,T) is a Hilbert space.■

One can prove (cf. Lions-Magenes [1], referred to as L.M. [1] in what follows) :

$$\left.\begin{array}{l}\text{every function } v \in W(0,T) \text{ is a.e. equal to a conti-} \\ \text{nuous function from } [0,T] \to H.\end{array}\right\} \tag{1.12}$$

We have the integration by parts formula : if $u,v \in W(0,T)$, then

$$\left.\begin{array}{l}\int_o^T (\frac{\partial u}{\partial t}, v)dt = (u(T),v(T)) - (u(o),v(o)) - \\ \qquad - \int_o^T (u, \frac{\partial v}{\partial t})dt.\end{array}\right\} \tag{1.13}$$

We also remark that :

$$\text{if } v \in L^2(0,T;V) \text{ then } A(t)v \in L^2(0,T;V') \tag{1.14}$$

and the mapping $v \to A(t)v$ is continuous from $L^2(0,T;V) \to L^2(0,T;V')$.

1.3. Abstract problem of "parabolic" type.

With the notations of sections 1.1 and 1.2, we now consider the following problem :

we look for $u \in L^2(0,T;V)$ such that

$$\frac{\partial u}{\partial t} + A(t)u = f \text{ , f given in } L^2(0,T;V'), \tag{1.15}$$

$$u(o) = u_o \text{ , } u_o \text{ given in H.} \tag{1.16}$$

Remark 1.1. : By virtue of (1.14), if u belongs to $L^2(0,T;V)$ and satisfies (1.15) then

$$\frac{\partial u}{\partial t} = f - A(t)u \in L^2(0,T;V')$$

i.e. $u \in W(0,T)$ so that (1.16) makes sense.

Remark 1.2. : All the remarks which follow readily extend to the equation

$$E\,\frac{\partial u(t)}{\partial t} + A(t)u = f \tag{1.17}$$

where $E \in \mathcal{L}(H;H)$, $E^* = E$, E invertible positive definite, and also to cases when $E(t)$ depends smoothly on t with similar hypotheses. The situation changes radically when E is not invertible. For an example of such a "singular" situation, let us give

$$x\,\frac{\partial u}{\partial t} - \frac{\partial^2 u}{\partial x^2} = f \text{ in } \Omega\times]0,T[\,,\ \Omega =]-1,1[\ , \tag{1.18}$$

$$u(\pm 1,t)=0\ , \tag{1.19}$$

and where the "initial condition" (1.16) has to be replaced by a "partly initial, partly final" condition, namely

$$\left.\begin{aligned} &u(x,o) \text{ given for } x > 0,\\ &u(x,T) \text{ given for } x < 0. \end{aligned}\right\} \tag{1.20}$$

For the solution, cf. Baouendi-Grisvard [1].■

Remark 1.3 : We do not restrict the generality by assuming that

$$a(t;v,v) \geq \alpha\|v\|^2\ . \tag{1.21}$$

Proof : change u into $e^{kt}u$ and choose $k \geq \lambda$, λ being the constant which appears in (1.5).■

We shall give in what follows a number of methods for proving the

Theorem 1.1 - We assume that (1.4)(1.5) (or(1.21)) hold true. Then problem (1.15)(1.16) admits a unique solution in $W(0,T)$ (or in $L^2(0,T;V)$, it amounts to the same thing). The mapping

$$f,u_o \to u$$

is continuous from $L^2(0,T;V') \times H \to L^2(0,T;V)$ (or $W(0,T)$).

2. EXAMPLES.

The rather abstract presentation of Section 1 is justified by the very large number of examples which all fit in the preceding framework.

We confine ourselves here to a few typical examples. For other examples we refer to L.M. [1], Lions [1] (L. [1]) and to the bibliography therein.

Example 2.1.

Let Ω be a bounded open set of $\mathbb{R}^n$, and let us define the usual Sobolev space of order 1 :

$$H^1(\Omega) = \{v \mid v, \frac{\partial v}{\partial x_i} \in L^2(\Omega)\} \tag{2.1}$$

equipped with its standard Hilbertian norm.
We define next

$$H_o^1(\Omega) = \text{closure of } \mathcal{D}(\Omega) \text{ in } H^1(\Omega) \tag{2.2}$$
(where $\mathcal{D}(\Omega)$ denotes the space of C^∞ scalar functions, with compact support in Ω).

If $\Gamma = \partial\Omega$ is smooth enough (cf. L.M. [1]) then

$$H_o^1(\Omega) = \{v \mid v \in H^1(\Omega) , v=0 \text{ on } \Gamma\} . \tag{2.3}$$

We take

$$V = H_o^1(\Omega) , H = L^2(\Omega) , \tag{2.4}$$

$$a(u,v) = \int_\Omega \sum_i \frac{\partial u}{\partial x_i} \frac{\partial v}{\partial x_i} dx. \tag{2.5}$$

We have : $V' = H^{-1}(\Omega)$, and (1.15)(1.16) is equivalent to the classical problem for the heat equation given by

$$\left.\begin{aligned} &\frac{\partial u}{\partial t} - \Delta u = f \quad \text{in } \Omega\times]0,T[, \\ &u = 0 \quad \text{on } \Gamma\times]0,T[\;= \Sigma, \\ &u(x,o) = u_o(x) \text{ in } \Omega. \end{aligned}\right\} \tag{2.6}$$

Example 2.2.

Let us take now

$$V = H^1(\Omega) \ , \ H = L^2(\Omega) \ . \tag{2.7}$$

and a(u,v) still given by (2.5)

Since V' cannot be identified with a space of distribution over Ω, it is better to formulate (1.15) in the (equivalent) "variational form".

$$(\frac{\partial u}{\partial t}, v) + a(t;u,v) = (f,v) \quad \forall v \in V. \tag{2.8}$$

We define

$$(f,v) = \int_\Omega f_o(x,t)v(x)dx + \int_\Gamma g(x,t)v(x)d\Gamma \tag{2.9}$$

where

$$f_o \in L^2(\Omega \times (0,T)), \ g \in L^2(0,T;L^2(\Gamma)) = L^2(\Sigma) . \tag{2.10}$$

It is easy to check that, assuming (2.10), (2.9) defines an element f of $L^2(0,T;V')$.

Remark 2.1 : If we use (L.M. [1]) the trace theorem in $H^1(\Omega)$, we see that we can take

$$g \in L^2(0,T,H^{-1/2}(\Gamma)) . \ \blacksquare \tag{2.11}$$

The problem (1.15)(1.16) is equivalent to the following :

$$\frac{\partial u}{\partial t} - \Delta u = f_o \ \text{in} \ \Omega\times]0,T[, \tag{2.12}$$

$$\frac{\partial u}{\partial \nu} = g \ \text{on} \ \Sigma \ , \tag{2.13}$$

where $\frac{\partial}{\partial \nu}$ = normal derivative to Γ directed toward the exterior of Ω, and with the initial condition (1.16). ■

Example 2.3.

Let us consider in $Q = \Omega\times]0,T[$ functions $a_{ij}(x,t) \in L^\infty(Q)$ such that

$$\sum_{i,j=1}^{n} a_{ij}(x,t)\xi_i\xi_j \geq \alpha \sum_{i=1}^{n} \xi_i^2 \quad , \ \alpha > 0 \ , \text{ a.e. in } Q. \tag{2.14}$$

We take, <u>for instance</u>

$$V = H_o^1(\Omega) \ , \tag{2.15}$$

and we define

$$a(t,u,v) = \sum_{i,j} \int_\Omega a_{ij}(x,t) \frac{\partial u}{\partial x_j} \frac{\partial v}{\partial x_i} dx \ . \tag{2.16}$$

We can apply Theorem 1.1. We have therefore existence and uniqueness of the solution of

$$\left. \begin{aligned} &\frac{\partial u}{\partial t} + A(t)u = f \quad \text{in } \Omega \times]0,T[, \\ &u=0 \text{ on } \Sigma, \\ &u(x,o) = u_o(x) \text{ on } \Omega \end{aligned} \right\} \tag{2.17}$$

where

$$A(t)u = - \sum_{i,j=1}^{u} \frac{\partial}{\partial x_i} (a_{ij}(x,t) \frac{\partial u}{\partial x_j}). \tag{2.18}$$

One has to be careful in the interpretation of (2.18) when - as it can be the case ! - coefficients a_{ij} are <u>discontinuous</u>.

Let us suppose (as it is sketchy indicated on Fig. 1 for u = 1) that

$$Q = Q_o \cup S \cup Q_1 \tag{2.19}$$

$$a_{ij} = \begin{cases} a_{ij}^o & \text{in } Q_o \\ a_{ij}^1 & \text{in } Q_1 \end{cases} \quad , \tag{2.20}$$

a_{ij}^o , a_{ij}^1 being continuous functions in $\overline{Q_o}$ and in $\overline{Q_1}$ but which take <u>different values</u> on S. If we denote by

$$u_k = \text{restriction of } u \text{ to } Q_k \ , \ k=0,1$$

$$A(t)^k u_k = - \sum \frac{\partial}{\partial x_i} (a_{ij}^k \frac{\partial u_k}{\partial x_j}) \text{ in } Q_k \ ,$$

then the 1st equation in (2.17) is <u>equivalent</u> to

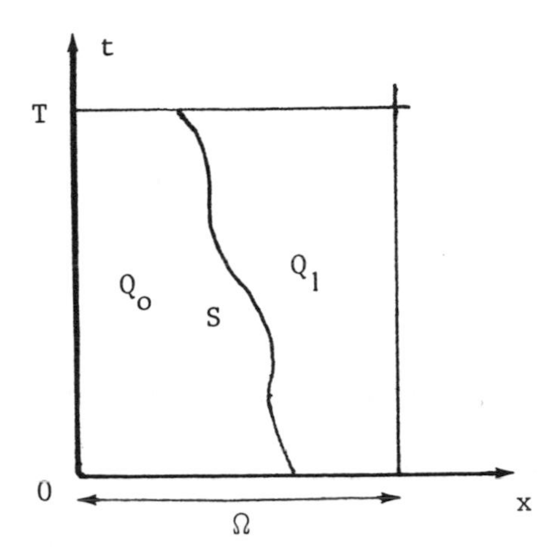

Fig. 1

$$\left.\begin{aligned} &\frac{\partial u_o}{\partial t} + A(t)^o\, u_o = f \text{ in } Q_o \\ &\frac{\partial u_1}{\partial t} + A(t)^1\, u_1 = f \text{ in } Q_1, \end{aligned}\right\} \tag{2.21}$$

with the transmission conditions

$$\left.\begin{aligned} &u_o = u_1 \text{ on } S, \\ &\sum_{i,j} a^o_{ij} \frac{\partial u_o}{\partial x_j} \nu_i = \sum_{i,j} a^1_{ij} \frac{\partial u_1}{\partial x_j} \nu_i \text{ on } S \end{aligned}\right\} \tag{2.22}$$

where $\{\nu_i\}$ = normal to S. The interpretation of (2.21) is formal, the correct meaning being through the variational formulation. ■

Example 2.4.

$$V = H^2_o(\Omega) \text{ , } H = L^2(\Omega), \tag{2.23}$$

$$a(u,v) = (\Delta u, \Delta v). \tag{2.24}$$

Then problem (1.15)(1.16) is equivalent to

$$\left.\begin{aligned} &\frac{\partial u}{\partial t} + \Delta^2 u = f \text{ in } Q, \\ &u = 0 \ , \ \frac{\partial u}{\partial \nu} = 0 \text{ on } \Sigma, \\ &u(x,o) = u_o(x) \text{ in } \Omega. \blacksquare \end{aligned}\right\} \tag{2.25}$$

Example 2.5. (Changing of "Pivot space").

In all of the examples taken so far, we have $H = L^2(\Omega)$. But it can be useful to make significantly different other choices. We take (cf. L.M. [1], Chapter 3, N° 4.7.5.) :

$$H = H_o^1(\Omega), \tag{2.26}$$

$$V = \{v| \quad v \in H, \frac{\partial \Delta v}{\partial x_i} \in L^2(\Omega)\} \tag{2.27}$$

$$a(u,v) = \Sigma \int_\Omega \frac{\partial \Delta u}{\partial x_i} \frac{\partial \Delta v}{\partial x_i} dx. \tag{2.28}$$

We can apply Theorem 1.1.
<u>The interpretation of the problem</u> is as follows : u satisfies

$$\left(\frac{\partial u}{\partial t}, v\right)_H + a(u(t),v) = (f(t),v)_H \quad \forall v \in V \tag{2.29}$$

where $(\ ,\ )_H$ denotes the scalar product between V' and V <u>compatible with the scalar product in H</u>. If we assume that

$$f \in L^2(0,T;H_o^1(\Omega)) \ , \tag{2.30}$$

then (2.29) can be written

$$\left(\frac{\partial u}{\partial t}, -\Delta v\right) + a(u(t),v) = (f(t), -\Delta v) \tag{2.31}$$

where $(\ ,\ )$ denotes the scalar product between $H_o^1(\Omega)$ and $H^{-1}(\Omega)$. If we set $-\Delta v = \phi$, then (2.31) is equivalent to

$$\left(\frac{\partial u}{\partial t}, \phi\right) + \Sigma \int \frac{\partial}{\partial x_i}(-\Delta u) \frac{\partial \phi}{\partial x_i} dx = (f,\phi) \ \forall \phi \in H^{-1}(\Omega) \tag{2.32}$$

hence it follows that

$$\frac{\partial u}{\partial t} + \Delta^2 u = f \text{ in } Q \tag{2.33}$$

subject to Boundary conditions

$$u = 0 \ , \ \frac{\partial \Delta u}{\partial \nu} = 0 \text{ on } \Sigma \tag{2.34}$$

and of course subject to the usual initial condition

$$u(x,o) = u_o(x) \ , \text{ where now } u_o \in H_o^1(\Omega). \ \blacksquare \tag{2.35}$$

Remark 2.2. : We refer to L. [1], L.M. [1] for examples for systems. ■

3. PROOF OF UNIQUENESS IN THEOREM 1.1.

Assume that u satisfies (1.15)(1.16) with $f = 0$, $u_o = 0$. Taking the scalar product of (1.15) with u gives

$$\left(\frac{\partial u}{\partial t} , u\right) + a(t;u,u) = 0 \tag{3.1}$$

hence it follows, by using (1.13), that

$$\frac{1}{2} |u(t)|^2 + \int_o^t a(s;u(s),u(s))ds = 0. \tag{3.2}$$

Since $a(s;u(s),u(s)) \geq \alpha||u(s)||^2$, it follows that $u = 0$.

Remark 3.1. : If we have not made the transformation as in Remark 1.3, we obtain from (3.2) that

$$\frac{1}{2} |u(t)|^2 \leq \lambda \int_o^t |u(s)|^2 ds$$

which is enough to conclude that $u=0$ by virtue of Gronwall's inequality. ■

4. A REVIEW OF EXISTENCE PROOFS IN THEOREM 1.1.

4.1. Projection theorem.

Let us recall (cf. L. [1], Chapter 3) the following "projection theorem" : let F be a real Hilbert space (and let $|| \ ||_F$ be its norm) and let Φ be a pre-hilbertian subspace of F :

$$\Phi \subset F \ , \tag{4.1}$$

$|||\phi|||$ = norm of $\phi \in \Phi$ in Φ (and Φ is not necessarily complete for this (pre-hilbertian) norm);
we assume that

$$||\phi||_F \leq C|||\phi||| \ . \tag{4.2}$$

Let $E(u,\phi)$ be a bilinear form on $F \times \Phi$ which satisfies :

$$\forall\phi \in \Phi \ , \ u \to E(u,\phi) \text{ is continuous on } F \ ; \tag{4.3}$$

$$E(\phi,\phi) \geq \alpha|||\phi|||^2 \ , \ \alpha > 0. \tag{4.4}$$

Let $\phi \to L(\phi)$ be a continuous linear form on Φ.

<u>Then there exists</u> $u \in F$ <u>such that</u>

$$E(u,\phi) = L(\phi) \quad \forall\phi \in \Phi \ . \ \blacksquare \tag{4.5}$$

<u>Application</u> : we take

$$F = L^2(0,T;V) \ , \tag{4.6}$$

$$\left.\begin{aligned} &\Phi = \{\phi|\phi \in F, \frac{\partial\phi}{\partial t} \in L^2(0,T;V') \ , \ \phi(T) = 0\} \ , \\ &|||\phi|||^2 = ||\phi||_F^2 + |\phi(o)|^2 \ \text{(so that } \Phi \text{ is indeed not complete !)} \end{aligned}\right. \tag{4.7}$$

$$E(u,\phi) = \int_o^T [-(u,\frac{\partial\phi}{\partial t}) + a(t;u,\phi)] \, dt \tag{4.8}$$

$$L(\phi) = \int_o^T (f,\phi)dt + (u_o,\phi(o)). \tag{4.9}$$

We leave as an Exercise to the reader to check that this proves Theorem 1.1.■

4.2. Approximation method (I). Semi-discretization in space.

We introduce a family of subspaces $V_h \subset V$ where h is, say, a scalar parameter $\in]0,1]$, and we assume that

$$\forall h, \ V_h \text{ is finite dimensional,} \tag{4.10}$$

$$\left.\begin{aligned} &V_h \text{ "converges" to } V \text{ is the following sense : } \forall v \in V \\ &\text{there exists } v_h \in V_h \text{ such that } ||v-v_h|| \to 0 \text{ as } h \to 0. \end{aligned}\right\} \tag{4.11}$$

We say that V_h is an <u>internal approximation</u> of V. We easily check that

$$\left.\begin{array}{l}\underline{\text{there exists a family } V_h \text{ of subspaces of } V \text{ satisfying}}\\ \underline{(4.10) \text{ and } (4.11)}\end{array}\right\} \quad (4.12)$$

Remark 4.1. : We can take an increasing family of spaces V_h, i.e. such that

$$V_h \subset V_{h'} \text{ , if } h > h' \text{ .}$$

This is the case in the Galerkin's method but it is not the case if we take for V_h "finite element subspaces". cf. CIARLET [1]. ∎

We consider now the following problem :

$$\left.\begin{array}{l} u_h(t) \in V_h, \\ (\frac{\partial}{\partial t} u_h(t), v) + a(t; u_h, v) = (f, v) \quad \forall v \in V_h, \\ u_h(o) = u_{oh} \text{ ,} \end{array}\right\} \quad (4.13)$$

where

$$u_{oh} \in V_h \text{ , } u_{oh} \to u_o \text{ in } H \text{ as } h \to 0. \quad (4.14)$$

(A sequence of u_{oh} satisfying (4.14) exists).

We remark that (4.13) is a set of differential equations (in number equal to the dimension of V_h), which admits a unique solution.

One easily checks (by taking $v = u_h$ in (4.13)) that

$$\|u_h\|_{L^2(0,T;V)} \leq C. \quad (4.15)$$

One can extract a subsequence, still denoted by u_h, such that

$$u_h \to u \text{ in } L^2(0,T;V) \text{ weakly,} \quad (4.16)$$

and one verifies that u satisfies $E(u,\phi) = L(\phi) \quad \forall \phi \in \Phi$ (in the notations of 4.1) hence the result follows.

4.3. Approximation method (II). Semidiscretization in time.

We introduce a "small" increment Δt and we replace $\frac{\partial u}{\partial t}$ by $\frac{u(t+\Delta t) - u(t)}{\Delta t}$. We define (we simplify somewhat the exposition by assuming here that $u_o \in V$) :

$$u^o_{\Delta t} = u_o \in V \; ; \tag{4.17}$$

assuming $u^{n-1}_{\Delta t} \in V$ to be computed, we define $u^n_{\Delta t} \in V$ as the solution of

$$\left.\begin{aligned} &\left(\frac{u^n_{\Delta t} - u^{n-1}_{\Delta t}}{\Delta t}, v\right) + \frac{1}{\Delta t}\int_{(n-1)\Delta t}^{n\Delta t} a(s;u^n_{\Delta t},v)ds = \\ &\qquad = \left(\frac{1}{\Delta t}\int_{(n-1)\Delta t}^{n\Delta t} f(s)ds, v\right) \quad \forall v \in V \end{aligned}\right\} \tag{4.18}$$

and we define the step function

$$u_{\Delta t} = u^n_{\Delta t} \text{ in } [n\Delta t,(n+1)\Delta t[\; . \tag{4.19}$$

One easily checks that

$$\|u_{\Delta t}\|_{L^2(0,T;V)} \leq C \tag{4.20}$$

and one can then extract a subsequence, still denoted by $u_{\Delta t}$, such that $u_{\Delta t} \to u$ in $L^2(0,T;V)$ weakly as $\Delta t \to 0$, and one verifies that u is again a solution of our problem.

Remark 4.2. : In numerical analysis one uses simultaneously the methods of Section 4.2. and of Section 4.3. (the method being then an implicit method). We refer to books in numerical analysis for explicit methods. ∎

4.4. Approximation method (III). Elliptic regularization.

We remark first that we do not restrict the generality by assuming that $u_o = 0$. Indeed, given u_o, one can find w in $V(0,T)$, which depends linearly and continuously on u_o, and which is such that

$$w(o) = u_o. \tag{4.21}$$

Then one replaces u by u-w. ■

We then consider, $\forall\, \varepsilon > 0$, the problem :

$$-\varepsilon \frac{\partial^2 u_\varepsilon}{\partial t^2} + \frac{\partial u_\varepsilon}{\partial t} + A(t)u_\varepsilon = f \ , \ 0 < t < T, \tag{4.22}$$

$$u_\varepsilon(o) = 0 \ , \ \frac{\partial u_\varepsilon}{\partial t}(T) = 0. \tag{4.23}$$

This problem is equivalent to the following. We introduce

$$\widetilde{W}(0,T) = \{v \mid v \subset L^2(0,T;V), \frac{\partial v}{\partial t} \subset L^2(0,T;H), v(o)=0\}, \tag{4.24}$$

provided with its natural Hilbertian structure.

For $u_\varepsilon, v \in \widetilde{W}(0,T)$, we define

$$\left.\begin{aligned} \pi_\varepsilon(u_\varepsilon,v) = \varepsilon\int_o^T \left(\frac{\partial u_\varepsilon}{\partial t}, \frac{\partial v}{\partial t}\right)dt + \int_o^T\left(\frac{\partial u_\varepsilon}{\partial t}, v\right)dt \,+ \\ + \int_o^T a(t;u_\varepsilon,v)dt. \end{aligned}\right\} \tag{4.25}$$

Problem (4.22)(4.23) is equivalent to

$$\pi_\varepsilon(u_\varepsilon,v) = \int_o^T (f,v)dt \quad \forall v \in \widetilde{W}(0,T) \tag{4.26}$$

and by virtue of Lax-Milgram's lemma, (4.26) admits a unique solution. One verifies that

$$\|u_\varepsilon\|_{L^2(0,T;V)} \leq C \tag{4.27}$$

and one checks that, as $\varepsilon \to 0$,

$$u_\varepsilon \to u \text{ in } L^2(0,T;V) \text{ weakly} \tag{4.28}$$

where u is the solution of our problem.

4.5. Other methods.

(i) - When $A(t) = A$ one can use the semi-group approach. Cf. Hille-Phillips [1], Yosida [1], and also for the application of this method to the case when $A(t)$ depends on t, cf. Yosida [1] and the references to Kato, Tanabe therein.

(ii) - Spectral calculus. A systematic use of Spectral calculus (Dunford's integrals) is made in Grisvard [1], together with application of interpolation theory, for the solution of general equations

$$Au + Bu = f$$

where A and B are unbounded operators.

Other results along these lines have been given by Dubinski [1], Da Prato-Grisvard [1] [2].

(iii) - Laplace transform. This method is useful when $A(t) = A$ does not depend on t and allows more general hypothesis than in the semi group approach. It leads to the theory of distributions semi groups (cf. Lions [2] [3], L.M. [1], Chazarain[1]).

(iv) - We present now, in Section 5 below, the mixed formulation.

5. MIXED FORMULATION.

5.1. Orientation.

We give now an adaptation to the evolution case of an approach (Brezzi [1]) used in hybrid methods for finite elements in stationary problems.

5.2. Setting of the problem.

Let Φ_1 and Φ_2 be two Hilbert spaces on $\mathbb{C}$, whose scalar product (resp. norm) is denoted by $(\, , \,)_i$ (resp. $\| \; \|_i$), $i=1,2$. We denote by Φ_i' the dual of Φ_i, and we denote by $\langle \, , \, \rangle$ the duality between Φ_i' and Φ_i. We are given

$$\left.\begin{aligned} a &= \text{continuous bilinear form on } \Phi_1 \times \Phi_1 \,, \\ b &= \quad '' \qquad '' \qquad '' \quad '' \;\; \Phi_1 \times \Phi_2 \,; \end{aligned}\right\} \tag{5.1}$$

We define A, B, B^* by

$$\left.\begin{aligned} & a(\phi_1,\psi_1) = \langle A\phi_1,\psi_1\rangle \quad , \quad A \in \mathcal{L}(\Phi_1;\Phi_1') \ , \\ & b(\phi_1,\psi_2) = \langle B\phi_1,\psi_2\rangle = \langle \phi_1,B^*\psi_2\rangle \quad , \\ & B \in \mathcal{L}(\Phi_1;\Phi_2') \ , \ B^* \in \mathcal{L}(\Phi_2;\Phi_1') \end{aligned}\right\} \tag{5.2}$$

and we denote by Λ the canonical isomorphism from Φ_2 into Φ_2' defined by

$$\langle \Lambda\phi_2,\psi_2\rangle = (\phi_2,\psi_2)_2 . \ \blacksquare \tag{5.3}$$

We are looking for a function ϕ with values in

$$\Phi = \Phi_1 \times \Phi_2 \quad , \quad \phi(t) = \{\phi_1(t),\phi_2(t)\} \ , \tag{5.4}$$

such that

$$\phi \in L^2(O,T;\Phi) \tag{5.5}$$

$$\frac{\partial\phi_2}{\partial t} \in L^2(O,T;\Phi_2) \tag{5.6}$$

$$\left(\frac{\partial\phi_2}{\partial t} , \psi_2\right)_2 + \pi(\phi,\psi) = \langle f,\psi_2\rangle \quad \forall \psi \in \Phi \tag{5.7}$$

where

$$\pi(\phi , \psi) = a(\phi_1,\psi_1) + b(\psi_1,\phi_2) \quad b(\phi_1,\psi_2) \tag{5.8}$$

and with

$$\phi_2(o) = \phi_{o2} \in \Phi_2 . \tag{5.9}$$

5.3 Laplace transform method.

If we denote by $\hat{\phi}(p)$ the Laplace transform of ϕ, we have to solve

$$p(\hat{\phi}_2(p),\psi_2)_2 + \pi(\hat{\phi}(p),\psi) = \langle g,\psi_2\rangle \quad , \tag{5.10}$$

where

$$\langle g,\psi_2\rangle = \langle \hat{f}(p),\psi_2\rangle + (\phi_{o2},\psi_2)_2 \ .$$

By standard techniques (cf. L [2]) it suffices to consider the case when g does not depend on t.

We make the hypothesis made in Brezzi [1] for the stationary case :

$$\left.\begin{aligned} & a(\phi_1,\phi_1) \geq 0 \quad \forall\phi_1 \in \Phi_1 , \\ & a(\phi_1,\phi_1) \geq \alpha\|\phi_1\|_1^2 \quad , \alpha > 0 , \quad \forall\phi_1 \in \operatorname{Ker} B, \end{aligned}\right\} \tag{5.11}$$

$$\sup_{\phi_1} \frac{|b(\phi_1,\psi_2)|}{\|\phi_1\|_1} \geq C\,\|\psi_2\|_2 \quad , C > 0. \tag{5.12}$$

Equation (5.10) is equivalent to

$$A\hat{\phi}_1 + B^*\hat{\phi}_2 = 0 \tag{5.13}$$

$$-B\hat{\phi}_1 + p\Lambda\hat{\phi}_2 = g. \tag{5.14}$$

We obtain from (5.14) :

$$p\hat{\phi}_2 = \Lambda^{-1}g + \Lambda^{-1}B\hat{\phi}_1 \tag{5.15}$$

hence (5.13) gives

$$\left.\begin{aligned} & pA\hat{\phi}_1 + B^*\Lambda^{-1}B\hat{\phi}_1 = -B^*\Lambda^{-1}g \in \Phi_1' , \\ & \hat{\phi}_1 \in \Phi_1 . \end{aligned}\right\} \tag{5.16}$$

But under hypothesis (5.11)(5.12), (5.16) is an "elliptic" equation.

Indeed, by virtue of (5.12)

$$B \text{ is an isomorphism from } \phi_1/\operatorname{Ker} B \text{ onto } \Phi_2'. \tag{5.17}$$

If we introduce

$$c(p;\phi_1,\psi_1) = p\ a(\phi_1,\psi_1) + (B\phi_1,B\psi_1)_2 \tag{5.18}$$

we see that (5.16) is equivalent to

$$c(p;\hat{\phi}_1,\psi_1) = \langle -B^*\Lambda^{-1}g,\psi_1\rangle \quad \forall\,\psi_1 \in \Phi_1 \tag{5.19}$$

and by virtue of (5.11) (5.17) we have

$$\left.\begin{array}{l} \operatorname{Re} c(p;\phi_1,\phi_1) \ge \alpha_o \|\phi_1\|_1^2 \qquad \phi_1 \in \Phi_1 \ , \ \alpha_o > 0 \ , \\ \text{if } \operatorname{Re} p \ge \xi_o > 0. \end{array}\right\} \tag{5.20}$$

Hence it follows that $\hat{\phi}_1(p), \hat{\phi}_2(p)$ exists, it is unique and we can compute ϕ_1, ϕ_2 by inverse Laplace transform.

5.4. Example.

We give here a very simple example. But what we say here readily extends to all examples considered in Brezzi [1], P.A. Raviart [1], M. Bercovier [1]. We take

$$\Phi_1 = \{\phi_1 | \ \phi_1 \in (L^2(\Omega))^n \ , \ \operatorname{div} \phi_1 \in L^2(\Omega)\} \ , \ \Phi_2 = L^2(\Omega) \ ,$$

$$a(\phi_1,\psi_1) = \int_\Omega \phi_1 \cdot \psi_1 \, dx \ , \ b(\phi_1,\psi_2) = (\operatorname{div} \phi_1, \psi_2) \ ;$$

we identify $\Phi_2' = \Phi_2$.

We have hypothesis (5.11)(5.12). Indeed Ker B = $\{\phi_1 | \operatorname{div} \phi_1 = 0\}$ and one has (5.11). Let us verify (5.12) ; let ψ_2 be given in $L^2(\Omega)$; we define $w \in H_o^1(\Omega)$ as the solution of

$$- \Delta w = \psi_2 \ ,$$

and we define ϕ_1 = grad w. Then

$$\frac{|b(\phi_1,\psi_2)|}{\|\phi_1\|_1} = \frac{\int_\Omega \psi_2^2 dx}{\|\phi_1\|_1} \ge c \|\psi_2\|_2 = c|\psi_2|_{L^2(\Omega)} \ .$$

The solution ϕ of (5.7) satisfies

$$(\phi_1,\psi_1) + (\operatorname{div} \psi_1,\phi_2) = 0 \ , \tag{5.21}$$

$$(\frac{\partial \phi_2}{\partial t} \ , \ \psi_2) - (\operatorname{div} \phi_1,\psi_2) = (f,\psi_2) \ . \tag{5.22}$$

But (5.21) is equivalent to

$$\phi_1 - \text{grad}\,\phi_2 = 0 \quad \text{and} \quad \phi_2 = 0 \text{ on } \Gamma = \partial\Omega,$$

and (5.22) is equivalent to

$$\frac{\partial\phi_2}{\partial t} - \text{div}\,\phi_1 = f$$

so that $\phi_2 \in L^2(0,T;H_o^1(\Omega))$ and

$$\frac{\partial\phi_2}{\partial t} - \Delta\phi_2 = f.$$

BIBLIOGRAPHY OF CHAPTER I

[1] M.S. Baouendi and P. Grisvard - J.F.A. 1969

[1] M. Bercovier - Thesis. 1976.

[1] F. Brezzi - On the existence, uniqueness and approximation of saddle point problems arising from Lagrangian multipliers. R.A.I.R.O. (1974), 129-151.

[1] J. Chazarain - Problèmes de Cauchy abstraits et applications à quelques problèmes mixtes. J.F.A. 7(1971), p. 386-446.

[1] Ph. Ciarlet - Lectures on finite element methods. Tata Institute Fund. Research - Bombay, 1975.

[1] G. Da Prato and P. Grisvard - Somme d'opérateurs linéaires et équations différentielles opérationnelles. J.M.P.A. 54 (1975), p. 305-387.

[2] G. Da Prato and P. Grisvard - Equations d'évolution abstraites non linéaires de type parabolique. C.R.A.S. September 1976.

[1] M. Dubinski - On an abstract theorem and its applications to boundary value problems for non classical equations. Math. Sbornik 79(121), 1969, p. 91-117 (Math. URSS Sbornik, 8, 1969, 87-113).

[1] E. Hille and R.S. Phillips - Functional analysis and semi groups. Coll. Pub. A.M.S. 1957.

[1] J.L. Lions - Equations différentielles opérationnelles et problèmes aux limites. Springer Verlag - 1961.

[2] J.L. Lions - Problèmes aux limites en théorie des distributions. Acta Math. 94(1955), p. 12-153.

[3] J.L. Lions - Sur les semi groupes distributions. Portugaliae Math. 19(1960), 141-164.

[1] J.L. Lions and E. Magenes - Problèmes aux limites non homogènes et applications - Dunod, Vol. 1, 2 (1968), vol. 3 (1970) English translation : Springer 1972, 1973.

[1] P.A. Raviart - Lectures Imperial College, June 1976.

[1] K. Yosida - Functional analysis. Fourth Ed. Springer 1974.

CHAPTER II

ASYMPTOTIC PROBLEMS - AN INTRODUCTION TO HOMOGENIZATION

1. SETTING C THE PROBLEMS.

1.1. Notations.

We shall consider in what follows <u>parabolic operators</u> of the type

$$\frac{\partial}{\partial t} - \frac{\partial}{\partial x_i}\left(a_{ij}\left(\frac{x}{\varepsilon}\right)\frac{\partial}{\partial x_j}\right), \tag{1.1}$$

$$\frac{\partial}{\partial t} - \frac{\partial}{\partial x_i}\left(a_{ij}\left(\frac{x}{\varepsilon}, \frac{t}{\varepsilon^k}\right)\frac{\partial}{\partial x_j}\right), \tag{1.2}$$

where - once for all in this Chapter - we adopt the summation convention with respect to indices which appear twice, and where the a_{ij}'s are <u>periodic</u> functions.

More precisely, let us introduce :

$$Y = \text{parallelepiped in } \mathbb{R}^n , \quad Y = \prod_j \,]0,y_j^o[\,, \tag{1.3}$$

$$\tilde{Y} = Y \times]0,\tau_o[\,, \subset \mathbb{R}^{n+1} . \tag{1.4}$$

We will consider functions $a_{ij}(y)$ or $a_{ij}(y,\tau)$ which satisfy :

$$\left.\begin{array}{l} a_{ij} \in L^\infty(\mathbb{R}^n) \,, \ a_{ij} \text{ is Y-periodic, i.e.} \\ a_{ij} \text{ admits period } y_k^o \text{ in the variable } y_k, k=1,\dots,n \end{array}\right\} \tag{1.5}$$

or

$$a_{ij} \in L^\infty(\mathbb{R}^{n+1}) \ , \ a_{ij} \text{ is } \tilde{Y} \text{ periodic.} \tag{1.6}$$

We assume that

$$\left.\begin{array}{l} a_{ij}\ \xi_i\ \xi_j \geq \alpha\xi_i\xi_i \ , \ \alpha > 0 \ , \ \forall\xi_i, \text{ a.e. in } y \\ \text{or a.e. in } y \text{ and } \tau. \end{array}\right\} \tag{1.7}$$

With these hypothesis, all the operators (1.1) or (1.2) where $\varepsilon > 0$ and $k=1,2,...$, are "uniformly parabolic" in ε. ■

Remark 1.1 : Operators of this type appear in the study of composite materials in mechanics, at least for (1.1). Operators of type (1.2) are studied here for mathematical reasons (we will see that they lead to interesting mathematical questions). ■

1.2. Problems.

Problem 1.1 - Let $u_\varepsilon = u_\varepsilon(x,t)$ be the solution of

$$\frac{\partial u_\varepsilon}{\partial t} - \frac{\partial}{\partial x_i} (a_{ij}(\frac{x}{\varepsilon}) \frac{\partial u_\varepsilon}{\partial x_j}) = f \text{ in } \Omega\times]0,T[\ , \tag{1.8}$$

$$u_\varepsilon = 0 \text{ on } \Sigma = \Gamma\times]0,T[\ , \ \Gamma = \partial\Omega \ , \tag{1.9}$$

$$u_\varepsilon(x,o) = u_o(x) \ , \ u_o \in L^2(\Omega). \tag{1.10}$$

By virtue of (1.7) this problem admits a unique solution

$$u_\varepsilon \in L^2(0,T;H_o^1(\Omega)) \ , \tag{1.11}$$

when f is given in $L^2(0,T;H^{-1}(\Omega))$.

The problem we want to study (Problem 1.1) is the behaviour of u_ε as $\varepsilon \to 0$.

We shall introduce also

Problem 1.2 - Let u_ε be the solution of

$$\frac{\partial u_\varepsilon}{\partial t} - \frac{\partial}{\partial x_i} (a_{ij} (\frac{x}{\varepsilon} , \frac{t}{\varepsilon^k}) \frac{\partial u_\varepsilon}{\partial x_j}) = f \text{ in } \Omega\times]0,T[\ , \tag{1.12}$$

f given in $L^2(0,T;H^{-1}(\Omega))$, u_ε being subject to (1.9)(1.10). By virtue of (1.7) this problem admits a unique solution which satisfies (1.11).

Problem 1.2. (k). consists in the study of u_ε as $\varepsilon \to 0$.
We shall see that there are three cases :

$$k=1 \ , \ k=2, \ k \text{ integer } \geq 3. \tag{1.13}$$

1.3. Orientation.

We study the above problems firstly by a (formal) asymptotic expansion : multiple scale expansion (cf. Section 2) and we give next a justification of this method by energy estimates (Section 3).

1.4. Remarks.

Very many extensions of what follows are possible. The following Chapter can be thought of as a mere introduction to the Book Bensoussan-Lions-Papanicolaou (B.L.P. [1]).

We confine ourselves here to a few remarks about some of the extensions.

The type of result we shall obtain is as follows : there exists an operator $\mathcal{A}$ (which is a second order elliptic operator with constant coefficients), which will be different in Problems 1.1 and 1.2 (k), k=1,2, and k ≥ 3, such that, if u denotes the solution of the "homogenized problem".

$$\frac{\partial u}{\partial t} + \mathcal{A}u = f \text{ in } \Omega\times]0,T[, \tag{1.14}$$

$$u = 0 \text{ on } \Sigma, \tag{1.15}$$

$$u(x,o) = u_o(x) \text{ in } \Omega, \tag{1.16}$$

then, as $\varepsilon \to 0$,

$$u_\varepsilon \to u \text{ in } L^2(0,T;H_o^1(\Omega)) \text{ weakly}. \tag{1.17}$$

Remark 1.2 : We have similar results for other Boundary conditions.

Remark 1.3 : All the results extend to all parabolic operators of any order or to systems of such operators.

Remark 1.4 : For Problem 1.1 (this is not the case for problem 1.2 (k)) the results extend to hyperbolic operators.

Remark 1.5 : For non linear problems we refer to B.L.P. [1].

2. MULTI-SCALE METHODS.

2.1. Notations. Principle of the method.

We introduce

$$y = x/\varepsilon \ , \quad \tau = t/\varepsilon^k. \tag{2.1}$$

Applied to a function $\Phi(x,y,t)$ or $\Phi(x,y,t,\tau)$ the operator $\frac{\partial}{\partial x_j}$ becomes $\frac{\partial}{\partial x_j} + \frac{1}{\varepsilon}\frac{\partial}{\partial y_j}$ and the operator $\frac{\partial}{\partial t}$ remains $\frac{\partial}{\partial t}$ in the first case and becomes $\frac{\partial}{\partial t} + \frac{1}{\varepsilon^k}\frac{\partial}{\partial \tau}$ in the second case.

We shall set :

$$P^\varepsilon = \frac{\partial}{\partial t} - \frac{\partial}{\partial x_i}\left(a_{ij}\left(\frac{x}{\varepsilon}\right)\frac{\partial}{\partial x_j}\right), \tag{2.2}$$

$$P^{\varepsilon,k} = \frac{\partial}{\partial t} - \frac{\partial}{\partial x_i}\left(a_{ij}\left(\frac{x}{\varepsilon}, \frac{t}{\varepsilon^k}\right)\frac{\partial}{\partial x_j}\right). \tag{2.3}$$

Applied to a function $\Phi(x,y,t)$, P^ε becomes :

$$P^\varepsilon = \varepsilon^{-2}P_1 + \varepsilon^{-1}P_2 + \varepsilon^{o}P_3 , \tag{2.4}$$

$$\left.\begin{aligned}
P_1 &= -\frac{\partial}{\partial y_i}\left(a_{ij}(y)\frac{\partial}{\partial y_j}\right), \\
P_2 &= -\frac{\partial}{\partial y_i}\left(a_{ij}(y)\frac{\partial}{\partial x_j}\right) - \frac{\partial}{\partial x_i}\left(a_{ij}(y)\frac{\partial}{\partial y_j}\right), \\
P_3 &= \frac{\partial}{\partial t} - \frac{\partial}{\partial x_i}\left(a_{ij}(y)\frac{\partial}{\partial x_j}\right).
\end{aligned}\right\} \tag{2.5}$$

Applied to a function $\Phi(x,y,t,\tau)$, $P^{\varepsilon,k}$ becomes :

$$P^{\varepsilon,1} = \varepsilon^{-2}Q_1 + \varepsilon^{-1}Q_2 + \varepsilon^{o}Q_3, \tag{2.6}$$

$$\left.\begin{aligned} Q_1 &= -\frac{\partial}{\partial y_i}\left(a_{ij}(y,\tau)\frac{\partial}{\partial y_j}\right), \\ Q_2 &= \frac{\partial}{\partial \tau} - \frac{\partial}{\partial x_i}\left(a_{ij}(y,\tau)\frac{\partial}{\partial x_j}\right) - \frac{\partial}{\partial x_i}\left(a_{ij}(y,\tau)\frac{\partial}{\partial y_j}\right), \\ Q_3 &= \frac{\partial}{\partial t} - \frac{\partial}{\partial x_i}\left(a_{ij}(y,\tau)\frac{\partial}{\partial x_j}\right); \end{aligned}\right\} \quad (2.7)$$

$$P^{\varepsilon,2} = \varepsilon^{-2}R_1 + \varepsilon^{-1}R_2 + \varepsilon^{o}R_3, \quad (2.8)$$

$$\left.\begin{aligned} R_1 &= \frac{\partial}{\partial \tau} - \frac{\partial}{\partial y_i}\left(a_{ij}(y,\tau)\frac{\partial}{\partial y_j}\right), \\ R_2 &= -\frac{\partial}{\partial y_i}\left(a_{ij}(y,\tau)\frac{\partial}{\partial x_j}\right) - \frac{\partial}{\partial x_i}\left(a_{ij}(y,\tau)\frac{\partial}{\partial y_j}\right), \\ R_3 &= \frac{\partial}{\partial t} - \frac{\partial}{\partial x_i}\left(a_{ij}(y,\tau)\frac{\partial}{\partial x_j}\right); \end{aligned}\right\} \quad (2.9)$$

$$P^{\varepsilon,k} = \varepsilon^{-k}S_1 + \varepsilon^{-2}S_2 + \varepsilon^{-1}S_3 + \varepsilon^{o}S_4, \quad (k \geq 3) \quad (2.10)$$

$$\left.\begin{aligned} S_1 &= \frac{\partial}{\partial \tau}, \\ S_2 &= -\frac{\partial}{\partial y_i}\left(a_{ij}(y,\tau)\frac{\partial}{\partial y_j}\right), \\ S_3 &= -\frac{\partial}{\partial y_i}\left(a_{ij}(y,\tau)\frac{\partial}{\partial x_j}\right) - \frac{\partial}{\partial x_i}\left(a_{ij}(y,\tau)\frac{\partial}{\partial y_j}\right) \\ S_4 &= \frac{\partial}{\partial t} - \frac{\partial}{\partial x_i}\left(a_{ij}(y,\tau)\frac{\partial}{\partial x_j}\right). \end{aligned}\right\} \quad (2.11)$$

The principle of the method is now, <u>for Problem</u> 1.1, to expand u_ε in the form

$$u_\varepsilon = w_o(x,y,t) + \varepsilon w_1(x,y,t) + \varepsilon^2 w_2(x,y,t) + \dots \quad (2.12)$$

where

$$w_j(x,y,t) \text{ is Y-periodic in } y \quad (2.13)$$

(and where of course, we replace y by x/ε in (2.12)) ;

for Problem 1.2 (k), we expand u_ε in the form

$$u_\varepsilon = w_o(x,y,t,\tau) + \varepsilon w_1(x,y,t,\tau) + \varepsilon^2 w_2(x,y,t,\tau)+\ldots \quad (2.14)$$

where

$$w_j(x,y,t,\tau) \text{ is } \tilde{Y}\text{-periodic in } y,\tau. \quad (2.15)$$

We make, in each case, an identification in (1.8) or in (1.12) and, as we are going to show, the compatibility of the computation will lead to the homogenized equation.

2.2. Problem 1.1.

We use (2.4) and (2.12). We obtain

$$P_1 w_o = 0, \quad (2.16)$$

$$P_1 w_1 + P_2 w_o = 0, \quad (2.17)$$

$$P_1 w_2 + P_2 w_1 + P_3 w_o = f. \quad (2.18)$$

In (2.16) we look for a periodic solution $w_o(x,y,t)$, where x and t are parameters. But, the only Y-periodic solution of

$$P_1 \phi = 0 \text{ in } Y \quad (2.19)$$

is

$$\phi = \text{constant} \quad (2.20)$$

and therefore

$$w_o(x,y,t) = u(x,t). \quad (2.21)$$

Using (2.21), (2.17) reduces to

$$P_1 w_1 - \left(\frac{\partial a_{ij}}{\partial y_i}(y)\right) \frac{\partial u}{\partial x_j} = 0. \quad (2.22)$$

We then introduce $\chi^j = \chi^j(y)$ by

$$\left.\begin{aligned} & P_1 \chi^j = - \frac{\partial a_{ij}}{\partial y_i} = P_1(y_j) \\ & \chi^j \text{ being Y-periodic,} \end{aligned}\right\} \quad (2.23)$$

which defines χ^j up to an additive constant. Then (2.22) gives

$$w_1 = -\chi^j(y)\frac{\partial u}{\partial x_j}(x,t) + u_1(x,t). \tag{2.24}$$

It remains to solve (2.18). We remark that the equation

$$P_1\phi = f \text{ in } Y,\ \phi \text{ Y-periodic} \tag{2.25}$$

admits a solution iff

$$\int_Y f(y)dy = 0. \tag{2.26}$$

(Notice that this condition is satisfied in (2.23)).

Then (2.18) admits a solution w_2 which is Y-periodic iff

$$\int_Y (P_2w_1 + P_3w_o)dy = \int_Y f dy = |Y|\ f(x) \tag{2.27}$$

where $|Y|$ = measure of Y.

Using (2.5), (2.24), we have :

$$\int_Y P_2w_1dy = -\frac{\partial}{\partial x_i}\int_Y a_{ik}(y)\frac{\partial w_1}{\partial y_k}dy =$$

$$= \frac{\partial}{\partial x_i}\left(\int_Y a_{ik}(y)\frac{\partial\chi^j}{\partial y_k}dy\frac{\partial u}{\partial x_j}\right)$$

and

$$\int_Y P_3w_o dy = |Y|\frac{\partial u}{\partial t} - \frac{\partial}{\partial x_i}\left(\int_Y a_{ij}(y)dy\frac{\partial u}{\partial x_j}\right).$$

Therefore (2.27) becomes :

$$\left.\begin{aligned} &\frac{\partial u}{\partial t} + \mathcal{A}u = f,\\ &\mathcal{A}u = -\frac{1}{|Y|}\frac{\partial}{\partial x_i}\left[\int_Y a_{ij}(y)dy - \int_Y a_{ik}(y)\frac{\partial\chi^j}{\partial y_k}dy\right]\frac{\partial u}{\partial x_j}, \end{aligned}\right\} \tag{2.28}$$

where χ^j is defined by (2.23) (and where the formula does not depend on the choice of the additive constant in (2.23)). <u>We will</u>

see in Section 3 below that one has (1.14)...(1.17) for this construction of $\mathcal{A}$.

2.3. Problem 1.2 (1).

We use now (2.6) and (2.14). We obtain :

$$Q_1 w_o = 0, \tag{2.29}$$

$$Q_1 w_1 + Q_2 w_o = 0, \tag{2.30}$$

$$Q_1 w_2 + Q_2 w_1 + Q_3 w_o = f. \tag{2.31}$$

In (2.9) x,t and τ play the role of parameters, so that

$$w_o(x,y,t,\tau) = \tilde{w}_o(x,t,\tau). \tag{2.32}$$

But (2.30) admits a solution w_1 which is Y-periodic iff

$$\int_Y Q_2 w_o dy = 0$$

i.e.

$$|Y| \frac{\partial \tilde{w}_o}{\partial \tau} = 0$$

and therefore

$$\tilde{w}_o = u(x,t). \tag{2.33}$$

Then (2.30) reduces to

$$Q_1 w_1 - \frac{\partial a_{ij}(y,\tau)}{\partial y_j} \frac{\partial u}{\partial x_j} = 0. \tag{2.34}$$

We proceed as for (2.22). We introduce $\mathcal{O}^j = \mathcal{O}^j(y,\tau)$ as the solution (defined up to an additive constant) of

$$\left.\begin{array}{l} Q_1 \mathcal{O}^j = Q_1(y_j), \\ \mathcal{O}^j \text{ being Y-periodic (and also automatically } \tau_o \\ \qquad\qquad \text{periodic in } \tau). \end{array}\right\} \tag{2.35}$$

Then

$$w_1 = -\mathcal{O}^j(y,\tau)\frac{\partial u}{\partial x_j} + \tilde{w}_1(x,t,\tau). \tag{2.36}$$

We can find w_2 Y-periodic solution of (2.31) iff

$$\int_Y (Q_2 w_1 + Q_3 w_o)dy = |Y| f. \tag{2.37}$$

But

$$\begin{aligned}\int_Y Q_2 w_1 dy = |Y| \frac{\partial \tilde{w}_1}{\partial \tau} - \int_Y \frac{\partial \mathcal{O}^j}{\partial \tau}(y,\tau)dy \frac{\partial u}{\partial x_j} + \\ + \frac{\partial}{\partial x_i} \int_Y a_{ik}(y,\tau) \frac{\partial \mathcal{O}^j}{\partial y_k}(y,\tau)dy \frac{\partial u}{\partial x_j})\end{aligned} \tag{2.38}$$

and the computation of $\tilde{w}_1$ is possible iff

$$\int_o^{\tau_o}\int_Y (Q_2 w_1 + Q_3 w_o)dy\, d\tau = |Y|\, \tau_o f. \tag{2.39}$$

But

$$\int_o^{\tau_o} d\tau \int_Y Q_2 w_1 dy = \frac{\partial}{\partial x_i}\int_o^{\tau_o}\int_Y a_{ik}(y,\tau)\frac{\partial \mathcal{O}^j}{\partial y_k}(y,\tau)dy\, d\tau \frac{\partial u}{\partial x_j}$$

so that (2.39) gives :

$$\left.\begin{aligned} &\frac{\partial u}{\partial t} + \mathcal{A}u = f, \\ &\mathcal{A}u = -\frac{1}{|Y|\tau_o}\frac{\partial}{\partial x_i}\int_o^{\tau_o}\int_Y \left(a_{ij} - a_{ik}\frac{\partial \mathcal{O}^j}{\partial y_k}(y,\tau)\right)dy d\tau \frac{\partial u}{\partial x_j}\end{aligned}\right\} \tag{2.40}$$

This computation will be justified in Section 3 below.

2.4. Problem 1.2(2).

We use now (2.8) and (2.14). We obtain :

$$R_1 w_o = o, \tag{2.41}$$

$$R_1 w_1 + R_2 w_o = 0, \tag{2.42}$$

$$R_1 w_2 + R_2 w_1 + R_3 w_o = f. \tag{2.43}$$

But let us remark that the only periodic solution in y and in τ of (2.41) is

$$w_o = u(x,t). \tag{2.44}$$

Then (2.42) reduces to

$$R_1 w_1 - \frac{\partial a_{ij}}{\partial y_i}(y,\tau)\frac{\partial u}{\partial x_j} = 0. \tag{2.45}$$

We then introduce $\phi^j = \phi^j(y,\tau)$ as the solution (defined up to an additive constant) of

$$\left.\begin{array}{l} R_1\,\phi^j = R_1\,y_j\,, \\ \phi^j \text{ is } Y \text{ and } \tau_o \text{ periodic in } y \text{ and in } \tau; \end{array}\right\} \tag{2.46}$$

then

$$w_1 = -\,\phi^j(y,\tau)\frac{\partial u}{\partial x_j}(x,t) + u_1(x,t). \tag{2.47}$$

The equation

$$R_1\phi = f\ ,\ \phi\ Y\text{-}\tau_o \text{ periodic} \tag{2.48}$$

admits a solution iff

$$\int_o^{\tau_o}\int_Y f(y,\tau)dy\,d\tau = 0.$$

Therefore we can obtain w_2 from (2.43) iff

$$\int_o^{\tau_o}\int_Y (R_2 w_1 + R_3 w_o)dy\,d\tau = |Y|\ \tau_o f. \tag{2.49}$$

But

$$\int_o^{\tau_o}\int_Y R_2 w_1 dy\,d\tau = \frac{\partial}{\partial x_i}\int_o^{\tau_o}\int_Y a_{ik}(y,\tau)\frac{\partial\phi^j}{\partial y_k}(y,\tau)dy\,d\tau\,\frac{\partial u}{\partial x_j}$$

so that we finally obtain

$$\left.\begin{aligned} &\frac{\partial u}{\partial t} + \mathcal{A}u = f, \\ &\mathcal{A}u = -\frac{1}{|Y|\tau_o}\frac{\partial}{\partial x_i}\left[\int_o^{\tau_o}\int_Y (a_{ij} - a_{ik}\frac{\partial \phi^j}{\partial y_k})\,dy\,d\tau\right]\frac{\partial u}{\partial x_j} \quad . \end{aligned}\right\} \tag{2.50}$$

2.5. Problem 1.2.(3).

We restrict ourselves to the case k=3. We leave it to the reader to verify that one obtains the <u>same</u> result for k arbitrary >3.

Using (2.8) and (2.10) we obtain

$$S_1 w_o = 0, \tag{2.51}$$

$$S_1 w_1 + S_2 w_o = 0, \tag{2.52}$$

$$S_1 w_2 + S_2 w_1 + S_3 w_o = 0, \tag{2.53}$$

$$S_1 w_3 + S_2 w_2 + S_3 w_1 + S_4 w_o = f. \tag{2.54}$$

But (2.51) is equivalent to

$$w_o(x,y,t,\tau) = \tilde{w}_o(x,y,t) \tag{2.55}$$

and (2.52) reduces to

$$\frac{\partial w_1}{\partial \tau} + S_2 \tilde{w}_o = 0. \tag{2.56}$$

A <u>periodic</u> solution in τ exists iff

$$\int_o^{\tau_o} S_2 \tilde{w}_o \, d\tau = 0$$

i.e.

$$\left.\begin{aligned} &-\frac{\partial}{\partial y_i}\left(\int_o^{\tau_o} a_{ij}(y,\tau)d\tau \, \frac{\partial \tilde{w}_o}{\partial y_j}\right) = 0 \;; \\ &\tilde{w}_o \text{ Y-periodic.} \end{aligned}\right\} \tag{2.57}$$

But (2.57) implies

$$\tilde{w}_o(x,y,t) = u(x,t) \tag{2.58}$$

and (2.56) reduces to $\frac{\partial w_1}{\partial \tau} = 0$ i.e.

$$w_1 = \tilde{w}_1(x,y,t). \tag{2.59}$$

We can obtain w_2 from (2.53) iff

$$\int_o^{\tau_o} (S_2 w_1 + S_3 w_o) d\tau = 0. \tag{2.60}$$

If we introduce

$$\bar{A} = - \frac{\partial}{\partial y_i} \left(\int_o^{\tau_o} a_{ij}(y,\tau) d\tau \frac{\partial}{\partial y_j} \right) \tag{2.61}$$

then (2.60) reduces to

$$\bar{A} w_1 - \int_o^{\tau_o} \frac{\partial a_{ij}}{\partial y_i} (y,\tau) d\tau \frac{\partial u}{\partial x_j} = 0. \tag{2.62}$$

We then introduce ψ^j by

$$\bar{A}\, \psi^j = \bar{A}\, y_j \ , \ \psi^j = \psi^j(y) \quad \text{Y-periodic} \tag{2.63}$$

which defines ψ^j up to an additive constant. Then (2.62) gives

$$w_1 = - \psi^j(y) \frac{\partial u}{\partial x_j} (x,t) + u_1(x,t). \tag{2.64}$$

Then (2.54) admits a solution w_3 iff

$$\int_o^{\tau_o} (S_2 w_2 + S_3 w_1 + S_4 w_o) d\tau = \tau_o f. \tag{2.65}$$

This equation admits a solution w_2 Y-periodic iff

$$\int_o^{\tau_o} \int_Y (S_3 w_1 + S_4 w_o) dy\, d\tau = \tau_o |Y| f. \tag{2.66}$$

But

$$\int_Y \int_0^{\tau_o} S_3 w_1 \, dy \, d\tau = \frac{\partial}{\partial x_i} \int_Y \int_0^{\tau_o} a_{ik}(y,\tau) \frac{\partial \psi^j}{\partial y_k}(y) dy \, d\tau \frac{\partial u}{\partial x_j}$$

and we finally obtain

$$\left.\begin{aligned} &\frac{\partial u}{\partial t} + \mathcal{A} u = f, \\ &\mathcal{A} u = - \frac{1}{|Y|\tau_o} \frac{\partial}{\partial x_i} \left[\int_Y \int_0^{\tau_o} \left[a_{ij} - a_{ik}(y,\tau) \frac{\partial \psi^j}{\partial y_k}(y) \right] dy \, d\tau \right] \frac{\partial u}{\partial x_j}. \end{aligned}\right\} \quad (2.67)$$

3. JUSTIFICATION BY ENERGY METHODS OF THE ASYMPTOTIC CALCULATIONS.

3.1. Problem 1.1.

We denote by u_ε the solution of (1.8)(1.9)(1.10).
We set

$$A^\varepsilon = - \frac{\partial}{\partial x_i} \left(a_{ij}\left(\frac{x}{\varepsilon}\right) \frac{\partial}{\partial x_j} \right) \quad (3.1)$$

and we denote by a^ε the bilinear form associated to A^ε on $H_o^1(\Omega)$:

$$a^\varepsilon(u,v) = \int_\Omega a_{ij}\left(\frac{x}{\varepsilon}\right) \frac{\partial u}{\partial x_j} \frac{\partial v}{\partial x_i} dx. \quad (3.2)$$

We have

$$\left(\frac{\partial u_\varepsilon}{\partial t}, v \right) + a^\varepsilon(u_\varepsilon, v) = (f,v) \quad \forall v \in H_o^1(\Omega). \quad (3.3)$$

By virtue of (1.7) we have

$$\left.\begin{aligned} &a^\varepsilon(v,v) \geq \alpha_1 \|v\|^2, \quad \alpha_1 > 0, \quad \forall v \in H_o^1(\Omega) \\ &\alpha_1 \ \underline{\text{independent of}}\ \varepsilon. \end{aligned}\right\} \quad (3.4)$$

Therefore

$$\|u_\varepsilon\|_{L^2(0,T;V)} \leq C, \quad V = H_o^1(\Omega) \quad (3.5)$$

where here and in what follows, the C's denote various constants <u>which do not depend on</u> ε.

Since $\frac{\partial u_\varepsilon}{\partial t} = f - A^\varepsilon u_\varepsilon$, one has

$$\left\|\frac{\partial u_\varepsilon}{\partial t}\right\|_{L^2(0,T;V)} \leq C. \tag{3.6}$$

Remark 3.1 : We shall obtain the same estimates for u_ε solution of Problem 1.2.(k). ■

Remark 3.2 : One obtains the analogous estimates for all variational boundary conditions.■

Notations :

$$a_{ij}^\varepsilon = a_{ij}(x/\varepsilon),$$

$$\xi_i^\varepsilon = a_{ij}^\varepsilon \frac{\partial u_\varepsilon}{\partial x_j} . \tag{3.7}$$

Equation (1.8) becomes :

$$\frac{\partial u_\varepsilon}{\partial t} - \frac{\partial}{\partial x_i} \xi_i^\varepsilon = f \text{ in } \Omega \times]0,T[. \tag{3.8}$$

By virtue of (3.5)(3.6), one can extract a sub-sequence, still denoted by u_ε , such that

$$u_\varepsilon \to u \text{ in } L^2(0,T;V) \text{ weakly}, \tag{3.9}$$

$$\frac{\partial u_\varepsilon}{\partial t} \to \frac{\partial u}{\partial t} \text{ in } L^2(0,T;V') \text{ weakly}, \tag{3.10}$$

and since $\|\xi_i^\varepsilon\|_{L^2(Q)} \leq C$, one can also assume that

$$\xi_i^\varepsilon \to \xi_i \text{ in } L^2(Q) \text{ weakly}. \tag{3.11}$$

It follows from (3.8) that

$$\frac{\partial u}{\partial t} - \frac{\partial \xi_i}{\partial x_i} = f. \tag{3.12}$$

Moreover, it follows from (3.9)(3.10) and from standard compactness arguments that

$$u_\varepsilon \to u \quad \text{in } L^2(Q) \text{ strongly.} \tag{3.13}$$

Remark 3.3 : If Ω is unbounded, it follows from (3.9)(3.10) that

$$u_\varepsilon \to u \text{ in } L^2(\mathcal{O}\times]0,T[) \text{ strongly}$$

where $\mathcal{O}$ is any bounded open set $\subset \Omega$, and this suffices for the proof which follows. ∎

Remark 3.4 : If we change $a^\varepsilon_{ij} = a_{ij}(x/\varepsilon)$ into

$$a^\varepsilon_{ij} = a_{ij}\left(\frac{x}{\varepsilon}, \frac{t}{\varepsilon^k}\right), \tag{3.14}$$

all what has been said till now immediately extends to Problem 1.2.(k) and it will not be repeated. ∎

All the problem is now reduced to computing $\mathcal{E}_i$.

We introduce :

$$\left.\begin{aligned} A_y &= -\frac{\partial}{\partial y_i}\left(a_{ij}(y)\frac{\partial}{\partial y_j}\right) = P_1 \ (\mathrm{cf}(2.5)), \\ \hat{A}_y &= -\frac{\partial}{\partial y_i}\left(\hat{a}_{ij}(y)\frac{\partial}{\partial y_j}\right), \ \hat{a}_{ij} = a_{ji} \end{aligned}\right\} \tag{3.15}$$

and we introduce

$$P(y) = \text{homogeneous polynomial in } y \text{ of degree } 1. \tag{3.16}$$

We now define $w = w(y)$ as the solution (defined up to an additive constant) of

$$\left.\begin{aligned} &A^*_y w = 0 \text{ in } Y, \\ &w-P \text{ is } Y\text{-periodic.} \end{aligned}\right\} \tag{3.17}$$

Therefore if

$$w-P = -\chi \tag{3.18}$$

we have :

$$A_y^* \chi = A_y^* P \text{ , } \chi \text{ is Y-periodic} \tag{3.19}$$

(which admits a solution defined up to an additive constant).

We introduce next

$$w_\varepsilon(x,t) = w_\varepsilon(x) = \varepsilon w(x/\varepsilon) = P(x) - \varepsilon\chi(x/\varepsilon). \tag{3.20}$$

We have

$$(A^\varepsilon)^* w_\varepsilon = 0. \tag{3.21}$$

We consider $\phi \in \mathcal{D}(Q)$ and we multiply (1.8) by ϕw_ε and (3.21) by ϕu_ε. After substraction and simplification, we obtain :

$$\left.\begin{aligned} &\left(\frac{\partial u_\varepsilon}{\partial t}, \phi w_\varepsilon\right) + \left(\xi_i^\varepsilon, \frac{\partial \phi}{\partial x_i} w_\varepsilon\right) - \int_\Omega a_{ij}^\varepsilon \frac{\partial w_\varepsilon}{\partial x_i} \frac{\partial \phi}{\partial x_j} u_\varepsilon \, dx = \\ &\qquad\qquad = (f, \phi w_\varepsilon). \end{aligned}\right\} \tag{3.22}$$

We integrate (3.22) in $t \in (0,T)$; we obtain :

$$\left.\begin{aligned} &- \int_Q u_\varepsilon w_\varepsilon \frac{\partial \phi}{\partial t} dx\, dt + \int_Q \xi_i^\varepsilon \frac{\partial \phi}{\partial x_i} w_\varepsilon \, dx\, dt - \\ &- \int_Q \left(a_{ij}(y) \frac{\partial w}{\partial y_i}(y)\right)_{y=x/\varepsilon} \frac{\partial \phi}{\partial x_j} u_\varepsilon \, dx\, dt = \int_0^T (f, \phi w_\varepsilon) dt. \end{aligned}\right\} \tag{3.23}$$

But by virtue of (3.13) and of the fact that $w_\varepsilon \to P$ in $L^2(Q)$ strongly the first two terms in (3.23) converge respectively towards

$$- \int_Q uP \frac{\partial \phi}{\partial t} dx\, dt \qquad \text{and} \qquad \int_Q \xi_i \frac{\partial \phi}{\partial x_i} P \, dx\, dt.$$

We observe that

$$\left.\begin{aligned} &\left(a_{ij} \frac{\partial w}{\partial y_i}\right)_{x/\varepsilon} \to \mathcal{M}\left(a_{ij} \frac{\partial w}{\partial y_i}\right) \text{ in } L^2(Q) \text{ weakly} \\ &\text{if } \mathcal{M}(\phi) = \frac{1}{|Y|} \int_Y \phi(y) dy \end{aligned}\right\} \tag{3.24}$$

so that (3.23) gives at the limit

$$\left. \begin{aligned} & - \int_Q uP \frac{\partial \phi}{\partial t} \, dx \, dt + \int_Q \xi_i \, P \frac{\partial \phi}{\partial x_i} \, dx \, dt - \\ & - \mathcal{M}(a_{ij} \frac{\partial w}{\partial y_i}) \int_Q \frac{\partial \phi}{\partial x_j} u \, dx \, dt = \int_Q f\phi P \, dx \, dt. \end{aligned} \right\} \quad (3.25)$$

It follows from (3.12) that

$$\int_Q f\phi P \, dx \, dt = - \int_Q uP \frac{\partial \phi}{\partial t} \, dx \, dt - \int_Q \frac{\partial \xi_i}{\partial x_i} \phi P \, dx \, dt$$

and using this formula in (3.25) gives

$$- \mathcal{M}(a_{ij} \frac{\partial w}{\partial y_i}) \int_Q \frac{\partial \phi}{\partial x_j} u \, dx \, dt = \int_Q \xi_i \phi \frac{\partial P}{\partial x_i} \, dx \, dt$$

$\forall \phi \in \mathcal{D}(Q)$, i.e.

$$\xi_i \frac{\partial P}{\partial x_i} = \mathcal{M}(a_{ij} \frac{\partial w}{\partial y_i}) \frac{\partial u}{\partial x_j} . \quad (3.26)$$

We now choose

$$P(y) = y_i \quad (3.27)$$

and we observe that the corresponding value of w equals

$$- (\chi^{i*} - y_i)$$

if we define χ^{i*} by

$$A_y^*(\chi^{i*} - y_i) = 0 \, , \; \chi^{i*} \text{ Y-periodic.} \quad (3.28)$$

Therefore

$$\xi_i = + \tilde{q}_{ij} \frac{\partial u}{\partial x_j} \quad (3.29)$$

where

$$\tilde{q}_{ij} = \mathcal{M}(a_{kj} \frac{\partial}{\partial y_k} (y_i - \chi^{i*})). \quad (3.30)$$

Replacing ξ_i by its value (3.29) in (3.12) gives the homogenized equation satisfied by u. ■

Remark 3.5 : The idea of this proof is due to L. Tartar [1].■

It remains to verify that we obtain the same result than in (2.28). We have to show that

$$\int_Y a_{ij}(y)dy - \int_Y a_{ik}(y) \frac{\partial \chi_j}{\partial y_k} dy = \int_Y a_{ij}(y)dy -$$

$$- \int_Y a_{kj}(y) \frac{\partial \chi^i}{\partial y_k} dy.$$

If we set

$$P_1(\phi,\psi) = \int_Y a_{ij}(y) \frac{\partial \phi}{\partial y_j} \frac{\partial \psi}{\partial y_i} dy$$

it amounts to showing that

$$P_1(\chi^j, y_i) = P_1^*(\chi^{i*}, y_i). \tag{3.31}$$

But (3.28) gives $P_1^*(y_i,\chi^j) = P_1(\chi^j,y_i) = P_1^*(\chi^{i*},\chi^j) = P_1(\chi^j,\chi^{i*}) =$ $= P_1(y_j,\chi^{i*})$ (by using (2.23)) $= P_1^*(\chi^{i*},y_j)$.■

3.2 Problem 1.2.(1).

We now justify (2.40).
We use a method similar to the method of Section 3.1.
We use now Q_1 and Q_1^* (cf.(2.7)) and we define $w = w(y,\tau)$ (τ = parameter) such that

$$\left.\begin{aligned} &Q_1^* w = 0 , \\ &w - P(y) \text{ is Y-periodic} \end{aligned}\right\} \tag{3.32}$$

where P is as in (3.16). We remark that w is automatically periodic in τ (period τ_0), provided we choose the "additive constant" conveniently (for instance we uniquely define w by $\int_Y w(y,\tau)dy=0$).

We introduce next

$$w_\varepsilon = \varepsilon w \left(\frac{x}{\varepsilon} , \frac{t}{\varepsilon} \right) \tag{3.33}$$

and we observe that

$$(A^\varepsilon)^* w_\varepsilon = 0. \tag{3.34}$$

We obtain again (3.22). After integration by parts, it follows that :

$$\left.\begin{aligned} &-\int_Q u_\varepsilon w_\varepsilon \frac{\partial\phi}{\partial t}\, dx\, dt - \int_Q u_\varepsilon \phi \left.\frac{\partial w}{\partial\tau}\right|_{x/\varepsilon, t/\varepsilon} dx\, dt + \\ &+ \int_Q \xi_i^\varepsilon \frac{\partial\phi}{\partial x_i} w_\varepsilon\, dx\, dt - \int_Q (a_{ij}(y,\tau) \frac{\partial w}{\partial y_i}(y,\tau)|_{x/\varepsilon, t/\varepsilon} \\ &\qquad \ldots \frac{\partial\phi}{\partial x_j} u_\varepsilon\, dx\, dt = \int_Q f\phi w_\varepsilon\, dx\, dt. \end{aligned}\right\} \quad (3.35)$$

But let us denote by $\mathcal{M}_{y,\tau}$ the mean value in y in τ :

$$\mathcal{M}_{y,\tau}(\phi) = \frac{1}{|Y|\tau_o} \iint_{Y\times(o,\tau_o)} \phi(y,\tau)dy\, d\tau .$$

We have

$$\left.\frac{\partial w}{\partial\tau}\right|_{x/\varepsilon, t/\varepsilon} \to \mathcal{M}_{y,\tau}\left(\frac{\partial w}{\partial\tau} \right) = 0 \text{ in } L^2(Q) \text{ weakly} \quad (3.36)$$

so that (3.35) gives

$$-\int_Q u\, P \frac{\partial\phi}{\partial t}\, dx\, dt + \int_Q \xi_i \frac{\partial\phi}{\partial x_i} P\, dx\, dt -$$

$$-\mathcal{M}_{y,\tau}(a_{ij} \frac{\partial w}{\partial y_i}) \int_Q \frac{\partial\phi}{\partial x_j} u\, dx\, dt = \int_Q f\,\phi\, P\, dx\, dt$$

and we obtain, as in (3.26) :

$$\xi_i \frac{\partial P}{\partial x_i} = \mathcal{M}_{y,\tau}(a_{ij} \frac{\partial w}{\partial y_i}) \frac{\partial u}{\partial x_j} . \quad (3.37)$$

We introduce $\mathcal{O}^{i*}$ as the solution (such as <u>for instance</u> $\int_Y \mathcal{O}^{i*} dy = 0$) of

$$Q_1^*(\mathcal{O}^{i*} - y_i) = 0. \quad (3.38)$$

Then

$$\xi_i = \mathcal{M}_{y,\tau}(a_{kj}(y,\tau) \frac{\partial}{\partial y_k} (y_i - \mathcal{O}^{i*})) \frac{\partial u}{\partial x_j} \quad (3.39)$$

and one verifies, as in the end of Section 3.1, that the formula so obtained is in fact identical to (2.40).

3.3. Problem 1.2.(2).

We now use R_1 and R_1^* (cf. (2.9)). We still consider $P = P(y)$ as in (3.16) and we define $w = w(y,\tau)$ as the solution (defined up to an additive constant) of

$$\left.\begin{aligned} & R_1^* \, w = 0, \\ & w - P(y) \text{ is } Y\text{-}\tau_o \text{ periodic} \end{aligned}\right\} \tag{3.40}$$

i.e., if

$$w - P = - \chi, \tag{3.41}$$

then

$$\left.\begin{aligned} & - \frac{\partial \chi}{\partial \tau} - \frac{\partial}{\partial y_i} \left(a_{ij}^*(y,\tau) \frac{\partial \chi}{\partial y_j}\right) = - \frac{\partial}{\partial y_i} \left(a_{ij}^*(y,\tau) \frac{\partial P}{\partial y_j}\right), \\ & \chi \text{ is } Y\text{-periodic}, \\ & \chi|_{\tau=0} = \chi|_{\tau=\tau_o} . \end{aligned}\right\} \tag{3.42}$$

We introduce next

$$w_\varepsilon(x,t) = \varepsilon w \left(\frac{x}{\varepsilon} , \frac{t}{\varepsilon^2} \right) \tag{3.43}$$

so that

$$- \frac{\partial w_\varepsilon}{\partial t} + (A^\varepsilon)^* \, w_\varepsilon = 0. \tag{3.44}$$

We take $\phi \in \mathcal{D}(Q)$ and we multiply (1.8) by ϕw_ε and (3.44) by $+ \phi u_\varepsilon$; after substracting, we obtain

$$\left.\begin{aligned} & \int_Q \phi \frac{\partial}{\partial t} (u_\varepsilon \, w_\varepsilon) dt + \int_Q \xi_i^\varepsilon \frac{\partial \phi}{\partial x_i} w_\varepsilon \, dx \, dt - \\ & - \int_Q \left(a_{ij}(y,\tau) \frac{\partial w}{\partial y_i} (y,\tau)\right)_{x/\varepsilon, t/\varepsilon^2} \frac{\partial \phi}{\partial x_j} u_\varepsilon \, dx \, dt = \\ & \qquad = \int_Q f \, \phi \, w_\varepsilon \, dx \, dt. \end{aligned}\right\} \tag{3.45}$$

The 1st integral in (3.45) equals $-\int_Q u_\varepsilon w_\varepsilon \frac{\partial \phi}{\partial t} dx\, dt$ so that (3.45) gives in the limit

$$-\int_Q u\, P \frac{\partial \phi}{\partial t} dx\, dt + \int_Q \xi_i \frac{\partial \phi}{\partial x_i} P\, dx\, dt -$$

$$- \mathcal{M}_{y,\tau}(a_{ij} \frac{\partial w}{\partial y_i}) \int_Q \frac{\partial \phi}{\partial x_j} u\, dx\, dt = \int_Q f\, \phi\, P\, dx\, dt$$

and we obtain, as in (3.37)

$$\xi_i \frac{\partial P}{\partial x_i} = \mathcal{M}_{y,\tau}(a_{ij} \frac{\partial w}{\partial y_i}) \frac{\partial u}{\partial x_j} . \tag{3.46}$$

We introduce ϕ^{i*} as the solution (defined up to an additive constant) of

$$R_1^*(\phi^{i*} - y_i) = 0 , \quad \phi^{i*} \quad Y\text{-}\tau_o \text{ periodic} \tag{3.47}$$

and we obtain

$$\xi_i = \mathcal{M}_{y,\tau}(a_{kj}(y,\tau) \frac{\partial}{\partial y_k} (y_i - \phi^{i*})) \frac{\partial u}{\partial x_j} . \tag{3.48}$$

<u>One verifies</u>, as in the end of Section 3.1, <u>the identity of this formula with</u> (2.50).

3.4. Problem 1.2.(3).

We are now going to justify (2.67). This justification seems to be much more delicate than the previous ones. Actually, we shall need [1] an additional assumption :

$$a_{ij} \in C^o(\tilde{\overline{Y}}) = C^o(\overline{Y} \times [o,\tau_o]) \quad \forall i,j. \tag{3.49}$$

We first verify :

<u>Lemma 3.1.</u> - <u>If we assume that</u> (1.7) <u>and</u> (3.49) <u>hold true, one does not restrict the generality in proving</u> (1.14)...(1.17) <u>and</u> (2.67), <u>by assuming that</u>

$$a_{ij} \in C^\infty(\overline{Y} \times [o,\tau_o]) , \quad a_{ij} \quad Y\text{-}\tau_o \text{ \underline{periodic} (in } C^\infty \text{ sense)} \tag{3.50}$$

(and the a_{ij}'s satisfy (1.7)).

[1] At least in <u>the proof</u>. This assumption is probably not necessary. Cf. remark at the end of the Chapter.

Proof : Let us consider a sequence of functions $a_{ij}^{(\beta)}$ which satisfy (3.50)(1.7) (may be with a slightly smaller ellipticity constant) and which are such that

$$\|a_{ij}^{(\beta)} - a_{ij}\|_{C^o(Y)} \to 0 \quad \text{as } \beta \to \infty . \tag{3.51}$$

We denote by $u_\varepsilon^{(\beta)}$ the solution of Problem 1.2(3) corresponding to $a_{ij}^{(\beta)}$.

We denote by $A^{\varepsilon\beta}$ the operator corresponding to $a_{ij}^{(\beta)}(\frac{x}{\varepsilon}, \frac{t}{\varepsilon^3})$ and by $a^{\varepsilon\beta}$ the corresponding bilinear form. We have

$$(\frac{\partial u_\varepsilon}{\partial t}, v) + a^\varepsilon(u_\varepsilon, v) = (f,v) \quad \forall v \in V ,$$

$$(\frac{\partial u_\varepsilon^{(\beta)}}{\partial t}, v) + a^{\varepsilon\beta}(u_\varepsilon^{(\beta)}, v) = (f,v) \quad \forall v \in V.$$

If we set

$$u_\varepsilon^{(\beta)} - u_\varepsilon = m_\varepsilon ,$$

we have

$$(\frac{\partial m_\varepsilon}{\partial t}, v) + a^{\varepsilon\beta}(m_\varepsilon, v) = \int_\Omega (a_{ij}^\varepsilon - a_{ij}^{\varepsilon\beta}) \frac{\partial u_\varepsilon}{\partial x_j} \frac{\partial v}{\partial x_i} dx$$

hence it follows that

$$\frac{1}{2}|m_\varepsilon(t)|^2 + \alpha_1 \int_o^t \|m_\varepsilon(s)\|^2 ds \leq$$

$$\leq C_{i,j} \|a_{ij} - a_{ij}^{(\beta)}\|_{C^o(\tilde{\bar{Y}})} \int_o^t \|u_\varepsilon\| \, \|m_\varepsilon\| ds$$

$$\leq C \sup_{i,j} \|a_{ij} - a_{ij}^{(\beta)}\|_{C^o(\tilde{\bar{Y}})} (\int_o^t \|m_\varepsilon(s)\|^2 ds)^{1/2}$$

(since we have (3.5)), so that

$$\|u_\varepsilon^{(\beta)} - u_\varepsilon\|_{L^2(0,T;V)} \leq C \sup_{i,j} \|a_{ij} - a_{ij}^{(\beta)}\|_{C^o(\tilde{\bar{Y}})} . \tag{3.52}$$

If we now introduce $\overline{A}^{(\beta)}$ by (compare to (2.61))

$$\overline{A}^{(\beta)} = - \frac{\partial}{\partial y_i} \left(\int_o^{\tau_o} a_{ij}^{(\beta)}(y,\tau)d\tau \frac{\partial}{\partial y_j} \right)$$

and if we define $\psi^{j(\beta)}$ as in (2.63) (with $\overline{A}$ replaced by $\overline{A}^{(\beta)}$), we introduce $\mathcal{A}^{(\beta)}$ as in (2.67) ; if we denote by q_{ij} (resp. $q_{ij}^{(\beta)}$) the coefficients of $\mathcal{A}$ (in (2.67)) and of $\mathcal{A}^{(\beta)}$ respectively, we have

$$|q_{ij}-q_{ij}^{(\beta)}| \to 0. \tag{3.53}$$

If we denote by $u^{(\beta)}$ the solution of

$$\frac{\partial u^{(\beta)}}{\partial t} + \mathcal{A}^{(\beta)} u^{(\beta)} = f$$

subject to $u^{(\beta)} = 0$ on Σ and $u^{(\beta)}(x,o) = u_o(x)$, we have as in (3.52) :

$$\|u^{(\beta)}-u\|_{L^2(0,T;V)} \leq C \sup. |q_{ij}-q_{ij}^{(\beta)}| . \tag{3.54}$$

By virtue of (3.52)(3.54), it suffices to show that, when $\varepsilon \to 0$,

$$u_\varepsilon^{(\beta)} \to u^{(\beta)} \text{ in } L^2(0,T;V) \text{ weakly,}$$

β fixed,

and the lemma is proved. ■

We assume from now on in this Section that the a_{ij}'s satisfy (1.7) and (3.50).

The general idea is the same that in the preceding sections, but the construction of functions analogous to functions w_ε as before is now more complicated. We are going to construct, for every $i = 1,\ldots,n$, functions M_ε^i, g_ε^i , such that

$$\left.\begin{array}{l} M_\varepsilon^i = \varepsilon\alpha^i(y) + \varepsilon^2 \beta^i(y,\tau) + \varepsilon^3 \gamma^i(y,\tau), \\ g_\varepsilon^i = h^i(y,\tau) + \varepsilon k^i(y,\tau) + \varepsilon^2 \ell^i(y,\tau), \\ \alpha^i,\beta^i,\ldots,\ell^i \text{ functions which are } C^\infty \text{ in } y \in \bar{Y} \text{ and} \\ \text{in } y,\tau \in \bar{Y}\times[o,\tau_o] \text{ and which are periodic (in the} \\ C^\infty \text{ sense on the Torus),} \end{array}\right\} \quad (3.55)$$

$$(-\frac{\partial}{\partial t} + (A^\varepsilon)^*)(M_\varepsilon^i - \varepsilon y_i) = \varepsilon g_\varepsilon^i \text{ in } \Omega\times]0,T[\;; \quad (3.56)$$

in a second step, we will use the function M_ε^i to derive the necessary formula for ξ_i. ■

With notations (2.10)(2.11), we have

$$(- \frac{\partial}{\partial t} + (A^\varepsilon)^*) = (P^{\varepsilon,3})^* = \varepsilon^{-3}S_1^* + \varepsilon^{-2}S_2^* + \varepsilon^{-1}S_3^* + S_4^* \quad (3.57)$$

and we identify the different powers of ε in (3.56).

We obtain in this manner for the ε^{-2} term

$$S_1^*(\alpha^i - y_i) = 0 \quad \text{i.e.}$$

α^i does not depend on τ, an hypothesis already made in (3.55). Next terms in the identification give

$$S_1^*\beta^i + S_2^*(\alpha^i - y_i) = 0, \quad (3.58)$$

$$S_1^*\beta^i + S_2^*\beta^i + S_3^*(\alpha^i - y_i) = 0, \quad (3.59)$$

$$S_2^*\gamma^i + S_3^*\beta^i + S_4^*(\alpha^i - y_i) = h^i , \quad (3.60)$$

$$S_3^*\gamma^i + S_4^*\beta^i = k^i, \quad (3.61)$$

$$S_4^*\gamma^i = \ell^i. \quad (3.62)$$

The computation of β^i from (3.58) is possible iff

$$\int_o^{\tau_o} S_1^*\beta^i \, d\tau = 0.$$

But using notation (2.61),

$$\int_0^{\tau_0} S_2^* \, d\tau = \overline{A}^* \tag{3.63}$$

so that we obtain

$$\overline{A}^*(\alpha^i - y_i) = 0, \; \alpha^i \text{ Y-periodic}, \; \alpha^i = \alpha^i(y), \tag{3.64}$$

which defines α^i up to an additive constant chosen independent of x and t. We have

$$\alpha^i \in C^\infty(\overline{Y}), \; \alpha^i \text{ periodic in the } C^\infty \text{ sense.} \tag{3.65}$$

We obtain then from (3.58) that

$$\beta^i(x,y,t,\tau) = \int_0^\tau S_2^*(\sigma)d\sigma \; (\alpha^i - y_i) + \tilde{\beta}^i(x,y,t). \tag{3.66}$$

The compatibility condition in (3.59) is $\int_0^{\tau_0} S_1^* \, \gamma^i \, d\tau = 0$; it gives, since $S_3^*(\alpha^i - y_i) = 0$:

$$\overline{A}^* \, \tilde{\beta}^i + \int_0^{\tau_0} S_2^*(\tau)d\tau \int_0^\tau S_2^*(\sigma)d\sigma \; (\alpha^i - y_i) = 0 \tag{3.67}$$

which defines $\tilde{\beta}^i = \tilde{\beta}^i(y)$ up to an additive constant (we choose all these additive constants independent of x and t).

Therefore

$$\beta^i = \beta^i(y,\tau) \in C^\infty(\overline{Y} \times [0,\tau_0]) \; , \; \text{Y-}\tau_0 \text{ periodic}, \tag{3.68}$$

and

$$\alpha^i = \int_0^\tau S_2^*(\sigma) \; \beta^i(y,\sigma)d\sigma + \tilde{\gamma}^i(x,y,t). \tag{3.69}$$

We <u>choose</u> $\tilde{\gamma}^i = 0$ so that

$$\left.\begin{aligned} &\gamma^i = \gamma^i(y,\tau) = \int_0^\tau S_2^*(\sigma) \; \beta^i(y,\sigma)d\sigma \in C^\infty(\overline{Y} \times [0,\tau_0]), \\ &\text{Y-}\tau_0 \text{ periodic.} \end{aligned}\right\} \tag{3.70}$$

We have then (3.60)(3.61)(3.62) as <u>definitions</u> for h^i, k^i, ℓ^i and by virtue of (3.65)(3.68)(3.70) we obtain (3.55)(3.56). ■

Let us set for a moment

$$w_\varepsilon = M_\varepsilon^i - \varepsilon y_i . \qquad (3.71)$$

We take $\phi \in \mathcal{D}(Q)$, we multiply (1.8) by ϕw_ε and we multiply (3.56) by ϕu_ε ; after substraction we obtain

$$\begin{aligned} &\int_Q \phi \frac{\partial}{\partial t}(u_\varepsilon w_\varepsilon) dx\, dt + \int_Q \xi_i^\varepsilon \frac{\partial \phi}{\partial x_i} w_\varepsilon \, dx\, dt - \\ &- \int_Q a_{ij}^\varepsilon \frac{\partial w_\varepsilon}{\partial x_i} \frac{\partial \phi}{\partial x_j} u_\varepsilon \, dx\, dt = \int_Q f\, \phi\, w_\varepsilon \, dx\, dt. \end{aligned} \qquad (3.72)$$

In order to avoid confusion of indices, we set for a moment

$$\left.\begin{aligned} P(y) = y_i \ , \ P = P(x), \ \alpha^i(y) = \alpha(y), \ \beta^i(y) = \beta(y) \ , \\ \gamma^i(y) = \gamma(y). \end{aligned}\right\} \qquad (3.73)$$

We observe that

$$\frac{\partial w_\varepsilon}{\partial x_i} = \left(\frac{\partial(\alpha - P)}{\partial y_i} + \varepsilon \frac{\partial \beta}{\partial y_i} + \varepsilon^2 \frac{\partial \gamma}{\partial y_i} \right) \left(\frac{x}{\varepsilon}, \frac{t}{\varepsilon^3} \right)$$

so that

$$\left.\begin{aligned} &\int_Q a_{ij}^\varepsilon \frac{\partial w_\varepsilon}{\partial x_i} \frac{\partial \phi}{\partial x_j} u_\varepsilon \, dx\, dt \rightarrow \\ &\rightarrow \mathcal{M}_{y,\tau}\left(a_{ij} \frac{\partial}{\partial y_i}(\alpha - P)\right) \int_Q \frac{\partial \phi}{\partial x_j} u \, dx\, dt \end{aligned}\right\} \qquad (3.74)$$

and (3.72) gives (since $w_\varepsilon \rightarrow -P$ in $L^2(Q)$)

$$\begin{aligned} &+ \int_Q u\, P \frac{\partial \phi}{\partial t} dx\, dt - \int_Q \xi_i \frac{\partial \phi}{\partial x_i} P \, dx\, dt - \\ &- \mathcal{M}_{y,\tau}\left(a_{ij} \frac{\partial}{\partial y_i}(\alpha - P)\right) \int_Q \frac{\partial \phi}{\partial x_j} u \, dx\, dt = - \int_Q f\, \phi\, P \, dx\, dt \end{aligned}$$

hence it follows that

$$\xi_i \frac{\partial P}{\partial x_i} = - \mathcal{M}_{y,\tau}\left(a_{ij} \frac{\partial}{\partial y_i}(\alpha - P)\right) \frac{\partial u}{\partial x_j} .$$

Using now (3.73) we obtain :

$$\xi_i = \left[\mathcal{M}_{y,\tau}(a_{ij}) - \mathcal{M}_{y,\tau}(a_{kj}(y,\tau) \frac{\partial \alpha^i}{\partial y_k}(y)) \right] \frac{\partial u}{\partial x_j} \quad \blacksquare \qquad (3.75)$$

It only remains to see the identity of this formula with (2.67). But this amounts to showing that

$$\overline{A}^*(\alpha^i, y_j) = \overline{A}(\psi^j, y_i) \qquad (3.76)$$

where $\overline{A}(\phi,\psi)$ = bilinear form on $H^1(Y)$ associated to $\overline{A}$. The equality (3.76) is verified as for (3.31).■

Remark 3.6 : Correctors in the asymptotic expansions and non linear problems for evolution operators with highly oscillating coefficients are studied in Bensoussan-Lions-Papanicolaou [1]. ■

Remark 3.7 : The situation of Problem 1.2.(3) is a variant of Colombini-Spagnolo [1], Th. 5.13., where the methods are entirely different. (Regularity hypothesis are stronger in Colombini--Spagnolo). ■

Remark 3.8 : One can study the "general case of coefficients

$$a_{ij}\left(\frac{x}{\varepsilon}, \frac{x}{\varepsilon^2}, \dots, \frac{x}{\varepsilon^N}, \frac{t}{\varepsilon}, \frac{t}{\varepsilon^2}, \dots, \frac{t}{\varepsilon^M}\right).$$

Cf. B.L.P. [1], where one will also find the case of systems of operators or of operators of any order.

We also refer to this book for the study of related problems for hyperbolic systems.■

BIBLIOGRAPHY OF CHAPTER 2.

[1] A. Bensoussan, J.L. Lions and G. Papanicolaou - Book, to appear (North Holland).
[2] Notes CRAS, Cf. CRAS, 282 (1976), p. 143-147 and the bibliography therein.
[1] F. Colombini and S. Spagnolo - Sur la convergence de solutions d'équations paraboliques. To appear.
[1] L. Tartar - To appear.

Additional remark.

Using an idea of Magenes (personal communication) and estimates of Meyers and Pulvirenti one can show the result of Section 3.4 under the only hypothesis that $a_{ij} \in L^\infty(y \times (0,\tau_0))$ (by an improvement of Lemma 3.1). This proof will be given in the book of Bensoussan, Lions and Papanicolaou.

CHAPTER III

OPTIMAL CONTROL AND GENERALIZED SOLUTIONS

1. BOUNDARY CONTROL.

1.1. Orientation.

In this Chapter we briefly indicate how optimal control of distributed parameter systems leads to the necessary introduction of "generalized solutions" (or solutions with "unbounded energy") of evolution equations.

We confine ourselves in this Chapter to three families of examples ; for technical details and many other examples we refer to Lions--Magenes [1], Lions [1].

1.2. An example of boundary control.

We consider the state equation of the system to be given by the heat equation

$$\frac{\partial y}{\partial t} - \Delta y = f \text{ in } Q = \Omega \times]0,T[, \tag{1.1}$$

$$y = v \text{ on } \Sigma = \Gamma \times]0,T[, \tag{1.2}$$

$$y(x,o) = y_o(x) \text{ on } \Omega. \tag{1.3}$$

In (1.2) v is the control function.

In general in the applications v is submitted to constraints of the type

$$0 \le v \le M \quad \text{a.e. on } \Sigma, \tag{1.4}$$

which do not assume any regularity on v.

Let us assume only that

$$v \in L^2(\Sigma) \tag{1.5}$$

and let us denote, in a formal manner for the time being, by

$$y(v) = y(x,t;v) = \text{solution of (1.1)(1.2)(1.3) when } v \text{ satisfies (1.5).} \tag{1.6}$$

Let the cost function be given by [1]

$$J(v) = \int_Q |y(v) - z_d|^2 \, dx\, dt + N \int_\Sigma v^2 d\Sigma \tag{1.7}$$

where z_d is given in $L^2(Q)$ and where N is given > 0.

The problem of optimal control is to find

$$\left.\begin{array}{l} \inf. J(v) \;,\; v \in \mathcal{U}_{ad}, \\ \mathcal{U}_{ad} = \text{closed convex subset of } L^2(\Sigma) \text{ (for instance} \\ \mathcal{U}_{ad} \text{ can be given by (1.4)).} \end{array}\right\} \tag{1.8}$$

In order to make this problem precise, one has to prove the following properties :

(i) to define y(v) as the solution of (1.1)(1.2)(1.3) when v satisfies (1.5) and to prove that $y(v) \in L^2(Q)$;
(ii) that $v \to y(v)$ is continuous from $L^2(\Sigma) \to L^2(Q)$. ∎

Once (ii) is proved, it is clear that problem (1.8) admits a unique solution, since $v \to J(v)$ is then a strictly convex, continuous function on $\mathcal{U}_{ad}$ and since $J(v) \to +\infty$ as $||v||_{L^2(\Sigma)} \to \infty$.

In order to define y(v) we use the transposition method that we now briefly recall (cf. L.M. [1] and the exposition of Magenes [1] in these proceedings, for transposition in Gevrey classes).

---

(1) This is still formal, for the time being.

1.3. Transposition method.

Let us consider the adjoint equation

$$\left.\begin{aligned} & -\frac{\partial\phi}{\partial t} - \Delta\phi = g \text{ in } Q, \\ & \phi = 0 \text{ on } \Sigma, \\ & \phi(x,T) = 0 \text{ on } \Omega. \end{aligned}\right\} \quad (1.9)$$

If $g \in L^2(Q)$ then (1.9) admits a unique solution which satisfies :

$$\left.\begin{aligned} & \phi \in L^2(0,T;H^2(\Omega) \cap H_o^1(\Omega)) \ , \ \frac{\partial\phi}{\partial t} \in L^2(Q), \\ & \phi(x,T) = 0 \ . \end{aligned}\right\} \quad (1.10)$$

The proof of (1.10) is quite simple ; one uses results of Chapter 1 and another "energy estimate" obtained by multiplying (1.9) by $\frac{\partial\phi}{\partial t}$; then $\frac{\partial\phi}{\partial t} \in L^2(Q)$ and $\phi \in L^2(0,T;H^2(\Omega))$ follows from classical estimates for elliptic equations.

We denote by X the space of functions ϕ satisfying (1.10) ; it is a Hilbert space for the norm

$$\left(\int_o^T \left[\|\phi(t)\|^2_{H^2(\Omega)} + \|\frac{\partial\phi}{\partial t}\|^2_{L^2(\Omega)} \right] dt \right)^{1/2} = \|\phi\|_X .$$

We have then :

$$\left.\begin{aligned} & \phi \to -\frac{\partial\phi}{\partial t} - \Delta\phi \text{ is an isomorphism from} \\ & X \to L^2(Q) \end{aligned}\right\} \quad (1.11)$$

and by transposition we obtain :

$$\left.\begin{aligned} & \text{given } L \in X' = \text{dual space of } X, \text{ there exists a unique} \\ & \text{function } y = y(L) \in L^2(Q) \text{ such that} \\ & (y \ , -\frac{\partial\phi}{\partial t} - \Delta\phi) = L(\phi) \quad \forall \phi \in X \end{aligned}\right\} \quad (1.12)$$

where (,) denotes the scalar product in $L^2(Q)$), and

$$L \to y(L) \text{ is a continuous mapping from } X' \to L^2(Q). \blacksquare \quad (1.13)$$

<u>Application</u> : We now <u>choose</u> L in the following form :

$$L(\phi) = \int_Q f\,\phi\,dx\,dt - \int_\Sigma v\,\frac{\partial\phi}{\partial n}\,d\Sigma + \int_\Omega y_o(x)\phi(x,o)dx \tag{1.14}$$

where

$$f \in L^2(0,T;H^{-1}(\Omega)), \tag{1.15}$$

$$v \in L^2(\Sigma)\ , \tag{1.16}$$

$$y_o \in H^{-1}(\Omega). \tag{1.17}$$

Since $\phi \to \frac{\partial\phi}{\partial n}$ is a continuous mapping from $X \to L^2(\Sigma)$ (one has even a better result, cf. Remark 1.1 below), it follows that L given by (1.14) is indeed in X' and therefore <u>there is a unique function</u> $y = y(f,v,y_o) \in L^2(Q)$ which satisfies (1.12) (1.14) and the mapping

$$f,v,y_o \to y(f,v,y_o)$$

is continuous from $L^2(0,T;H^{-1}(\Omega)) \times L^2(\Sigma) \times H^{-1}(\Omega) \to L^2(Q)$.

Il we fix f and y_o, we write

$$y(f,v,y_o) = y(v) \tag{1.18}$$

and $v \to y(v)$ is (affine) continuous from $L^2(\Sigma) \to L^2(Q)$.

<u>It remains to show that</u> y(v) <u>satisfies</u> - in some sense - (1.1) (1.2) (1.3). ■

If in (1.12) we take $\phi \in \mathcal{D}(Q)$, we obtain (1.1). Therefore $y \in L^2(Q)$ and $\frac{\partial y}{\partial t} - \Delta y \in L^2(0,T;H^{-1}(\Omega))$. One can show that, under these hypothesis, (cf. L.M. [1]), one can define the <u>traces</u> $y|_\Sigma$ and $y|_{t=0}$ and that one can apply Green's formula. A formal application of Green's formula gives :

$$(y(T),\phi(T)) - (y(o),\phi(o)) - \int_\Sigma \frac{\partial y}{\partial n}\,\phi\,d\Sigma + \int_\Sigma y\,\frac{\partial\phi}{\partial n}\,d\Sigma\ +$$
$$+\ (y, -\frac{\partial\phi}{\partial t} - \Delta\phi) = (f,\phi)$$

and since $\phi(T) = 0$, $\phi|_\Sigma = 0$:

$$(y,-\frac{\partial\phi}{\partial t} - \Delta\phi) = (f,\phi)+(y(o),\phi(o)) - \int_\Sigma y\,\frac{\partial\phi}{\partial n}\,d\Sigma \tag{1.19}$$

and by comparison with (1.14) we obtain (formally, but this can be justified, cf. L.M. [1]) (1.2) (1.3). ■

Remark 1.1 : When ϕ spans χ, $\frac{\partial\phi}{\partial n}$ spans a space which is strictly smaller than $L^2(\Sigma)$ so that the information obtained

"$y(v) \in L^2(Q)$"

is not the best possible. In order to obtain the best possible results, one has to use interpolation theory as in L.M. [1]. ■

2. GEOMETRICAL CONTROL. HADAMARD'S TYPE FORMULAES.

2.1. The domain as "control variable".

In many applications (Optimum design theory) the "control variable" is the domain itself and this leads to many open problems and to the need of working with generalized solutions.

Let us consider $\Omega_o = \Omega \subset \mathbf{R}^n$, with smooth boundary Γ ; let us denote by $\nu(x)$ = unitary normal to $x \in \Gamma$ directed towards the exterior of Ω ; for $\lambda \geq 0$ small enough we define

$$\left.\begin{array}{l} \Gamma_\lambda = \text{variety described by } x+\lambda\alpha(x)\nu(x) \text{ when } x \\ \text{spans } \Gamma, \\ \text{where } \alpha \text{ is a given smooth function on } \Gamma, \end{array}\right\} \tag{2.1}$$

and we define

$$\Omega_\lambda = \text{open set "interior" to } \Gamma_\lambda. \tag{2.2}$$

Given $\lambda > 0$, we denote by y_λ the solution of

$$\frac{\partial y_\lambda}{\partial t} - \Delta y_\lambda = f \text{ in } \Omega_\lambda \times]0,T[, \tag{2.3}$$

where f is given in $L^2(\mathcal{O} \times]0,T[)$, $\Omega_\lambda \subset \mathcal{O}$ for $\lambda \geq 0$ (and, say, $\lambda \leq 1$), y_λ being subject to

$$y_\lambda = 0 \text{ on } \Sigma_\lambda = \Gamma_\lambda \times]0,T[, \tag{2.4}$$

$$y_\lambda(x,o) = y_o(x) \text{ on } \Omega_\lambda, \tag{2.5}$$

y_o given in $L^2(\mathcal{O})$;

$y_\lambda = y_\lambda(x,t)$ is the state of the system and problems of optimum design lead to the need of computing - if it exists - the derivative

$$\frac{d}{d\lambda} y_\lambda|_{\lambda=o} = \dot{y}. \tag{2.6}$$

This is a classical problem, going back to Hadamard [1], for elliptic problems.

2.2. Formal computation of $\dot{y}$.

A formal computation of $\dot{y}$ is easy. If we write (2.4) in the explicit form

$$y_\lambda(x+\lambda\alpha(x)\nu(x)) = 0 \ , \ x \in \Gamma \ ,$$

we have - assuming y_λ smooth enough -

$$y_\lambda(x)+\lambda\alpha(x) \frac{\partial y_\lambda}{\partial\nu}(x) + \ldots = 0$$

and since $y(x) = 0$ if $x \in \Gamma$ (we set $y_{\lambda=o} = y$) :

$$\lambda^{-1} [y_\lambda(x) - y(x)] + \alpha(x) \frac{\partial y_\lambda}{\partial\nu}(x) + \ldots = 0$$

and therefore, letting $\lambda \to 0$,

$$\dot{y} + \alpha \frac{\partial y}{\partial\nu} = 0 \text{ on } \Sigma.$$

Consequently :

$$\frac{\partial\dot{y}}{\partial t} - \Delta\dot{y} = 0 \text{ in } Q = \Omega \times]0,T[, \tag{2.7}$$

$$\dot{y} = -\alpha\frac{\partial y}{\partial\nu} \text{ on } \Sigma, \tag{2.8}$$

$$\dot{y}|_{t=o} = 0. \tag{2.9}$$

In general one has, for solving (2.7) (2.8) (2.9), to use solutions with unbounded energy.

Remark 2.1. : The preceding calculation can indeed be justified.

Remark 2.2. : One has similar formulaes for other parabolic equations and also for hyperbolic equations. If

$$\frac{\partial^2 y_\lambda}{\partial t^2} - \Delta y_\lambda = f \text{ in } \Omega_\lambda \times]0,T[, \tag{2.10}$$

$$y_\lambda = 0 \text{ on } \Sigma_\lambda , \tag{2.11}$$

$$y_\lambda(x,o) = y_o(x) \ , \ \frac{\partial y_\lambda}{\partial t}(x,o) = y_1(x) \text{ in } \Omega_\lambda \tag{2.12}$$

then, with notation (2.6) :

$$\frac{\partial^2 \dot{y}}{\partial t^2} - \Delta \dot{y} = 0 \text{ in } \Omega \times]0,T[, \tag{2.13}$$

$$\dot{y} = -\alpha \frac{\partial y}{\partial \nu} \text{ on } \Sigma , \tag{2.14}$$

$$\dot{y}|_{t=o} = 0 \ , \ \frac{\partial \dot{y}}{\partial t}\Big|_{t=o} = 0 \text{ in } \Omega . \tag{2.15}$$

Remark 2.3 : We refer to Pironneau [1] and to the bibliography therein for problems of optimum design.

BIBLIOGRAPHY TO CHAPTER III

[1] J. Hadamard - Leçons sur le Calcul des Variations. Paris, Hermann 1912.

[1] J.L. Lions - Contrôle Optimal de Systèmes gouvernés par des équations aux dérivées partielles. Dunod, 1968.

[1] J.L. Lions and E. Magenes - Problèmes aux limites non homogènes et applications. Dunod, Vol. 1, 2 (1968), Vol. 3 (1970).

[1] O. Pironneau - Thesis, Paris, 1976.

CHAPTER IV

SINGULAR PROBLEMS OF EVOLUTION

1. AN EQUATION OF MIXED TYPE.

1.1. Orientation.

We recall, in Section 1.2 below, a result due to Baouendi-Grisvard [1] which is relative to an equation of mixed type. We show, in Section 2, how this equation - which is not of the evolution type - can be transformed in an evolution equation of singular type.

1.2. Mixed problem.

We shall use the following notations :

$$\Omega =]-1,1[\quad , \; t \in]0,T[\; ;$$

$$Q = \Omega \times]0,T[\; ;$$

we consider the equation

$$x \frac{\partial u}{\partial t} - \frac{\partial^2 u}{\partial x^2} = f \text{ in } Q \; , \; f \in L^2(Q) \; , \tag{1.1}$$

subject to the boundary conditions of Dirichlet type

$$u(\pm 1,t) = 0 \; ; \tag{1.2}$$

the "initial" conditions are of different type in the region $x > 0$ (where the operator is "parabolic upward in time") and in the region $x < 0$ (where the operator is "parabolic backward in time") :

$$u(x,o) = 0 \text{ for } x > 0 \text{ , } u(x,T) = 0 \text{ for } x < 0. \tag{1.3}$$

It is proven in Baouendi-Grisvard [1] that the problem (1.1)(1.2)(1.3) admits a unique solution which is such that

$$u \in L^2(0,T;H_o^1(\Omega)) \text{ ,} \tag{1.4}$$

$$x \frac{\partial u}{\partial t} \in L^2(0,T;H^{-1}(\Omega)). \tag{1.5}$$

We show in Section 2 below how this problem can be transformed into <u>an evolution equation of singular type</u>.

<u>Remark 1.1</u> : The preceding result readily extends to the equation

$$x \frac{\partial u}{\partial t} + (-1)^m \frac{\partial^{2m} u}{\partial x^{2m}} = f \text{ ,} \tag{1.6}$$

$$\left.\begin{aligned} & u \in L^2(0,T;H_o^m(\Omega)), \\ & x \frac{\partial u}{\partial t} \in L^2(0,T;H^{-m}(\Omega)), \end{aligned}\right\} \tag{1.7}$$

and u satisfying (1.3).

The results of Section 2 also extend to this situation.

2. A SINGULAR EQUATION OF EVOLUTION.

2.1. Invariant imbedding.

We are going to use an idea derived from the invariant imbedding, technique due to Ambarzumian, Chandrasekhar, Bellman (cf. Bellman -Kalaba-Wing [1]), and somewhat similar to the technique used in Lions [1] for obtaining the integro-differential equation of Riccati's type arising in optimal control of distributed systems.

We shall denote by u_+ (resp. u_-) the restriction of u to $x > 0$ (resp. to $x < 0$).

Let h be given in $L_+^2 = L^2(0,1)$; we consider the equation

$$x \frac{\partial \phi}{\partial t} - \frac{\partial^2 \phi}{\partial x^2} = f \ , \ x \in \Omega \ , \ t \in]s,T[\tag{2.1}$$

$$\left.\begin{aligned} &\phi(x,s) = h(x) \text{ if } x > 0, \\ &\phi(x,T) = 0 \quad \text{ if } x < 0, \end{aligned}\right. \tag{2.2}$$

$$\phi(\pm 1,t) = 0 \tag{2.3}$$

This equation admits a unique solution, and therefore $\phi(x,s)$, $x < 0$, is uniquely defined ; we denote by ϕ_+ (resp. ϕ_-) the restriction of ϕ to $x > 0$ (resp. to $x < 0$), and by $\phi_+(s)$ $(\phi_-(s))$ the function $x \to \phi_+(x,s)$ $(x \to \phi_-(x,s))$. Then

$$h \to \phi_-(s)$$

is an affine continuous mapping from $L^2_+ \to L^2_- = L^2(-1,0)$, i.e.

$$\left.\begin{aligned} &\phi_-(s) = P(s)\phi_+(s) + r(s), \\ &P(s) \in \mathcal{L}(L^2_+;L^2_-), \ r \in L^2. \end{aligned}\right\} \tag{2.4}$$

If we take now

$$h(x) = u_+(x,s)$$

then the solution ϕ of (2.1)(2.2)(2.3) is the restriction of u to $\Omega \times]s,T[$ and therefore (2.4) can be written

$$u_-(s) = P(s)u_+(s)+r(s)$$

and since s is arbitrary, we have in fact the identity

$$\left.\begin{aligned} &u_-(t) = P(t)u_+(t)+r(t), \text{ for a.e. } t \in]0,T[\\ &P(t) \in \mathcal{L}(L^2_+;L^2_-) \end{aligned}\right. \tag{2.5}$$

and if we define

$$\left.\begin{aligned} &F_+ = \{v | v \in H^1(0,1), \ v(1) = 0\} \ , \\ &F_- = \{v | v \in H^1(-1,0), v(-1) = 0\}, \end{aligned}\right. \tag{2.6}$$

then

$$P(t) \in \mathcal{L}(F_+;F_-) \tag{2.7}$$

$$r \in L^2(0,T;F_-). \tag{2.8}$$

If x is arbitrarily fixed in $]-1,0[$,

$$v \to P(t)v(x)$$

is a continuous linear form on F_+, so that

$$P(t)v(x) = \int_0^1 P(x,\xi,t)v(\xi)d\xi \ , \tag{2.9}$$

$\xi \to P(x,\xi,t)$ being an element of F'_+.

Our goal is now to obtain an equation satisfied by $P(x,\xi,t)$.

2.2. An identification procedure.

We write ϕ' instead of $\frac{\partial\phi}{\partial t}$, $A\phi = -\frac{\partial^2\phi}{\partial x^2}$; we have

$$xu'_- + Au_- = f_- . \tag{2.10}$$

Assumming that the computation is valid (verifications are quite long) we obtain from (2.10), using (2.5) :

$$x(P'u_+ + Pu'_+ + r') + APu_+ + Ar = f_- . \tag{2.11}$$

But

$$xu'_+ + Au_+ = f_+$$

so that (2.11) becomes

$$xP'u_+ + xP(-\frac{1}{\xi}Au_+ + \frac{1}{\xi}f_+) + xr' + APu_+ + Ar = f_- . \tag{2.12}$$

But for fixed t=s, this is valid with $u_+(s) = h$ arbitrary so that (2.12) is equivalent to

$$xP' + xP\ (-\frac{1}{\xi}A) + AP = 0, \tag{2.13}$$

$$xr' + Ar + xP(\frac{1}{\xi}f_+) = f_- . \tag{2.14}$$

If we introduce $M(x,\xi,t)$ defined by

$$P(x,\xi,t) = \xi M(x,\xi,t) \tag{2.15}$$

then (2.13) is equivalent to

$$x\xi\frac{\partial M}{\partial t} + x\frac{\partial^2 M}{\partial\xi^2} - \xi\frac{\partial^2 M}{\partial x^2} = 0 \tag{2.16}$$

and to boundary conditions on which we shall return.

In (2.16) M is defined for $x < 0$, $\xi > 0$. If we change x in -x :

$$N(x,\xi,t) = M(-x,\xi,t), \tag{2.17}$$

then

$$\left.\begin{aligned} & -x\xi \frac{\partial N}{\partial t} - \left(x \frac{\partial^2 N}{\partial \xi^2} + \xi \frac{\partial^2 N}{\partial x^2}\right) = 0 , \\ & x,\xi \in]0,1[^2 , \quad t \in]0,T[. \ \blacksquare \end{aligned}\right\} \tag{2.18}$$

Let us now prove that the kernel N is <u>symmetric</u> :

$$N(x,\xi,t) = N(\xi,x,t). \tag{2.19}$$

Indeed, let us consider the analogous of (2.1)(2.2)(2.3) with $f = 0$, i.e.

$$\begin{aligned} & x \frac{\partial \phi}{\partial t} + A\phi = 0, \\ & \phi_+(s) = h, \ \phi_-(T) = 0, \\ & \phi(\pm 1,t) = 0 ; \end{aligned} \tag{2.20}$$

let us denote by $\hat{\phi}$ the solution of the anologous equation corresponding to $\hat{h}$ instead of h. Then

$$\phi_-(s) = P(s)h , \ \hat{\phi}_-(s) = P(s)\hat{h}. \tag{2.21}$$

We have

$$x\phi'_- + A\phi_- = 0 \tag{2.22}$$

and we multiply (2.22) by $\hat{\phi}_+(-x,t)$ (defined therefore for $x < 0$). We obtain :

$$0 = (x\phi_-(T),\hat{\phi}_+(-x,T))_{L^2_-} - (x\phi_-(s),\hat{\phi}_+(-x,s))_{L^2_-} -$$

$$- \int_s^T \frac{\partial \phi_-}{\partial x}(o,t)\hat{\phi}_+(o,t)dt - \int_s^T \phi_-(o;t) \frac{\partial \hat{\phi}_+}{\partial x}(o,t)dt +$$

$$+ \int_s^T \left(\phi_-, -x \frac{\partial}{\partial t} \hat{\phi}_+(-x,t) - \frac{\partial^2}{\partial x^2} \hat{\phi}_+(-x,t)\right)dt$$

hence it follows that

$$(x\phi_-(s),\hat{\phi}_+(-x,s)_{L^2_-} = -\int_s^T \left[\frac{\partial\phi_-}{\partial x}(o,t)\hat{\phi}_+(o,t)+\phi_-(o,t)\frac{\partial\hat{\phi}_+}{\partial x}(o,t)\right]dt$$

i.e.

$$(xP(s)h\ ,\ \hat{h}(-x))_{L^2_-}\ \underline{\text{is symmetric in}}\ h,\hat{h} \tag{2.23}$$

i.e.

$$\int_{-1}^{o}\int_{o}^{1} P(x,\xi,s)h(\xi)\ x\hat{h}(-x)dx\ d\xi =$$

$$\int_{-1}^{o}\int_{o}^{1} P(x,\xi,s)\hat{h}(\xi)xh(-x)dx\ d\xi \quad \forall h,\hat{h}$$

i.e.

$$xP(-x,\xi,s) = \xi P(-\xi,x,s)$$

i.e. (2.19). ■

Boundary conditions. We should have $P(t)v(-1) = 0\ \forall v \in F_+$ i.e.

$$P(-1,\xi,t) = 0$$

i.e.

$$N(1,\xi,t) = 0 \tag{2.24}$$

and therefore, by virtue of (2.19) :

$$N(x,1,t) = 0. \tag{2.25}$$

On the other hand,

$$P(t)\phi_+(0) = \phi_-(0) = \phi_+(0)$$

i.e.

$$\int_o^1 P(0,\xi,t)\phi_+(\xi)d\xi = \phi_+(0)$$

i.e.

$$\xi N(o,\xi,t) = \delta_\xi(o) \text{ (Dirac measure at the origin)} \tag{2.26}$$

and, by virtue of (2.19)

$$xN(x,o,t) = \delta_x(o). \tag{2.27}$$

Final condition :

$$N(x,\xi,T) = 0. \tag{2.28}$$

CONCLUSION.

The kernel $P(x,\xi,t)$ is given by (2.15)(2.17), where $N(x,\xi,t)$ is the solution of the singular backward parabolic equation (2.18) subject to conditions (2.24)...(2.28).

Remark 2.1 : The function r is given (once M is known) by

$$x\frac{\partial r}{\partial t} + Ar = f_- - x\int_0^1 M(x,\xi,t)f_+(\xi,t)d\xi \,, \tag{2.29}$$

$$r(o,t) = r(1,t) = 0, \tag{2.30}$$

$$r(x,T) = 0 \text{ for } x < 0. \tag{2.31}$$

Remark 2.2 : A direct study of (2.18) subject to (2.24)...(2.28) seems to be an open problem.

Remark 2.3 : One could also consider, by the same kind of technique, the singular evolution equation associated with the equation

$$|x|^p \operatorname{sign} x \frac{\partial u}{\partial t} - \frac{\partial^2 u}{\partial x^2} = f \,, \quad p > -1 \tag{2.32}$$

(we have of course the case already studied if p=1), studied in Talenti [1]. Cf. other examples in Beals [1], Cooper [1].

Remark 2.4 : For operators of evolution with coefficients which are singular for t=0 we refer to Baiocchi-Baouendi [1], Bernardi [1].

BIBLIOGRAPHY OF CHAPTER IV

[1] C. Baiocchi and M.S. Baouendi - Singular evolution equations. To appear.
[1] M.S. Baouendi and P. Grivard - Sur une équation d'évolution changeant de type. J.F.A. (1969).
[1] R. Bellman, R. Kalaba and G.M. Wing - Invariant imbedding and Mathematical Physics. J. Math. Phy. Vol. 1, 1960, p. 280-308.
[1] J.L. Lions - Contrôle optimal de systèmes gouvernés par des équations aux dérivées partielles. Dunod, 1968.
[1] G. Talenti - Equations paraboliques "en avant-en arrière". Colloque Int. CNRS, Pub. SMF, Astérique 2,3, 1973, p. 292-304.
[1] M.B. Bernardi - Su alcune equazioni d'evoluzione Singolari. Boll. U.M.I. 13-B (1976), p. 1-20.
[1] R. Beals - An abstract approach to some scattering and transport problems. To appear.
[1] J. Cooper - J.M.A.A. (1969).
In connection with this reference cf. also.
[1] T. Kato - Proc. Japan Acad. (1973).
and
[1] Mc Intosh. To appear.

TOPICS IN PARABOLIC EQUATIONS: SOME TYPICAL FREE BOUNDARY PROBLEMS

E. Magenes

Università di Pavia and Laboratorio di Analisi Numerica del C.N.R.

INTRODUCTION. It is well-known that in the applications one is frequently faced with free boundary problems for linear partial differential equations (in particular for parabolic type equations). A classical example is that of the "Stefan problem" for the heat equation arising in the ice melting phenomenon. The interest on this sort of problems has recently increased (particularly, after the discovery that they are closely related to the theory of variational inequalities).

The aim of these lectures is to study the relation existing between free boundary value problems and variational inequalities. We do this by means of simple but rather significant models such as free boundary problems for the heat equation in one dimensional space arising from the oxygen diffusion in time, from ice melting and from fluid filtration in porous media respectively.

In section 1 we introduce formally the mathematical models of the three above mentioned physical phenomena. In sections 2, 3 and 4 we specify the mathematical formulations of the problem and reduce them to parabolic variational inequalities. We also show what type of results can be obtained in this way. In sections 5, 6 and 7 we develop the proofs for the third problem that seems to be the most interesting and difficult. In section 8 we deal with the question about the regularity of the free boundary. We close with some remarks and some references to the literature on the subject.

For Table of Contents see page 312.

Garnir (ed.), Boundary Value Problems for Linear Evolution Partial Equations. 239-312.

1. MATHEMATICAL MODELS OF SOME PHYSICAL PROBLEMS

a) First of all let us consider the oxygen diffusion in an absorbing tissue; $u(x,t)$ represents the oxygen concentration (suitably normalized) in the tissue at the time t and at the point x of the tissue (usually the problem is set in one space dimension); we suppose that the oxygen is absorbed at unit rate wherever it is present and there is no diffusion at $x=0$; the initial concentration $g(x)$ is the steady-state concentration, i.e. the concentration when at the surface $x=o$ of the tissue the concentration is maintained constant; then $g(x)$, suitable scaled, is given by $g(x)=\frac{1}{2}(b-x)^2$ where b is the furthest depth of oxygen penetration in the above steady condition; we can generalize g, by assuming as "compatibility conditions" the following hypothesis:

$$(1.1)\quad \begin{cases} g(x)>0 \quad 0\leq x<b,\ g(b)=0 \\ g'(x)\leq 0,\ 0\leq x\leq b,\ g'(b)=0,\ g''(x)\leq 1,\ 0\leq x\leq b \end{cases}$$

If $s(t)$ denotes the location of the "interface" between the region where u is positive and the region where u is zero, at this "interface" also the oxygen flux must be zero. Then we can state the mathematical model of the problem "formally" as follows:

PROBLEM 1: Given $b>0$ and $g(x)$ satisfying (1.1), find $\{T, s(t), u(x,t)\}$ such that

$$(1.2)\quad s(t)>0 \ ,\quad 0\leq t<T\ ;\quad s(0)=b,\ s(T)=0$$

$$(1.3)\quad u_{xx}-u_t=1 \ ,\quad 0<x<s(t)\ ,\ 0<t<T$$

$$(1.4)\quad u(s(t),t)=0 \ ,\quad 0<t\leq T$$

$$(1.5)\quad u_x(s(t),t)=0 \ ,\quad 0<t<T$$

$$(1.6)\quad u_x(0,t)=0 \ ,\quad 0<t<T$$

$$(1.7)\quad u(x,0)=g(x) \ ,\quad 0\leq x\leq b.$$

b) Let us consider now a particular case of the so called Stefan problem: a physical system composed of a "segment" of water, denoted in the mathematical model by an interval $[0,a]$, of the real axis $\mathbb{R}$, and of a thin block of ice, occupying in the mathematical model the interval $]a\ +\infty[$. The system is described by the distribution of the water temperature $u(x,t)$ in the space-time; let $g(x)$, $0\leqslant x\leqslant a$, be the initial distribution of the temperature with the "compatibility condition"

$$(1.8) \quad g(x)>0 \ , \ g(a)=0 \ ; \ g'(x)\leqslant 0;$$

and let us suppose the thermal flux to vanish at the point $x=0$ of our system for every t. Then the ice will begin to melt and for every time $t>0$ the water will occupy an unknown interval $0\leqslant x<s(t)$. The water temperature $u(x,t)$ must satisfy the heat equation in the domain $0<t<T$, $0<x<s(t)$, and the obvious condition $u(s(t),t)=0$ on the "free boundary" $x=s(t)$; moreover on the same free boundary an additional condition is given namely the law of conservation of energy. With a suitable normalization of certain physical constants, the mathematical model of the problem is "formally" the following:

PROBLEM II: Given $T>0$, $a>0$ and $g(x)$ satisfying (1.8) find $\{s(t),u(x,t)\}$ such that

$$(1.9) \quad s(t)>0 \ , \quad 0\leqslant t\leqslant T \ , \quad s(0)=a,$$

$$(1.10) \quad u_{xx}-u_t=0 \ , \quad 0<x<s(t), \quad 0<t<T,$$

$$(1.11) \quad u(s(t),t)=0 \ , \quad 0<t\leqslant T,$$

$$(1.12) \quad u_x(s(t),t)= -s'(t) \ , \quad 0<t\leqslant T,$$

$$(1.13) \quad u_x(0,t)=0 \ , \quad 0<t\leqslant T,$$

$$(1.14) \quad u(x,0)=g(x), \quad 0\leqslant x\leqslant a.$$

REMARK 1.1 Condition (1.12) frequently is formulated in the equivalent form

$$u_t(s(t),t)=u_x^2(s(t),t) \ , \qquad 0<t<T.$$

c) Let us consider finally a compressible fluid moving in an underground vertical pipe, the interior of the pipe consisting of a homogeneous porous medium. The variable x represents the height and $x=0$ and $x=a$, $a>0$ are respectively the bottom and the top of the pipe; the variable t represents always the time, $0\leqslant t\leqslant T$. The function $u(x,t)$ is the piezometric head and $-u_x(x,t)$ is the velocity of the fluid (using Darcy's law); the level of the fluid in the pipe is denoted by $s(t)$ and is the "free boundary". We suppose that the fluid is moving through the bottom of the pipe upward (if $\ell(t)\geqslant 0$), or downward (if $\ell(t)\leqslant 0$) at the rate $|\ell(t)|$, where $\ell(t)$, $0\leqslant t\leqslant T$ is a given function. The potential or piezometric head $u(x,t)$

is supposed known at the time $t=0$: $u(x,0)=g(x)$, $0\leqslant x\leqslant a$. Since $u(x,t)$ is the sum $p(x,t)+x$ where p is the inner pressure and x comes from the gravity, we should have $u(x,t)>x$ if $0\leqslant x<s(t)$; we have then the physical condition on g:

$$(1.15)\quad g(x)>x \quad , \; 0\leqslant x<a \quad , \quad g(a)=a.$$

Physical reasons, depending on the porosity of the medium, suggest to assume also that

$$(1.16)\quad \ell(t)>-1 \quad , \; 0\leqslant t\leqslant T.$$

Then the mathematical model of the phenomenon can be "formally" stated as follows.

PROBLEM III: Given $T>0$, $a>0$, $g(x)$ and $\ell(t)$ satisfying respectively (1.15) and (1.16), find $\{s(t),u(x,t)\}$ such that

$$(1.17)\quad s(t)>0 \quad , \quad 0\leqslant t\leqslant T \quad , \quad s(0)=a,$$

$$(1.18)\quad u_{xx}-u_t=0 \quad , \quad 0<x<s(t), \quad 0<t<T,$$

$$(1.19)\quad u(s(t),t)=s(t) \quad , \quad 0<t\leqslant T,$$

$$(1.20)\quad u_x(s(t),t)=-s'(t) \quad , \quad 0<t\leqslant T,$$

$$(1.21)\quad u_x(0,t)=-\ell(t) \quad , \quad 0<t\leqslant T,$$

$$(1.22)\quad u(x,0)=g(x) \quad , \quad 0\leqslant x\leqslant a.$$

REMARK 1.2 Condition (1.20) frequently is formulated in the equivalent form:

$$u_t(s(t),t)=u_x^2(s(t),t)-u_x(s(t),t).$$

d) Problems I,II,III seem similar; but we shall see that, even if it is possible to solve them by the same methods, they present different difficulties in increasing order from the first to the last. Before giving a precise mathematical formulation of the problems and their reduction to variational inequalities, let us introduce some notations. If A is an open set in $\mathbb{R}^1$ or in $\mathbb{R}^2$, k a positive integer and $1\leqslant p\leqslant\infty$, we denote by $W_p^k(A)$ the usual Sobolev space of the real functions v such that v and their derivatives (in the sense of distributions on A) until the order k are in $L^p(A)$. Denoting by $\bar{A}$ the closure of A, we shall use also the usual Banach space $C^k(\bar{A})$, $k=0,1,\ldots$ of the real functions v which are continuous in $\bar{A}$ with their derivatives until the order k;

$C^{\infty}(\bar{A})$ shall be the space of infinitely derivable functions in $\bar{A}$. In the case of $\mathbb{R}^2$ we shall use also the spaces $W_p^{k,h}(Q)$, where Q is a rectangle, k,h are positive integers, $1\leq p\leq\infty$, of the functions $v(x,t)$ (x,t denoting the coordinates in $\mathbb{R}^2$) belonging to $L^p(Q)$ together with the derivatives (always in the sense of distributions in Q) $\frac{\partial^j v}{\partial x^j}$, $j=1,\dots,k$ and $\frac{\partial^i v}{\partial t^i}$, $i=1,\dots,h$. The space $W_p^k(A)$, $C^k(\bar{A})$, $W_p^{k,h}(A)$ are Banach spaces with respect to their natural norms. In the case of an interval I of $\mathbb{R}^1$ we shall also use the Sobolev spaces $W_p^k(I)$ with k real and positive; if $0<k<1$ it is the space of the functions $v(x)$ which are in $L^p(I)$ and such that

$$\int_I\int_I \frac{|v(x)-v(y)|^p}{|x-y|^{1+kp}}\,dxdy <+\infty$$

In the case of $k>1$ it is the space of the functions v which are in $W_p^{[k]}(I)$, $[k]$=maximum integer less then k and such that $v^{([k])}(x)\in W_p^{k-[k]}(I)$. Let us remember that frequently $W_2^k(A)$, $W_2^{k,h}(Q)$ are designed by $H^k(A)$, $H^{k,h}(Q)$. $D(A)$ and $D'(A)$ denote respectively the usual space of infinitely differentiable functions with compact support in A and its dual, the space of distributions in A. Moreover, if B is a Banach space we shall denote by $L^p(\Theta,T;B)$, $\Theta\leq T$, $1\leq p\leq\infty$, (resp. by $C^0([\Theta,T];B)$) the Banach space of the functions $t\to v(t)$ defined in $[\Theta,T]$ with values in B, strongly measurables and such that $||v(t)||_B$ is a real function in $L^p(]\Theta,T[)$, (resp. in $C^0([\Theta,T])$) with the natural norm. Let us remember that $W_p^{k,h}(Q)$ if $Q=]0,b[\times]\Theta,T[$ may be identified with the space of functions $t\to v(t)$ such that

$$v\in L^p(\Theta,T;W^k(]0,b[)),\ \frac{d^j v}{dt^j}\in L^p(\Theta,T;L^p(]0,b[)),\ j=1,\dots,h.$$

the derivatives being taken in the sense of distributions in $]\Theta,T[$ with values in $L^p(]0,b[)$. Finally we shall denote frequently by $D_x v$, $D_{xx}v$, $D_t v,\dots$ or by v_x, v_{xx}, $v_t,\dots$ the derivatives $\frac{\partial v}{\partial x}$, $\frac{\partial^2 v}{\partial x^2}$, $\frac{\partial v}{\partial t}$, ...; and we denote by $E(v)$ the "heat operator": $E(v)=v_{xx}-v_t$.

In all previous Problems I,II,III, we shall denote by Ω

$$(1.23)\qquad \Omega=\{(x,t);\ 0<x<s(t),\ 0<t<T\}$$

2. THE OXYGEN DIFFUSION IN AN ABSORBING TISSUE

a) Let us define "classical solutions" of Pr. 1:

DEFINITION 2.1: Under the assumption that

(2.0) $g \in C^2([0,b])$ and verifies (1.1)

$\{T,s(t),u(x,t)\}$ is a classical solution of Pr. I if: $T>0$, $s \in C^o([0,T])$ and verifies (1.2), $s'(t)$ is continuous for $0<t<T$, u $C^o(\bar{\Omega}) \cap H^1(\Omega)$, where Ω is given by (1.23), u_x, u_t, u_{xt} are is continuous for $0\leq x\leq s(t)$, $0<t<T$, and (1.3), (1.4), (1.5), (1.6), (1.7) are verified((1.3) in the sense of $D'(\Omega)$) and $u\geq 0$ in Ω.

Definition 2.1 seems to be the "good" definition for the problem of the oxygen diffusion, in the case that the datum $g(x)$ does not permit more regularity for the function u; in fact if for instance we look for u belonging also to $C^1(\bar{\Omega})$ then we have to add a "compatibility condition" on g, since by (1.6) and 1.7) we must have:

$$(2.1)\quad g'(0) = 0$$

This condition is verified in certain diffusion problems. But in the case of the oxygen diffusion we have $g(x) = \frac{1}{2}(b-x)^2$, then (2.1) is not verified. Nevertheless we will suppose in this section a), in order to explain better the relations between Pr. I and variational inequalities, the validity of (2.1) and of

$$(2.2)\quad u \in C^1(\bar{\Omega})$$

Then if $\{T,s(t), u(x,t)\}$ is a classical solution of Pr. 1, setting $v=u_t$ it is easy to prove

$$(2.3)\quad E(v)=0 \qquad \text{in } \Omega$$

$$(2.4)\quad v(s(t),t)=0 \quad , \quad 0<t<T$$

$$(2.5)\quad v_x(s(t),t)=-s'(t) \quad , \quad 0<t<T$$

$$(2.6)\quad v_x(0,t)=0 \quad , \quad 0<t<T$$

$$(2.7)\quad v(x,0)=g''(x)-1 \quad , \quad 0<x<a$$

In fact we can derive with respect to t the equations (1.3), (1.4) and (1.5) and, using also (1.6) and (1.7), we obtain

$$E(u_t)=0 \qquad \text{in } \Omega$$

$$u_x(s(t),t)s'(t)+u_t(s(t),t)=u_t(s(t),t)=0, \quad 0<t<T$$

$$u_{xx}(s(t),t)=1+u_t(s(t),t)=1 \qquad 0<t<T$$

$$u_{xx}(s(t),t)s'(t)+u_{xt}(s(t),t)=s'(t)+u_{xt}(s(t),t)=0, \quad 0<t<T$$

$$u_{xt}(0,t)=0 \qquad 0<t<T$$

$$u_t(x,0)=u_{xx}(x,0)-1=g''(x)-1 \qquad 0<x<a$$

REMARK 2.1: Problem (2.3).....(2.7) is of the same kind as the Stefan problem considered in n. I, b); we shall come back to this remark later.

We can then prove the following
PROPOSITION 2.1 If $\{T,s(t),u(x,t)\}$ is a classical solution satisfying (2.2) (in the hypothesis (2.1))of Pr. I, then

(2.8) $u_t(x,t)\leq 0$, $u_x(x,t)\leq 0$ in $\bar{\Omega}$

(2.9) $u(x,t)>0$ in Ω

(2.10) s is strictly decreasing in $[0,T]$

(2.11) $T\leq g(0)$

PROOF: We have already noted that $v=u_t$ is a solution of the Problems (2.3).....(2.7); moreover $v\in C^0(\bar{\Omega})$, then we can apply the classical "maximum principle" (see [30]): v must take its maximum on the "parabolic boundary" of Ω, i.e. $\partial\Omega - \{(x,t);\ t=T\}$. But this maximum can't be positive: in fact $v(x,0)=g''(x)-1\leq 0$, $0\leq x\leq a$, $v(s(t),t)=0$, $0\leq t\leq T$ so if it would be positive it ought be such on the segment$\{(x,t),\ x=0,\ 0<t<T\}$; but by a strong form of maximum principle (cf.[30]) in the maximum point $(0,\bar{t})$ one should have $v_x(0,\bar{t})<0$, contrary to (2.6). Then $v(x,t)=u_t(x,t)\leq 0$ in $\bar{\Omega}$ and more precisely, noting that v cannot be identically zero, we conclude that $v(x,t)<0$ in Ω, again by the maximum principle. Let us consider now $v=u_x$; we have similarly $E(v)=0$ in Ω, $v(0,t)=0$, $v(s(t),t)=0$, $0\leq t\leq T$, $v(x,0)=g'(x)\leq 0$, $0\leq x\leq a$; then from the maximum principle we deduce that $v(x,t)=u_x(x,t)\leq 0$ in $\bar{\Omega}$ and (2.8) are proved. Now from the Definition 2.1 we have $u(x,t)\geq 0$ in $\bar{\Omega}$; more precisely nothing that $E(u)=1$ in Ω again from the maximum principle we have (2.9). In order to prove (2.10) let us fi rst note that, setting $u_t=v$, we have (2.5); but $v_x(s(t),t)\geq 0$, since $v(s(t),t)=0$ $0<t<T$ (cf. (2.4)) and we know that $v(x,t)\leq 0$ in $\bar{\Omega}$ (cf. (2.8));then $s'(t)\leq 0$, $0<t<T$, and $s(t)$ is decreasing in $[0,T]$. Let us prove now that s is strictly decreasing; indeed, in the contrary case there exist two points t', t'' $(t'<t'')$ such that $s(t')=s(t)$, $t'\leq t\leq t''$. We can now apply the strong form of the maximum principle on the segment $x=s(t)$, $t'\leq t\leq t''$ and, recalling that $v(x,t)<0$ in Ω we obtain $v_x(s(t),t)>0$, i.e. $s'(t)<0$, in contradiction with the assumption that $s(t')=s(t)=s(t'')$, $t'\leq t\leq t''$. Finally, in order to prove (2.11) it is enough to remark that (1.6) and (2.8) imply that $u_{xx}(0,t)\leq 0$, $0<t<T$; then $u_t(0,t)\equiv u_{xx}(0,t)-1\leq -1$, $0<t<T$, so that $u(0,T)-u(0,0)=-g(0)=\int_0^T u_t(0,t)dt\leq -T$, that is, (2.11).

REMARK 2.2 Let us note that the estimate found for the unknown T (cf. (2.11)) depends only on g(0).

Now let us define Q as follows:

$$Q=\{(x,t);\ 0<x<b,\ 0<t<T'=g(0)\} \tag{2.13}$$

and again in the assumptions of Prop. 2.I let us define

$$\tilde{u}(x,t)=u(x,t) \text{ in } \bar{\Omega},\quad \tilde{u}(x,t)=0 \text{ in } \bar{Q}-\bar{\Omega} \tag{2.14}$$

We have

$$\tilde{u}\in C^1(\bar{Q}),\quad \tilde{u}_{xx}\in L^\infty(Q) \tag{2.15}$$

We can compute $E(\tilde{u})$ in the sense of $D'(Q)$ and we have $\forall\phi\in D(Q)$: using (1.5) and (1.3)

$$<E(\tilde{u}),\phi>=-\iint_Q \tilde{u}_x\phi_x dxdt-\iint_Q \tilde{u}_t\phi dxdt=-\iint_\Omega u_x\phi_x dxdt -$$

$$-\iint_\Omega u_t\phi dxdt=\int_0^T\int_0^{s(t)} u_{xx}\phi dxdt-\int_0^T u_x(s(t),t)\phi(s(t),t)dt$$

$$-\iint_\Omega u_t\phi dxdt=\iint_\Omega (u_{xx}-u_t)\phi dxdt=\iint_Q \phi\chi(\Omega)dxdt$$

then

$$E(\tilde{u})=\chi(\Omega)\qquad \text{in } Q\,. \tag{2.16}$$

Finally we have that $\tilde{u}$ satisfies the following system

$$(2.17)\quad \begin{cases} \tilde{u}(x,t)\geq 0 \text{ in } Q, & \tilde{u}(x,t)>0 \text{ in } \Omega, \quad \tilde{u}(x,t)=0 \text{ in } Q-\Omega \\ E(\tilde{u})\in L^{\infty}(Q)\ , & E(\tilde{u})\leq 1 \quad \text{in } Q \\ \{E(\tilde{u})-1\}\tilde{u}=0 & \text{in } Q \\ \tilde{u}_x(0,t)=0\ , & 0<t<T' \\ \tilde{u}(b,t)=0\ , & 0<t<T' \\ \tilde{u}(x,0)=g(x), & 0\leq x\leq b \end{cases}$$

The system (2.17) is a so called "parabolic variational inequality". Before studying it, let us remark that the system (2.17) can be considered "formally" equivalent to Pr.I, in the sense that if we can solve it and prove that the set $\Omega=\{(x,t);\ \tilde{u}(x,t)>0\}$ is of the form (1.23) with s verifying (1.2), then setting $u(x,t)=\tilde{u}(x,t)$ in Ω, we obtain a solution of Pr.I.

b) Now let us introduce a precise formulation of our variational inequality, in a "semistrong" form. Denote by V the Hilbert space

$$(2.18)\quad V=\{v; v\in H^1(]0,b[),\quad v(b)=0\}$$

and set

$$(2.19)\quad (u,v)=\int_0^b u(x)v(x)dx \qquad \forall u,v\in L^2(]0,b[),$$

$$(2.20)\quad a(u,v)=\int_0^b u_x(x)v_x(x)dx \qquad \forall u,v\in V.$$

$$(2.21)\quad J(v)=\int_0^b v^+(x)dx \qquad \forall v\in V$$

where, as usual, $v^+=\sup(v,0)$, $v^-=\sup(-v,0)$. Let us consider again the function $\tilde{u}$ and prove that it verifies the inequality for almost every t in $]0,T'[$

$$(\tilde{u}_t,v-\tilde{u})+a(\tilde{u},v-\tilde{u})+J(v)-J(\tilde{u})\geq 0,\quad \forall v\in V$$

In fact from (2.16) we have a.e. in $]0,T'[$ and $\forall v\in V$

$$(E(\tilde{u}),v-\tilde{u})=-(\tilde{u}_x,(v-\tilde{u})_x)+[\tilde{u}_x\cdot(v-\tilde{u})]_{x=0}^{x=b}-(\tilde{u}_t,v-\tilde{u})=$$
$$=(\chi(\Omega),v-\tilde{u})$$

Then using (2.17) we have

$$(\tilde{u}_t, v-\tilde{u})+a(\tilde{u}, v-\tilde{u})+(\chi(\Omega), v)-(\chi(\Omega), \tilde{u})=0$$

But $(\chi(\Omega),\tilde{u})=J(\tilde{u})$ and $(\chi(\Omega),v)=(\chi(\Omega),v^+-v^-)=$

$= \int_0^{s(t)} v^+(x)dx - \int_0^{s(t)} v^-(x)dx \leq J(v)$; so we have proved the inequality before. We have thus in a position to formulate

PROBLEM I': Given g with

$$(2.22) \quad g \in V, \quad g(x)>0, \quad 0\leq x<b$$

find w such that

$$(2.23) \quad w \in H^1(Q),\ w(b,t)=0 \quad \text{in } [0,T'].$$

$$(2.24) \quad \begin{cases} \text{for almost every } t \text{ in }]0,T'[\\ (w_t, v-w)+a(w,v-w)+J(v)-J(w)\geq 0 \quad \forall v \in V \end{cases}$$

$$(2.25) \quad w(x,0)=g(x) \quad \text{on } [0,b]$$

REMARK 2.3: The conditions $w(x,0)=g(x)$ and $w(b,t)=0$ are meaningful in consequence of the trace theorem of the space $H^1(Q)$.

We shall give in a moment an existence and uniqueness theorem for Pr.I'; now let us first study the relation between Pr. I' and the system (2.17); we have

PROPOSITION 2.2: If w is a solution of Pr.I', then

$$(2.26) \quad w\geq 0 \quad \text{in } Q$$

$$(2.27) \quad E(w) \in L^\infty(Q), \quad E(w)\leq 1 \quad \text{in } Q$$

$$(2.28) \quad \{E(w)-1\}\tilde{w}=0 \quad \text{in } Q$$

$$(2.29) \quad w \in H^{2,1}(Q)$$

$$(2.30) \quad w_x(0,t)=0 \ , \quad 0<t<T$$

$$(2.31) \quad w(b,t)=0 \ , \quad 0<t<T$$

$$(2.32) \quad w(x,0)=g(x), \quad 0\leq x\leq b,$$

where (2.26),(2.31),(2.32) are in the sense of $C^\circ(\bar{Q})$,

(2.27) (2.28) almost everywhere in Q, (2.30) in the sense of $H^{2,1}(Q)$.

PROOF: Let us take $v=w^+$ in (2.24), as it is possible, and remember that $w=w^+-w^-$; and we have

$$0\leq (w_t,w^-)+a(w^+-w^-,w^-)+J(w^+)-J(w)=$$

$$= -((w^-)_t,w^-)-((w^-)_x,(w^-)_x)$$

Then in particular

$$\frac{d}{dt}\int_0^b |w^-(x,t)|^2dx\leq 0 \qquad \text{a.e. in }]0,T'[$$

from which, since $w^-(x,0)=g^-(x)=0$, we have (2.26). Let us take now $v=w+\phi$ in (2.24), where $\phi\in D(]0,b[)$ and $\phi\geq 0$; we have, since (by (2.26)) $w^+=w$ and $(w+\phi)^+=w+\phi$,

$$(w_t,\phi)+a(w,\phi)+(1,\phi)=(w_t,\phi)-\langle w_{xx},\phi\rangle+(1,\phi) =$$

$= \langle 1-E(w),\phi\rangle\geq 0$, where $\langle,\rangle$ is the pairing between $D'(]0,b[)$ and $D(]0,b[)$; from which we have $1-E(w)\geq 0$ in the sense of $D'(]0,b[)$ for a.e. t in $]0,T'[$; and consequently $E(w)\leq 1$ in the sense of $D'(Q)$, since $w\in H^1(Q)$. Taking now $v=w-\phi$ with $\phi\in D(]0,b[)$ and $\phi\geq 0$, we have similarly:

$$-(w_t,\phi)+\langle w_{xx},\phi\rangle\geq J(w)-J(w-\phi)\geq 0$$

and then $E(w)\geq 0$; thus (2.27) is proved , and from it we can also deduce (2.29), using $E(w)=w_{xx}-w_t$ and (2.23). Now $w\in C^0(\bar{Q})$ and (2.26), (2.31),(2.32) have meaning in classical sense and (2.30) has meaning in the sense of the trace theorem for the functions belonging to $H^{2,1}(Q)$(cf. e.g. [48]): $w_x(0,t)\in H^{1/4}(]0,T'[)$. Finally let us introduce the set

$$(2.33)\qquad \Omega=\{(x,t)\in Q,\ w(x,t)>0\},$$

wich is a non void open set since $w\in C^0(\bar{Q})$ and by virtue of (2.32) and (2.22). Let us take now $\psi\in D(Q)$ with support contained in Ω; then there exists $\lambda_\psi>0$ such that for each real λ with $|\lambda|\leq\lambda_\psi$ one has $w+\lambda\psi\geq 0$ in Q, hence $(w+\lambda\psi)^+=w+\lambda\psi$; moreover for each $t\in]0,T'[$ $w(.,t)+$ $+\lambda\psi(.,t)\in V$, then we can take $v=w(.,t)+\lambda\psi(.,t)$ in (2.24) and we obtain

$$\lambda\iint_\Omega w_t\phi dxdt+\lambda\iint_\Omega w_x\psi_x dxdt+\iint_\Omega (w+\lambda\psi)dxdt-\iint_\Omega wdxdt\geq 0$$

i.e.

$$\lambda\iint_{\Omega}(E(w)-1)\psi dxdt\leqslant 0$$

from which, as the sign of λ is arbitrary,

$$\iint_{\Omega}(E(w)-1)\psi dxdt=0 \qquad \text{i.e.}$$

$$(2.34) \qquad E(w)=1 \qquad \text{in } \Omega$$

and we can deduce (2.28) almost everywhere in Q.

REMARK 2.4: If we denote by $H(\lambda)$ the Heaveside function (as a monotone grapf: $H(\lambda)=0$, $\lambda<0$, $H(0)=[0,1]$, $H(\lambda)=1$ $\lambda>0$) (2.26), (2.27), (2.28) can be condensed into the single relation

$$(2.35) \qquad E(w)\in H(w)$$

Now we can state the following

THEOREM 2.1: There exists one and only one solution w of Pr.I'; moreover

$$(2.36) \qquad w\in H^{2,1}(Q)$$

$$(2.37) \qquad w\in C^{0}(\bar{Q})$$

$$(2.38) \qquad w_t\in L^{\infty}(\Theta,T';L^2(]0,b[))\cap L^2(\Theta,T';V); \quad \forall\,\Theta>0,\Theta<T$$

if g satisfies moreover

$$(2.39) \qquad \begin{cases} g\in H^2(]0,b[);\ g'(x)\leqslant 0,\ 0\leqslant x\leqslant b;\ g'(b)=0; \\ g''(x)\leqslant 1.\ \text{a.e. in } [0,b]. \end{cases}$$

then we have

$$(2.40) \qquad w\in W_p^{2,1}(Q), \quad 1\leqslant p<3$$

$$(2.4I) \qquad w_x\leqslant 0,\ w_t\leqslant 0 \quad \text{a.e. in } Q;$$

finally if g satisfies moreover (2.1) we have

$$(2.42) \qquad \begin{cases} w\in W_p^{2,1}(Q) \quad \forall p\ 1\leqslant p\leqslant 6 \\ (\text{and consequentely } w_x\in C^{0}(\bar{Q})) \end{cases}$$

c) Now we come back to Pr. I using the solution w of Pr.I', given by Theor. 2.1. Using (2.41) it can be proved that Ω defined by (2.33) is actually given by

(2.43) $\Omega=\{(x,t);\ 0<x<s(t),\quad 0<t<T\}$

where

(2.44) $T\leqslant T'=g(0)$, and

(2.45) $\begin{cases} s\in C^{o}([0,T]), s(t)>0,\ 0\leqslant t<T,\ s(0)=b, s(T)=0, \\ s \text{ is strictly decreasing in } [0,T]. \end{cases}$

Moreover setting

(2.46) $u(x,t)=w(x,t)$ in $\bar{\Omega}$

the triplet $\{T,s(t),\ u(x,t)\}$ can be considered as a "weak solution" of Pr.I in the following sense:

DEFINITION 2.2: Under the assumptions (2.22) and (2.39) for g, $\{T,s(t),u(x,t)\}$ is a "weak solution" of Pr. I if T is >0, s verifies (2.45), $u\in C^{o}(\bar{\Omega})$, where Ω is defined by (2.43), and u verifies (1.4), (1.7) and (1.3) (this one in the sense of $D'(\Omega)$) and moreover the function $\tilde{u}$ defined by $\tilde{u}(x,t)=u(x,t)$ in $\bar{\Omega}$, $\tilde{u}(x,t)=0$ in $\bar{Q}'-\bar{\Omega}$ where $Q'=\{(x,t);\ 0<x<b,\ 0<t<\ T\ \}$, belongs to $H^{2,1}(Q')$ and $\tilde{u}_x(0,t)=0$ for a.e. $t\in[0,T\ ]$; moreover $u\geqslant 0$ in Ω.

REMARK 2.5: The boundary conditions (1.5) and (1.6) are contained in Definition 2.2 in the fact that $\tilde{u}\in H^{2,1}(Q)$ and $\tilde{u}_x(0,t)=0$. Let us remark also that we do not use condition (2.1) in Def. 2.1; so we can apply this definition and the following Theorem 2.2 also to the oxygen diffusion. It is interesting to note that althought the problem of oxygen diffusion (i.e. when $g'(0)\neq 0$) seems to be not reducible to a problem of Stefan type (remember indeed that if $g'(0)\neq 0$ we don't have for instance $u\in C^{1}(\bar{\Omega})$ and the proofs of the Proposition 2.1 are not valid), nevertheless it can be studied by the same variational inequality as for the case $g'(0)=0$.

Then we can state the following

THEOREM 2.2: There exists one and only one weak solution of Pr. I.

PROOF: The existence follows by Theor. 2.1, as we have just seen. For the uniqueness let us note, that if $\{T,s(t),u(x,t)\}$ is a weak solution, then it is easy to prove that, if we take a rectangle $Q_o=]0,b_o[\times]0,T_o[$ such that $\bar{\Omega}\subset\bar{Q}$ and if we extend u in $\bar{Q}_o$ setting $\tilde{u}=u$ in Ω,

$\tilde{u}=0$ in $\bar{Q}_o-\bar{\Omega}$, then $\tilde{u}$ is the solution of the variational inequality (2.23),(2.24),(2.25), written with Q_o instead of Q and $\tilde{g}(x)$ instead of $g(x)$, where $\tilde{g}(x)=g(x)$ in $[0,b]$, $\tilde{g}(x)=0$ in $[b,b_o]$. Then $\tilde{u}$ satisfies also the analogous of Theor. 2.1 and we have in particular $\tilde{u}_x\leq 0$ in Q_o, i.e. $u_x\leq 0$ in Ω. Then it is possible to repeate the proof given in the Prop.2.1, in order to obtain that $T\leq g(0)$. Now we can take $Q_o=Q$ and it is immediate to see that $\tilde{u}$ solves Pr.I'. Then the uniqueness follows from the uniqueness of the solution of Pr.I'.

REMARK 2.6: By a similar type of proof we can also obtain the uniqueness of the classical solution of Pr.I.

3. A STEFAN PROBLEM

a) First of all let us define "classical solutions" of Pr. II:

DEFINITION 3.1: Under tha assumptions

(3.0) $g \in C^1([0,a])$ and verifies (1.8)

the pair $\{s(t),u(x,t)\}$ is a "classical solution" of Pr. II if: $s \in C^o([0,T])$ and verifies (1.9), $s'(t)$ is continuous for $0<t\leq T$, $u \in C^o(\bar{\Omega})$, where Ω is given by (1.23), $u_x \in L^2(\Omega)$, u_x, u_t are continuous for $0\leq x\leq s(t)$, $0<t\leq T$, and u verifies (1.10), (1.11), (1.12), (1.13), (1.14) (the equation (1.10) in the sense of $D'(\Omega)$, then also in clas sical sense). We have

PROPOSITION 3.1: If $\{s(t),u(x,t)\}$ is a classical solution of Pr. II, then

(3.1) $u(x,t)>0$ in Ω,

(3.2) s is strictly increasing in $[0,T]$.

PROOF: The proof is the same used in the Proposition 2.1 for the function there denoted by u_t (let us remember the Remark 2.1); the only difference is that now $u(x,0)=g(x)>0$ and then $u(x,t)>0$ in Ω and $u_x(s(t),t)<0$, $0<t<T$.

PROPOSITION 3.2: If $\{s(t),u(x,t)\}$ is a classical solution of Pr. II, then

$$s(t)+\int_0^{s(t)} u(x,t)dx=a+\int_0^a g(x)dx \quad 0\leq t\leq T. \tag{3.3}$$

PROOF: First let us remark that $u_{xx}=u_t$ is also continuous for $0\leq x\leq s(t)$, $0<t\leq T$.

Then for every t in $]0,T[$ we have, using (1.13), (1.10), (1.11),

$$\begin{aligned} 0 &= s'(t)+u_x(s(t),t)=s'(t)+\int_0^{s(t)} u_{xx}(x,t)dx-u_x(0,t)= \\ &s'(t)+\int_0^{s(t)} u_t(x,t)dx=s'(t)+\frac{d}{dt}\int_0^{s(t)} u(x,t)dx- \\ &- s'(t)u(s(t),t) \end{aligned} \tag{3.4}$$

$$=\frac{d}{dt}\left[s(t)+\int_0^{s(t)} u(x,t)dx\right]$$

Then the function

$$s(t)+\int_0^{s(t)} u(x,t)dt$$

is constant in the interval $[0,T]$ and finally by (1.9) and (1.14) we have (3.3).

COROLLARY 3.1: There exists $b>0$ such that

(3.5) $\quad s(t)<b \quad , \quad 0<t<T$

In fact from (3.3) and (3.1) we have for instance $s(t)\leqslant =a+\int_0^a g(x)dx$; we can then take for instance $b=a+$ $+\int_0^a g(x)dx+1$.

Let us try to put Problem II in the form of a variational inequality, like in section 2; setting

(3.6) $\quad Q=\{(x,t);\ \ 0<x<b,\ \ 0<t<T\}$

with b given by the Corollary 3.1, we are suggested by (1.11) and the physical situation, to set

(3.7) $\quad \tilde{u}(x,t)=u(x,t)$ in $\bar{\Omega}$; $\tilde{u}(x,t)=0$ in $\bar{Q}-\bar{\Omega}$

But now u is not "sufficiently smooth"(we have $\tilde{u}\in C^0(\bar{\Omega})$ and $\tilde{u}_x\in L^2(Q)$) for being a solution of a variational inequality of the kind of (2.24); in fact, if this were possible, we could deduce from (1.12) $s'(t)=0$, $0<t<T$, which is in contradiction with (3.2). On the other hand let us try to compiute $E(\tilde{u})$ in the sense of $D'(Q)$; first we shall prove the following

PROPOSITION 3.3: If $\{s(t),u(x,t)\}$ is a classical solution of Pr.II, then we have

(3.8) $$\iint_\Omega (u_x\phi_x-u\phi_t)dxdt=\iint_\Omega \phi_t dxdt \quad \forall\phi\in C_*^\infty(\bar{Q})$$

where

(3.9) $$\begin{cases} C_*^\infty(\bar{Q})=\{\phi\in C^\infty(\bar{Q});\ \phi\equiv 0 \text{ in a neighbourough of the} \\ \text{set } \partial Q-\{(x,t);\ x=0,\ 0<t<T\} \end{cases}$$

PROOF: Let us recall that u_{xx} is continuous for $0 \leq x \leq s(t)$, $0<t<T$ and let us note that every $\phi \in C^\infty_*(\bar{Q})$ has its support in $[0,b[\times[\xi,T-\xi]$, for some $\xi>0$; the following computations are valid

$$\iint_\Omega u_x\phi_x dxdt - \iint_\Omega u\phi_t dxdt = -\iint_\Omega u_{xx}\phi dxdt + \int_0^T [u_x\phi]_0^{s(t)} dt +$$

$$+\iint_\Omega u_t\phi dxdt = -\iint_\Omega (u_{xx}-u_t)\phi dxdt - \int_0^T s'(t)\phi(s(t),t)dt$$

$$= -\int_0^T \left(\frac{d}{dt}\int_0^{s(t)} \phi(x,t)dx\right)dt + \int_0^T\int_0^{s(t)} \phi_t dxdt = \iint_\Omega \phi_t dxdt$$

Then we deduce

PROPOSITION 3.4: If $\{s(t),u(x,t)\}$ is a classical solution and we define $\tilde{u}$ by means of (3.7), we have

$$E(\tilde{u}) = D_t\chi(\Omega) \quad \text{in the sense of } D'(Q) \tag{3.10}$$

PROOF: For each $\phi \in D(Q)$ we have, using (3.8) and remarking that the "free boundary" $\partial\Omega \cap Q$ is of zero measure in $\mathbb{R}^2$ and that $\tilde{u}_x \in L^2(Q)$:

$$\langle E\tilde{u},\phi\rangle = \langle\tilde{u}_x,\phi_x\rangle + \langle\tilde{u},\phi_t\rangle = -\iint_\Omega u_x\phi_x dxdt + \iint_\Omega u\phi_t dxdt =$$

$$= -\iint_\Omega \phi_t dxdt = -\iint_Q \chi(\Omega)\phi_t dxdt = \langle D_t\chi(\Omega),\phi\rangle$$

where $\langle , \rangle$ denotes the pairing between $D'(Q)$ and $D(Q)$.

By looking at the equation (3.10) it is natural to introduce a new unknown function

$$w(x,t) = \int_0^t u(x,\tau)d\tau \quad \text{in } \bar{Q} \tag{3.11}$$

This will be actually the "good" unknown for the problem, because w satisfies a good variational inequality and, if we know w, we can come back to the solution of Pr.II; in fact we have

PROPOSITION 3.5: If $\{s(t),u(x,t)\}$ is a classical solution and w and $\tilde{u}$ are defined by (3.11) and (3.7), then we have

$$w(x,t) \geq 0 \quad \text{in } Q \tag{3.12}$$

(3.13) $w(x,t)>0$ in Ω , $w(x,t)=0$ in $Q-\Omega$

(3.14) $E(w)\in L^\infty(Q)$, $E(w)\leqslant 1-f$ in Q

where f is defined by

(3.15) $f(x,t)=1+g(x), 0<x<a, 0<t<T;\ f(x,t)=0, a<x<b, 0<t<T;$

(3.16) $(E(w)-1+f)w=0$ in Q

(3.17) $w(b,t)=0$, $0<t<T$

(3.18) $w_x(0,t)=0$, $0<t<T$

(3.19) $w(x,0)=0$, $0\leqslant x\leqslant b$

where (3.14),(3.16) have a meaning almost everywhere in Q and all the other relations in the classical sense.

PROOF: The conditions (3.12),(3.13),(3.17),(3.18),(3.19), together with the properties $w, w_t \in C^0(\bar{Q})$, $w_x, w_{xt} \in L^2(Q)$ are an easy consequence of the definition of w and of the properties of $\tilde{u}$. Now let us consider in Q the distribution $E(w)-\chi(\Omega)$; we have from (3.11) and (3.10), in the sense of $D'(Q)$

$$D_t(E(w)-\chi(\Omega))=E(D_t w)-D_t\chi(\Omega)=E(\tilde{u})-D_t\chi(\Omega)=0$$

Then $E(w)-\chi(\Omega)$ is constant with respect to t i.e. is of the form

$$E(w)-\chi(\Omega)=\lambda(x)$$

but we know that $w=0$ in $Q-\Omega$, so we obtain that

$$\lambda(x)=E(w)-\chi(\Omega)=0 \quad \text{for} \quad a<x<b\ ,\ 0<t<T$$

Secondly, in the rectangle $Q_a=\{(x,t);\ 0\leq x<a, 0<t<T\}$, we have $\chi(\Omega)=1$, and $w_t=\tilde{u}$ is continuous in $\bar{Q}_a$ and so also for $t=0$, and by (1.14) we have $w_t(x,0)=\tilde{u}(x)=u(x)=g(x)$ for $0\leqslant x\leqslant a$. Moreover $w_{xx}=\lambda(x)+1+w_t$, then w_{xx} is continuous with respect to t in $\bar{Q}_a$ and we have from (3.11) $w_{xx}(x,0)=0$; then $E(w)-\chi(\Omega)$ can evaluated for $t=0$, $0<x\leqslant a$, obtaining as value $-g(x)-1$. Then

$$E(w)-\chi(\Omega)=-1-g(x) \quad \text{in } \bar{Q}_a$$

Finally we have

(3.20) $\quad E(w)=1-f \quad \text{in } \Omega$

and (3.14),(3.16) are proved.

Suppose now "formally" that we can solve the system of the inequalities (3.12),(3.14),(3.16),(3.17),(3.18), (3.19) with w "sufficiently smooth" and such that the set $\Omega=\{(x,t)\in Q;\ w(x,t)>0\}$ be of the form (1.21) with s verifying (1.9); then setting $u(x,t)=w_t(x,t)$ in $\bar{\Omega}$, we obtain a solution of Pr.II. The system (3.12),(3.14),... ...,(3.19)is similar to the variational inequalities (2.26), ...,(2.32) of n.2; then, in the same way as in n.2,b), we shall now introduce a precise formulation for the system (3.12),(3.14),...,(3.19).

b) With the same definitions (2.18),(2.19),(2.20), (2.21) for V, (u,v), a(u,v), J(v), let us consider the following

PROBLEM II': Given g with

(3.21) $\quad g(a)=0,\ g(x)>0,\ 0\leqslant x<a, \quad g\in H^1(]0,a[)$

find w such that

(3.22) $\quad w\in H^1(Q), \quad w(b,t)=0 \quad \text{on } [0,T]$

and

(3.23) $\quad \begin{cases} \text{for almost every } t \text{ in } [0,T], \\ (w_t, w-w)+a(w,v-w)+J(v)-J(w)\geqslant(f,v-w) \\ \forall v\in V \\ \text{where } f \text{ is defined by (3.15)} \end{cases}$

(3.24) $\quad w(x,0)=0 \quad \text{on } [0,b].$

With the same proofs used for Prop.2.3 and using also the fact that f, defined by (3.15) is $\geqslant 0$ in Q, we can now prove the following

PROPOSITION 3.6: If w is a solution of Pr.II', then w satisfies (3.12),(3.14),(3.16),(3.17),(3.18),(3.19), together with

(3.25) $\quad w\in H^{2,1}(Q)$;

in particular

(3.26) $\quad E(w)=1-f \quad \text{in } \Omega$

where

(3.27) $\Omega=\{(x,t)\in Q;\ w(x,t)>0\}.$

REMARK 3.1: (3.12),(3.14),(3.16) can be condensed into the relation

(3.28) $E(w)+f\in H(w)$

Finally we have the

THEOREM 3.1: There exists one and only one solution w of Pr.II': moreover

(3.29) $w_t\in C^0([0,T];\ L^2(]0,b[)\cap L^2(0,T;V)$

(3.30) $w\in W_p^{2,1}(Q) \quad \forall p>1$

(3.31) $w, w_x\in C^0(\bar{Q})$

(3.32) $w_t, w_{xx}\in L^\infty(Q)$

(3.33) $w_t\geq 0$ a.e. in Q;

if moreover g verifies

(3.34) $g'(x)\leq 0$ a.e. in $[0,a]$

then we have

(3.35) $w_x\leq 0$ in Q

c) We can back to Pr.II, using the solution w of Pr.II' given by Theor.3.1; using in particular (3.33) and (3.35) it is possible to prove that the set Ω defined by (3.27) is actually given by

(3.36) $\Omega=\{(x,t);\ 0<x<s(t),\ 0<t<T\}$

where

(3.37) $\begin{cases} s\in C^0([0,T]),\ s(t)>0,\ s(0)=a,\ s \text{ is strictly} \\ \text{increasing in } [0,T]. \end{cases}$

Setting

(3.38) $u(x,t)=w_t(x,t)$ in $\bar{\Omega}$,

the pair $\{s(t),u(x,t)\}$ can be considered a "weak solution" of Pr. in the following sense:

DEFINITION 3.2: Under the assumptions (3.21) and (3.34)

for g, $\{s(t),u(x,t)\}$ is a "weak solution" of Pr.II if: s verifies (3.37), $u\in L^{\infty}(\Omega)$, $u_x\in L^2(\Omega)$, where Ω is defined by (3.36), $u\geqslant 0$ a.e. in Ω, u verifies (3.8) and the function $\tilde{u}$ defined by $\tilde{u}(x,t)=u(x,t)$ in Ω , $\tilde{u}(x,t)=0$ in $\bar{Q}-\Omega$, belongs to $H^{1,0}(Q)\cap C^0([0,T];L^2(]0,b[)$ and verifies $\tilde{u}(x,0)=\tilde{g}(x)$ in the sense of $C^0([0,T];L^2(]0,b[))$, where $\tilde{g}(x)=g(x)$, $0\leqslant x\leqslant a$, $\tilde{g}(x)=0$, $a\leqslant x\leqslant b$.

REMARK 3.2: Definition 3.2 is really a "weak formulation" of Pr.II: in fact the boundary condition (1.11) is contained (in the sense of $H^{1,0}(Q)$, then e.g. also for a.e. t in $]0,T[$) in the condition $\tilde{u}\in H^{1,0}(Q)$ the initial condition (1.14) is contained in the condition $\tilde{u}(x,0)=\tilde{g}(x)$; the equation (1.10) follows in the sense of $D'(\Omega)$ from (3.8) (where we take ϕ with support in Ω); and the boundary conditions (1.12),(1.13) are contained in a "weak sense" in the relation (3.8); moreover condition $u\geqslant 0$ a.e. in Ω, together with $E(u)=0$ in Ω, assures us, using the maximum principle in the interior of Ω , that $u(x,t)>0$ in Ω.

REMARK 3.3: Another "weaker" definition of "weak solution" of Pr. II has been introduced by Kamenonostkaja [40] and studied also by Olenik [52] , Friedman [29] ; it is not difficult to see that a "weak solution" in the sense of Def.3.2 is also a "weak solution" in the other sense.

Using Theor.3.1 and Def.3.2, it is possible to prove the following

THEOREM 3.2: There exists one and only one weak solution of Pr.II.

REMARK 3.4: We shall come back in section 8 to the regularity of the weak solution of Pr.II and to the existence of a classical solution; in any case since "classical solution are also weak solutions(cf. in particular Prop.3.1, and 3.3) (or since, as we have seen, if $\{s(t),u(x,t)\}$ is a classical solution then $\tilde{u}$ defined by (2.14) is a solution of Pr.I'), we obtain from Theor. 3.2 the uniqueness of classical solution.

4. A PROBLEM OF FLUID FLOWS THROUGH A POROUS MEDIUM.

a) Let us consider first "classical solutions" of Pr. III:

DEFINITION 4.I: Under the assumptions

$$(4.0)\quad \begin{array}{l} g \in C^1([0,a]) \text{ and verifies (I.I5),} \\ \ell \in C^1([0,T]) \text{ and verifies (I.I6)} \end{array}$$

the pair $\{s(t),u(x,t)\}$ is a "classical solution" of Pr. III if: $s \in C^0([0,T])$ and verifies (I.I7), $s'(t)$ is continuous for $0<t\leqslant T$, $u \in C^0(\bar{\Omega})$, where Ω is given by (I.23), $u_x \in L^2(\Omega)$, u_x, u_t are continuous for $0\leqslant x\leqslant s(t)$, $0<t\leqslant T$, and u verifies (I.I8), (I.I9), (I.20), (I.2I), (I.22) (the equation (I.I8) in the sense of $D'(\Omega)$).

We have

PROPOSITION 4.I: If $\{s(t),u(x,t)\}$ is a classical solution of Pr. III, then

$$(4.I)\quad u(x,t)>x \qquad \text{in } \Omega$$

$$(4.2)\quad \begin{cases} s'(t)\geqslant -I,\ 0<t\leqslant T \text{ and the "free boundary"} \\ \partial\Omega \cap Q \text{ does not contain segments parallel to the} \\ \text{line } x+t=0 \end{cases}$$

PROOF: The same proof of Proposition 2.I, applied to the function $v(x,t)=u(x,t)-x$.

PROPOSITION 4.2: If $\{s(t),u(x,t)\}$ is a classical solution of Pr. III, then, denoting by $L(t)$ the function

$$(4.3)\quad L(t)=\int_0^t \ell(\tau)d\tau+\int_0^a (I+g(x)-x)dx$$

we have

$$(4.4)\quad s(t)+\int_0^{s(t)} \{u(x,t)-x\}dx=L(t) \qquad 0\leqslant t\leqslant T$$

PROOF: The same proof of Proposition 3.2.

Now by (4.4) we have $s(t)<L(t)$, $0\leqslant t\leqslant T$ and consequentely:

$$(4.5)\quad 0<\lambda_o = \inf_{0\leqslant t\leqslant T} L(t)$$

$$(4.6)\quad s(t)\leqslant\lambda_1 = \sup_{0\leqslant t\leqslant T} L(t)$$

Denote by b a fixed number greater than λ_1, for instance

(4.7) $b = \lambda_1 + I$

and Q

(4.8) $Q = \{(x,t) : 0<x<b,\ 0<t<T\}$

We are suggested by (I.I9) to define also

(4.9) $$\tilde{u}(x,t) = \begin{cases} u(x,t) & \text{if } (x,t) \in \bar{\Omega} \\ x & \text{if } (x,t) \in \bar{Q}-\bar{\Omega} \end{cases}$$

and, remarking that by virtue of (I.20) we are in a situation similar to the Stefan problem one, we try to introduce a new unknown function by a transformation similar to (3.7). Let us follow, first of all, Friedman and Jensen [34] and define

(4.I0) $$w(x,t) = \int_x^b \{\tilde{u}(\xi,t)-\xi\} d\xi$$

We shall now see "formally" that w verifies the following system of inequalities:

(4.II) $w \geq 0$ in Q

(4.I2) $w > 0$ in Ω, $w = 0$ in $Q-\Omega$

(4.I3) $E(w) \in L^{\infty}(Q)$, $E(w) \leq I + s'(t)$ in Q

(4.I4) $\{E(w)-(I+s'(t))\}w = 0$ in Q

(4.I5) $w(b,t) = 0$, $0<t<T$

(4.I6) $w(0,t) = h(0) + \int_0^t \ell(\tau)d\tau - s(t) + a$, $0<t<T$

(4.I7) $w(x,0) = h(x)$ $0 \leq x \leq b$

where

(4.I8) $$h(x) = \int_x^a (g(\xi)-\xi)d\xi, \quad 0 \leq x \leq a; \quad h(x) = 0, \quad a \leq x \leq b$$

In fact we evaluate

$w_x(x,t) = x - u(x,t)$ in Ω; $w_x(x,t) = 0$ in $Q-\Omega$

$w_{xx}(x,t) = I - u_x(x,t)$ in Ω; $w_{xx}(x,t) = 0$ in $Q-\Omega$

$w_t(x,t)=\int_x^b \tilde{u}_t(\xi,t)d\xi$ in Ω; $w_t(x,t)=0$ in $Q-\Omega$

since $\tilde{u}(x,t)-x=0$ if $x=s(t)$. Consequently we have in Ω

$$E(w)=I-u_x(x,t)-\int_x^b \tilde{u}_x(\xi,t)d\xi=I-u_x(x,t)-$$

$$(4.I9)\qquad -\int_x^{s(t)} u_{\xi\xi}(\xi,t)d\xi=I-u_x(x,t)-u_x(s(t),t)+u_x(x,t)=$$

$$=I-u_x(s(t),t)=I+s'(t).$$

and in $Q-\Omega$, by (4.2)

$$(4.20)\qquad E(w)=0\leqslant I+s'(t)$$

Furthermore (4.II),(4.I2), (4.I5), (4.I7) are immediate. In order to prove (4.I6), let us take $x=0$ in (4.I9) and let us remember that $w_{xx}(0,t)=I-u_x(0,t)=I+\ell(t)$; we obtain $w_t(0,t)=\ell(t)-s'(t)$, then integrating and using (4.I7), we have (4.I6). Suppose now "formally" that we can solve the system (4.II)....(4.I7) in the unknown s and w; then taking $u(x,t)=x-w_x(x,t)$ in $\bar{\Omega}$, with Ω given by (I.23), we obtain a solution of Pr. III. The system (4.II).....(4.I7) (I.23) in the unknown w and s is a so called"quasi-variational inequality".If $s(t)$ is a known function then (4.II) (4.I3).....(4.I7) form a variational inequality of the same kind as the variational inequality of Sections 2 and 3, and if we denote by $w^{(s)}$ its solution then,if $w_x^{(s)}\leqslant 0$,the curve $x=\sigma(t)$ given by

$$\sigma(t)=\sup\{x;\ w^{(s)}(x,t)>0\}$$

is the corresponding "free boundary". If $\sigma(t)$ coincides with $s(t)$, then the pair $\{s,w^{(s)}\}$ is a solution of the system (4.II)(4.I2)....(4.I7)(I.23). We can then think that a method for solving quasi-variational inequalities of type (4.II)....(4.I7)(I.23) may be obtained by combining fixed point theorems with existence theorems of the variational inequalities. And actually this idea is used to solve general quasi-variational inequalities. But quasi-variational inequalities being a more sophisticated tool than variational inequalities we shall adopt here a different approach to Pr. III, following an idea of Torelli, who by a change of unknown, different form (4.I0), was able to reduce Pr. III to a variational inequalities of the same kind as those of Pr. I' and II" (see [62]).

b) First let us remark some other properties of classical solution. If $C^\infty_*(\bar{Q})$ is still defined by (3.9), we have

PROPOSITION 4.3: If $\{s(t), u(x,t)\}$ is a classical solution of Pr. III, then

$$(4.2I)\quad \iint_\Omega \{(u_x-I)\phi_x-(u-x)\phi_t\}dxdt = \iint_\Omega (\phi_t-\phi_x)dxdt + \int_0^T \ell(t)\phi(0,t)dt \qquad \forall\phi\in C^\infty_*(\bar{Q})$$

PROOF: It is the same as for the Prop. 3.3; we have to use the condition $u_x(0,t)=-\ell(t)$ instead of the condition $u_x(0,t)=0$.

Moreover as in the Proof. 3.4 we deduce immediately from (4.2I) that

$$(4.22)\quad E(\tilde{u})=(D_t-D_x)\chi(\Omega) \qquad \text{in the sense of } D'(Q)$$

where $\tilde{u}$ is the function defined by (4.9). Let us compare (4.22) with (3.I0) (and with (2.II)!): Torelli's idea is to integrate with respect to the oblique direction y making with the x-axis an angle of measure $\frac{3}{4}\pi$. Then let us introduce the set

$$(4.23)\quad \tilde{Q}=\{(x,t); 0<x<b+T-t,\quad 0<t<T\}$$

and let us define by continuation $\tilde{u}$ in $\bar{\tilde{Q}}-\bar{Q}$, i.e. setting

$$(4.24)\quad \tilde{u}(x,t)=x \qquad \text{in } \bar{\tilde{Q}}-\bar{Q}$$

Let us note also the following properties of the set Ω defined by (I.23) if $\{s(t),u(x,t)\}$ is a classical solution: the first is a part of (4.2), which we rewrite here

$$(4.25)\quad \begin{cases}\text{the "free boundary " } \partial\Omega\cap Q \text{ does'nt contain} \\ \text{segments parallel to the line } x+t=0;\end{cases}$$

the second is:

$$(4.26)\quad \begin{cases}\text{if } (x,t)\in\tilde{Q}-\bar{\Omega}, \text{ then the segment } \{(\xi,\tau); \xi=x+\theta t, \\ \tau=t-\theta t,\ 0\leq\theta<I\} \text{ belongs to } \tilde{Q}-\bar{\Omega}\end{cases}$$

and follows immediately from (4.2). Now we are in a position to define the function

$$(4.27)\quad \tilde{w}(x,t)=\int_0^t \tilde{u}(t+x-\zeta,\zeta)-(t+x-\zeta)\}d\zeta \qquad (x,t)\in\bar{Q}$$

for which we have the

PROPOSITION 4.4: The function $\tilde{w}$, given by (4.27),(4.24), (4.9), satisfies, in the sense of $D'(Q)$:

(4.28) $\tilde{w}_t - \tilde{w}_x = \tilde{u} - x$

(4.29) $\tilde{w}_x(x,t) = \int_0^t \{\tilde{u}_x(t+x-\zeta,\zeta) - I\} d\zeta$

(4.30) $\tilde{w}_t(x,t) = \int_0^t \{\tilde{u}_x(t+x-\zeta,\zeta) - I\} d\zeta + \tilde{u}(x,t) - x$

(4.3I) $E(\tilde{w}) = \chi(\Omega - Z) - \chi(Z)\{g(x+t) - (x+t)\}$

Were

(4.32) $Z = \{(x,t) \in Q;\ \ x+t < a\}$

PROOF: Let us first note that by (4.2)

(4.33) $Z \subset \Omega$

Secondly, recalling that $\tilde{u} \in C^\circ(\bar{Q})$ and $\tilde{u}_x \in L^2(Q)$, we can obtain directly (4.29) (4.30) and consequently (4.28); from (4.28),(4.29), (4.30) we deduce that

(4.34) $\tilde{w}, \tilde{w}_t - \tilde{w}_x \in C^\circ(\bar{Q})$; $\tilde{w}_t, \tilde{w}_x \in L^2(Q)$

Then from (4.28) and (4.22) we have

(4.35) $(D_t - D_x)(E(\tilde{w}) - \chi(\Omega)) = 0$ in the sense of $D'(Q)$

The distributions $E(\tilde{w}) - \chi(\Omega)$ is then constant on the parallels to the line $x+t=0$, i.e. it is of the form

(4.36) $E(\tilde{w}) - \chi(\Omega) = \lambda(x+t)$

Now using (4.25) and (4.26) we can deduce that $\tilde{w}(x,t)=0$ in $Q-\Omega$; then remarking that $s(0)=a$ and $s(t)<b$ for $0 \leqslant t \leqslant T$, we have

(4.37) $E(\tilde{w}) - \chi(\Omega) = 0$ in $Q-Z$

In $\bar{Z} \cap Q$ we have $\chi(\Omega)=I$ and $\tilde{w}_t - \tilde{w}_x = \tilde{u} - x$ is a continuous function; then by (4.36) we have

$$\tilde{w}_{xx} - \tilde{w}_x = \lambda(x+t) + I + \tilde{w}_t - \tilde{w}_x$$

and then $\tilde{w}_{xx} - \tilde{w}_x$ is continuous in $\bar{Z} \cap Q$ on the parallels to the line $x+t=0$; then we can evaluate

$$\lambda(x+t)=E(\tilde{w})-\chi(\Omega)\quad (=(\tilde{w}_{xx}-\tilde{w}_x)-(\tilde{w}_t-\tilde{w}_x)-I \quad \text{in } \bar{Z}\cap Q)$$

for $t=0$, $0<x\leqslant a$ and we obtain from (4.27)

$$\lambda(x)=(\tilde{w}_{xx}-\tilde{w}_x)(x,0)-(\tilde{w}_t-\tilde{w}_x)(x,0)-I=0-\tilde{g}(x)+x-I =$$
$$= -g(x)+x-I$$

Then we have

$$(4.38)\quad E(\tilde{w})-\chi(\Omega)=-g(x+t)+(x+t)-I \quad \text{in } \bar{Z}\cap Q$$

Finally (4.36),(4.38) can be written in the form (4.3I)

Now we make another changement of unknown; let us consider the function $z(x,t)$ defined by

$$(4.42)\quad \begin{cases} z(x,t)=0 \text{ in } \bar{Q}-Z\ , \quad z(x,t)= \\ \qquad =\int_{x+t}^{a}(I+g)(\xi)-\xi)(I-e^{x+t-\xi})d\xi \text{ in } Z \end{cases}$$

We have

$$(4.43)\quad z_x(x,t)=z_t(x,t)=\begin{cases} 0 \quad \text{in } Q-Z \\ -\int_{x+t}^{a}(I+g(\xi)-\xi)e^{x+t-\xi}d\xi \quad \text{in } Z \end{cases}$$

$$(4.44)\quad z_{xx}(x,t)=\begin{cases} 0 \qquad \text{in } Q-Z \\ -\int_{x+t}^{a}(I+g(\xi)-\xi)e^{x+t-\xi}d\xi+I+g(x+t)-(x+t) \text{ in } Z \end{cases}$$

and consequently

$$(4.45)\quad z_{xx}(x,t)-z_t(x,t)=\chi(Z)(I+g(x+t)-(x+t) \quad \text{in } Q \quad \text{and}$$

$$(4.46)\quad z,z_x,z_t,z_{xx}\in C^{\circ}(\bar{Q}).$$

We introduce now the new unknown $w(x,t)$, defined by

$$(4.47)\quad w(x,t)=\tilde{w}(x,t)+z(x,t) \quad \text{in } \bar{Q}$$

and we prove the

PROPOSITION 4.5: The function w defined by (4.47)(4.42) verifies

$$(4.48)\quad w(x,t)\geqslant 0 \quad \text{in } \bar{Q}$$

(4.49) $w(x,t)>0$ in Ω , $w(x,t)=0$ in $\bar{Q}-\Omega$

(4.50) $E(w)\in L^{\infty}(Q)$, $E(w)\leqslant I$ in Q

(4.5I) $\{E(w)-I\}w=0$ in Q

(4.52) $w_x(0,t)-w(0,t)=-L(t)$, $0<t<T$

(4.53) $w(b,t)=0$, $0<t<T$

(4.54) $w(x,0)=G(x)$, $0\leqslant x\leqslant b$

where $L(t)$ is given by (4.3) and $G(x)$ by

(4.55) $G(x)=\int_x^a (I+g(\xi)-\xi)(I-e^{x-\xi})d\xi$, $0\leqslant x\leqslant a$; $G(x)=0, a<x\leqslant b$

and where (4.50),(4.5I) have meaning almost everywhere in Q and (4.48),(4.49),(4.52),(4.53),(4.54) in classical sense.

PROOF : First we note that from (4.46) and (4.34) we have

(4.56) $w, w_t-w_x\in C^{\circ}(\bar{Q})$; $w_t, w_x\in L^2(Q)$

Secondly using the properties of w and z it is easy to prove (4.48),(4.49),(4.50),(4.5I),(4.53),(4.54). We have thus to verify only (4.52). Now from (4.28)and (4.43) we have

(4.57) $w_t-w_x=\tilde{u}-x$ in Q

Deriving,in the sense of $D'(Q)$, we obtain, using also $E(w)=\chi(\Omega)$(which is contained in (4.48) and (4.5I))

$$w_{tx}-w_{xx}=w_{tx}-(w_t+\chi(\Omega))=(w_x-w)_t-\chi(\Omega)=\tilde{u}_x-I$$

Writing the last equation for $x=0$, $0<t<T$ (which is possible because $\tilde{u}_x=u_x$ in Ω and u_x is continuous for $x=0$ $0<t<T$), we have

$$(w_x(0,t)-w(0,t))_t=-\ell(t)$$

from which, by (4.54)

$$w_x(0,t)-w(0,t)=-\int_0^t \ell(\tau)d\tau+G'(0)-G(0)=-\int_0^t \ell(\tau)d\ -$$
$$-\int_0^a (I+g(\xi)-\xi)d\xi=-L(t)$$

The system (4.49)....(4.54) is of the same kind as the other systems(2.I2)...(2.I7) and (3.I3)....(3.I8).And similarly as in §. 2 and 3 if we suppose "formally"to have solved it with w "sufficiently smooth" and such that the set $\Omega=\{(x,t)\in Q;\ w(x,t)>0\}$ is of the form (I.2I), then setting $u(x,t)=w_t(x,t)-w_x(x,t)+x$ in Ω , we obtain a solution of Pr. III.

c) Then let V, (u,v), J(v) be defined as in S. 2 and 3 by (2.I8),(2.I9),(2.2I) and set now

$$(4.58)\quad a(u,v)=\int_0^b u_x(x)v_x(x)dx+u(0)v(0)\qquad \forall u,v\in V.$$

Let us consider the following

PROBLEM III': Given g and ℓ such that

$$(4.59)\quad g\subset H^1(]0,a[);\ g(x)>x,\ 0\leqslant x<a,\ g(a)=0$$

$$(4.60)\quad \ell\in C^0([0,T]);\ \ell(t)>-I,\ 0\leqslant t\leqslant T;\ \ell\in H^{\frac{1}{4}}(]0,T[)$$

and with G and L defined by (4.55) and (4.53), find w such that

$$(4.6I)\quad w\in H^1(Q);\quad w(b,t)=0\quad \text{on } [0,T];$$

for almost every t in $[0,T]$

$$(4.62)\quad (w_t,v-w)+a(w,v-w)+J(v)-J(w)\geqslant L(t)(v(0)-w(0,t)),\forall v\in V$$

$$(4.63)\quad w(x,0)=G(x)\quad \text{on } [0,b].$$

REMARK 4.I: As we have seen in n. 2 b) it is easy to prove that if $\{s(t),u(s,t)\}$ is a classical solution and we define w by (4.47), then w is a solution of Pr. III'.

By means of the same proof used for Prop. 2.3, using in addition the positivity of L(t) (cf. (4.5)), we can prove the

PROPOSITION 4.6: If w is a solution of Pr. III then w satisfies (4.48),(4.50).....(4.54) together with

$$(4.64)\quad w\in H^{2,1}(Q);$$

in particular

$$(4.65)\quad E(w)=I\quad \text{in } \Omega$$

where

$$(4.66) \quad \Omega=\{(x,t)\in Q;\quad w(x,t)>0\}$$

REMARK 4.2: (4.48),(4.50),(4.5I) can be condensed into

$$(4.67) \quad E(w)\in H(w)$$

Finally we have the

THEOREM 4.I: There exists one and only one solution of Pr. III'; moreover

$$(4.68) \quad w_t \in C^0(0,T;L^2]0,b[)\cap L^2(0,T;V)$$

$$(4.69) \quad w\in W_p^{2,1}(Q) \quad \forall p\geq I;\ w,w_x\in C^0(\bar{Q})$$

$$(4.70) \quad w_t,w_{xx}\in L^\infty(Q)$$

$$(4.7I) \quad w_t-w_x\geq 0\ ,\ w_x\leq 0 \quad \text{a.e. in } Q$$

d) Using the solution w given by Theor. 4.I we can now come back to Pr. III. We shall prove using in particular (4.7I) that the set Ω defined by (4.66) is actually of the form

$$(4.72) \quad \Omega=\{(x,t);\ 0<x<s(t),\ 0<t<T\},$$

where

$$(4.73) \quad s\in C^0([0,T])\ ,\quad s(t)>0,\ 0\leq t\leq T,\ s(0)=a,$$

and Ω satisfies the condition

$$(4.74) \quad \begin{cases} \forall P\in\bar{\Omega},\ \text{then } \Lambda^-(P)\subset\Omega \\ \forall P\in Q-\Omega,\ \text{then } \Lambda^+(P)\subset Q-\bar{\Omega}, \end{cases}$$

where if $P=(\bar{x},\bar{t})$

$$(4.75) \quad \begin{cases} \Lambda^+(P)=\{(x,t)\in Q;\ t\leq\bar{t},\ x+t\geq\bar{x}+\bar{t}\}-\{P\} \\ \Lambda^-(P)=\{(x,t)\in Q;\ t\geq\bar{t},\ x+t\leq\bar{x}+\bar{t}\}-\{P\} \end{cases}$$

Setting

$$(4.76) \quad u(x,t)=w_t(x,t)-w_x(x,t)+x \quad \text{in } \bar{\Omega}$$

the pair $\{s(t),u(x,t)\}$ can be considered a "weak solution" of Pr. III in the following sense:

DEFINITION 4.2: Under the hypothesis for g and ℓ(4.59) and (4.60), $\{s(t),u(x,t)\}$ is a "weak solution" of Pr. III if: s verifies (4.73), Ω, defined by (4.72), verifies (4.74),(4.75), $u \in L^{\infty}(\Omega)$, $u_x \in L^2(\Omega)$, $u(x,t) \geq x$ in Ω, u verifies (4.2I) and, setting $\tilde{u}(x,t)=u(x,t)$ in Ω, $\tilde{u}(x,t)=x$ in $Q-\Omega$, $\tilde{u} \in H^{1,0}(Q) \cap C^{0}([0,T];L^2(]0,b[)$ and verifies $\tilde{u}(x,o)=\tilde{g}(x)$ in the sense of $C^{0}([0,T];L^2(0,b))$

REMARK 4.3: Def. 4.2 is really a "weak formulation" of Pr. III; in fact the boundary condition (I.I9) is contained in the condition $\tilde{u} \in H^{1,0}(Q)$ (hence e.g. for a.e. t in $[0,T]$); (I.22) is contained in the condition $\tilde{u}(x,0)=\tilde{g}(x)$; the equation (I.I8)follows in the sense of $D'(\Omega)$ from (4.2I); and (4.2I) contains in a "weak sense" also the conditions (I.20),(I.2I); moreover the condition $u(x,t) \geq x$ in Ω, gives us, together with the equation (I.I8), the property $u(x,t)>0$ in Ω.

REMARK 4.4.: Condition (4.74) (4.75) gives us that $\partial\Omega \cap Q$ is a graph which is lipschitzian with respect to the new axis (x',t) defined by $x'=\alpha t+\beta x$, $t'=-\beta t+\alpha x$ where

$\alpha=\cos \frac{\pi}{8}$, $\beta=\text{sen } \frac{\pi}{8}$. In particular we have also that $Z \subset \Omega$, where Z u defined by (4.32).

Finally we have

THEOREM 4.2: There exists one and only one weak solution of Pr. III.

REMARK 4.5: We shall come back in S. 8 to the regularity of the weak solution and to the existence of a classical solution; in any case remark that classical solutions are also weak solutions and that , as a consequence of that or of the Prop. 3.5 and the uniqueness of solution of Pr. III', we have the uniqueness of the classical solution.

5. PROOF OF THEOREMS OF S.4: APPROXIMATION

We shall prove now Theor.4.1.

a) Uniqueness: Let w_1 and w_2 be two solutions of Pr.III'. Setting $w=w_1-w_2$ and taking in (4.62) written for w_1(resp. w_2) $v=w_2$ (resp. $v=w_1$) and adding the two inequalities we have

$$w(x,0)=0 \quad 0\leqslant x\leqslant a \quad \text{and}$$

$$-(w'(t),w(t))-a(w(t),w(t))\geqslant 0 \quad \text{a.e. in } [0,T]$$

i.e.

$$\frac{1}{2}\frac{d}{dt}\int_0^b |w(x,t)|^2dx+\int_0^b |w_x(x,t)|^2dx+|w(0,t)|^2\leqslant 0$$

a.e. in $[0,T]$

from which we obtain $w\equiv 0$.

b) We shall give the existence proof by regularization. Let us recall that Pr.III' is "equivalent" to find $w\in H^{2,1}(Q)$ such that

$$E(w)\in H(w) \tag{5.1}$$

$$w(b,t)=0 \ , \quad 0<t<T \tag{5.2}$$

$$w_x(0,t)-w(0,t)=-L(t), \quad 0<t<T \tag{5.3}$$

$$w(x,0)=G(x) \quad 0\leqslant x\leqslant b \tag{5.4}$$

We choose now an adequate approximation of H, L and G.

c) For n=1,2,..., let $H_n(\lambda)$ be a function satisfying the following properties:

$$H_n(\lambda)\in C^0(\mathbb{R}) \ , \quad H_n(\lambda)\in C^\infty([0,+\infty[) \tag{5.5}$$

$$H_n(\lambda)=\begin{cases}0 & \text{if} \quad -\infty<\lambda\leqslant 0\\ \frac{3}{2}n\lambda & \text{"} \quad 0\leqslant\lambda\leqslant\frac{1}{2n}\\ 1 & \text{"} \quad \frac{1}{n}\leqslant\lambda<+\infty\end{cases} \tag{5.6}$$

$$H_n'(\lambda)\geqslant 0, \quad H_n''(\lambda)<0 \ , \quad 0<\lambda<+\infty \tag{5.7}$$

$$H_1(\lambda)\leqslant H_2(\lambda)\leqslant\ldots\leqslant H_n(\lambda)\leqslant\ldots\leqslant H(\lambda) \tag{5.8}$$

If $u(x,t)$ is a function defined in Q let us consider $H_n(u)$; we have

(5.9) $\quad H_n(u)\in L^\infty(Q) \quad (\text{resp.}\, C^0(\bar{Q}))$ if $u\in L^\infty(Q)(\text{resp.}\, C^0(\bar{Q}))$

Stronger smoothness properties of u are not always transferred to $H_n(u)$, since H_n is only a lipschitzian function in $\mathbb{R}$; but we have a better situation if $u\geq 0$ in Q because $H_n\in C^\infty([0,+\infty[)$. In particular we have:

(5.10) $\quad$ If $u\in W_p^{2,1}(Q)$ and $u\geq 0$, then $H_n(u)\in W_p^{2,1}(Q)$, $\frac{3}{2}\leq p<+\infty$

and moreover

(5.11) $\quad$ the usual derivation rules are valid

In fact let ϕ_i, $i=1,2,\ldots$, be functions such that $\phi_i\in C^\infty(\bar{Q})$ and ϕ_i converges to u in $W_p^{2,1}(Q)$; let us remember that then $u\in C^0(\bar{Q})$, and ϕ_i converges to u in $C^0(\bar{Q})$ and $u_x\in L_{2p}(Q)$, for $p\geq\frac{3}{2}$, and $D_x\phi_i$ converges to u_x in $L_{2p}(Q)$, as a consequence of Sobolev's imbedding theorems in the space $W_p^{2,1}(Q)$ (cf.e.g.[43]). Let $K_n(\lambda)$ be a function such that $K_n(\lambda)\in C^\infty(\mathbb{R})$ and $K_n(\lambda)=H_n(\lambda)$ for $0\leq\lambda<+\infty$. Then $K_n(\phi_i)\in C^\infty(\bar{Q})$ and we have

$$(5.12)\quad \begin{cases} D_xK_n(\phi_i)=K_n'(\phi_i)D_x\phi_i, \quad D_tK_n(\phi_i)=K_n'(\phi_i)D_t\phi_i, \\ D_{xx}K_n(\phi_i)=K_n''(\phi_i)(D_x\phi_i)^2+K_n'(\phi_i)D_{xx}\phi_i \end{cases}$$

We can pass to the limit, for $i\to+\infty$, in (5.12) and find that $K_n(u)\in W_p^{2,1}(Q)$ and

$$(5.13)\quad \begin{cases} D_xK_n(u)=K_n'(u)D_xu, \quad D_tK_n(u)=K_n'(u)D_tu \\ D_{xx}K_n(u)=K_n''(u)(D_xu)^2+K_n'(u)D_{xx}u \end{cases}$$

But now $u\geq 0$ in Q, then $K_n(u)=H_n(u)$ and we have (5.10) and (5.11).

d) Let us set now

(5.14) $\quad \tilde{g}(x)=g(x),\quad 0\leq x\leq a;\quad \tilde{g}(x)=x \quad a\leq x\leq b$

(5.15) $\quad \rho_n(x)=H_n(a+\frac{1}{n}-x) \quad 0\leq x\leq b$

so that by (5.5),(5.6),(5.7) we have

$$(5.16)\qquad \rho_n(x)=\begin{cases} 1 & 0\leqslant x\leqslant a \\ -\frac{3}{2}n(x-(a+\frac{1}{n})) & a+\frac{1}{2n}\leqslant x\leqslant a+\frac{1}{n} \\ 0 & a+\frac{1}{n}\leqslant x\leqslant b \end{cases}$$

and $\rho_n\in C^0[0,b], \rho_n\in C^\infty([0,a+\frac{1}{n}])$, $\rho_n'(x)\leqslant 0$, $\rho_n''(x)\leqslant 0$;

$$(5.17)\qquad G_n(x)=\int_x^b(\rho_n(\xi)+\tilde{g}(\xi)-\xi)(1-e^{x-\xi})d\xi,\ 0\leqslant x\leqslant b$$

$$(5.18)\qquad L_n(t)=\int_0^t \ell_n(\tau)d\tau+\int_0^b(\rho_n(\xi)+\tilde{g}(\xi)-\xi)d\xi,\ 0\leqslant t\leqslant T$$

where $\{\ell_n\}$ is a sequence of functions such that

$$(5.19)\qquad \begin{cases} \ell_n\in C^\infty([0,T]),\ \ell_1(t)\geqslant\ell_2(t)\geqslant\ldots\geqslant\ell_n(t)\geqslant\ldots\geqslant\ell(t) \\ \text{and } \lim_{n\to\infty}\ell_n=\ell \text{ in } C^0[0,T]\cap H^{\frac{1}{4}}(]0,T[) \end{cases}$$

Let us remark that we have

$$(5.20)\qquad L_n\in C^\infty([0,T])$$

$$(5.21)\qquad G_n\in H^3(]0,b[)$$

with

$$(5.22)\qquad \begin{aligned} G_n'(x)&=-\int_x^b(\rho_n(\xi)+\tilde{g}(\xi)-\xi)e^{x-\xi}d\xi \\ G_n''(x)&=-\int_x^b(\rho_n(\xi)+g(\xi)-\xi)e^{x-\xi})d\xi+\rho_n(x)+\tilde{g}(x)-x \end{aligned}$$

We have also if $\chi([0,a])$ is the caracteristic function of $[0,a]$.

$$(5.23)\qquad \rho_1(x)\geqslant\rho_2(x)\geqslant\ldots\geqslant\rho_n(x)\ldots\geqslant\chi([0,a]),$$

$$(5.24)\qquad G_1(x)\geqslant G_2(x)\geqslant\ldots\geqslant G_n(x)\ldots\geqslant G(x)$$

$$(5.25)\qquad L_1(t)\geqslant L_2(t)\geqslant\ldots\geqslant L_n(t)\geqslant\ldots\geqslant L(t).$$

Let us define also

$$(5.26)\qquad j_n(\lambda)=\int_{-\infty}^{\lambda}H_n(\xi)d\xi\qquad \forall\lambda\in\mathbb{R}$$

$$(5.27)\qquad J_n(v)=\int_0^b j_n(v)dx\qquad \forall v\in V$$

and that

(5.28) $\quad j_1(\lambda)\leqslant j_2(\lambda)\leqslant\ldots\leqslant j_n(\lambda)\ldots$

PROPOSITION 5.1: If $n=2,3,\ldots$, then

(5.29) $\quad H_n(G_n(x))\leqslant\rho_n(x),\quad 0\leqslant x\leqslant b$

PROOF: Firstly (5.29) is valid if $0\leqslant x\leqslant a$, because $\rho_n(x)=1$ and $H_n(G_n(x))\leqslant 1$; (5.29) is also valid if $a+\frac{1}{n}\leqslant x\leqslant b$, since $\rho_n(x)=0$ and $H_n(G_n(x))=0$ (because $G_n(x)=0$). It remains to prove (5.29) if $a\leqslant x\leqslant a+\frac{1}{n}$; we have for any such x

(5.30) $$G_n(x)=\int_x^{a+\frac{1}{n}}\rho_n(\xi)(1-e^{x-\xi})d\xi\leqslant\int_x^{a+\frac{1}{n}}\rho_n(\xi)(\xi-x)d\xi$$

If $a\leqslant x\leqslant a+\frac{1}{3n}$, we have

(5.31) $$G_n(x)\leqslant\int_x^{a+\frac{1}{3n}}\rho_n(\xi)(\xi-x)d\xi+\int_{a+\frac{1}{3n}}^{a+\frac{1}{n}}\rho_n(\xi)(\xi-x)d\xi\leqslant$$

$$\leqslant\int_x^{a+\frac{1}{3n}}(\xi-x)d\zeta+\int_{a+\frac{1}{3n}}^{a+\frac{1}{n}}(-\tfrac{3}{2}n(\xi-(a+\tfrac{1}{n}))(\xi-x)d\xi\leqslant$$

$$\leqslant\frac{1}{2}(a+\frac{1}{3n}-x)^2+\frac{3n}{2}\int_x^{a+\frac{1}{n}}|\xi-x|\,|\xi-(a+\tfrac{1}{n})|d\xi\leqslant$$

$$\leqslant\frac{1}{2}(\frac{1}{3n})^2+\frac{3n}{2}\,\frac{1}{4}(a+\frac{1}{n}-x)^3\leqslant\frac{1}{18n^2}+\frac{3n}{4}(\frac{1}{n})^3\leqslant\frac{1}{2n},\quad\text{if } n\geqslant 2$$

If $\quad a+\frac{1}{3n}\leqslant x\leqslant a+\frac{1}{n}$, we have

(5.32) $$G_n(x)\leqslant\frac{3n}{2}\int_x^{a+\frac{1}{n}}|\xi-x|\,|\xi-(a+\tfrac{1}{n})|d\xi\leqslant\frac{3}{4n^2}\leqslant\frac{1}{2n},\quad\text{if } n\geqslant 2$$

In any case we have $G_n(x)\leqslant\frac{1}{2n}$ if $a\leqslant x\leqslant a+\frac{1}{n}$; then by definition of $H_n(\lambda)$, we have:

(5.33) $$H_n(G_n(x))=\tfrac{3}{2}nG_n(x)=\tfrac{3}{2}n\int_x^{a+\frac{1}{n}}\rho_n(\xi)(1-e^{x-\xi})d\xi\leqslant$$

$$\leqslant\tfrac{3}{2}n\int_x^{a+\frac{1}{n}}\rho_n(\xi)(\xi-x)d\zeta$$

Let us again distinguish two cases

i) if $a\leq x\leq a+\frac{1}{2n}$ we have

$$\int_x^{a+\frac{1}{n}}\rho_n(\xi)(\xi-x)d\xi=\int_x^{a+\frac{1}{2n}}\rho_n(\xi)(\xi-x)d\xi+\int_{a+\frac{1}{2n}}^{a+\frac{1}{n}}\rho_n(\xi)(\xi-x)d\xi\leq$$

$$\leq\rho_n(x)\int_x^{a+\frac{1}{2n}}(\xi-x)d\xi+\rho_n(a+\frac{1}{2n})\int_{a+\frac{1}{2n}}^{a+\frac{1}{n}}(\xi-x)d\xi\leq$$

$$\leq\rho_n(x)\frac{1}{2}(\frac{1}{2n})^2+\rho_n(x)\frac{1}{2}(\frac{1}{n})^2\leq\rho_n(x)\frac{1}{n^2}(\frac{1}{8}+\frac{1}{2})\leq\rho_n(x)\frac{2}{3n} \text{ if } n\geq 2$$

ii) if $a+\frac{1}{2n}\leq x\leq a+\frac{1}{n}$ we have

$$\int_x^{a+\frac{1}{n}}\rho_n(\xi)(\xi-x)d\xi\leq\rho_n(x)\int_x^{a+\frac{1}{n}}(\xi-x)d\xi\leq\rho_n(x)\frac{1}{2}(\frac{1}{2n})^2\leq\rho_n(x)\frac{2}{3n}$$

From (5.33) and i) and ii) we deduce again

$$H_n(G_n(x))\leq\rho_n(x)$$

e) Let us consider now the following

PROBLEM 5.1: For any fixed n=1,2,...., find $w^{(n)}$ such that

(5.34) $E(w^{(n)})=H_n(w^{(n)})$ in Q

(5.35) $w^{(n)}(b,t)=0 \quad 0<t<T$

(5.36) $w_x^{(n)}(0,t)-w^{(n)}(0,t)=-L_n(t), \quad 0<t<T$

(5.37) $w^{(n)}(x,0)=G_n(x), \quad 0\leq x\leq b$

Pr.5.1 is a "good" non linear problem for a parabolic equation, due to the definition of H_n; and it can be studied by different points of view: in the classical spaces of Hölder continuous functions (cf.e.g. [30], [43]) or in the spaces $H^{2,1}(Q)$ or $W_p^{2,1}(Q)$ (cf.e.g. [45], [25], [43]). It can be formulated also in an "equivalent" form as a variational inequality; more precisely in the following

PROBLEM 5.2: For n=1,2,... find $w^{(n)}$ such that

$$(5.38)\quad \begin{cases} w^{(n)} \in L^2(0,T;V),\ w_t^{(n)} \in L^2(0,T;V)\ (\text{i.e. } w^{(n)} \in H^1(Q), \\ \text{with } w_{xt}^{(n)} \in L^2(Q))\ ;\ w_t^{(n)} \in L^\infty(0,T;\ L^2(]0,b[)) \end{cases}$$

and

$$(5.39)\quad \begin{cases} \text{for almost every } t \text{ in } [0,T]: \\ (D_t w^{(n)}, v-w^{(n)}) + a(w^{(n)}, v-w^{(n)}) + J_n(v) - J_n(w^{(n)}) \geq \\ \geq L_n(t)(v(0)-w^{(n)}(0,t)) \quad \forall v \in V \end{cases}$$

$$(5.40)\quad w^{(n)}(x,0) = G_n(x) \qquad 0 \leq x \leq b.$$

Following Duvaut-Lions[25] ch.I § 5.6 we can solve Pr.5.2 using the Faedo-Galerkin method; we refer to that book for the proof. For every n we have one and only one solution $w^{(n)}$ of Probl.5.2, which satisfies in addition the following uniform extimates

$$(5.41)\quad \begin{aligned} &||w^{(n)}||_{L^2(0,T;V)} + ||w_t^{(n)}||_{L(0,T;V)} + \\ &+ ||w_t^{(n)}||_{L^\infty(0,T;L^2(]0,b[)} \leq C, \ \forall n \end{aligned}$$

with C indipendent of n. Moreover $w^{(n)}$ solves also Problem 5.1 (this can be easily obtained by the same proof of Propositions 2.3, 3.6, 4.6). We can now apply the regularity theorems in the L^p Sobolev spaces for the linear heat equation; we know (cf. [38] theor. 9.13) that the problem

$$(5.42)\quad \begin{cases} Ew = \psi \\ w(b,t) = 0 \quad , \quad 0<t<T \\ w_x(0,t) - w(0,t) = \psi_1(t) \ , \quad 0<t<T \\ w(x,0) = \psi_0(x) \quad , \quad 0 \leq x \leq b \end{cases}$$

has one and only one solution in $W_p^{2,1}(Q)$ depending by continuity from the data if

$$(5.43)\quad \psi \in L^p(Q),\ \psi_1 \in W_p^{\frac{1}{2}-\frac{1}{2p}}(]0,T[),\ \psi_0\ W_p^{2-\frac{2}{p}}(]0,b[)\ p \geq 2$$

and the following "compatibility" conditions at the corners (0,0) and (b,0) of Q are satisfied

$$(5.44)\quad \begin{cases} \psi_0(b) = 0, \quad \text{if } p \leq 3 \\ \psi_0(b) = 0 \quad \text{and } \psi_0'(0) - \psi_0(0) = \psi_1(0) \text{ if } p > 3 \end{cases}$$

In our case the "compatibility" conditions on the corners are satisfied because $G_n(b)=0$ and $G'_n(0)-G_n(0)=-L_n(0)$; from (5.20) and (5.21) and noting that

$$(5.45) \quad H^3(]0,b[)\in W_p^{2-\frac{2}{p}}(]0,b[) \quad \forall p\geq 1$$

(by virtue of imbedding theorems in "fractionary" Sobolev Spaces; see e.g. [51]); we have

$$(5.46) \quad L_n\in W_p^{\frac{1}{2}-\frac{1}{2p}}(]0,T[) \quad \forall p\geq 1$$

$$(5.47) \quad G_n\in W_p^{2-\frac{2}{p}}(]0,b[) \quad \forall p\geq 1$$

$$(5.48) \quad E(w^{(n)})\in L^{\infty}(Q)\subset L^p(Q) \quad \forall p\geq 1$$

We conclude that our solution $w^{(n)}$ of Pr.5.2 verifies

$$(5.49) \quad w^{(n)}\in W_p^{2,1}(Q) \quad \forall p\geq 2$$

and the norm of $w^{(n)}$ in $W_p^{2,1}(Q)$ is uniformly bounded because $|E(w^{(n)})|\leq 1$ in Q $\forall n$ and $L_n\to L$ in $W_p^{\frac{1}{2}-\frac{1}{2p}}(]0,T[)$ and $G_n\to G$ in $W_p^{2-\frac{2}{p}}(]0,b[)$; then we have

$$(5.50) \quad ||w^{(n)}||_{W_p^{2,1}(Q)}\leq C_p; \quad \forall p\geq 2, \ \forall n$$

where C_p depends only on p. From Sobolev's imbedding theorem we deduce from (5.50) that $w^{(n)}$, $w_x^{(n)}\in C^0(Q)$ and

$$(5.51) \quad ||w^{(n)}||_{C^0(\bar{Q})}+||w_x^{(n)}||_{C^0(\bar{Q})}\leq C$$

with C independent of n.

6. FURTHER PROPERTIES OF THE APPROXIMATIONS

a) Let us first prove that actually $w^{(n)}$ satisfies further smoothness properties. We shall use sometimes for the sake of simplicity the following notations:

$$\Gamma_o=\{(x,t);\ 0<x<b,\ t=0\}$$
(6.1)
$$B_o=\{(x,t);\ x=0, 0<t\leq T\},\quad B_b=\{(x,t);\ x=b,\ 0<t\leq T\}$$

PROPOSITION 6.1: There exists C_o, $0<C_o<a$, such that

(6.2) $$0\leq \Lambda_o(x)\leq w^{(n)}(x,t) \quad \text{in } \bar{Q} \qquad \forall n$$

where

(6.3) $$\Lambda_o(x)= \frac{(C_o-x)^2}{2} \text{ for } 0\leq x\leq C_o,\quad \Lambda_o(x)=0 \text{ for } C_o\leq x\leq b$$

PROOF: First let us remark that

(6.4) $$J_n(v_1)-J(v_2)\geq J_n(\sup(v_1,v_2))+J(\inf(v_1,v_2))\ \forall v\in V$$

Where J_n and J are defined respectively by (5.27) and (2.21); (6.4) follows easily from the definitions of J_n and J, recalling in particular (5.8). Moreover we can choose C_o such that

(6.5) $$C_o+\frac{C_o^2}{2}<\lambda_o \quad (\lambda_o \text{ given by } (4.5))$$

(6.6) $$\Lambda_o(x)\leq G(x)\leq G_n(x) \quad 0\leq x\leq b$$

since λ_o is positive and $G(x)$ is a continuous function with $G(0)>0$. Now using the notations (2.19) and (4.58) it is easy to see that

(6.7) $$a(\Lambda_o,v-\Lambda_o)+J(v)-J(\Lambda_o)\geq(\frac{C_o^2}{2}+C_o)(v(0)-\Lambda_o(0))\ \forall v\in V$$

Let us now prove that $\Lambda_o(x)\leq w^{(n)}(x,t)$; indeed let us set

(6.8) $$v=\sup(w^{(n)},\Lambda_o)=w^{(n)}+(\Lambda_o-w^{(n)})^+ \text{ in the inequality (5.39)}$$

(6.9) $$v=\inf(w^{(n)},\Lambda_o)=\Lambda_o-(\Lambda_o-w^{(n)})^+ \text{ in the inequality (6.7)}$$

(6.10)
$$(D_t w^{(n)},(\Lambda_o-w^{(n)})^+)+a(w^{(n)},(\Lambda_o-w^{(n)})^+)+$$
$$+J_n(\sup(w^{(n)},\Lambda_o)-J_n(w^{(n)})\geq L_n(t)(\Lambda_o(0)-w^{(n)}(0,t)),$$

$$-a(\Lambda_o,(\Lambda_o-w^{(n)})^+)+J(\inf(w^{(n)},\Lambda_o)-J(\Lambda_o)\geqslant$$
$$-(\frac{C_o^2}{2}+C_o)(\Lambda_o(0)-w^{(n)}(0,t))^+$$

from which, adding the two inequalities and recalling (6.4),(6.6) and (4.5), we obtain

$$-a(\Lambda_o-w^{(n)},(\Lambda_o-w^{(n)})^+)+(D_t(w^{(n)}-\Lambda_o),(\Lambda_o-w^{(n)})^+)\geqslant 0$$

or

$$-a((\Lambda_o-w^n)^+,(\Lambda_o-w^{(n)})^+)-\frac{1}{2}\frac{d}{dt}||(-w^{(n)}+\Lambda_o)^+||^2_{L^2(]0,b[)}\geqslant 0$$

from which

$$(6.11)\qquad \frac{d}{dt}||(-w^{(n)}+\Lambda_o)^+||^2\leqslant 0;$$

then, using also (5.37) and (6.6), we have $(-w^{(n)}+\Lambda_o)^+=0$ i.e. $w^{(n)}\geqslant\Lambda_o$.

b) We come back now to the smoothness properties of $w^{(n)}$. From (6.2) and (5.10) we deduce that in particular

$$(6.12)\qquad H_n(w^{(n)})\in H^{2,1}(Q)$$

Now using (5.20),(5.21) we can apply the regularity theorems for heat equation in L^2 spaces (see again [38] but also [48] chap.IV); the problem (5.42) has now one and only one solution in $H^{4,2}(Q)$, if $\psi\in H^{2,1}(Q)$, $\psi_1\in H^{5/4}(]0,T[)$, $\psi_o\in H^3(]0,b[)$ if the following compatibility conditions are satisfied

$$(6.13)\qquad \begin{cases}\text{there exists } z(x,t)\in H^{4,2}(Q) \text{ such that:}\\ z(x,0)=\psi_o(x), 0\leqslant x\leqslant b;\ z(b,t)=0, z_x(0,t)-z(0,t)=\\ =\psi_1(t), 0<t<T;\quad z_{xx}(x,0)-z_t(x,0)=\psi(x,0).\end{cases}$$

Taking $\psi=H_n(w^{(n)})$, $\psi_1=-L_n$, $\psi_o=G_n$ $\forall n$ fixed it is not difficult, using the trace theorems in the space $H^{4,2}(Q)$ (cf. Lions-Magenes [48] chap.IV), to prove that the compatibility condition is verified by $\{H_n(w^{(n)}),-L_n,G_n\}$. Then we obtain that

$$(6.14)\qquad w^{(n)}\in H^{4,2}(Q)\quad \forall n$$

(but let us remark that the norms of $w^{(n)}$ in $H^{4,2}(Q)$ are not uniformly bounded with respect to n). From (6.14) and the Sobolev imbedding theorem for the space $H^{4,2}(Q)$ (cf.e.g. [43]) we obtain

(6.15) $\quad w^{(n)} \in C^1(\bar{Q}), \; w^{(n)}_{xx} \in C^0(\bar{Q}), \quad \forall n$

Finally let us note some other <u>local regularity</u> of the functions $w^{(n)}$: for every point (x_o,t_o) such that

(6.16) $\quad (x_o,t_o) \in Q$

(6.17) $\quad$ or $(x_o,t_o) \in B_b$, $\;$ or $a+\frac{1}{n}<x_o<b, \; t_o=0$

we have

(6.18) $\begin{cases} \text{there exists a neighbourhood } I\,(x_o,t_o;n) \text{ of} \\ (x_o,t_o) \text{ (depending on } (x_o,t_o) \text{ and on } n) \text{ such} \\ \text{that } w^{(n)} \in W^{4,2}_p(I(x_o,t_o;n) \quad p \geq 2 \end{cases}$

We can apply indeed the local regularity results in the L^p spaces (see again [38]) since from (5.49) and (5.10) we have

(6.19) $\quad H_n(w^{(n)}) \in W^{2,1}_p(Q) \quad \forall p \geq \frac{3}{2} \quad \forall n$

and the boundary data are C^ω in a neighbourhood of these points: recall (5.36) and (5.20), (5.37) and (5.17) from the last of which we have $G_n(x)=0$, $a+\frac{1}{n}<x\leq b$.(On the contrary we have not the global regularity of $w^{(n)}$ in $W^{4,2}_p(Q)$ since G_n belongs only to $H^3(]0,b[)$). From (6.17) (6.18) and from the imbedding theorems in the space $W^{4,2}_p$ (see [43]) for $p>3$ we obtain:

(6.20) $\begin{cases} w^{(n)}_{xt} \text{ and } w^{(n)}_{xxx} \text{ are continuous in every point} \\ (x_o,t_o) \text{ satisfying (6.16) or (6.17)} \end{cases}$

Finally from (6.2) we have that for n sufficiently large $H_n(w^{(n)})=1$ in a neighbourhood of B_o; then recalling (5.36) and (5.20), from the local regularity results on the boundary for heat equation we have

(6.21) $\begin{cases} w^{(n)} \text{ is infinitely differentiable in a} \\ \text{neighbourhood of every point of } B_o \end{cases}$

In conclusion we can summarize the results obtained until now on $w^{(n)}$ by the

THEOREM 6.1: For every n there exists one and only one solution $w^{(n)}$ of Pr.5.1 (and Pr.5.2), satisfying:

i) the uniform (with respect to n) global regularity conditions: (5.38)-(5.41) and (5.49)-(5.50)

ii) the global regularity conditions (6.14),(6.15)

iii) the local regularity conditions (6.16),(6.17),(6.18), (6.20) and (6.21).

c) PROPOSITION 6.2: For every n sufficiently large, we have

(6.22) $$w_t^{(n)}-w_x^{(n)}\geq 0 \quad \text{in } Q$$

(6.23) $$w_t^{(n)}-w_x^{(n)}\leq \lambda^{(n)}(x,t) \quad \text{in } Q$$

where

(6.24) $$\begin{cases} \lambda^{(n)}(x,t)=\mu_2+\mu_1^{(n)}b-\mu_1^{(n)}(x,t) \text{ with} \\ \mu_1^{(n)}=\sup\limits_{0\leq t\leq T}(\ell_n(t)+1),\ \mu_2=\sup\limits_{0\leq x\leq a}(g(x)-x+1) \end{cases}$$

PROOF: Firstly, let us remark that, by (5.19) and (4.60), $\mu_1^{(n)}>0$, and, by (4.59) $\mu_2\geq 1$. For any fixed n, setting

$$z(x,t)=w_t^{(n)}(x,t)-w_x^{(n)}(x,t)$$

we have, by (6.2) and (5.35):

(6.25) $$z=w_t^{(n)}-w_x^{(n)}=-w_x^{(n)}\geq 0 \quad \text{on } B_b$$

Moreover, by (5.34),(5.37), we have

$$z(x,0)=w_t^{(n)}(x,0)-w_x^{(n)}(x,0)=w_{xx}^{(n)}(x,0)-H_n(w^{(n)}(x,0))-$$
$$-w_x^{(n)}(x,0)=G_n''(x)-H_n(G_n(x))-G_n'(x)=\rho_n(x)+\tilde{g}(x)-x-$$
$$-H_n(G_n(x));$$

then by (5.29) and (4.59):

(6.26) $$z=\rho_n+\tilde{g}-x-H_n(G_n)\geq 0 \quad \text{on } \Gamma_o$$

Moreover by (5.34),(5.36)

$$z_x(0,t)=w_{xt}^{(n)}-w_{xx}^{(n)}=w_{xt}^{(n)}-w_t^{(n)}-H_n(w^{(n)})=$$

$=-\ell_n(t)-H_n(w^{(n)}(0,t))=$ (by (6.2) and the definition of $H_n(\lambda)$, if n is sufficiently large) $=-\ell_n(t)-1$

and by (5.19) and (4.60) we obtain

(6.27) $\quad z_x=-\ell_n-1<0 \quad$ on B_o .

We also have by (5.34)

(6.28) $\quad E(z)=(D_t-D_x)H_n(w^{(n)})=H_n'(w^{(n)})z \quad$ in Q

Let us apply now the maximum principle to the operator $E(z)+cz$, with $c=-H_n'(w^{(n)})\leq 0$ (by (5.51) and (5.7)). Suppose that z has a negative minimum in $\bar{Q}$; then it must be on $B_o \cup \bar{\Gamma}_o \cup B_b$ because of (6.28) and the maximum principle; but it can t be on $\bar{\Gamma}_o \cup B_b$ since (6.25) and (6.26); then it must be on B_o, but on B_o by (6.27), $z_x<0$ and this is impossible in a point of B_o of negativ minimum. In order to obtain (6.23) let us consider the function

$z-\lambda^{(n)}$, where $z=w_t^{(n)}-w_x^{(n)}$ and $\lambda^{(n)}$ is defined by (6.24)

From (6.28),(6.22),(6.2),(5.7) we have

(6.29) $\quad E(z-\lambda^{(n)})=H_n'(w^{(n)})z+\mu_1^{(n)}\geq 0$

Then from the maximum principle the $\max(z-\lambda^{(n)})$ in $\bar{Q}$ must be on $B_o \cup \Gamma_o \cup B_b$. But we have, using (6.24),(5.35)

$$(6.30) \quad \begin{aligned} &(D_t+D_x)(z-\lambda^{(n)})=w_{tt}^{(n)}-w_{xx}^{(n)}-(D_t+D_x\lambda^{(n)}=-w_{xx}^{(n)}= \\ &=-w_t^{(n)}-H_n(w^{(n)})=0 \quad \text{on} \quad B_b \end{aligned}$$

Moreover, by (6.27)

(6.31) $\quad D_x(z-\lambda^{(n)})=-\ell_n(t)-1+\mu_1^{(n)}\geq 0 \quad$ on B_o

and, by (6.26)

$$(6.32) \quad \begin{cases} z-\lambda^{(n)}=\rho_n(x)+\tilde{g}(x)-x-H_n(G_n(x))-\mu_2-\mu_1(b-x)\leq \\ \leq\rho_n(x)+\tilde{g}(x)-x-H_n(G_n(x)-\mu_2\leq 1+\tilde{g}(x)-x-\mu_2\leq 0 \text{ on } \bar{\Gamma}_o \end{cases}$$

Then if the maximum would be positive it ought be on $B_o \cup B_b$; but this is in contradiction with the strong maximum principle and (6.30),(6.31) D_t+D_x and $-D_x$ are non tangential inward derivative. Then the maximum of

$z-\lambda^{(n)}$ is ≤ 0 and we have (6.23)

PROPOSITION 6.3: For n sufficiently large we have

$$(6.33) \qquad -\lambda_1^{(n)} \leq w_x^{(n)} \leq 0 \quad \text{in } Q$$

where

$$(6.34) \qquad \lambda_1^{(n)} = \sup_{0\leq t\leq T} L_n(t) + \frac{1}{n}$$

PROOF: The proof is similar to the proof of Prop.6.2; let us set now

$$z = w_x^{(n)} \quad \text{in } Q;$$

we have

$$(6.35) \qquad E(z) = H_n'(w^n)z \quad \text{in } Q$$

$$(6.36) \qquad z_x = w_{xx}^{(n)} = w_t^{(n)} + H_n(w^{(n)}) = 0 \quad \text{on } B_b$$

thanks to (5.34) and (5.35);

$$(6.37) \qquad z = -L_n + w^{(n)} \geq -L_n > -\lambda_1^{(n)} \quad \text{on } B_o$$

thanks to (5.36) and (6.2);

$$(6.38) \qquad z = G_n' \quad \text{on } \overline{\Gamma}_o$$

But using (6.34) and (5.18)

$$\lambda_1^{(n)} - x \geq L_n(0) + \frac{1}{n} - x = \int_o^b (\rho_n(\xi) + \tilde{g}(\xi) - \xi)\,d\xi + \frac{1}{n} - x =$$

$$= \int_o^x (\rho_n(\xi) + \tilde{g}(\xi) - \xi)\,d\xi + \int_x^b (\rho_n(\xi) + \tilde{g}(\xi) - \xi)\,d\xi + \frac{1}{n} - x \geq x - \frac{1}{n} +$$

$$+ \int_x^b (\rho_n(\xi) + \tilde{g}(\xi) - \xi)\,d\xi + \frac{1}{n} - x \geq \int_x^b (\rho_n(\xi) + \tilde{g}(\xi) - \xi)e^{x-\xi}\,d\xi = -G_n'(x)$$

and finally

$$(6.39) \qquad z = G_n' \geq -\lambda_1^{(n)} \quad \text{on } \Gamma_o.$$

Then we can conclude that the minimum of z must be on $\overline{\Gamma}_o \cup B_o$; so we have the condition

$$w_x^{(n)} \geq -\lambda_1^{(n)} \quad \text{in } \overline{Q}$$

For the second condition of (6.33) we have to evaluate the derivative $z_t - z_x$ on B_o; but this is exactly done by the formula (6.27), which may be written now with the present notation in the following form, for n sufficiently large

(6.40) $\quad z_t - z_x = w_{xt}^{(n)} - w_{xx}^{(n)} = -\ell_n - 1 < 0 \quad$ on B_o

Thus the maximum of z, which must be taken on $\bar{\Gamma}_o \cup B_o \cup B_b$ by (6.35) and the maximum principle for the operator $E(v)+cv$ with $c=-H_n'(w^{(n)})$, cannot be positive, because of (6.37),(6.40) and (6.39).

PROPOSITION 6.4: For n sufficiently large we have

(6.41) $\quad w^{(n)}(x,t) \leq L_n(t) \quad$ in $\bar{Q}$

PROOF: Immediately by the Proposition (6.3) and by (5.36).

Finally we have

PROPOSITION 6.5: There exists a constant $K>0$, independent of n, such that for n sufficiently large, we have

$$|w^{(n)}| \leq K, \ |w_x^{(n)}| \leq K, \ |w_t^{(n)}| \leq K, \ |w_{xx}^{(n)}| \leq K.$$

PROOF: It is enough to remark that for $n \to +\infty$, ℓ_n and L_n converge unformly in $[0,T]$ to ℓ and L respectively;

then $\lim_{n\to\infty} \mu_1^{(n)} = \mu_1 = \sup_{0\leq t\leq T} (\ell(t)+1)$; $\lim_{n\to\infty} \lambda_1^{(n)} = \lambda_1$

and Prop. 6.5 follows from Propositions 6.1, 6.2, 6.3, 6.4 and from (5.34).

7. CONVERGENCE OF THE APPROXIMATION AND CONCLUSIONS

a) We have all the tools for passing to the limit from $n\to\infty$ on the $w^{(n)}$; the extimate (5.41) is sufficient to pass to the limit in (5.39),(5.40) and to obtain that a subsequence of $\{w^{(n)}\}$, which we shall denote again by $\{w^{(n)}\}$converges to a function w solution of Pr. III'; we refer for the proof to the book of Duvaut--Lions [25] pag. 56-57. The same proof gives us also that

$$(7.0)\qquad w_t \in L^\infty(0,T;\ L^2(]0,b[))\cap L^2(0,T;V)$$

Moreover from (5.50) and from the Propositions(6.1), (6.2), (6.3), (6.4) and Sobolev's imbedding theorems we obtain that w verifies also (4.69),(4.70),(4.71); moreover we can prove by the same proof of Prop. 6.1, applied to the fonction w and Λ_1 that

$$(7.1)\qquad \begin{aligned} &0\leq\Lambda_0(x)\leq w(x,t)\leq\Lambda_1(x) \quad \text{in } \bar{Q}\ , \text{ where} \\ &\Lambda_1(x)=\frac{(\lambda_1-x)^2}{2},\ 0\leq x\leq\lambda_1,\ \ \Lambda_1(x)=0,\ \lambda_1\leq x\leq b,\ \lambda_1=\text{defined by (4.6)} \end{aligned}$$

Then Theorem 4.1 is proved except for the property $w_t\in C^0([0,T];\ L^2(]0,b[))$ about which we will come back later.

b) We have

PROPOSITION 7.1: Let $(x_0,t_0)\in\bar{Q}$; then

$$(7.2)\quad \begin{cases} \text{if } w(x_0,t_0)=0 \text{ then } w(x,t_0)=0 & \text{for } x_0\leq x\leq b \\ \text{if } w(x_0,t_0)>0 \text{ then } w(x,t_0)>0 & \text{for } 0\leq x\leq x_0 \end{cases}$$

$$(7.3)\quad \begin{cases} \text{if } w(x_0,t_0)=0 \text{ then } w(x,t)=0 \text{ for } x=x_0+t_0-t,\ 0\leq t\leq t_0 \\ \text{if } w(x_0,t_0)>0 \text{ then } w(x,t)>0 \text{ for } (x,t)\in\bar{Q} \text{ and} \\ \qquad\qquad x=x_0+t_0-t,\ \ t_0\leq t\leq T. \end{cases}$$

PROOF: It follows immediately from (4.71)

PROPOSITION 7.2: If Ω is defined by (4.66) then

(7.4) $\partial\Omega\cap Q$ does not contain segments parallel to the line $x+t=0$

(7.5) $\partial\Omega\cap Q$ does not contain segments parallel to the axis $t=0$

(7.6) $\partial\Omega\cap\{(x,t):\ t=0\}=\{(x,t):\ 0<x<a,\ t=0\}$

(7.7) $\{(x,t);\ 0<x<C_0,\ 0<t<T\}\subset\Omega$

PROOF: i) Suppose first that $\partial\Omega \cap Q$ contains a segment $\overline{P_1 P_2}$ parallel to the line $x+t=0$, with $P_1=(x_1,t_1)$ and $P_2=(x_2,t_2)$ $x_1+t_1=K$, $x_2+t_2=K$, $t_1<t_2$; then using (7.2) and (4.69), we obtain that w is solution of the following Cauchy problem

$$(7.8)\qquad \begin{cases} E(w)=I & \text{in } M \\ w=w_x=0 & \text{on } \overline{P_1 P_2} \end{cases}$$

where M is defined by

$$(7.9)\qquad M = \{(x,t):\ 0<x<K-t,\ t_1<t<t_2\}$$

From the uniqueness of this Cauchy problem we obtain that

$$(7.10)\qquad w(x,t)=K-x-t+e^{x+t-K}-T \qquad \text{in } \bar{M},$$

so that

$$w_x(0,t)-w(0,t)=t-K \qquad t_1\leq t\leq t_2;$$

then by Proposition 4.6 and (4.52) we have

$$-L(t)=t-K;\quad \text{hence } -\ell(t)=I \qquad t_1\leq t\leq t_2$$

which is contrary to our assumption (4.6). Thus (7.4) is proved.

ii) Suppose now that $\partial\Omega \cap Q$ contains an horizontal segment $\overline{P_1P_2}=\{(x,t):\ x_1<x<x_2,\ t=t_1\ ,\ 0<t<T\}$. First let us remark that by (7.3) Ω contains "upper neighbourhood" of $\overline{P_1P_2}$, that is for every $P_o \in \overline{P_1P_2}$ there exists a disk $B(P_o)$ with center P_o such that $\Omega \cap B(P_o)$ is composed by points (x,t) with $t\geq t_1$; then we have that w is a solution of the Cauchy problem

$$(7.11)\qquad \begin{cases} E(w)=I & \text{in } B(P_o)\cap\Omega \\ w=0 & \text{on } B(P_o)\cap \overline{P_1P_2} \end{cases}$$

Then w must be regular in particular $w\in C^1(\overline{\Omega\cap B(P)})$ and we must have $w_t=-I$ on $B(P_o)\cap \overline{P_1P_2}$ and consequentely w must be negativ in a neighborhood of the point P_o, which is impossible by (7.1). Thus (7.5) is proved

iii) In order to prove (7.6) let us remark that the segment $\{(x,t);\ 0\leq x\leq a,\ t=0\}$ must belong to $\partial\Omega$ because

$w(x,0)=G(x)>0$, $0\leq x\leq a$. Suppose that a point $P_o=(x_o,0)$ with $x_o>a$ belongs to $\partial\Omega$; then by (7.2) we obtain that the whole segment $I=\{(x,t);\ a\leq x\leq x_o,\ t=0\}\subset\partial\Omega$ and then we have a contradiction, because in a neighbourhood of I w satisfies the Cauchy problem $E(w)=I$, $w(x,0)=0$ and, as before for problem (7.II), we have $w_t(x,0)=-I$, and then w must be negative for some $t>0$ near $t=0$; and this is in contradiction with (7.I).

iii) Finally (7.7) is an immediate consequence of (7.I).

PROPOSITION 7.3: The set Ω verifies (4.74).

PROOF: It follows immediately from Prop. 7.I and 7.2.

Now let us define for every $t\subset]0,T[$ the set

$$\Omega_t = \{x\in]0,b[:\ (x,t)\in\Omega\}$$

Since we have (7.7) Ω_t is non empty for $t\in]0,T[$ and we can define

$$(7.I2)\qquad s(t)=\sup_{x\in]0,b[}\Omega_t\ ,\qquad 0<t<T$$

Thanks to the Prop. 7.3 (see also the Remark 4.4) s(t) is "lipschitzian graph" with respect to the axis (x',t') with $x'=\alpha t+\beta x$ $t'=-\beta t+\alpha x$ where $\alpha=\cos\frac{\pi}{8}$, $\beta=\operatorname{sen}\frac{\pi}{8}$. Then we can define

$$(7.I3)\quad s(0)=\lim_{t\to0+}s(t)\ ,\quad s(T)=\lim_{t\to T-}s(t)$$

By (7.6) we have $s(0)=a$, by (7.7) we have $s(t)>0$, $0\leq t\leq T$. Then we can conclude with the following

PROPOSITION 7.4: The set Ω and the function s defined by (7.I2),(7.I3) satisfy (4.73)(4.74).

Now we can conclude the proof of Theor. 4.I and give further regularity properties for w. In fact now we know that w solves the boundary value problem.

$$(7.I4)\quad\begin{cases}E(w)=\chi(\Omega)\ \text{in}\ Q,\ w=G(x)\ \text{on}\ \bar{\Gamma}_o,\quad w=0\ \text{on}\ B_b\\ w_x-w=-L\quad\text{on}\ B_o\end{cases}$$

where $\chi(\Omega)$ is the caracteristic function of a "good set", since Ω and s verify (4.73) (4.74). We can use, as in Section 6 b) the regularity results for the heat equation and we obtain in particular

$$(7.15)\quad \begin{cases} w \text{ is } C^{\infty} \text{ in } Q-\partial\Omega\cap Q,\ w\equiv 0 \text{ in } \bar{Q}-\bar{\Omega}, \\ w\in H^{4,2}(I(x_o,t_o)) \text{ where } I(x_o,t_o) \text{ is a neighbour-} \\ \text{hood of } (x_o,t_o), \text{ depending on } (x_o,t_o), \text{ for all} \\ (x_o,t_o)\in \bar{B}_o\cup (x,t);\ 0\leqslant x<a,\ t=0\};\ \text{then in parti-} \\ \text{cular } w\in C^1(\overline{I(x_o,t_o)}) \end{cases}$$

From (7.15) we have that w_t is continuous in $\overline{Q-\partial\Omega\cap Q}$, i. e. everywhere in Q except in the "free boundary"; moreover by (4.70) w_t is bounded in $\bar{Q}$; so we can deduce that the function $t\to w(t,x)$ is continuous on $[0,T]$, with values in $L^2(]0,b[)$ and the proof of Theor. 4.1 is complete.

c) Now let us define

$$(7.16)\quad \tilde{u}(x,t)=w_t(x,t)-w_x(x,t)+x \quad \text{in } \bar{Q}$$

$$(7.17)\quad u(u,t)=\tilde{u}(x,t) \quad \text{in } \bar{\Omega}$$

We have the following

THEOREM 7.1: The pair $\{s(t),u(x,t)\}$ defined by (7.12), (7.13),(7.17) is a"weak solution" of Pr.III in the sense of the Def.4.2.

PROOF: By 7.3 and 7.4 s verifies (4.72),(4.73),(4.74), (4.75). Moreover by (4.69),(4.70) we have $u\in L^{\infty}(\Omega)$, and by (4.60),(4.69), $u_x\in L^2(\Omega)$ and $\hat{\tilde{u}}\in H^{1,0}(Q)\cap C^{o}([0,T];L^2(]0,b[))$; moreover by Prop.4.6 $\tilde{u}(x,t)=x$ in $Q-\Omega$ and $u(x,t)\geqslant x$ in Ω. Next we have also in the sense of $C^0([0,T];L^2(]0,b[))$ $\tilde{u}(x,0)=w_t(x,0)-w_x(x,0)+x=w_{xx}(x,0)-\chi([0,a[)-G'(x)+x=g(x)$. Then we have only to prove that u verifies (4.21). Let us consider for every n the function

$$(7.18)\quad z=w_t^{(n)}-w_x^{(n)} \quad \text{in } \bar{Q}$$

which we have already considered in the proof of Prop. 6.2; we remember (see (6.27),(6.28) that we have

$$(7.19)\quad z_x(0,t)=-\ell_n(t)-1 \quad 0<t<T$$

(7.20) $E(z)=(D_t-D_x)H_n(w^n)$ in Q

Let us multiply (7.20) by $\phi\in C_*^\infty(\bar{Q})$ (see (3.9)) and integrate over Q: the first member gives us

$$(7.2I)\quad \iint_Q (z_{xx}-z_t)\phi dxdt=\int_0^T [z_x\phi]_{x=0}^{x=b} dt-\iint_Q z_x\phi_x dxdt - \\ -\int_0^b [z\phi]_{t=0}^{t=T} dx +\iint_Q z\ \phi_t dxdt=\int_0^T (I+\ell_n(t))\phi(0,t)dt - \\ -\iint_Q z_x\phi_x dxdt+\iint_Q z\phi_t dxdt$$

The second member gives us similarly

$$(7.22)\quad \begin{aligned} &\iint_Q (D_t-D_x)H_n(w_n)\phi dxdt=-\iint_Q H_n(w^n)\phi_t dxdt + \\ &+\int_0^T H_n(w^{(n)}(0,t))\phi(0,t)dt+\iint_Q H^n(w^n)\phi_x dxdt \end{aligned}$$

Finally, if we remark also that for n sufficiently large we have $H_n(w^{(n)}(0,t))=1$ we obtain

$$(7.23)\quad \begin{aligned} &\int_0^T (1+\ell_n(t))\phi(0,t)dt-\iint_Q z_x\phi_x dxdt+\iint_Q z\phi_t dxdt= \\ &-\iint_Q H_n(w^n)\phi_t dxdt-\int_0^T \phi(0,t)dt+\iint_Q H^n(w^n)\phi_x dxdt \end{aligned}$$

Let us remember now that from the a priori estimates on the $w^{(n)}$, we have in particular (see again [25]) for $\{w^{(n)}\}$(or for a subsequence still denoted by $\{w^{(n)}\}$)

$$\lim_{n\to\infty} w_t^{(n)}-w_x^{(n)}=w_t-w_x=\tilde{u}-x \qquad \text{in } L^2(Q) \text{ weakly}$$

$$\lim_{n\to\infty} (w_t^{(n)}-w_x^{(n)})_x=\tilde{u}_x+1 \qquad \text{" " "}$$

$$\lim H_n(w^{(n)})=\chi(\Omega) \qquad \text{in } L^\infty(Q) \text{ weak star}$$

then we obtain for $n\to+\infty$ from (7.23) and (7.18)

$$(7.24)\quad \iint_Q (\tilde{u}_x-1)\phi_x dxdt-\iint_Q (\tilde{u}-x)\phi_t dxdt=$$

$$=\iint_Q \chi(\Omega)(\phi_t-\phi_x)\,dxdt+\int_0^T \ell(t)\phi(0,t)\,dt$$

But $\tilde{u}_x-1=u_x-1$ in Ω, $\tilde{u}-x=0$ in $Q-\Omega$; then (7.24) becames exactly (4.21).

Theorem 7.1 proofs the existence part of Theorem 4.1; for the uniqueness we have to prove that if $\{s(t),u(x,t)\}$ is a weak solution in the sense of Def.4.2, and we construct the function w by the formulas (4.47),(4.42), (4.27), then w is a solution of Pr.III'. We do not give here the details of the proof (see |62|); but we have already given this proof by the Prop.4.4 and 4.5 when we proved the same property for the function w constructed starting by a "classical solution $\{s(t),u(x,t)\}$ of Pr.III.

8. FURTHER RESULTS OF REGULARITY

a) We will consider in this Section the regularity of the weak solution of Pr.III and the existence of a classical solution.
First let us define the function $\xi_n(x)$, for every fixed n, by

$$(8.1)\quad \begin{cases} \xi_n(x)=\frac{1}{n}\mathrm{sh}(\sqrt{n}(b-x)), & b-x_n\leq x\leq b \quad (\mathrm{sh}t=\frac{e^t-e^{-t}}{2}) \\ \xi_n(x)=\frac{1}{2}x_n^2-\sqrt{\frac{2}{n}}x_n+\frac{1}{n}+(\sqrt{\frac{2}{n}}-x_n)(b-x)+\frac{1}{2}(b-x)^2, & 0\leq x\leq x_n \end{cases}$$

where

$$(8.2)\quad x_n=\frac{1}{\sqrt{n}}\ \mathrm{arcsh}\ 1 \quad (\text{i.e. } \mathrm{sh}(\sqrt{n}x_n)=1)$$

We can suppose that b is so large that

$$(8.3)\quad \mathrm{arcsh}\ 1<b$$

and, for n sufficiently large,

$$(8.4)\quad a+\frac{1}{n}<b-x_n; \qquad \sup_{0\leq t\leq T} L_n(t)\leq b-1$$

It is easy verify, by the definition, that

$$(8.5)\quad \xi_n\in C^1([0,b])$$

and

$$(8.6)\quad \xi_n''(x)=1,\ 0\leq x<b-x_n;\ \xi_n''(x)=n\xi_n(x),\ b-x_n<x\leq b$$

Let us define

$$(8.7)\quad \tilde{H}_n(\lambda)=\begin{cases} 0, & -\infty<\lambda\leq 0 \\ n\lambda, & 0\leq\lambda\leq\frac{1}{n} \\ 1, & \frac{1}{n}\leq\lambda<+\infty \end{cases}$$

From (8.7),(5.5),(5.6),(5.7) we have

$$(8.8)\quad \tilde{H}_n(\lambda)\leq H_n(\lambda) \qquad \lambda\in\mathbb{R}$$

Now we can write for $\xi_n(x)$ the equation

$$(8.9)\quad \xi_n''(x)=\tilde{H}_n(\xi_n(x)),\quad 0<x<b$$

and the boundary conditions

$$(8.10)\quad \xi_n(b)=0,\ \xi_n'(b)=-\frac{1}{\sqrt{n}},\ \xi_n'(0)=-\sqrt{\frac{2}{n}}+x_n-b$$

Since $x_n \le \sqrt{\frac{2}{n}}$, we have

$$(8.11) \qquad \xi_n'(0) < -b$$

and we verify also by the definition (8.1) that

$$(8.12) \qquad \xi_n'(x) < -b+x, \; 0 \le x \le b-x_n; \; \xi_n'(x) < 0, \; b-x_n \le x \le b.$$

Then by the first of (8.10) we have

$$(8.13) \qquad \xi_n(x) \ge 0 \;, \; 0 \le x \le b$$

Moreover we can prove that

$$(8.14) \qquad G_n(x) \le \xi_n(x) \;, \; 0 \le x \le b$$

Indeed, since $G_n(b)=\xi_n(b)=0$, it is sufficient to prove that

$$(8.15) \qquad \xi_n'(x) \le G_n'(x) \;, \quad 0 \le x \le b.$$

Inequality (8.15) is obvious for $a+\frac{1}{n} \le x \le b$; in order to prove it for $0 \le x \le a+\frac{1}{n}$ let us remark that from (5.22) we have

$$-G_n'(x) = \int_x^{a+\frac{1}{n}} (\rho_n(\xi)+\tilde{g}(\xi)-\xi)e^{x-\xi}d\xi \le \int_x^{a+\frac{1}{n}} (\rho_n(\xi)+\tilde{g}(\xi)-\xi)d\xi \le$$

$$\le \int_x^{a+\frac{1}{n}} \rho_n(\xi)d\xi + \int_x^{a+\frac{1}{n}} (\tilde{g}(\xi)-\xi)d\xi \le \int_0^{a+\frac{1}{n}} \rho_n(\xi)d\xi - \int_0^x \rho_n(\xi)d\xi +$$

$$+ \int_0^{a+\frac{1}{n}} (\tilde{g}(\xi)-\xi)d\xi \le L_n(0) - \int_0^x \rho_n(\xi)d\xi \le L_n(0)+1-x \le (\text{by } (8.4),$$

$$\text{for } n \text{ sufficiently large}) \le b-x \le (\text{by}(8.12)) \le -\xi_n'(x)$$

Now let us define

$$(8.16) \qquad \tilde{j}_n(\lambda) = \int_{-\infty}^{\lambda} \tilde{H}_n(\xi)d\xi \qquad \forall \lambda \in \mathbb{R},$$

$$(8.17) \qquad \tilde{J}_n(v) = \int_0^b \tilde{j}_n(v)dx \qquad \forall v \in V.$$

We can verify easily, using (8.8), that $\tilde{j}_n(\lambda)$ and $j_n(\lambda)$, defined by (5.26), satisfy

$$(8.18) \qquad \tilde{j}_n(\sup(\lambda,\mu)) + j_n(\inf(\lambda,\mu)) \le \tilde{j}_n(\lambda) + j_n(\mu), \; \lambda,\mu \in \mathbb{R}$$

Now we can prove the

PROPOSITION 8.1: For n sufficiently large we have

(8.19) $\quad w^{(n)}(x,t)\leq\xi_n(x)$ in Q

where $\xi_n(x)$ is defined by (8.1).

PROOF: Using (8.9),(8.10) it easy to verify the inequality

(8.20) $\quad a(\xi_n,v-\xi_n)+\tilde{J}_n(v)-\tilde{J}_n(\xi_n)\geq(\xi_n-D_x\xi_n)(v-\xi_n)(0)\ \forall v\in V$

where a(u,v) is defined by (4.58) and $\tilde{J}_n$ by (8.17). Then let us take $v=\inf(\xi_n,w^{(n)})=w^{(n)}-(w^{(n)}-\xi_n)^+$ in the inequality (5.39) and $v=\sup(\xi_n,w^{(n)})=\xi_n+(w^{(n)}-\xi_n)^+$ in the inequality (8.20); we obtain

$$-a(w^{(n)},(w^{(n)}-\xi_n)^+)-(w_t^{(n)},(w^{(n)}-\xi_n)^+ +J_n(\mathrm{Inf}(\xi_n,w^{(n)})-$$

$$-J_n(w^{(n)})\geq-L_n(t)(w^{(n)}(0,t)-\xi_n(0))^+,$$

$$a(\xi_n,(w^{(n)}-\xi_n)^+)+\tilde{J}_n(\sup(\xi_n,w^{(n)}))-\tilde{J}_n(\xi_n)\geq$$

$$\geq(\xi_n(0)-D_x\xi_n(0))(w^{(n)}(0,t)-\xi_n(0))^+$$

from which, adding the two inequalities and recalling (8.17),(8.18),(5.27), we obtain

$$-a((w^{(n)}-\xi_n)^+,(w^{(n)}-\xi_n)^+)-(D_t(w^{(n)}-\xi_n),(w^{(n)}-\xi_n)^+)\geq$$

$$\geq(\xi_n(0)-D_x\xi_n(0)-L_n(t)(w^{(n)}(0,t)-\xi_n(0))^+;$$

then, using (8.13),(8.11),(8.14), we have

$$\frac{d}{dt}\int_0^b|(w^{(n)}(x,t)-\xi_n(x))^+|^2dx\leq0$$

and finally, by (8.14), we obtain $(w^{(n)}(x,t)-\xi_n(x))^+=0$ i.e. (8.19).

From (8.19),(6.2),(5.35),(8.10) we have the

COROLLARY 8.1: For n sufficiently large we have

(8.21) $-\frac{1}{\sqrt{n}}\leq w_x^{(n)}(b,t)\leq 0$, $0\leq t\leq T$.

Now let us introduce a new assumption on the function ℓ:

(8.22) $\ell\in H^1(]0,T[)$

Then we can suppose that ℓ_n verify (5.39) and also

(8.22') $\lim_{n\to\infty} \ell_n=\ell$ in $H^1(]0,T[)$.

PROPOSITION 8.2: Under the assumption (8.22) there exists a positive number C, such that, for n sufficiently large and for $0<\sigma\leq t\leq T$:

(8.23) $$\int_0^b |w_{xt}^{(n)}(x,t)|^2dx+\int_\sigma^T\int_0^b |w_{tt}^{(n)}(x,t)|^2dxdt\leq\frac{C}{\sigma}$$

PROOF: With the following notation

$$D_y v=v_y=D_t v-D_x v,$$

let us derive with respect to y the equation (5.34); we obtain

(8.24) $w_{xxy}^{(n)}-w_{ty}^{(n)}=H_n'(w^{(n)})w_y^{(n)}$

Setting, for fixed n, $z=w_y^{(n)}$, from Proposition 6.2 and 6.5 we have

(8.25) $0\leq z\leq 2K$ in Q, for n sufficiently large

Multiplying (8.24) by z_y we have

(8.26) $z_{xx}z_y-z_tz_y=H_n'(w^{(n)})zz_y$

Let us integrate (8.26) over $[0,b]$ for fixed t, $0<t<T$; the first term of (8.26) gives

(8.27) $$\int_0^b z_{xx}z_ydx-\int_0^b z_tz_ydx=-\int_0^b z_xz_{yx}+z_x(b,t)z_y(b,t)-$$

$$-z_x(0,t)z_y(0,t)-\int_0^b z_tz_ydx=-\frac{1}{2}\int_0^b D_yz_x^2dx+z_x(b,t)z_y(b,t)-$$

$$-z_x(0,t)z_y(0,t)-\int_0^b z_tz_ydx$$

From (6.27) $z_x(0,t)=-1-\ell_n(t)$, and from (5.36),(6.27)

$$z_y(0,t)=z_t(0,t)+1+\ell_n(t)=(w_t^{(n)}-w^{(n)}+w^{(n)}-w_x^{(n)})_t(0,t)+1+\ell_n(t)=w_{tt}^{(n)}(0,t)-w_t^{(n)}(0,t)+\ell_n(t)+1+\ell_n(t).$$

Then we have

$$z_x(0,t)z_y(0,t)=-(1+\ell_n(t))(w_{tt}^{(n)}(0,t)-w_t^{(n)}(0,t))-$$
$$-(1+\ell_n(t))(1+2\ell_n(t))$$

Similarly, using (5.35),(5.34),(5.6) we have

$$w_{xx}^{(n)}(b,t)=w_t^{(n)}(b,t)-H_n(w^{(n)}(b,t))=0;$$

then we obtain

$$(8.29)\qquad z_x(b,t)z_y(b,t)=(w_t^{(n)}-w_x^{(n)})_x\Big[(w_t^{(n)}-w_x^{(n)})_x-$$
$$-(w_t^{(n)}-w_x^{(n)})_x\Big](b,t)=(w_{tt}^{(n)}w_{tx}^{(n)})-2(w_{tx}^{(n)})^2-(w_{xx}^{(n)})^2+$$
$$+3w_{tx}^{(n)}w_{xx}^{(n)}-w_{tt}^{(n)}w_{xx}^{(n)})(b,t)=-2(w_{tx}^{(n)}(b,t))^2$$

The second term of (8.26) integrated over $[0,b]$, gives

$$(8.30)\qquad \int_0^b H_n'(w^{(n)})zz_y\,dx=\frac{1}{2}\int_0^b D_y\big[H_n'(w^{(n)})z^2\big]dx-\frac{1}{2}\int_0^b H_n''(w^{(n)})z^3dx$$

where we have, from (8.25),(6.22),(5.7)

$$(8.31)\qquad H_n''(w^{(n)})z^3\leq 0$$

Now from (8.26)...(8.31) we obtain

$$(8.32)\qquad \int_0^b D_y\{z_x^2+H_n'(w^{(n)})z^2\}dx+2\int_0^b z_tz_y\,dx+2(w_{tx}^{(n)}(b,t))^2\leq$$
$$\leq(1+\ell_n(t))(w_{tt}^{(n)}(0,t)-w_t^{(n)}(0,t))+(1+2\ell_n(t))(1+\ell_n(t))$$

On the other hand

$$(8.33)\qquad \int_0^b D_y\{\dots\}dx=\int_0^b D_t\{\dots\}dx-\int_0^b D_x\{\dots\}dx=$$

$$\int_0^b D_t\{\dots\}dx-z_x^2(b,t)-(H_n'(w^{(n)})z^2)(b,t)+z_x^2(0,t)+$$

$$+H_n'(w^{(n)}(0,t))z^2(0,t)$$

Then from (8.32),(8.33), recalling that $H_n'(w^{(n)})\geq 0$ and that $z_x(b,t)=w_{tx}^{(n)}(b,t)$, we have

$$\int_0^b D_t\{z_x^2+H_n'(w^{(n)})z^2\}dx+2\int_0^b z_t z_y dx\leq$$

$$(8.34)\qquad \leq(1+\ell_n(t))(1+2\ell_n(t))+(1+\ell_n(t))(w_{tt}^{(n)}(0,t)-w_t^{(n)}(0,t))+$$

$$+H_n'(w^{(n)}(b,t))z^2(b,t)$$

Integrating (8.34) over $[\tau,t]$, $0<\tau<t$ and recalling (5.19), (5.6), (8.22') and the Prop.6.5, we obtain

$$(8.35)\qquad \int_0^b z_x^2(x,t)dx+\int_0^b (H_n'(w^{(n)})z^2)(x,t)dx-\int_0^b z_x^2(x,\tau)dx-$$

$$-\int_0^b (H_n'(w^{(n)})z^2)(x,\tau)dx+2\int_\tau^t\int_0^b z_t z_y dxdt'\leq$$

$$\leq C'+\int_\tau^t (1+\ell_n(t))w_{tt}^{(n)}(0,t')dt'+\int_\tau^t \tfrac{3}{2}nz^2(b,t')dt'\leq$$

$$\leq(\text{since } z_x^2(b,t)=(w_x^{(n)}(b,t))^2\leq\tfrac{1}{n} \text{ by virtue of } (8.21))$$

$$\leq C'+(1+\ell_n(t))w_t^{(n)}(0,t)-(1-\ell_n(\tau))w_t^{(n)}(0,\tau)-$$

$$-\int_\tau^t \ell_n'(t')w_t^{(n)}(0,t')dt'+\tfrac{3}{2}(t-\tau)\leq C''$$

where C' and C" are positive numbers indipendent of n; t and τ. Recalling again that $H_n'(w^{(n)})\geq 0$, we obtain

$$(8.36)\qquad \int_0^b z_x^2(x,t)dx+2\int_\tau^t\int_0^b z_t z_y dxdt'\leq C''+\int_0^b z_x^2(x,\tau)dx+$$

$$+\int_0^b (H_n'(w^{(n)})z^2)(x,\tau)dx$$

Now let us integrate (8.36) with respect to τ over the interval $]0,\sigma[$, $0<\sigma<t$; we have

$$(8.37)\qquad \sigma\int_0^b z_x^2(x,t)dx+2\int_0^\sigma\int_\tau^t\int_0^b z_t z_y dxdt'd\tau \leq \sigma C''+$$

$$+\int_0^\sigma\int_0^b z_x^2(x,\tau)dxd\tau+\int_0^\sigma\int_0^b (H_n'(w^{(n)})z^2)(x,\tau)dxd\tau \leq$$

$$\leq (\text{from } (5.41),(5.50)) \leq C'''+\int_0^\sigma\int_0^b zD_yH_n(w^{(n)})dxd\tau \leq$$

$$\leq (\text{because } D_yH_n(w^{(n)})=zH_n'(w^{(n)})\geq 0 \text{ and we have}$$

$$(8.25)) \leq C'''+2K\int_0^\sigma\int_0^b D_yH_n(w^n)dxd\tau \leq C'''+$$

$$+2K\{\int_0^b H_n(w^{(n)}(x,\sigma)dx-\int_0^b H_n(w^{(n)}(x,0))dx-$$

$$-\int_0^\sigma H_n(w^{(n)}(b,t))dt+\int_0^\sigma H_n(w^{(n)}(0,t))dt\} \leq C^{IV}$$

where C'', C''', C^{IV} are positive number indipendent of n, ,t. Now putting in (8.37) the definition of z, we have

$$(8.38)\qquad \sigma\int_0^b |w_{tx}^{(n)}(x,t)-w_{xx}^{(n)}(x,t)|^2dx+$$

$$+2\int_0^\sigma\int_\tau^t\int_0^b (w_{tt}^{(n)}-w_{tx}^{(n)})(w_{tt}^{(n)}-2w_{tx}^{(n)}+w_{xx}^{(n)})dxdt'd\tau \leq C$$

from which, recalling (5.41),(5.50), the Prop.6.5, we deduce by a standard elementary estimate that there exists C indipendent of n,σ,t such that

$$(8.39)\qquad \sigma\int_0^b |w_{tx}^{(n)}(x,t)|^2dx+\int_0^\sigma\int_\tau^t\int_0^b |w_{tt}^{(n)}|^2dxdt'd\tau \leq C$$

for every n sufficiently large, which contains in particular (8.23).
From Prop.8.2 we deduce obviously, passing to the limit as $n\to\infty$, the following

PROPOSITION 8.3: Under the assumption (8.22), the solution w of Pr.III' satisfies

$$(8.40)\qquad \int_0^b |w_{tx}(x,t)|^2 dx + \int_\sigma^T\int_0^b |w_{tt}(x,t)|^2 dxdt \leq \frac{C}{\sigma}$$

for every $0<\sigma\leq t\leq T$, where C is indipendent of σ and t

b) Inequality (8.40) is very important, in order to obtain the regularity properties of w and s. First we can deduce the following

PROPOSITION 8.4: Under the assumption (8.22), the function w_t is continuous at every point $(s(t),t)$ for $0<t<T$.

PROOF: Let be fixed, $0<\sigma<T$; using (8.40) we obtain

$$(8.41)\qquad |w_t(x_1,t)-w_t(x_2,t)|\leq C'|x_2-x_1|^{\frac{1}{2}},\ \forall t,\quad \sigma\leq t\leq T,$$

where C' not depends on t in $[\sigma,T]$; then w_t is continuous in x on $[0,b]$, uniformely with respect to t in the interval $[\sigma,T]$. We have also that

$$(8.42)\qquad w_t(s(t),t)=0 \quad 0<t\leq T$$

Moreover for any continuous function $\psi(x)$, $0\leq x\leq b$, and $0<\sigma\leq t'<t''\leq T$, we have from (8.40)

$$(8.43)\qquad \left|\int_0^b (w_t(x,t'')-w_t(x,t'))\psi(x)dx\right| \leq \left|\int_{t'}^{t''}\int_0^b w_{tt}(x,t)\cdot\right.$$

$$\left.\cdot\psi(x)dxdt\right| \leq C_\psi |t''-t'|^{\frac{1}{2}}\left(\int_{t'}^{t''}\int_0^b |w_{tt}|^2 dxdt\right) \leq C_\psi \frac{C}{\sigma}|t''-t'|^{\frac{1}{2}}$$

where C_ψ depends only on ψ. Let us suppose that w_t is not continuous with respect to (x,t) in a point $\bar{P}=(\bar{s}(\bar{t}),\bar{t})$ $0<\bar{t}<T$. Then, in view of (8.41) there exists a sequence $\{t_i\}$ such that

$$(8.44)\qquad \lim_{i\to+\infty} t_i=\bar{t},\ \lim_{i\to+\infty} w_t(s(\bar{t}),t_i)=\gamma=0$$

Suppose $\gamma>0$ (similarly if $\gamma<0$). From (8.41) we get, for some $\delta>0$,

$$(8.45) \quad w_t(x_i t_i)-w_t(x,\bar{t})>\gamma/2 \quad \text{if} \quad |x-s(t)|<\delta$$

Taking now in (8.43), $t''=t_i$, $t'=\bar{t}$, $\psi(x)\geq 1$ if $(x-s(\bar{t}))<\delta/2$, $\psi(x)=0$ if $|x-s(\bar{t})|\geq\delta$, $\psi(x)\geq 0$ elsewhere, we then get a contradiction, because from (8.44),(8.43) we have

$$\lim_{i\to\infty}\int_0^b (w_t(x,t_i)-w_t(x,\bar{t}))\psi(x)dx=0$$

and that is impossible since we have (8.45).
Now we want to prove that the restriction of w_{xt} at $\bar{\Omega}$ is continuous for $0\leq x\leq s(t)$, $0<t\leq T$. In order to do that we can follow two methods: the first is suggested by the proof given by Kinderlehrer and Niremberg (see [41][42]) of the analogous assertion in the case of the Stefan problem. By applying a method of Bernstern to a suitable approximation of w_t it is possible to obtain some estimates from which the continuity of w_{xt} follows directly. The second methods, which we follows here, is suggested by the proof of the analoguous assertion in a free boundary problem studied by Friedman [32], Theor.4.6 (see also Friedman-Jensen [34] lemma 9.2). First of all let us prove the following

PROPOSITION 8.5: Under the assumption (8.22), for any $\sigma>0$ the function $s(t)$ is Hölder continuous with exponent 3/4 in $[\sigma,T]$.

PRROF: Let us take t',t'' with $0<\sigma\leq t'\leq t''\leq T$, and let us note that from Prop.7.3 we obtain that if $t''-t'$ is sufficiently small then the closed segment $\overline{p'p''}$ with $p''=(s(t')+t'-t'',t'')$ and $p'=(s(t')+t'-t'',t')$ belongs to Ω. Since $w_x(s(t),t)=0$, we have $0=w_x(s(t''),t'')-w_x(s(t'),t')=$
$=w_x(s(t''),t'')-w_x(s(t')+t'-t'',t'')+w_x(s(t')+t'-t'',t'')-$
$-w_x(s(t')+t'-t'',t')+w_x(s(t')+t'-t'',t')-w_x(s(t'),t')$
So we have

$$(8.46) \quad \int_{s(t')+t'-t''}^{s(t'')} w_{xx}(x,t'')dx+\int_{t'}^{t''} w_{xt}(s(t')+t'-t'',t)dt- -\int_{s(t')+t'-t''}^{s(t')} w_{xx}(x,t')dx=0$$

From (8.42) and from the equation $E(w)=\chi(\Omega)$ we can define by continuity from the left w_{xx} on the free boundary $(s(t),t)$ and we obtain $w_{xx}(s(t),t)=1$, $0<t<T$. Hence

$$(8.47)\quad \begin{cases} \displaystyle\int_{s(t')+t'-t''}^{s(t'')} w_{xx}(x,t'')dx=\{(s(t'')-s(t')+t''-t'\} \\ \{1+O(t''-t')\} \\ \displaystyle\int_{s(t')+t'-t''}^{s(t')} w_{xx}(x,t)dx=(t''-t')\{1+O(t''-t')\} \end{cases}$$

where $O(\lambda)$ denotes a function of λ which tends to zero if $\lambda\to 0$. We have also

$$w_{xt}(s(t')+t'-t'',t)-w_{xt}(s(t')+t'-t''-\eta,t)+$$

$$+\int_{s(t')+t'-t''-\eta}^{s(t')+t'-t''} w_{xtx}(x,t)dx=w_{xt}(s(t')+t'-t''-\eta,t)+$$

$$+\int_{s(t')+t'-t''-\eta}^{s(t')+t'-t''} w_{tt}(x,t)dx$$

Then

$$(8.48)\quad \begin{aligned} &\int_{t'}^{t''} w_{xt}(s(t')+t'-t'',t)dt=\int_{t'}^{t''} w_{xt}(s(t')+t'-t''-\eta,t)dt+ \\ &+\int_{t'}^{t''}\int_{s(t')+t'-t''-\eta}^{s(t')+t'-t''} w_{tt}(x,t)dxdt \end{aligned}$$

From (8.46),(8.47),(8.48) after integrating with respect to η, $0<\eta<\eta_o$, we get

$$\eta_o|s(t'')-s(t')|\leq\int_{t'}^{t''}\int_0^{\eta_o} w_{xt}(s(t')+t'-t''-\eta,t)|d\eta dt+$$

$$+\eta_o\int_{t'}^{t''}\int_{s(t')+t'-t''-\eta_o}^{s(t')+t'-t''} |w_{tt}(x,t)dt+\eta_o O(t'-t'')$$

So using (8.40) we obtain

$$\eta_o|s(t")-s(t')|\leq\tilde{C}\{\eta_o^{\frac{1}{2}}|t"-t'|+\eta_o|t"-t'|^{\frac{1}{2}}\eta_o^{\frac{1}{2}}+\eta_o|t"-t'|\}$$

where $\tilde{C}$ depends on σ only. Taking $\eta_o=|t"-t'|^{\frac{1}{2}}$ we have

$$|s(t")-s(t')|\leq\tilde{C}\{|t"-t'|+\{t"-t'|^{\frac{3}{4}}+|t"-t'|\}$$

and Prop. 8.5 follows.
Now let us recal a lemma of Cannon, Henry, Kotlov [20] (see also [21]) useful in this type of situation

LEMMA 8.1: Let z be a solution of $E(z)=0$ for $0<x<s(t)$, $\sigma<t<T$, where $s(t)$ is Hölder continuous in $[\sigma,T]$ with exponent $\alpha>\frac{1}{2}$ and let z continuous up to the boundary $x=s(t)$, with $z(s(t),t)=0$, $\sigma\leq t\leq T$. Then z_x is continuous up to the boundary $x=s(t)$, $\sigma\leq t\leq T$. From this Lemma and Prop.8.4 and 8.5 we obtain that w_{tx} is continuous for $0<x\leq s(t)$, $\sigma\leq t\leq T$, for every $\sigma>0$. Then from (8.46),(8.47) we have

$$(8.49)\qquad \frac{s(t")-s(t')}{t"-t'}=-\frac{1}{t"-t'}\int_{t'}^{t"}w_{xt}(s(t'),t)dt+O(t"-t')$$

and the second term of (8.49) converges to $-w_{xt}(s(t'),t')$ when $t"\to t'$; and the analogous assertion is valid if $t'\to t"$. Moreover under the assumption (8.22) using the regularity results near the boundary for heat equation (see. e.g. [48] chap.IV) and Sobolev imbedding theorem, we have that w_{xt} is continuous also on B_o. Then we proved the following

PROPOSITION 8.6: Under the assumption (8.22) the function w_{xt} is continuous for $0\leq x\leq s(t)$, $0<t\leq T$ and the function $s(t)$ is continuous differentiable for $0<t\leq T$ and we have

$$(8.50)\qquad s'(t)=-w_{xt}(s(t),t),\quad 0<t\leq T.$$

Now we can conclude that the "weak solution" of Pr.III $\{s(t),u(x,t)\}$ defined by (7.12),(7.13),(7.17), satisfies all the conditions of the Definition 4.1 of "classical solution", except for the continuity of u at the point (a,0). Indeed we have $u=w_t-w_x+x$ in Ω and we proved by Theor.4.1 that $w_x\in C^o(\bar{Q})$ and by (7.15) and Prop.8.4 that w_t is continuous in $\bar{Q}-\{(a,0)\}$. Actually we can prove a a little more. We have indeed, using (8.42),(8.40), for $t>0$

$$(8.51)\qquad \begin{aligned} &|w_t(x,t)|=|w_t(x,t)-w_t(s(t),t)|\leq|x-s(t)|^{\frac{1}{2}}\cdot\\ &\cdot\left(\int_x^{s(t)}|w_{tx}(\xi,t)|^2 d\xi\right)^{\frac{1}{2}}\leq|x-s(t)|^{\frac{1}{2}}C\ t^{-\frac{1}{2}}\end{aligned}$$

Then

$$(8.52)\qquad \begin{cases} w_t(x,t)\to 0 \text{ if } (x,t)\to(a,0) \text{ with the constraint}\\ t^{-1}|x-s(t)|=o(t)\end{cases}$$

In any case the continuity of u (and consequently of w_t) at the point (a,0) also holds, as a consequence of the results obtained by Friedman and Jensen [34] in their approach of Pr.III. In order to obtain this continuity by the method which we developed here, it is sufficient, for instance, to improve the estimate (8.40). Namely

$$(8.53)\qquad \int_0^b|w_{xt}(x,t)|^2dx+\iint_Q|w_{tt}(x,t)|^2dxdt\leq C,\ \text{for } 0\leq t\leq T,$$

with C indipendent of t. Then the same proof of Prop.8.4 gives the continuity of w_t also for t=0. But in order to prove (8.53) we have to modify the approximations given in Section 5 by H_n,ρ_n,G_n,L_n so that we can integrate (8.34) over $[0,t]$ and we can estimate the term

$$\int_0^b z_x^2(x,0)dx+\int_0^b H_n'(w^{(n)}(x,0))z^2(x,0)dx.$$

I believe that this is possible but I shall not attempt do it here. In any case we can refer to the paper of Friedman and Jensen [34] for the complete proof of the following

THEOREM 8.1: Under the assumption (8.22) there exists one and only one classical solution of Pr.III.

c) Now we can look for the infinite differentiability of the free boundary. One can prove that the function s(t) is infinite differentiable for 0<t<T and moreover that the function u is infinite differentiable for $0<x\leq s(t)$, 0<t<T. In order to do that, we can follow, for instance, two methods. We sketch here only the ideas of the methods.
The first method is due to Schaeffer [55], who developed it for the Stefan problem. We use the transformation

(8.54) $\eta = x - s(t)$, $z(\eta,t) = u(x,t) - s(t)$

in a neighborood of the free boundary. The heat equation for u is transformed in the following equation for z:

(8.55) $-z_{\eta\eta} + z_t - z_\eta s'(t) = -s'(t)$ for $-s(t) < \eta < 0$, $0 < t < T$.

Using step by step the standard estimates in the spaces of Hölder continuous functions for parabolic equations, one can prove that s(t) and u(x,t) have continuous derivatives of any order for $0 < t \leq T$, $0 < x \leq s(t)$. This method is followed by Friedman and Jensen [34], to whom we refer for the details of the proof.
The second method is due to Kinderlehrer([41],[42]), who developed it for the Stefan problem in the case of several space dimension and is based on the Legendre transform. For every fixed point $P_o(s(t_o),t_o)$ of the free boundary, with $t_o > 0$ we apply the transformation of $\bar{\Omega} \cap B$ (where B is a neighborood of P_o contained in Q) defined by

(8.56) $\xi = -w_x(x,t)$, $\tau = t$, $(x,t) \in \bar{\Omega} \cap B$

This mapping is C^1 by Prop.8.4 (note that in Ω we have $w_{xx} = w_t + 1$) and Prop.8.6, with Jacobian=1 at the point P_o. Then the mapping is non singualr and maps a neighborood of P_o, say $\Omega \cap B$, onto a region $\mathcal{U} \subset \{(\xi,\tau);\ \xi > 0\}$.Since $w_x(s(t),t) = 0$, the set $B \cap \partial\Omega$ is mapped onto a subset Σ of $\{(\xi,\tau);\xi = 0\}$. Then the Legendre transform of w is defined by

(8.57) $z(\xi,\tau) = x\xi + w(x,t)$

and verifies the equation

(8.58) $\frac{1}{z_{\xi\xi}} + z_\tau = -1$ in $\mathcal{U}$

with the boundary condition z=0 on Σ . This equation is a non linear parabolic equation to which it is possible to apply known results of regularity at the boundary and to obtain that z is infinite differentiable in $\mathcal{U} \cup \Sigma$. Thus

(8.59) $x = z_\xi(0,t)$ t varying in a neighborood of t_o

is a C^∞ parametrization of $\Omega \cap B$. Then s(t) is infinite differentiable in a neighborood of t_o.

We wish to remark that in order to obtain this result we do not need the continuity of w_t (or equivalently of u) also at the point (a,0). We can therefore state the following

THEOREM 8.2: Under the assumption (8.22) the weak solution $\{s(t),u(x,t)\}$ of Pr.III, defined by (7.12),(7.13), (7.17), satisfies the conditions

s(t) is infinite differentiable for $0<t<T$

u(x,t) is infinite differentiable for $0<x<s(t)$, $0<t<T$

Finally, we remark that the problem of the analyticity of the free boundary, under the assumption that $\ell(t)$ is analytic in $[0,T]$, was solved recently by Friedman ([68] Theor.4.3), who proved by very sharp and quite technical estimates that s(t) is analytic for $0<t\leq T$. With the limited terme at our disposal we cannot developed here its proof.

d) Further interesting properties of the classical solution of Pr.III has been proved by Friedman and Jensen in [34] n.10,11. In particular they provod that

(8.60) if $g'(x)\geq 0$ and $\ell(t)\leq 0$, then s(t) is decreasing,

(8.61) if $g'(x)\leq 0$ and $\ell(t)\geq 0$, then s(t) is increasing

(8.62) if $\ell(t)\equiv 0$ and $g'(x)$ changes sign a finite number m of times, then s(t) is piecewise monotone and the direction of monotocity changes at most in times

The proofs of these assertions is based on the use of the maximum principle for the function w_t, where w is defined, starting from a classical solution, by (4.9), (4.27),(4.42),(4.47). For instance if $g'(x)\geq 0$ and $\ell(t)\leq 0$ we get $w_t\leq 0$ in Q, then s(t) is decreasing. It would be interesting to prove these properties also for our "weak solution", using the approximations $w^{(n)}$ (as for the Propositions 6.2 and 6.3). But here we find the same type of difficulties as for the continuity of w_t at the point (a,0), where w is now the solution of Pr. III' (see the discussion at the end of this Section, part.b)).

COMMENTS AND REFERENCES

a) The theory and the proofs, given in Sections 5, 6,7,8 for Problems III and III' can be developed also for Problems II and II', in order to prove Theor. 3.1 and 3.2 and the properties given in Section 8. In this case we have to take H_n defined by (5.5)....(5.8) and $G_n(x)=0$, $L_n(t)=0$ and to define a decreasing sequence $\{f_n\}$ of sufficient regular functions f_n, for instance belonging to $C^\infty([0,b])$, uniformly bounded and converging, as $n \to +\infty$, to the function f, defined by (3.15), in $L^p(]0,b[)$ for every p>1. The proofs are simpler than in the case of Pr. III', in particular for Prop. 8.2. We find also the same difficulty in order to obtain the continuity of w_t at the point (a,0) as for the case of Pr.III. In any case the existence and the uniqueness of the classical solution of Pr.II are well known and they where proved by classical direct methods (see e.g. the books [30], [54] and the report [53] and the references therein; see also the following n.d)). For the approach with variational inequalities see Duvaut [23], [24] and in particular Friedman-Kinderlehrer [36], in which Theor. 3.1 and the analogous of Prop. 8.2 and 8.3 are proved essentially by the same method which we followed here.

For Pr.I the method works if g'(0)=0, more precisely under the assumptions (2.23), (2.39) and (2.1), defining H_n by (5.5)...(5.8), $G_n(x)=g(x)$, $L_n(t)=0$. There is only a difficulty in order to prove the inequality $w_t \leqslant 0$ in Q of the type already found and discussed in Section 8 b) and d) (i.e. how to get the inequality $w_t^{(n)}(x,0)= G_n''(x)-H_n(G_n(x)) \leqslant 0$). However, in the general case and in particular in the case of oxygen diffusion (where $g'(0)\neq 0$) our approach does not work. In order to prove the existence and the uniqueness of the solution of Pr.I' and the validity of (2.36), (2.37), (2.38) we can use (instead of Théor. 5.1 and 5.2, Chap. I of Duvaut-Lions [25]) the existence and uniqueness theorem of weak solutions of parabolic variational inequality of Brezis [14] (see Théor.II 9, Rémarque II.15 and Paragraphe II.2.4; see also Brezis-Friedman [16]). But for the complete proof of Théor. 2.1 we have to refer to the paper of Baiocchi and Pozzi [6] where also the validity of (2.41) and the proof of Theor.2.2 are proved, using a different approach of the variational inequality (2.24), which is based on "semidiscretization" by finite differences with respect to time variable (let us note only that (2.40) and (2.42) are not explicity written in [6] ; but their proof is an easy consequence of the regularity results of linear boundary value problems for heat equation in the space $W_p^{2,1}(Q)$, already used and quoted in Section 5 (see [38]).

By the way I wish to emphazise that for every problem a suitable approximation is needed. In order to prove existence and uniqueness of the solution of problems of the type as Pr.I',II',III', it is possible to apply the general results for abstract parabolic variational inequality in a weak or strong form (see [14],[25],[45],[49]). In particular for Problem as Pr. II', III', it is useful to use Théor. 5.1 and 5.2, chap.I, of [25], which give sufficient conditions on the data and the approximations of the data in order to obtain existence and uniqueness of the solution satisfying the following properties

$$w \in L^2(0,T;V),\ w_t \in L^2(0,T;V) \cap L^\infty(0,T;L^2(]0,b[))$$

On the other hand, if one need further regularity properties for the solution, one has to choose more convenient approximations. This is, e.g., the case of our approximations H_n, ρ_n, G_n, L_n, which we introduced here following Torelli [62], but with a modification in the definition of $H_n(\lambda)$, suggested by the approximations used by Friedman-Kinderlehrer [36] for the Stefan problem. This modification allowed us to obtain Prop. 8.2. I also whish to emphasize the importance of estimates for $w^{(n)}$ as in Prop. 8.1 (which is due to Torelli), not only from the theoretical point of view, but also for the numerical point of view.

b) The relations existing between variational inequality (briefly v.i.) and free boundary value problems (briefly f.b. v.p.), of elliptic and of evolution type, have been particularly studied in these recent years, starting from the work of Lewy-Stampacchia [44] who have shown that the solution of a v.i. with "obstacle" solves also a f.b.v.p. On the other hand we have seen (Pr.II and Pr.III) that there are f.b.v.p. which are not "directly" reducible to a v.i. The idea of introduce a changement of the unknown function as e.g. (3.11), (4.10), (4.27) ("regularizing" in a certain sense the original unknown) in order to obtain a "good" v.i. is due to Baiocchi, who introduced it in a steady-state f.b.v.p. for fluid flows through porous media [1]. This idea has been taken over by several authours for many other problems (see [2],[15],[17],[23],[24],[36],[58]...[62], the references given in [2]). Recently Baiocchi ([3][4]) was able to characterize all the f.b.v.p. for second order linear operators, that can be reduced to a v.i. (Pr.II and Pr. III,as we have seen, are of this type). On the other hand there exist also f.b.v.p. which can be solved using quasi-variational inequalities. This type of new inequalities was introduced by Bensoussan and Lions [9],[10], [11] in order to study problem of optimal stopping times or of impulsive controls and it turned out to be useful also in f.b.v.p. of fluid flows ([2]) or of the type of Pr.I ([37],[33]). For the relations between f.b.v.p., v.i. and control theory we refer also to the lectures of Lions [46],[47]. Finally let us remark that there also f.b.v.p. which have not yet been reduced to v.i. or quasi-variational inequalities [69].

c) J would like also remark that v.i. are particularly useful in f.b.v.p. in the case of several space dimension, which can not be treated by classical methods used for the case of one space dimension (see later d)). Thus for example in a Stefan problem which generalize our Pr.II, the reduction to a v.i. was quite useful in order to obtain very deep results on this problem. We refer to the papers of Duvaut [23], [24] (in which the reduction and the existence theorem for the solution of the v.i. are given), to the paper of Friedman and Kinderlehrer [36](where stronger properties of this solution are proved, of the same type of Theor. 3.1 and where also many interesting properties of the free boundary surface are obtained) and to the recent papers of Caffarelli [18][19], Kinderlehrer and Nirenberg [41][42] (who obtained the infinite differentiability of the free boundary surface in a favorable case).

Also Pr.III and an analogous problems for an incompressible fluid, were studied in two space dimension in a sequence of papers of Torelli ([58],[59],[60],[61], [63], and of Friedman-Torelli [64] (see also Friedman-Jensen [35]). Is is exactly in these papers that Torelli introduced the transformation of the type (4.47) (4.27) and a definition of weak solution that suggested our Def. 3.2 and 4.2.

From the numerical point of view the use of v.i. or quasi variational inequalities in f.b.v.p. seems to give very good results. I refer to the papers concerned with flows trough porous media, in particular with Pr.III obtained at the Laboratorio di Analisi Numerica in Pavia (see [5] and the references therein), and to the papers [39],[70],[12],[13],[22],[7].

d) Finally, we have to remark that in the one space dimensional case the use of v.i. for f.b.v.p. seems to be not so important as it is in the several space dimensional case. Indeed, it is well known that the Stefan problem and its generalizations to two phase and to some non linear boundary value conditions has been considered in many papers, using different methods (as heat potential theory and non linear integral equations). J refer to the books of Friedman [30] and Rubinstein [54] and to the report of Primicerio [53], which contains a very com prehensive literature on the subject until 1973. For mo re recent papers J wish to refer in particular to [20], [21],[26],[27],[55],[65],[66].

Similarly, it is possible to apply these methods al so to Pr.III, as Friedman-Jensen suggested in a final remark of their paper [34]. Problem I, in the case g'(0)=0, can be reduce, by derivation (see Section 2 a)) to a Stefan problem and from this point of view it has been studied using the alove methods (see [56]). However, in the case g'(0)≠0 (oxygen diffusion) J believe that the first correct proof of an existence and uniqueness theorem was obtained by Baiocchi and Pozzi [6], by means of the v.i. (2.24).

Concluding, I hope that these lectures (even though they were referred to the one space dimensional case) will be sufficient in order to get a picture of the im portance of v.i. in the treatment of f.b.v.p. of parabolic type and the increasing interest of f.b.v.p.

I also wish to thank my collegues C. Baiocchi, G.A. Pozzi and A. Torelli for the advices in preparing this paper.

REFERENCES

1. C. Baiocchi, Su un problema a frontiera libera connesso a questioni di idraulica, Ann. Mat. Pura Appl. (4) 92 (1972), 107-127; C.R.Acad. Sc. Paris, 273 (1971), 1215-1217
2. C. Baiocchi, Free boundary problems in the theory of fluid flow through porous media, International Congress of Mathematicians, Vancouver (1974), vl.2, 373-243
3. C. Baiocchi, Problèmes à frontière libre et inéquations variationnelles, C.R. Acad. Sc. Paris, 1976 seance 26 avril.
4. C. Baiocchi, Problèmes à frontière libre en hydraulique: milieux non homogènes. (to appear)
5. C. Baiocchi-F. Brezzi-V. Comincioli, Free boundary problems in fluid flow through porous media, Second Symp. on Finite Element Methods in Flow Problems, I.C.C.A.D., 1976, preprints, 409-420
6. C. Baiocchi-G.A. Pozzi, An evolution variational inequality related to a diffusion absorption problem (to appear on Appl. Math. and Optim).
7. C. Baiocchi-G.A. Pozzi, Error estimates and free boundary convergence for a finite difference discretization of a parabolic variational inequal. (to appear)
8. J. Bear, Dynamics of Fluids in Porous Media, Amer. Elsevier Publ. Comp. New York, 1972
9. A. Bensoussan-J.L. Lions, Contrôle impulsionnel et inéquations quasi-variationnelles d'évolution, C.R. Acad. Sc. Paris, A (1973), 1333-1338
10. A. Bensoussan-J.L. Lions, Problèmes de temps d'arrêt optimal et inéquations variationelles paraboliques, J. Appl. Anal.
11. A. Bensoussan-J.L. Lions - book in preparation
12. A.E. Berger, The truncation method for the solution of a class of variational inequalities, R.A.I.R.O. Anal. Numér., 10 (1976), 29-42
13. A.E. Berger-M. Ciment-J.C.W. Rogers, Numerical solution of a diffusion consumption problem with a free boundary, S.I.A.M. Journ. Num. Anal., 12 (1975), 646-672
14. H. Brezis, Problèmes unilateraux, J. Math. pures et appl., 51 (1972), 1-168
15. H. Brezis-G. Duvaut, Ecoulement avec sillage autour d'un symétrique sans incidence, C.R. Acad. Sc. Paris A 276 (1973), 875-878
16. H. Brezis-A. Friedman, Estimates on the support of solution of parabolic variational inequality, Illinois J. Math., 20 (1976), 82-97
17. H. Brezis-G. Stampacchia, Une nouvelle méthode pour

l'étude d'écoulement stationnaires, C.R. Acad. Sc. Paris A, 276 (1973), 129-132.

18. L.A. Caffarelli, The regularity of free boundaries in higher dimensions (to appear)

19. L.A. Caffarelli-N.M. Rivière, Smoothness and analiticity of free boundaries in variational inequalities, Ann. Sc. Norm. Sup. Pisa, IV, 3 (1976), 289-310

20. J.R. Cannon-D.B. Henry-D.B. Kotlow, Continuous differentiability of the free boundary of weak solutions of the Stefan problem - Bull. Amer. Math.Soc., 80 (1974), 45-48

21. J.R. Cannon-D.B. Henry-D.B. Kotlow, Classical Solutions of the One-dimensional Two-Phase Stefan Problem, Ann. Mat. pura e appl., IV, 107 (1975), 311-341

22. J. Crank-R.S. Gupta, A moving boundary problem arising from diffusion of oxigen in absorbing tissue, J. Inst. Math. Appl., 10 (1972), 19-33

23. G. Duvaut, Résolution d'un problème de Stéfan, C.R. Acad. Sci. Paris Sér. A, 276 (1973), 1461-1463

24. G. Duvaut, Résolution d'un problème de Stefan. New Variational Techniques in Math. Phys., C.I.M.E., Cremonese- 1974, 84-102

25. G. Duvaut-J.L. Lions, Les inéquations en mécanique et en physique, Dunod, Paris 1972 (english transl.: Springer, Berlin, 1975)

26. A. Fasano-M. Primicerio, General free-boundary problems for the heat equation . P.I,II,III, J. Math. Anal. and Appl.

27. A. Fasano-S.Kamin-M. Primicerio, Regularity of weak solutions of one dimensional two-phase Stefan problem, (to appear on J. Diff. Equat.)

28. A. Friedman, One dimensional Stephan problems with nonmonotone free boundary, Trans. Amer. Math. Soc., 133 (1968), 89-114

29. A. Friedman, The Stephan problem in several space variables, Trans. Amer. Math. Soc., 133 (1968), 51-87

30. A. Friedman, Partial Differential Equations of Parabolic Type, Prentice Hall, Englewood Cliffs 1964

31. A. Friedman, Regularity theorems for variational inequalities in unbounded domain and applications to stopping time problems, Arch. Rat. Mech. Anal. 52 (1973), 134-160

32. A. Friedman, Parabolic variational inequalities in one space dimension and smoothness of the boundary, J. Funct. Anal., 18 (1975), 151-176

33. A. Friedman, A class of parabolic quasi-variational inequalities - II (to appear)

34. A. Friedman-R. Jensen, A parabolic quasi-variational inequality arising in Hydraulics, Ann. Sc.Norm. Sup. Pisa, IV, 2 (1975), 421-468
35. A. Friedman-R. Jensen, Elliptic quasi-variational inequalities and application to a non-stationary problem in Hydraulics, Ann. Sc. Norm. Sup. Pisa, IV, 3 (1976), 47-88
36. A. Friedman-D. Kinderlehrer, A one phase Stefan Problem, Ind. Univ. Math. J., 24, 11 (1975), 1005-1035
37. A. Friedman-D. Kinderlehrer, A class of parabolic quasi variational inequalities, J. Diff. Equat.
38. P. Grisvard, Equations differentielles abstraites, Ann. Sc. Ec. Norm. Sup., 4, 2 (1969), 311-395
39. R. Jensen, Finite difference approximation to the free boundary of a parabolic variational inequality, to appear
40. S.L. Kamenostkaja, On Stefan's problem, Mat. Sbornik 53 (95), (1961), 488-514
41. D. Kinderlehrer-L. Nirenberg, The smoothness of the free boundary in the one phase Stefan problem, Séminaire Collège de France, mai 1976
42. D. Kinderlehrer-L. Nirenberg, A parabolic free boundary value problem (to appear)
43. O.A. Ladyzenskaja-V.A. Solonnikov-N.N. Ural'ceva, Linear and Quasi-linear Equations of Parabolic Type, Amer. Math. Soc. Translations, 23, Providence 1968
44. H. Lewy-G. Stampacchia, On the regularity of the solution of a variational inequality, Comm. Pure Appl. Math., 22 (1969), 153-188
45. J.L. Lions, Quelques Méthodes de résolution des problèmes aux limites non linéaires, Dunod-Gauthier-Villars Paris 1969
46. J.L. Lions, On free surface problems: methods of variational and quasi-variational inequalities, Int. Conf. on Comput. Methods in Non Linear Mechanics, Austin, 1974
47. J.L. Lions, Sur la théorie du contrôle, Proc. Intern. Congress of Math., Vancouver, 1974, I,139-154
48. J.L. Lions-E. Magenes, Problèmes aux limites non homogènes, vol. I,II,III, Dunod-1968-1970 (english transl.: Springer, Berlin, 1972)
49. J.L. Lions-G. Stampacchia, Variational inequalities, Comm. Pure Appl. Math., 20 (1967), 493-519
50. S.P. Neumann-P.A. Witherspoon, Variational principles for confined and nonconfined flow of ground water, Water Resources Res., 6 (1970), 1376-1382
51. S.M. Nikolskii, Approximation of functions of several variables and imbedding theorems, Springer, Berlin, 1975

52. O.A. Oleinik, A method of solution of the general Stefan problem, Sov. Math. Dokl. 1 (1960), 1350-1353

53. M. Primicerio, Problemi a contorno libero per equazione della diffusione, Rend. Sem. Mat. Torino, 32 (1973-74), 183-206

54. L.I. Rubinstein, The Stefan Problem, Amer. Math. Soc. Transl., 27 Providence, 1971

55. D. Schaeffer, A new proof of infinite differentiability of the free boundary in the Stefan problem, J. Diff. Equat., 20 (1976), 266-269

56. A. Schatz, Free boundary problems of Stefan type with prescribed flux, J. Math. Anal. Appl. 28 (1969) 569-580

57. B. Sherman, General one-phase Stefan problem and free boundary problems for the heat equation with Cauchy data prescribed on the free boundary, S.I. A.M.J. App. Math., 20 (1971) 555-570

58. A. Torelli, Un problème à frontière libre d'évolution en hydraulique, C.R. Acad. Sci. Paris, Ser.A, 280 (1975), 353-356

59. A. Torelli, Su un problema a frontiera libera di evoluzione, Boll. U.M.I. (4) 11 (1975), 559-570

60. A. Torelli, Su un problema non lineare con una condizione sulla frontiera, to appear on Ann. di Mat. pura e appl.

61. A. Torelli, On a free boundary value problem connected with a non steady filtration phenomenon (to appear on Ann. Sc. Norm. Sup. Pisa)

62. A. Torelli, Existence and uniqueness of the solution for a non-steady free boundary problem, (to appear)

63. A. Torelli, On a free boundary problem connected with a non steady filtration phenomenon of compressible fluid, (to appear)

64. A. Friedman-A. Torelli, A free boundary problem connected with non steady filtration in porous media (to appear)

65. P. Van Moerbeke, An optimal stopping problem for linear reward, Acta Math., 132 (1974), 1-41

66. P. Van Moerbeke, Optimal stopping and free boundary problems, Arch. Rat. Mech. Anal.

67. C. Li-Shang, Existence and differentiability of the solution of the two-please Stefan problem for quasi-linear parabolic equations, Chinese Math. Acta, 7 (1965), 481-496.

68. A. Friedman, Analyticity of the free boundary for the Stefan problem, Arch. Rat. Mech. and Anal., 61 (1976), 97-125.

69. S.L. Daniluk, Sur une classe de fonctionnelles intégrales à domain variable d'intégration, Act. Congrès Int. des Math. de Nice, vol. 2, Gauthier--Villars, Paris (1970) 703-715.

70. J.F. Ciavaldini-G.Tournemine, Une metode directe d'etude d'ecoulement subcritique par resolution numerique dans le plan de l'hodographe, Second Symp. on Finite Element Methods in Flow Problems, I.C.C.A.D., 1976, preprints, 421-431.

71. J.F. Ciavaldini, M. Pogu, G. Tournemine, Modele d'ecoulements autour de profils en presence de parois poreuses, Second Symp. on Finite Element Methods in Flow Problems, I.C.C.A.D., 1976, preprints, 463-469

CONTENTS:

CRITERIA OF HYPERBOLICITY FOR CONSTANT COEFFICIENT POLYNOMIALS

M. Münster

Université de Liège

1. INTRODUCTION

Let P_k $(k=0,\ldots,m)$ be a polynomial in n+1 variables, homogeneous of degree k and with constant coefficients.

We say that the polynomial

$$P = \sum_{k=0}^{m} P_{m-k}$$

is (c-)<u>hyperbolic</u> if $P_m(0,1) \neq 0$ and if

$$\left.\begin{array}{l} P(x,z) = 0 \\ x \in R^n,\ z \in C \end{array}\right\} \Rightarrow |\Im z| \leq c.$$

It is well known that the principal part of a hyperbolic polynomial is 0-hyperbolic.

Necessary and (or) sufficient conditions for a polynomial with hyperbolic principal part to be hyperbolic may be found in [2], [3] and [5]. We give here a new necessary and sufficient condition, which includes these results. Our proof is quite elementary : it does not rely on PUISEUX's series nor SEIDENBERG-TARSKI's lemma.

We give here all the details only for polynomials of degree 2 and 3. The general case can be treated in the same way (see [4], where other usefull criteria may be found).

Garnir (ed.), Boundary Value Problems for Linear Evolution Partial Equations. 313-318.

2. CRITERION OF HYPERBOLICITY

The polynomial P is hyperbolic if and only if P_m is hyperbolic and there exists a constant K such that

$$(1)\qquad |P_{m-k}| \leq K \max_{\substack{0\leq j\leq k \\ k\leq j'\leq m}} \{|D^j P_m|^{\frac{j'-k}{j'-j}} |D^{j'} P_m|^{\frac{k-j}{j'-j}}\} \text{ in } R^n \times C.$$

Here and in the sequel, D denotes differentiation with respect to the last variable.

PROOF

a) Sufficiency.

It is easily seen that if P_m is 0-hyperbolic, there exists a constant K' such that

$$(2)\qquad |D_\tau^j P_m| \leq \frac{K'|P_m|}{|\mathcal{R}\tau|^j} \text{ in } R^n \times \{\tau \in C, \mathcal{R}\tau \neq 0\}.$$

So, if condition (1) is satisfied, $|P_{m-k}|$ is dominated by $|P_m|$ for $|\mathcal{R}\tau|$ large enough (uniformly in $x \in R^n$) and this implies the hyperbolicity of P (see [5]).

b) Necessity, for $m = 2$.

For any $t > 0$, $\alpha > 0$, $x \in R^n$, $\tau \in C$, we have

$$\alpha^2 P(\frac{tx}{\alpha}, \frac{z+t\tau}{\alpha}) \equiv t^2 P_2(x,\tau) + tDP_2(x,\tau)z + \frac{D^2P_2}{2} z^2$$
$$+ \alpha[tP_1(x,\tau) + DP_1 z]$$
$$+ \alpha^2 P_0 = 0 \Rightarrow |\mathcal{R}(\frac{z+t\tau}{\alpha})| \leq c$$
$$\Rightarrow |\mathcal{R}z| \leq c\alpha + t|\mathcal{R}\tau|.$$

We shall first prove that it is impossible to find sequences $t > 0$, $\alpha > 0$, $x \in R^n$ and $\tau \in C$(*) such that $c\alpha + t|\mathcal{R}\tau| \to 0$, $|\alpha t P_1(x,\tau)| = 1$ and all the other coefficients tend to 0, except $\frac{D^2P_2}{2}$ (which is a constant $\neq 0$), i.e.

$$(3)\qquad \begin{cases} \alpha \to 0,\ t|\mathcal{R}\tau| \to 0,\ |\alpha t P_1(x,\tau)| = 1, \\ t^2 P_2(x,\tau) \to 0,\ t\, DP_2(x,\tau) \to 0. \end{cases}$$

(*) For short, we omit the indices for sequences.

Indeed, if it were possible, choosing a subsequence such that

$$\alpha\, t\, P_1(x,\tau) \to r \quad (|r| = 1),$$

we should obtain, passing to the limit,

$$\frac{D^2 P_2}{2} z^2 + r = 0 \Longrightarrow \quad z = 0,$$

or

$$\frac{r}{D^2 P_2} \geq 0.$$

But if we keep the same sequences t and α and change (x,τ) in $(-x,-\tau)$, we should also obtain

$$\frac{-r}{D^2 P_2} \geq 0,$$

which implies $r = 0$, a contradiction.
Now, (3) is equivalent to

$$(4) \quad P_1(x,\tau) \neq 0, \ \frac{1}{t\,|P_1(x,\tau)|} \to 0 \text{ and } t \max(|P_2(x,\tau)|^{\frac{1}{2}}, |DP_2(x,\tau)|) \to 0,$$

because

$$|D^2 P_2| \leq \frac{K' |P_2|}{|\mathcal{R}\tau|^2} \text{ in } R^n \times [\tau \in C : \mathcal{R}\tau \neq 0]$$

(see (2)) implies

$$t^2 |\mathcal{R}\tau|^2 \leq \frac{K'}{|D^2 P_2|} t^2 |P_2| \text{ in } R^n \times C$$

and therefore,

$$t^2 P_2(x,\tau) \to 0 \Longrightarrow t\,|\mathcal{R}\tau| \to 0.$$

Since (4) cannot hold for <u>any</u> sequences $t > 0$, $x \in R^n$ and $\tau \in C$, it is also impossible to find sequences $x \in R^n$ and $\tau \in C$ such that

$$P_1(x,\tau) \neq 0 \text{ and } \frac{1}{|P_1(x,\tau)|} \max(|P_2(x,\tau)|^{\frac{1}{2}}, |DP_2(x,\tau)|) \to 0. \quad (*)$$

(*) Here, we use the following simple property : <u>given two sequences</u> $a_m \geq 0$, $b_m \geq 0$ <u>such that</u> $a_m b_m \to 0$, <u>there exists a sequence</u> $t_m > 0$ <u>such that</u> $t_m a_m \to 0$ <u>and</u> $b_m/t_m \to 0$. Take, for instance, $t_m = \sqrt{b_m/a_m}$ if $b_m \neq 0$, $a_m \neq 0$; $t_m = m b_m$ if $a_m = 0$, $b_m \neq 0$; $t_m = 1/m a_m$ if $a_m \neq 0$, $b_m = 0$ and $t_m = 1$ if $a_m = b_m = 0$.

In other words, there exists a constant $\varepsilon > 0$ such that

$$P_1 \neq 0 \Longrightarrow \frac{1}{|P_1|} \max(|P_2|^{\frac{1}{2}}, |DP_2|) > \varepsilon \text{ in } R^n xC,$$

or, with $K = \frac{1}{\varepsilon}$,

$$|P_1| \leq K \max(|P_2|^{\frac{1}{2}}, |DP_2|) \text{ in } R^n xC,$$

q.e.d.

c) Necessity, for m = 3.

For any $t > 0$, $\alpha > 0$, $x \in R^n$, $\tau \in C$, we have

$$(5)\alpha^3 P(\frac{tx}{\alpha}, \frac{z+t\tau}{\alpha}) \equiv t^3 P_3(x,\tau) + t^2 DP_3(x,\tau)z + t\frac{D^2P_3(x,\tau)}{2}z^2 + \frac{D^3P_3}{6}z^3$$
$$+\alpha[t^2P_2(x,\tau) + t\,DP_2(x,\tau)z + \frac{D^2P_2}{2}z^2]$$
$$+\alpha^2[tP_1(x,\tau) + DP_1 z]$$
$$+\alpha^3 P_0 = 0 \Longrightarrow |\mathcal{R}z| \leq C\alpha + t|\mathcal{R}\tau|.$$

It can be proved as in b) that it is impossible to find sequences $t > 0$, $\alpha > 0$, $x \in R^n$, $\tau \in C$ such that $c\alpha + t|\mathcal{R}\tau| \to 0$, $|\alpha^2 tP_1(x,\tau)| = 1$ and all the other coefficients, except $\frac{D^3P_3}{6}$, tend to zero. From case b), applied to the hyperbolic polynomial DP, we have

$$|DP_2| \leq K \max\{|DP_3|^{\frac{1}{2}}, |D^2P_3|\} \text{ in } R^n xC.$$

Therefore,

$$\left.\begin{array}{l} \left.\begin{array}{l} t^2\,DP_3(x,\tau) \to 0 \\ t\,D^2P_3(x,\tau) \to 0 \end{array}\right\} \Longrightarrow t\,DP_2(x,\tau) \to 0 \\ \qquad \alpha \to 0 \end{array}\right\} \Longrightarrow \alpha t\,DP_2(x,\tau) \to 0$$

and we obtain, as in b), the inequality

$$(6) \qquad |P_1| \leq K \max\{|P_3|^{\frac{1}{3}}, |DP_3|^{\frac{1}{2}}, |D^2P_3|, |P_2|^{\frac{1}{2}}\} \text{ in } R^n xC.$$

In the same way, but taking now $|\alpha t^2 P_2(x,\tau)| = 1$, we find

$$(7)\ |P_2| \leq K \max\{|P_3|^{\frac{2}{3}}, |DP_3|, |D^2P_3|^2, |P_3|^{\frac{1}{2}}|P_1|^{\frac{1}{2}}, |DP_3|^{\frac{3}{4}}|P_1|^{\frac{1}{2}}, |D^2P_3|^{\frac{3}{2}}|P_1|^{\frac{1}{2}}\}$$

in $R^n xC$.

Replacing now, in (6), $|P_2|$ by the right member of (7), we obtain, with another constant K,

$$(8) \qquad |P_1| \le K \max\{|P_3|^{\frac{1}{3}}, |DP_3|^{\frac{1}{2}}, |D^2P_3|\} \text{ in } R^n \times C,$$

the desired inequality for P_1.
From (8), we see that

$$\left.\begin{array}{l} \left.\begin{array}{l} t^3 P_3(x,\tau) \to 0 \\ t^2 DP_3(x,\tau) \to 0 \\ t\ D^2P_3(x,\tau) \to 0 \end{array}\right\} \Rightarrow t\ P_1(x,\tau) \to 0 \\ \qquad\qquad \alpha \to 0 \end{array}\right\} \Rightarrow \alpha^2 t\ P_1(x,\tau) \to 0.$$

So, we can eliminate the terms including $|P_1|^{\frac{1}{2}}$ in (7), i.e., with another K,

$$(9) \qquad |P_2| \le K \max\{|P_3|^{\frac{2}{3}}, |DP_3|, |D^2P_3|^2\} \text{ in } R^n \times C.$$

To get a better inequality for $|P_2|$, note that it is impossible to find sequences $t>0$, $\alpha>0$, $x \in R^n$, $\tau \in C$ such that

$$c\alpha + t|P\tau| \to 0, \ |\alpha t^2 P_2(x,\tau)| = |tD^2P_3(x,\tau)| \ne 0,$$

and all the other coefficients in (5) divided by $tD^2P_3(x,\tau)$ tend to 0.
This leads to the existence of a constant K such that, in $R^n \times C$,

$$(10) \qquad \begin{cases} \text{either } |D^2P_3| \le K \max\{|P_3|^{\frac{1}{3}}, |DP_3|^{\frac{1}{2}}\} \\ \quad \text{or } \quad |P_2| \le K \max\{|DP_3|, |P_3|^{\frac{1}{2}}|D^2P_3|^{\frac{1}{2}}\}. \end{cases}$$

Combining (9) and (10), we find, with another constant K,

$$|P_2| \le K \max\{|P_3|^{\frac{2}{3}}, |DP_3|, |P_3|^{\frac{1}{2}}|D^2P_3|^{\frac{1}{2}}\},$$

the desired inequality for $|P_2|$.
The same proof could be used to prove the general case by induction on the degree of P (see [4]).

3. COROLLARIES

<u>Corollary</u> 1. (A. LAX [3]).

<u>If</u> P <u>is hyperbolic</u>, <u>then</u>, <u>any root of</u> P_m <u>with multiplicity</u> $\alpha+1$ $(\alpha>0)$ <u>is a root of</u> P_{m-k} <u>with multiplicity</u> $\ge \alpha+1-k$, <u>for any</u> $k \le \alpha$.

It is a root of P_{m-k} for any $k \leq \alpha$, because

$$D^j P_m(x_o,\tau_o)=0, \forall j \leq \alpha \Rightarrow D^j P_m(x_o,\tau_o)=0, \forall j \leq k \Rightarrow P_{m-k}(x_o,\tau_o)=0,$$

by condition (1). To obtain the good multiplicity, it remains to apply condition (1) to each τ-derivative of P.

Corollary 2. (L. HORMANDER [2] p. 136).

If P_m is hyperbolic and has only τ-roots of multiplicity $\leq k_o$ for any $x \neq 0$, then $P_m + \sum_{k=k_o}^{m} P_{m-k}$ is hyperbolic for any P_{m-k} homogenenous of degree $m-k$.

Indeed, for $k \geq k_o$, the two members of (1) are homogeneous of the same degree $m-k$ and the right member vanishes only at (0,0); so, there exists a constant K for which (1) is satisfied.

REFERENCES

[1] L. Garding, Linear hyperbolic partial differential equations with constant coefficients, Acta Math. 85, 1-62, (1950).

[2] L. Hörmander, Linear partial differential operators, Springer, Berlin, (1963).

[3] A. Lax, On Cauchy's problem for partial differential equations with multiple characteristics, Comm. Pure Appl. Math. 9, 135-169, (1956).

[4] M. Münster, On hyperbolic polynomials with constant coefficients, to appear.

[5] S.L. Svensson, Necessary and sufficient conditions for the hyperbolicity of polynomials with hyperbolic principal part, Arkiv för Mat. 8, 145-162, (1969).

STABILITY OF MOTION FOR SEMILINEAR EQUATIONS

Jeffrey Rauch*

University of Michigan, Ann Arbor, Michigan USA

§1. INTRODUCTION

The purpose of this paper is to discuss the asymptotic behavior as $t \to +\infty$ of solutions to semilinear equations of the form

$$\phi_t = A\phi + J(\phi) \tag{1.1}$$

where $\phi(t)$ takes values in a Banach space B, and A generates a C_0 semigroup on B. Of particular interest is the stability of equilibrium or periodic solutions of (1.1).

In section 2 we describe an abstract framework, similar to that in [12], where existence and uniqueness theorems for (1.1) can be obtained in essentially the same completeness (and with the same proofs) as for the ordinary differential equation $\phi_t = J(\phi)$. The main goal is to prove differentiable dependence on the initial values $\phi(0)$ and to show that the differential satisfies the linearized equations

$$v_t = Av + dJ_{<\phi(t)>}v . \tag{1.2}$$

With these facts one can attack the stability problem by Poincare's method of the first return map. If the differential of the nth return map is a contraction then one obtains asymptotic stability

* This research partially supported by the National Science Foundations under grant NSF GP 34260

Garnir (ed.), Boundary Value Problems for Linear Evolution Partial Equations. 319-349.

of the associated periodicorbit. Similar results are obtained for equilibria. The relationship with Floquet multipliers (spectrum of the differential of the first return map) is also discussed.

The main application of this technique is to the study of the asumptotic behavior of solutions to semilinear partial differential equations. As with ordinary differential equations, linearization yields limited information and when other methods work they are usually preferable. We present two applications to partial differential equations where either no other technique is available or linearization complements the information provided by other methods.

In §4 we discuss the scalar parabolic equation on $\Omega \subset \mathbb{R}^3$,

$$u_t - \Delta u + g(u) = f \quad \text{on } [0,\infty) \times \Omega \tag{1.3}$$

$$\frac{\partial u}{\partial \nu} = 0 \quad \text{on } [0,\infty) \times \partial\Omega \ . \tag{1.4}$$

By linearization one shows that if u_e is an equilibrium solution of (1.3),(1.4) with the property that the operator $-\Delta + g'(u_e)$ with domain $\{\psi \in H_2(\Omega) \,|\, \frac{\partial \psi}{\partial \nu} = 0 \text{ on } \partial\Omega\}$ is a <u>strictly</u> positive selfadjoint operator on $L_2(\Omega)$ then u_e is asymptotically stable. If one assumes the stronger condition that g is strictly increasing then under approriate conditions on f and g one can show that there is exactly one equilibrium, and using energy inequalities and some ideas from dynamical systems we show that every solution converges to u_e as $t \to +\infty$. Here is a case where three different techniques combine to give a fairly complete picture.

In §5 we discuss equilibrium solutions of the nonlinear wave equation on $\Omega \subset \mathbb{R}^3$,

$$u_{tt} - \Delta u + g(u) + a(x)u_t = f \quad \text{on } [0,\infty) \times \Omega \tag{1.5}$$

$$u = 0 \quad \text{on } [0,\infty) \times \partial\Omega \tag{1.6}$$

If u_e is an equilibrium solution of (1.5) - (1.6) the linearized equations at u_e are

$$v_{tt} - \Delta v + g'(u_e)v + a(x)v_t = 0 \quad \text{on } [0,\infty) \times \Omega \tag{1.7}$$

$$v = 0 \quad \text{on } [0,\infty) \times \partial\Omega \tag{1.8}$$

In contrast to the results in §4 the analysis of this linearized problem presents a challenge. If $a > 0$ and $-\Delta + g'(u_e)$ with domain $H_2(\Omega) \cap \overset{\circ}{H}_1(\Omega)$ is <u>strictly</u> positive then precise decay estimates for (1.7) - (1.8) have been obtained [9] and in particular one can conclude that u_e is asymptotically stable. We prove a less sharp but sufficient decay theorem by using a trick from classical mechanics to construct a Lyapunov function. Particularly striking is the fact that if g is strongly nonlinear then existence of global (in time) smooth solutions of (1.5) - (1.6) is an open problem. However, the asymptotic stability conclusion is that if $(u(0), u_t(0))$ is sufficiently close to $(u_e, 0)$ then there is a global smooth solution with these initial data and $(u(t), u_t(t)) \to (u_e, 0)$ is $t \to \infty+$. This is reminiscent of the low energy scattering theory of Strauss [14] which provides global solutions with special asymptotics for data in a restricted class even when global soluability is not known for arbitrary data. As an example if $g(u) = \gamma u + u|u|^{r-1}$ with $\gamma \geq 0$ and $r \geq 2$ then if $(u(0), u_t(0))$ is sufficiently small in $B \equiv H_2(\Omega) \cap \overset{\circ}{H}_1(\Omega)$ then there is a global solution, (u, u_t) is continuous with values in B and converges to zero in B at an exponential rate as $t \to +\infty$.

These results for the nonlinear wave equation generalize results of Sattinger [13] who assumes that g is real analytic and that a is not too large. His proof employs a perturbation series. Sattinger gives a beautiful interpretation of the positivity of $-\Delta + g'(u_e)$ as a continuum mechanics analogue of the classical results of Dirichlet and Lyapunov on stability of mechanical systems at a minimum of potential energy. This idea is also described in §5. A similar analogy between the parabolic equation (1.3) and gradient dynamical systems is mentioned in §4.

A third application of linearization, to the threshold problem for equations modelling the conduction of nerve impulses is described in [11, §4.1]. It was in that work that the author first realized the utility of these ideas.

In section six we describe some difficulties which arise when the techniques are applied to problems in high dimensions and/or with general boundary conditions. The solutions to some of these are sketched and some open problems remain. Specifically, we extend the results of §4 to all dimensions and those of §5 to dimensions less than eight.

I would like to thank Professor C. Dafermos , J. Smoller, W. Strauss, and M. Taylor for their advice and encouragement on this project. Without them the work could not have been done.

§2. Abstract Linearization Theorems.

We consider the initial value problem

$$\phi_t = A\phi + J(\phi) \tag{2.1}$$

$$\phi(t_o) = \phi_o \tag{2.2}$$

where $t \mapsto \phi(t)$ is a continuous function taking values in a Banach space B. The following assumptions are in force

(2.3) J: B → B is locally Lipshitzian, that is, there is a continuous function $c : \overline{\mathbb{R}}_+ \times \overline{\mathbb{R}}_+ \to \mathbb{R}$, monotonically increasing in each variable and such that for all ϕ, ψ B,

$$||J(\phi) - J(\psi)|| \le c(||\phi||, ||\psi||)\,||\phi - \psi||$$

(2.4) A is the infinitesimal generator of a C_o semi-group, e^{tA}, on B.

(2.5) J is reasonable smooth. Precisely we suppose that J is Fréchet differentiable, the map $\psi \to dJ_{\langle\psi\rangle}$ is continuous from B to Hom B, and, for any bounded set $\beta \subset B$ there is a constant c such that $\forall\, \phi, \psi \in \beta$

$$||J(\phi) - J(\psi) - dJ_{\langle\psi\rangle}(\phi - \psi)|| \le c||\phi - \psi||^2$$

Definition 1. A function $\phi \in C([t_o, t_1] : B)$ *is a solution of* (2.1) *if for every* $t \in [t_o, t_1]$

$$\phi(t) = e^{(t-t_o)A}\phi(t_o) + \int_{t_o}^{t} e^{(t-s)A} J(\phi(s))ds \tag{2.6}$$

Notice that the integrand in (2.6) is a continuous function of s with values in B so the integral is a Riemann integral. Second notice that a solution need *not* be strongly differentiable with respect to t and $\phi(t)$ need not be in $\mathcal{D}(A)$. The differential equation (2.1) is satisfied in a weak sense. With a few natural assumptions strongly differentiable solutions can be created, see [12]. The first order of business is to prove the existence and uniqueness of solutions. Since the methods are standard some of the arguments will only be outlined.

THEOREM 1. (Uniqueness of solutions) If ϕ and ψ in

$C([t_o, t_1]: B)$ are solutions of (2.1), (2.2) then $\phi = \psi$.

Proof. Choose $M \in \mathbb{R}$ so that

$$||e^{\tau A}|| + c(||\phi(t)||, ||\psi(t)||) \leq M . \qquad \begin{array}{l} t_o \leq t \leq t_1 \\ 0 \leq \tau \leq t_1 - t_o \end{array}$$

Since ϕ and ψ are solutions we have

$$\phi(t) - \psi(t) = \int_{t_o}^{t} e^{(t-s)A} (J(\phi(s)) - J(\psi(s)) \, ds .$$

Therefore $||\phi(t) - \psi(t)|| \leq M \int_{t_o}^{t} ||\phi(s) - \psi(s)||ds$ and Gronwall's inequality implies $\phi(t) - \psi(t) = 0$.

THEOREM 2 (Local existence). For any $\phi_o \in B$ and $t_o > 0$ there is a δ depending only on $||\phi_o||$ so that (2.1), (2.2) has a solution $\phi \in C([t_o, t_o + \delta] : B)$.

Proof (Picard iteration). For any $t_1 > t_o$ define $K : C([t_o, t_1] : B)\circlearrowleft$ by

$$K\phi(t) = e^{(t-t_o)} \phi_o + \int_{t_o}^{t} e^{(t-s)A} J(\phi(s)) \, ds .$$

We must find a fixed point of K . For $1 \geq \delta > 0$ define

$$\Omega = \{\phi \in C([t_o, t_o + \delta] : B) \ : \ ||\phi(t) - e^{(t-t_o)A} \phi_o|| \leq 1$$
$$\text{for } t \in [t_o, t_1]\}$$

$$M_1 = 1 + \sup_{0 \leq \tau \leq 1} ||e^{\tau A} \phi_o||$$

$$\sqrt{M} = \sup_{||z|| \leq M_1} ||J(z)|| + M_1 + c(M_1, M_1)$$

Then for $\phi, \psi \in \Omega$, $t \in [t_o, t_o + \delta]$

$$||(K\phi)(t) - e^{(t-t_o)A} \phi_o|| \leq (t - t_o)M$$

$$||K\phi(t) - K\psi(t)|| \leq (t - t_o)M$$

Choose $\delta \in (0, 1]$ so that $\delta M < 1$ then K is a contraction of Ω into itself so Banach's theorem implies that there is a fixed point in Ω.

THEOREM 3 (Patching together local solutions). If ψ_1 is a solution of (2.1) for $t_o \leq t \leq t_1$ and ψ_2 is a solution of (2.1) for $t_1 \leq t \leq t_2$ with $\psi_2(t_1) = \psi_1(t_1)$ then if

$$\phi(t) = \begin{cases} \psi_1(t) & \text{for } t \in [t_o, t_1] \\ \psi_2(t) & \text{for } t \in [t_1, t_2] \end{cases}$$

then ϕ is a solution of (2.1) for $t_o \leq t \leq t_2$.

Proof. We must show that (2.6) holds for $t_o \leq t \leq t_2$. For $t_o \leq t \leq t_1$ this is true since ψ_1 is a solution. For $t_1 \leq t \leq t_2$ one adds the following identities

$$\psi_2(t) = e^{(t-t_1)A}\psi_1(t_1) + \int_{t_1}^{t} e^{(t-t_1)A} J(\psi_2(s))\, ds$$

$$e^{(t-t_1)A}\psi_1(t_1) = e^{(t-t_1)A}\left[e^{t_1 A}\psi_1(t_1) + \int_{t_o}^{t_1} e^{(t_1-t_o)A} J(\psi_1(s))\, ds\right]$$

to complete the proof.

This patching result allows one to piece together a maximal solution.

THEOREM 4 (Maximal Orbits). For any $\phi_o \in B$ there is a $T_c \in \mathbb{R}_+ \cup \{\infty\}$ and a $\phi \in C([0, T_c) : B)$ which is a solution in every interval $[0, T] \subset [0, T_c)$ and such that either $T_c = \infty$ or $||\phi(t)|| \to \infty$ as $t \to T_c$.

The standard proof is omitted. Next we show that if one has a solution $\phi(t)$ for $0 \le t \le T$ then for initial data close to $\phi(0)$ solutions exist at least up to time T.

THEOREM 5 (Semiglobal existence). If $\phi \in C([0, T] : B)$ is a solution of (2.1) there is a neighborhood $\mathcal{O}$ of $\phi(0)$ such that for any $\psi_o \in \mathcal{O}$ there is a solution $\psi \in C([0, T] : B)$ with $\psi(0) = \psi_o$. In addition there is a constant c such that

$$||\psi(t) - \phi(t)|| \le c||\psi(0) - \phi(0)|| \tag{2.7}$$

for all $\psi_1 \in \mathcal{O}$, $0 \le t \le T$.

Proof. We derive an á priori estimate for solutions $\psi_1 \in C([0, \tau] : B)$ with $\tau \le T$. Subtracting the following identities

$$\psi(t) = e^{tA}\psi(0) + \int_0^t e^{(t-s)A} J(\psi(s))\, ds$$

$$\phi(t) = e^{tA}\phi(0) + \int_0^t e^{(t-s)A} J(\phi(s))\, ds$$

and letting $\delta(t) = \psi(t) - \phi(t)$ we find

$$\delta(t) = e^{tA}\delta(0) + \int_0^t e^{(t-s)A} (J(\psi(s)) - J(\phi(s)))\, ds \tag{2.8}$$

Choose $M \in \mathbb{R}$ such that

$$M > \sup_{0\le\tau\le T} (||e^{tA}|| + ||\phi(t)||) .$$

Then for δ we find the inequality

$$||\delta(t)|| \le M||\delta(0)|| + M\int_0^t c(||\psi(s)||, M)||\delta(s)||\, ds$$

Claim: If $||\delta(0)||M \exp(M\, c\, (M + 1, M)T) < 1$ then $||\delta(t)|| \le 1$ for $0 \le t \le \tau$.

Proof of Claim. If not there is a $T_1 \le \tau$ with $||\delta(T_1)|| = 1$ and $||\delta(s)|| < 1$ for $0 \le s \le T_1$. Then for $s \in [0, T_1]$, $||\psi(s)|| < M + 1$ so for $t \in [0, T_1]$

$$||\delta(t)|| \leq M||\delta(0)|| + M\ c\ (M+1,\ M) \int_0^t ||\delta(s)||\ ds\ . \quad (2.9)$$

Gronwall's inequality implies that for $t \in [0, T_1]$

$$||\delta(t)|| \leq ||\delta(0)||\ M \exp\ (M\ c\ (M+1,\ M)t)\ . \quad (2.10)$$

In particular $||\delta(T_1)|| < 1$ contradicting the choice of T_1.

We have now shown that if $||\psi(0) - \phi(0)||$ satisfies the inequality of the claim then $||\psi(t) - \phi(t)|| \leq 1$ for $0 \leq t \leq \tau$. In particular $||\psi(t)|| \leq M + 1$. This á priori estimate allows us to extend ψ to a solution for $0 \leq t \leq T$ and to conclude that $||\psi(t)|| \leq M + 1$ for $0 \leq t \leq T$. Then inequalities (2.9) and (2.10) hold for $t \in [0, T]$ which proves (2.8) with $c = M \exp\ (M\ c\ (M+1,\ M)T)$.

Let ϕ, ψ and $\mathcal{O}$ be as in the above theorem. Define the nonlinear solution operator by

$$S(t) : \mathcal{O} \to B \qquad 0 \leq t \leq T \quad (2.11)$$

$$S(t)\big(\psi(0)\big) = \psi(t)\ . \quad (2.12)$$

Our next goal is to show that, for each t, $S(t)$ is Fréchet differentiable and that the derivative $\gamma(t) = dS_{\langle\psi(0)\rangle}\gamma_o$ solves the linearized equations

$$\gamma_t = A\gamma + dJ_{\langle\psi(t)\rangle}\gamma \quad (2.13)$$

$$\gamma(0) = \gamma_o\ . \quad (2.14)$$

Notice that the continuity of ϕ and (2.5) imply that the map $t \to dJ_{\langle\psi(t)\rangle}$ is continuous with values in $\mathrm{Hom}(B)$. This is sufficient to insure that the linearized equations are solvable.

THEOREM 6 (Existence for linear equations). If $D \in C([o, T] : \mathrm{Hom}(B))$ and $\gamma_o \in B$ then there is a unique solution $\gamma \in C([0, T] : B)$ of $\gamma_t = A\gamma + D(t)\gamma$ with $\gamma(0) = \gamma_o$ in the sense that

$$\gamma(t) = e^{tA}\gamma_o + \int_0^t e^{(t-s)A} D(s)\ \gamma(s)\ ds \qquad 0 \leq t \leq T\ . \quad (2.15)$$

Proof. Local existence and uniqueness are entirely analogous to theorems one and two. In fact we could have let J depend on t in these results. To prove that solutions exist for $0 \leq t \leq T$ we need an á priori estimate. If $\tau \leq T$ and γ satisfies (2.15) for $0 \leq t \leq \tau$ we show that $||\gamma(t)|| \leq c||\gamma_o||$ with c independent of t, τ, γ_o. Choose M $\mathbb{R}$ with

$$M > \sup_{0\leq t\leq T} (||e^{tA}|| + ||D(t)||_{Hom(B)}) .$$

Then

$$||\gamma(t)|| \leq M||\gamma_o|| + M^2 \int_0^t ||\gamma(s)|| \, ds$$

and Gronwall's inequality completes the proof.

Given ψ a solution of (2.1) for $0 \leq t \leq T$ with $\psi(0) \in \mathcal{O}$ the linearized solution operator $S_L(t)$ is defined for $t \in [0, T]$ by

$$S_L(t) : B \to B \tag{2.16}$$

$$S_L(t)\gamma_o = \gamma(t) \tag{2.17}$$

where $\gamma(t)$ is the solution of (2.13) with $\gamma(0) = \gamma_o$. The main result is the following.

THEOREM 7 (Differentiable dependence on initial data). Suppose ϕ, $\mathcal{O}$, S are as above then for $t \in [0, T]$, $S(t)$ is a Fréchet differentiable map from $\mathcal{O}$ to B and for any $\psi_o \in \mathcal{O}$,

$$dS(t)_{\langle \psi_o \rangle} = S_L(t) .$$

Proof. We must show that $||S(t)(\psi_o + h) - S(t)\psi_o - S_L(t)h|| = o(||h||)$ as $h \to o$ in B. Let $\gamma(t) = S_L(t)h$, $\psi(t) = S(t)\psi_o$, $\tilde{\psi}(t) = S(t)(\psi_o + h)$, $\delta(t) = \tilde{\psi}(t) - \psi(t)$. Then as in the derivation of (2.8)

$$\delta(t) = e^{tA} h + \int_0^t e^{(t-s)A}[J(\tilde{\psi}(s)) - J(\psi(s))] \, ds .$$

By the smoothness (2.5) of J we have

$$\delta(t) = e^{tA} h + \int_0^t e^{(t-s)A} [dJ_{\langle\psi(s)\rangle} \cdot \delta(s) + \rho(s)]\, ds \quad (2.18)$$

where $||\rho(t)|| \le c||\delta(t)||^2$, $0 \le t \le T$ and c is uniform over all h with $||h|| \le 1$. Use the estimate (2.7) to conclude that for $t \in [0, T]$ there is a new constant c with

$$||\int_0^t e^{(t-s)A} \rho(s) ds|| \le c||\delta(0)||^2 = c||h||^2$$

Thus if $\nu(t) = \delta(t) - \gamma(t)$ we have

$$\nu(t) = \int_0^t e^{(t-s)A} dJ_{\langle\psi(s)\rangle} \nu(s) ds + O(||h||^2)$$

A, by now familiar, Gronwall estimate yields $||\nu(t)|| \le c||h||^2 = o(||h||)$. This is precisely the desired estimate.

<u>THEOREM 8</u>. <u>The map</u> $S(t)$ <u>is continuously Fréchet differentiable on</u> $\mathcal{O}$.

<u>Proof</u>. We must show that if $\psi_1(0)$ and $\psi_2(0)$ are nearby points of $\mathcal{O}$ and S_L^1, S_L^2 the associated linearized solution operators then $S_L^2(t)h - S_L^2(t)h$ is small uniformly for $||h|| \le 1$. Let $\gamma_1(t) = S_L^1(t)h$, $\gamma_2(t) = S_L^2(t)h$, $\delta = \gamma_1 - \gamma_2$. Then

$$\delta(t) = \int_0^t e^{(t-s)A} (dJ_{\langle\psi_1(s)\rangle} \gamma_1(s) - dJ_{\langle\psi_2(s)\rangle} \gamma_2(s)) ds . \quad (2.19)$$

For ψ_1 the lipshitz dependence (2.7) and smoothness of J allows us to choose $\eta > 0$ so that if $||\psi_1(0) - \psi_2(0)|| < \eta$, then $\psi_2(0) \in \mathcal{O}$ and

$$||dJ_{\langle\psi_1(s)\rangle} - dJ_{\langle\psi_2(s)\rangle}|| \le c||\psi_1(0) - \psi_2(0)||$$

for $0 \le s \le T$. In particular we have a uniform estimate $||dJ_{\langle\psi_i(s)\rangle}|| \le c'$. The difference of derivatives in the integral

2.19 is equal to

$$dJ_{\langle\psi_2(s)\rangle}\delta + (dJ_{\langle\psi_1(s)\rangle} - dJ_{\langle\psi_2(s)\rangle})\gamma_1(s)$$

so estimating crudely we have

$$||\delta(t)|| \leq \text{const} \int_0^t ||\delta(s)||ds + \text{const}\ ||\psi_1(0) - \psi_2(0)||$$

where the constants can be chosen uniformly for $||\psi_1(0) - \psi_2(0)|| < \eta$ and $||h|| \leq 1$. Gronwall's inequality yields

$$||\delta(t)|| \leq \text{const}\ ||\psi_1(0) - \psi_2(0)|| \quad \text{for} \quad 0 \leq t \leq T .$$

This estimate implies the continuous differentiability of S .

§3. INFINITESIMAL STABILITY AND STABILITY

In this section we show how Poincaré's method of the first return map and the theory of Floquet multipliers extends to the abstract setting.

Definition 2. Suppose $\phi \in C([0,\infty):B)$ is a solution of (2.1) periodic with period $p > 0$. Let $S_L(t)$ be the solution operator of (2.13), the linearized equations at ϕ. We say that ϕ is infinitesmally exponentially stable if there is a $t_0 > 0$ and a constant $\alpha > 0$ so that $||S_L(t)|| \leq e^{-\alpha t}$ for $t > t_0$.

There is a parallel definition for equilibria.

Definition 3. Suppose $\phi_e \in B$ is an equilibrium solution (independent of t) of (2.1). We say that ϕ_e is infinitesmally exponentially stable if there are positive numbers t_0, α with $||S_L(t)|| \leq e^{-\alpha t}$ for $t \geq t_0$, where S_L is the linearized solution operator at ϕ_e.

Each of these definitions has a variety of equivalent formulations. We summarize some of them below.

THEOREM 9. With ϕ, S_L and p as in definition 2 the following are equivalent.

1. ϕ is infinitesimally exponentially stable
2. $||S_L(t)|| \to 0$ as $t \to \infty$
3. There is an integer $n > 0$ such that $||S_L(np)|| < 1$
4. There is an integer $n > 0$ such that the spectrum of $S_L(np)$ is contained in $\{z \mid |z| < 1\}$.
5. The spectrum of $S_L(p) \subset \{z \mid |z| < 1\}$.

THEOREM 10. With ϕ_e and S_L as in Definition 3, the following are equivalent.

1. ϕ_e is infinitesimally asymptotically stable.
2. $||S_L(t)|| \to 0$ as $t \to \infty$
3. There is a $t > 0$ such that the spectrum of $S_L(t)$ is contained in $\{z \mid |z| < 1\}$.

Proof of Theorem 9. 1 => 2 => 3 => 4 are automatic. The periodicity of ϕ and the patching together principle (theorem 3) imply that $S_L(np) = S_L(p)^n$ so spectrum $S_L(np) = \{z^n \mid z \in \text{spectrum } S_L(p)\}$. Thus (4) and (5) are equivalent..

To see that 5 => 1 use the spectral radius formula.

$$1 > \text{spectral radius } S_L(p) = \lim_{n\to\infty} ||S_L(np)||^{1/p} .$$

choose $\rho \in \mathbb{R}$

$$1 > \rho > \text{spectral radius } S_L(p) .$$

Then $|| S_L(np) || \leq \rho^n$ for n large. Then since $S_L(t) = S_L(t - [t/p]p) S_L [t/p]p)$ where [ ] is the greatest integer symbol we have

$$||S_L(t)|| \leq \sup_{0 \leq t \leq p} ||S_L(t)|| \cdot \rho^{[t/p]}$$

which implies infinitesimal exponential stability. The proof of theorem 10 is analogous. In practice there is rarely a shortcut and criteria number 1 of these theorems must be proved directly. An exception is for equilibria of parabolic systems where $S_L(t)$ defines a C_o semigroup on B which is compact for $t > 0$.

THEOREM 11. Suppose that ϕ_e and S_L are as in Theorem 10. If $S_L(t)$ is compact for t large then the conditions of theorem 10 are equivalent to

4. There is a $\alpha < 0$ such that the spectrum of $A + dJ_{<\phi_e>}$ is contained in the halfspace $\{z | \text{Re} z \leq \alpha\}$

Proof. Under the hypothesis that the semigroup $S_L(t) = \exp t(A + dJ_{<\phi_e>})$ is eventually compact we have the spectral mapping theorem (see [5])

$$\text{spectrum } S_L(t) = \{e^{tz} | z \in \text{spectrum } A + dJ_{<\phi_e>}\}.$$

Thus (4) is equivalent to condition (3) of theorem 10.

The main results of this section assert that infinitesimal stability implies stability.

Definition 4. A periodic solution $\phi(t)$ of (2.1) is exponentially stable if there is a neighborhood $\mathcal{O}$ of $\phi(0)$ and positive numbers c, α such that if ψ is a solution of (2.1) with $\psi(0) \in \mathcal{O}$ then ψ exists in $0 \leq t < \infty$ and $||\psi(t) - \phi(t)|| \leq ce^{-\alpha t}||\psi(0) - \phi(0)||$ for $t \geq 0$.

Definition 5. An equilibrium solution ϕ_e of (2.1) is exponentially stable if there is a neighborhood $\mathcal{O}$ of ϕ_e and positive numbers c, α such that if ψ is a solution of (2.1) with $\psi(0) \in \mathcal{O}$ then ψ exists for $0 \leq t < \infty$ and $||\psi(t) - \phi_e|| \leq ce^{-\alpha t}||\psi(0) - \phi_e||$ for $t \geq 0$.

THEOREM 11. (1) If ϕ is an infinitesimally exponentially stable periodic solution of (2.1) then ϕ is exponentially stable.

(2) If ϕ is an infinitesmally exponentially stable equilibrium solution of (2.1) then ϕ is exponentially stable.

In both cases if $||S_L(t)|| \leq e^{-\alpha t}$ for t large then for any $0 < \alpha' < \alpha$ we have $||\psi(t) - \phi(t)|| \leq ce^{-\alpha' t}||\psi(0) - \phi(0)||$ for all $t \geq 0$ and $||\psi(0) - \phi(0)||$ sufficiently small.

Proof. We prove part (1). The second assertion is treated similarly. Choose an integer $n > 0$ so that $|| S_L(np)|| \leq e^{-\alpha(np)}$. If $0 < \alpha' < \alpha$ we may choose an open ball $\mathcal{O} = \{\psi \in B| || \psi - \psi(0)|| < r\}$ such that if $\psi(t)$ is a solution of (2.1) with $\psi(0) \in \mathcal{O}$ then

$$\psi \in C([0,np]:B) \text{ (theorem 5), and} \tag{3.1}$$

$$||S_L(np)|| \leq e^{-\alpha' np} \text{ (theorem 8)} . \tag{3.2}$$

where $S_L(np)$ is the Frechet derivative of $S(np)$ at $\psi(0)$. It follows from the mean value theorem ([4, 8.5]) that

$$||S(np)\psi - S(np)\tilde{\psi}|| \leq e^{-\alpha' np}||\psi - \tilde{\psi}|| \text{ for } \psi,\tilde{\psi} \in \mathcal{O} . \tag{3.3}$$

In addition, $S(np)\phi(0) = \phi(0)$ so $S(np)$ maps $\mathcal{O}$ onto itself. The patching theorem then shows that for $\psi \in \mathcal{O}$, $S(t)\psi$ exists for all $t \geq 0$. In addition, for any integer $k > 0$

$$\begin{aligned} ||S(knp)\psi - \phi(0)|| &= ||S(knp)\psi - S(knp)\phi(0)|| \\ &= ||S(np)^k\psi - S(np)^k\phi(0)|| \\ &\leq e^{-k\alpha' np}||\psi - \phi(0)|| \end{aligned} \tag{3.4}$$

from the contraction inequality (3.3). For any $t > 0$ let $k = [t/np]$ and $t_o = t - knp$ so $0 \leq t_o \leq np$. Since $\phi(t) = S(t)\phi(0) = S(to)S(knp)\phi(0) = S(to)\phi(0)$ we may apply the estimate (2.7) to obtain

$$\begin{aligned} ||S(t)\psi - \phi(t)|| &= ||S(to)S(knp)\psi - S(to)\phi_0|| \\ &\leq \text{const}||S(knp)\psi - \phi_0|| \end{aligned} \tag{3.5}$$

where the constant is independent of $\psi \in \mathcal{O}$ and $t \in \mathbb{R}+$. Inequalities (3.4) and (3.5) prove the exponential stability of ϕ.

§4. A PARABOLIC EQUATION WITH KERNEL.

In this section we study the local and global stability of equilibrium solutions of the boundary value problem

$$u_t - \Delta u + g(u) = f \quad \text{in} \quad [0,\infty) \times \Omega \tag{4.1}$$

$$\frac{\partial u}{\partial \nu} = 0 \quad \text{in} \quad [0,\infty) \times \partial\Omega \tag{4.2}$$

where Ω is a bounded open set in $\mathbb{R}^n$ lying on one side of its smooth boundary and $\frac{\partial}{\partial \nu}$ is differentiation is the direction normal to $\partial\Omega$. This problem is interesting because the associated linear problem $(g \equiv 0)$ may have no equilibria and when there is one there are many and none is asymptotically stable. This results from the fact that if u is a solution of (4.1), (4.2) then so is $u+$ constant. Any uniqueness and stability results must rely in an essential way on the function g. We first put this problem in the framework of §2.

Define a nonpositive self adjoint operator A on $L_2(\Omega)$ by

$$\mathcal{D}(A) = \{\psi \in H_2(\Omega) \,|\, \frac{\partial \psi}{\partial \nu} = 0 \quad \text{on} \quad \partial\Omega\}$$

$$A\psi = \Delta\psi \quad \text{for} \quad \psi \in \mathcal{D}(A) .$$

The space B is defined by

$$B = D(A) \tag{4.3}$$

$$||\psi||_B^2 = ||\psi||^2_{L_2(\Omega)} + ||A\psi||^2_{L_2(\Omega)} \quad \text{for} \quad \psi \in B .$$

Since $A \le 0$ it follows that $\exp tA$ is a Co contraction semigroup on B. We make the following assumptions.

$$f \subset B \tag{4.4}$$

$$\Omega \subset \mathbb{R}^n \text{ with } n \le 3 . \text{ Then } B \subset H_2(\Omega) \subset C(\overline{\Omega}) . \tag{4.5}$$

Generalizations to arbitrary n are discussed in §6.

$$g \in C^2(\mathbb{R}) . \text{ Then } J:\psi \to f - g(\psi) \text{ is a } C^2 \text{ map of } H_2(\Omega) \text{ to itself and } dJ_{\langle\psi\rangle}(h) = -g'(\psi)h . \tag{4.6}$$

In addition J maps B into itself for if $\psi \in B$ then, on $\partial\Omega$,

$$\frac{\partial}{\partial \nu}(g(\psi) - f) = g'(\psi)\frac{\partial \psi}{\partial \nu} - \frac{\partial f}{\partial \nu} = 0 - 0 .$$

Thus hypotheses (2.3) and (2.5) on J are satisfied, and (4.1) and (4.2) is equivalent to the abstract equation for ϕ with values in B ,

$$\phi_t = A\phi + J(\phi) . \tag{4.7}$$

Suppose $u_e \in B$ is an equilibrium solution of (4.1),(4.2) (equivalently 4.7) . The linearized equations at u_e are

$$v_t = -\Delta v + g'(u_e)v \quad \text{on} \quad [0,\infty) \times \Omega$$

$$\frac{\partial v}{\partial \nu} = 0 \quad \text{on} \quad [0,\infty) \times \partial\Omega \quad .$$

THEOREM 13. Suppose Ω, f, g, B are as in (4.3) - (4.6) and that $u_e \in B$ is a solution of the equilibrium equations $-\Delta u_e + g(u_e) = f$. If the self adjoint operator $-\Delta + g'(u_e)$ on $L_2(\Omega)$ with domain equal to B is strictly positive, say $-\Delta + g'(u_e) > aI > 0$ then there is a $c \in \mathbb{R}$ and a neighborhood $\mathcal{O} \subset B$ of u_e such that for any $b \in \mathcal{O}$ there is a unique solution $u \in C([0,\infty):B)$ of (4.1),(4.2) with $u(0) = b$ and $||u(t) - u_e||_B \leq ce^{-at}||u(0) - u_e||_B$ for all $t \geq 0$.

Remark 1 Under these hypotheses global solutions may fail to exist for some $b \in B$.

Remark 2 There is a geometric interpretation of the positivity of $-\Delta + g'(u_e)$. If $V: B \to \mathbb{R}$ is defined by

$$V(\psi) = \int_\Omega |\nabla \psi|^2 + G(\psi) - f\psi \, dx$$

Where $G(s) = \int_0^s g(\sigma)\, d\sigma$, then the fact that u_e is an equilibrium is equivalent to the statement that V has a critical point at u_e , that is $\delta V = 0$. The condition $-\Delta + g'(u_e)$ strict positive is equivalent to strict positivity of the second variation $\delta^2 V$. Thus $-\Delta + g'(u_e)$ positive forces V to have a strict minimum at u_e . Now the differential equation (4.1) is equivalent (see §5 for similar computations) to the gradient system $u_t = -\delta V_{<u>}$ and the conclusion of the theorem asserts that orbits starting near the strict minimum u_e converge to that minimum. After the proof of theorem 13 we will put further restrictions on g which imply $V(u_e) = \inf_{\psi \in B} V(\psi)$ and that every orbit approaches this absolute minimum as $t \to +\infty$.

Proof. We need only verify that $||S_L(t)||_{\text{Hom}(B)} \leq \text{const } e^{-a't}$ for some $a' > a$ and all $t \geq 0$. Now

$$S_L(t) = \exp t(-\Delta + g'(u_e)) .$$

Let $a' = \sup\{\lambda | \lambda \in \sigma(\Delta - g'(u_e))\}$ then $a' > a$ by hypothesis and

$$||S_L(t)||_{Hom(L_2(\Omega))} = e^{-a't} .$$

In addition the commutator $[S_L, -\Delta + g'(u_e)] = 0$ so for any $b \in B$

$$||(-\Delta + g'(u_e))S_L b||_{L_2} = ||S_L(-\Delta + g'(u_e))b||_{L_2}$$
$$\leq e^{-a't}||(-\Delta + g'(u_e))b||_{L_2} .$$

Thus $||S_L(t)||_{Hom(B)} \leq e^{-a't}$ and the proof is complete.

A simple sufficient condition for the positivity of $-\Delta + g'(u_e)$ is that $g'(u_e) \geq 0$ and $g'(u_e)$ is not identically zero. Then

$$((-\Delta + g'(u_e)\psi, \psi)_{L_2(\Omega)} = \int_\Omega |\nabla\psi|^2 + g'(u_e)\psi^2 \, dx$$

Let ω be an open set such that $g'(u_e) > c > 0$ on ω then

$$((-\Delta + g'(u_e))\psi, \psi)_{L_2(\Omega)} \geq \int_\Omega |\nabla\psi|^2 \, dx + c \int_\omega \psi^2 \, dx$$
$$\geq \text{const}||\psi||^2_{H_1(\Omega)} .$$

Specializing still further if g is strictly monotone in the sense that

$$g(s) > g(t) \quad \text{for any} \quad s > t \tag{4.8}$$

then any nonconstant equilibrium is exponentially stable.

In addition assuming (4.8) there is at most one equilibrium. for, if u_e and $\tilde{u}_e$ are equilibria the equation $-\Delta(u_e - \tilde{u}_e) + g(u_e) - g(\tilde{u}_e)$ in Ω implies

$$\int_\Omega |\nabla(u_e - \tilde{u}_e)|^2 + (u_e - \tilde{u}_e)(g(u_e) - g(\tilde{u}_e)) \, dx = 0$$

and it follows that $u_e = \tilde{u}_e$.

The existence of equilibria when g is monotone has recently been settled by Brezis (see [2],[8]). A necessary and sufficient condition for the existence of an equilibrium is that

$$g(-\infty) < |\Omega|^{-1} \int_{\Omega} f < g(+\infty) \quad . \tag{4.9}$$

Assuming (4.8) and (4.9) there is exactly one equilibrium, u_e, and it is stable. We next ask the more delicate question: Is it true that all solutions of (4.1),(4.2) converge to u_e as $t \to \infty$?

The first thing that we want to show is that for any $b \in B$ the solution of (4.1),(4.2) with $u(0) = b$ exists for all $t \geq 0$. To do this we must derive an á priori estimate for solutions. We sketch two proofs. First considering the second remark following Theorem 13 one sees that for $t \geq 0$, $V(u(t))$ is a decreasing function of t. Thus $V(u(t)) \leq V(u(0))$ for all $t \geq 0$. However, the basic estimate of McKenna and Rauch [8] asserts that if (4.9) holds then there are positive constants c_1 and c_2 such taht

$$V(\psi) \geq c_1 ||\psi||^2_{H_1} - c_2 \quad \forall \psi \in H_1(\Omega) \quad . \tag{4.10}$$

Thus $\{u(t)\}_{t \geq 0}$ is bounded in $H_1(\Omega)$. Some additional arguments are needed to show that $\{u(t)\}_{t \geq 0}$ is bounded in $H_2(\Omega)$. One such is presented below. Notice that when g is monotone G and therefore V are convex.

A second approach relies more heavily on monotonicity but does not use the inequality (4.10). The basic fact is that, for $t > 0$, $S(t)$ is a contraction on $L_2(\Omega)$. Precisely if u and v in $C([0,T]|B)$ are solutions of (4.1) then for $0 < t \leq T$

$$||u(t) - v(t)||_{L_2(\Omega)} < || u(0) - v(0)||_{L_2(\Omega)} \quad .$$

Equivalently,

$$||S(t)u(0) - S(t)v(0)||_{L_2} < ||u(0) - v(0)||_{L_2} \quad . \tag{4.11}$$

For the proof let $w = u - v$, then

$$w_t = \Delta w + g(u) - g(v)$$

Multiply by w and integrate over $[0,t] \times \Omega$ to obtain

$$\frac{1}{2}\int_\Omega w^2\, dx \Big|_0^t = -\iint_{[0,t]\times\Omega} |\nabla w|^2 + (u-v)(g(u)-g(v))\, dx\, dt < 0$$

which is the desired inequality.

Next we estimate $||\frac{\partial u}{\partial t}||_{L_2(\Omega)}$ when $u \in C([0,T]|B)$ satisfies (4.1). To do this we formally differentiate the equation with respect to t. If $z = \partial u/\partial t$ one finds that

$$z_t = \Delta z - g'(u)z \quad \text{on} \quad [0,\infty) \times \Omega \tag{4.12}$$

$$\frac{\partial z}{\partial \nu} = 0 \quad \text{on} \quad [0,\infty) \times \partial\Omega .$$

Multiply (4.12) by z and integrate over $[0,t]\times\Omega$ to obtain

$$\frac{1}{2}\int_\Omega z^2\, dx\Big|_0^t = -\iint_{[0,t]\times\Omega} |\nabla z|^2 + g'(u)z^2\, dx\, dt < 0 .$$

Formally this proves that

$$\left|\left| \frac{\partial u}{\partial t}(t)\right|\right|_{L_2(\Omega)} < \left|\left|\frac{\partial u}{\partial t}(0)\right|\right|_{L_2(\Omega)} \tag{4.13}$$

for all $t > 0$. To make this rigorous one uses the trick of smoothing first in t, making the above argument, then removing the smoothing. The conclusion is that $u \in C^1([0,T):L_2(\Omega))$ and (4.13) holds. Details of an entirely analogous proof can be found in [10].

Next we estimate $||u(t)||_{H_2(\Omega)}$. Let $h(t) = u_t(t) - f$. The above estimate shows that $||h(t)||_{L_2(\Omega)}$ is bounded independent of $t \geq 0$. The differential equation for u is

$$\Delta u - g(u) = h(t)$$

Multiply this identity by Δu and integrate over Ω to obtain

$$\int_\Omega |\Delta u|^2 + g'(u)|\nabla u|^2\, dx \leq c\left(\int_\Omega |\Delta u|^2\, dx\right)^{1/2} .$$

Since $g'(u) \geq 0$ this implies that $||\Delta u||_{L_2}$ is bounded independent of $t \geq 0$. We then use the coerciveness estimate

$$||u||_{H_2} \leq \text{const}\left(||\Delta u||_{L_2} + ||u||_{L_2} + \left|\left|\frac{\partial u}{\partial \nu}\right|\right|_{H_{3/2}(\partial\Omega)}\right).$$

Now $\frac{\partial u}{\partial \nu} = 0$ on $\partial\Omega$ and the other two terms are bounded independent of t for $0 \le t \le T$ so we obtain

$$||u(t)||_{H_2(t)} \le \text{const} \quad . \tag{4.14}$$

for $0 \le t \le T$ where the constant does not depend on T. This á priori estimate implies global existence with estimates (4.11) and (4.14) holding for all $t \ge 0$. We are halfway through the proof of

THEOREM 14. Suppose f,Ω,g,a and B are as in theorem 13 and in addition (4.8) and (4.9) are satisfied. Then there is exactly one equilibrium solution $u_e \in B$ and for any $b \in B$ there is a unique $u \in C([0,\infty): B)$, with $u(0) = b$, satisfying (4.1). In addition

$$||u(t) - u_e||_{H_2(\Omega)} = O(e^{-at})$$

as $t \to +\infty$.

Proof. We first use the smoothing property of the heat equation to show that $\{u(t)\}_{t \ge 1}$ lies in a precompact subset of $H_2(\Omega)$. Let $w = u - u_e$. Then,

$$w_t - \Delta w = g(u) - g(u_e) \overset{\text{def}}{\equiv} \chi(t) \quad \text{on} \quad [0,\infty) \times \Omega$$

$$\frac{\partial w}{\partial \nu} = 0 \qquad \text{on} \quad [0,\infty) \times \partial\Omega$$

Observe that $\chi \in C([0,\infty): B)$ and that $||\chi(t)||_B$ is bounded independent of $t \ge 0$. As before, let A be the nonpositive selfadjoint operator on $L_2(\Omega)$ given by the Laplacian with Neumann boundary condition, then for $t \ge 1$

$$w(t) = e^A w(t-1) + \int_{t-1}^{t} e^{(t-s)A}\chi(s)\, ds \tag{4.15}$$

To estimate $||w||_{H_3(\Omega)}$ we use the fact that $\mathcal{D}(A^{3/2}) \subset H_3(\Omega)$ and $||\psi||_{H_3} \le \text{const}(||A^{3/2}\psi||_{L_2} + ||\psi||_{L_2})$ for all $\psi \in \mathcal{D}(A^{3/2})$. Since $A \le 0$ it follows that e^A maps $L_2(\Omega)$ into $\mathcal{D}(A^{3/2})$ and we can estimate

$$||A^{3/2}e^A w(t-1)||_{L_2} \le \text{const}||w(t-1)||_{L_2}$$

$$\le \text{constant independent of } t \ge 1 \; .$$

The second term on the right of (4.15) is also in $\mathcal{D}(A^{3/2})$.

To see this let

$$\psi_\varepsilon = \int_{t-1+\varepsilon}^{t} e^{(t-s)A}\chi(s)\, dx$$

Then, for $\varepsilon > 0$, $\psi_\varepsilon \in (A^{3/2})$ and we must show that $\psi_0 \in \mathcal{D}(A^{3/2})$. As $\varepsilon \to 0$ $\psi_\varepsilon \to \psi_0$ in $L_2(\Omega)$, and for $\varepsilon > 0$

$$A^{3/2}\psi_\varepsilon = \int_{t-1+\varepsilon}^{t} A^{1/2}e^{(t-s)A}A\chi(s)\, dx\,.$$

For $\lambda \leq 0$ and $\tau > 0$, $\sqrt{\lambda}\, e^{\lambda\tau} \leq \text{const}\, \tau^{-1/2}$ so by the spectral theorem $||A^{1/2}e^{(t-s)A}||_{Hom(L_2)} \leq \text{const}(t - s)^{-1/2}$. Since

$||A\chi(s)||_{L_2}$ is bounded independent of s we may apply Lebesque's theorem to conclude that as $\varepsilon \to 0$

$$A^{3/2}\,\psi_\varepsilon \to \int_{t-1}^{t} A^{1/2}e^{(t-s)A}A\chi(s)\, ds \tag{4.16}$$

in $L_2(\Omega)$. Since $A^{3/2}$ has closed graph it follows that $\psi_0 \in \mathcal{D}(A^{3/2})$ and $A^{3/2}\psi_0$ is given by the integral in (4.16). Then

$$||A^{3/2}\psi_0||_{L_2} \leq \int_{t-1}^{t} \text{const}(t-s)^{-1/2} \sup_{s\geq 0}||A\chi(s)||\, ds$$

$$\leq \text{constant independet of } t \geq 1\,.$$

Thus, $\{w(t)\}_{t\geq 1}$ is a bounded subset of $H_3(\Omega)$ and therefore precompact in H_2. Since $u(t) = w(t) + u_e$, $\{u(t)\}_{t\geq 1}$ is precompact in $H_2(\Omega)$.

We next investigate the ω limit set of u. Let

$$K = \{k \in L_2(\Omega) | \exists t_1 < t_2 < \ldots \to +\infty \text{ such that } \lim_{n\to\infty} u(t_n) = k\}\,.$$

By the above remarks K is a nonempty subset of $B \subset H_2(\Omega)$. It follows immediately from the definition that K is invariant under S, that is

$$\text{For any } k \in K \text{ and } t \geq 0,\ S(t)k \in K\,. \tag{4.18}$$

In addition since $S(t)$ is a contraction on $L_2(\Omega)$ for any $t \geq 0$ it follows that (see [3, theorem 1]) S is an isometry in

K , that is

$$\text{For any } k_1 \text{ and } k_2 \text{ in } K \text{ and } t \geq 0 \quad ||S(t)k_1 - S(t)k_2||_{L_2} = ||k_1 - k_2||_{L_2} \quad . \tag{4.19}$$

However, the inequality in (4.11) is strict so 4.19 can only hold if K consists of exactly one point, $K = \{k\}$. The invariance of K under S implies that k must be an equilibrium. Thus $K = \{u_e\}$. Since $\{u(t)\}_{t \geq 0}$ is precompact in $H_2(\Omega)$ it follows that $||u(t) - u_e||_{H_2} \to 0$ as $t \to \infty$. In particular for t large $u(t) \in \mathcal{O}$, the neighborhood in theorem 13. Once in $\mathcal{O}$ the convergence to u_e at an exponential rate follows from theorem 13 and the proof is complete.

§5. EQUILIBRIA OF NONLINEAR WAVE EQUATIONS

We next turn our attention to equilibrium solutions of the nonlinear wave equation

$$u_{tt} - \Delta u + a(x)u_t + g(u) = f(x) \quad \text{on} \quad [0,\infty) \times \Omega \tag{5.1}$$

$$u = 0 \quad \text{on} \quad [0,\infty) \times \partial\Omega \tag{5.2}$$

The Dirichlet condition (5.2) could be replaced by a Neumann condition however we choose to complement the ideas of §4 as much as possible.

The wave equation (5.1) is a continuum mechanics analogue of the ordinary differential equation for $y = (y,\ldots,y_n)$,

$$\ddot{y} = -\text{grad}\, V(y) - a\dot{y} \,. \tag{5.3}$$

The analogy comes about as follows. Let

$$\mathcal{V}(\phi) = \int_\Omega \frac{|\nabla\phi|^2}{2} + G(\phi) - f\phi \, dx$$

where $G(s) = \int_0^s g(\sigma)\, d\sigma$. Then the first variation of $\mathcal{V}$ is given by

$$\delta V_{<\phi>}(\psi) = \int_\Omega \nabla\phi \cdot \nabla\psi + g(\phi)\psi - f\psi \, dx$$

so that if we admit only variations vanishing at $\partial\Omega$ the equations (5.1) is equivalent to

$$u_{tt} = -\delta\mathcal{V}_{<u>} - a(x)u_t \,.$$

In particular, a state u_e is an equilibrium if and only if it is a critical point of $\mathcal{V}$ (see remark 2 of § 4 for an analogue of $u_t = -\text{grad}V(y)$) .

As a guide for our intuition we consider the equations, (5.3), from classical mechanics. For these we have the energy identity

$$\frac{d}{dt}[|\dot{y}|^2/2 + V(y)] = -a|\dot{y}|^2 \leq 0 \,.$$

The energy E is the sum of the kinetic energy $|\dot{y}|^2/2$ and the potential energy $V(y)$. A state y_0 is an equilibrium if and only if $\text{grad}V(y_0) = 0$. It is a classical observation of Dirichlet that if $a \geq 0$ and y_0 is a strict local minimum of V then the equilibrium solution y_0 is stable in the sense that if $y(0) - y_0$ and $\dot{y}(0)$ are small then $y(t) - y_0$ and $\dot{y}(t)$ remain small for all $t \geq 0$. This is proved by observing first that

$$\begin{aligned} V(y(t)) &= E(t) - |\dot{y}(t)|^2/2 \\ &\leq E(0) = V(y(0)) + \frac{|\dot{y}(0)|^2}{2} \end{aligned} \tag{5.4}$$

since y_0 is a strict local minimum of V this forces $y(t) - y_0$ to be small provided $\dot{y}(0)$ and $y(0) - y_0$ are small. Then

$$\begin{aligned} |\dot{y}(t)|^2/2 &= E(t) - V(y(t)) \\ &\leq E(0) - V(y(t)) \\ &= |\dot{y}(0)|^2/2 + V(y(0)) - V(y(t)) \end{aligned} \tag{5.5}$$

which remains small as $t \to \infty$. This argument can be refined to show that if $a > 0$ the as $t \to \infty$ $(y(t),\dot{y}(t)) \to (y_0,0)$. A natural way to insure that V has a minimum at x_0 is to suppose that the Hessian $[V_{y_iy_j}]$ is a positive definite matrix. In this case the stability of y_0 can be proved by considering the linearized equations at y_0,

$$\ddot{z} = -[V_{y_iy_j}]z - a\dot{z}$$

whose solutions decay exponentially be virtue of the positivity $V_{y_iy_j}$ and a.

We next investigate to what extent these ideas are useful in analysing the the nonlinear wave equation. For solutions of (5.1),(5.2) we have the energy identity

$$\frac{d}{dt}\left[\int_\Omega \frac{u_t^2}{2}\,dx + \mathcal{V}(u(t))\right] = -\int_\Omega a(x)u_t^2\,dx .$$

Thus if $a \geq 0$ the energy is a sum of kinetic and potential energies and is a decreasing function of time. We have the estimate analogous to (5.4),

$$\mathcal{V}(u(t)) \leq \mathcal{V}(u(0)) + \frac{1}{2}\int_\Omega u_t^2(0,x)\,dx . \tag{5.6}$$

Thus if u_e is an equilibrium which furnishes a strict local minimum for V this indicates that if $u(0) - u_e$ and $u_t(0)$ are small then $u(t) - u(0)$ remains small for $t \geq 0$. The difficulty here is that the best one could hope for is that (5.6) implies that $||u(t) - u_e||_{H_1(\Omega)}$ and perhaps a functional of the form $\int_\Omega \Phi(u-u_e)\,dx$ remain small. If the number of

space dimension is 2 or 3 this is not strong enough to estimate $||u(t) - u_e||_{L_\infty}$. If the nonlinearity g is rapidly growing at infinity this failure prevents one from showing that $u(t)-u_e$ is small in any stronger topologies. In particular even if V has a strict absolute minimum at u_e and if $(u(0),u_t(0)) \in C_0^\infty(\Omega)^2$ is close to $(u_e,0)$ in the $C^\infty(\bar{\Omega})$ topology it is not known whether the solution of (5.1) with data $(u(0), u_t(0))$ can be continued for all $t \geq 0$ as an element of $C([0,\infty)|H_2(\Omega))$. As a result we abandon the approach of Dirichlet.

Again relying on our experience with the ordinary differential equation (5.3) we are lead to guess that if the second variation $\delta^2 V_{<u_e>}$ is positive definite then not only must V have a strict local minimum but the stability may be proved by linearization. Now,

$$\delta^2 V_{<u_e>}(\psi) = \int_\Omega |\nabla\psi|^2 + g'(u_e)\psi^2\,dx$$

which leads to the following guess: If $-\Delta + g'(u_e)$ with Dirichlet boundary conditions is a positive definite operator on $L_2(\Omega)$ and if $a > 0$ in $\bar{\Omega}$ then u_e is exponentially stable.

We begin the demonstration by putting the problem into the framework of sections two and three. Let H be the Hilbert space $\mathring{H}_1(\Omega) \oplus L_2(\Omega)$ and let A be the operator on H defined by

$$\mathcal{D}(A) = H_2(\Omega) \cap \mathring{H}_1(\Omega) \oplus \mathring{H}_1(\Omega) \tag{5.7}$$

$$A(\phi,\psi) = (\psi, \Delta\phi - a\psi) \quad \text{for} \quad (\phi,\psi) \in \mathcal{D}(A). \tag{5.8}$$

Then provided that Ω is reasonable A generates a Co contraction semigroup on H (see [6],[9]). Let

$$B = \mathcal{D}(A) \quad \text{with norm} \quad ||b||_B^2 = ||b||^2 + ||Ab||^2 \tag{5.9}$$

for $b \in \mathcal{D}(B)$

Define an operator A on B by

$$\mathcal{D}(A) = \{b \in B : \mathcal{A}b \in B\} \ , \ Ab = \mathcal{A}b \quad \text{for } b \in \mathcal{D}(A) \ . \tag{5.10}$$

Then A generates a Co contraction semigroup on B. Concerning g and we suppose that

$g \in C^2(\mathbb{R})$, $f \in H_1(\Omega)$ and $f|_{\partial\Omega} = g(0)$. Then $J:(\phi,\psi) \to (0, g(\phi) - f)$ maps B into itself, satisfies (2.3) and (2.5), and $dJ_{<\phi,\psi>}(\chi,\eta) = (0, g'(\phi)\chi)$ (5.11)

With these conventions if $u \in C([0,\infty)|H_2(\Omega) \cap \overset{\circ}{H}_1(\Omega))$ satisfies (5.1) then $U = (u,u_t) \in C([0,\infty)|B)$ and satisfies

$$U_t = AU + J(U) \ . \tag{5.12}$$

Conversely if $U = (u^1(t), u^2(t))$ satisfies (5.12) then $u^2(t) = \frac{\partial u^1}{\partial t}$ and $u^1 \in C([0,\infty)|H_2(\Omega) \cap \overset{\circ}{H}_1(\Omega))$ satisfies (5.1). Now suppose $U_e = (u_e, 0) \in B$ is an equilibrium solution of (5.12). Then $Z = (v,w)$ satisfies the linearized equations.

$$Z_t = AZ + dJ_{<U_e>}Z \tag{5.13}$$

if and only if $w = \frac{\partial v}{\partial t}$ and v satisfies the linearization of (5.1), namely,

$$v_{tt} - \Delta v + a(x)v_t + g'(u_e)v = 0 \quad \text{on} \quad [0,\infty) \times \Omega \tag{5.14}$$

$$v = 0 \quad \text{on} \quad [0,\infty) \times \partial\Omega \tag{5.15}$$

We now state the main result.

<u>THEOREM 15</u>. Suppose $\Omega \subset \mathbb{R}^n$ with $n \leq 3$ is open and lies on one side of its smooth compact boundary, $\partial\Omega$, and that f, g, and B, are as described in (5.7)-(5.11). In addition, suppose that $a \in C^2(\overline{\Omega})$ and $\min a > 0$. If $u_e \in H_2(\Omega) \cap \overset{\circ}{H}_1(\Omega)$ is a solution of the equilibrium equation $-\Delta u_e + g(u_e) = 0$ and if the self adjoint differential operator $-\Delta + g'(u_e)$ with domain $H_2(\Omega) \cap \overset{\circ}{H}_1(\Omega)$ is strictly positive then there is an open set $\mathcal{O} \subset B$ with $(u_e, 0) \in \mathcal{O}$ such that for any $b \in \mathcal{O}$ there is a unique $u \in C([0,\infty); H_2(\Omega) \cap \overset{\circ}{H}_1(\Omega)) \cap C^1([0,\infty); \overset{\circ}{H}_1(\Omega))$ which satisfies the differential equation (5.1) and the initial condition $(u(0), u_t(0)) = b$. In addition there are positive constants c_1, c_2 independent of b such that

$$\|u(t) - u_e\|_{H_2(\Omega)} + \|u_t(t)\|_{H_1(\Omega)} \leq c_1 e^{-c_2 t}(\|u(0) - u_e\|_{H_2(\Omega)} + \|u_t(0)\|_{H_1(\Omega)}) \ .$$

Proof. It suffices to prove that solutions of the linearized equations (5.13) decay exponentially in B . That is, we must find positive constants d_1 and d_2 such that for $t \geq 0$

$$||\exp t(A + dJ_{<U_e>}|| \leq d_1 e^{-d_2 t} \qquad (5.16)$$

where $|| \ ||$ is the norm in $\mathrm{Hom}(B)$. As in the proof of Theorem 13 in section 4 it suffices to prove estimate (5.16) where $|| \ ||$ is the norm in $\mathrm{Hom}(H)$. The brief details of this reduction are omitted. Since $-\Delta + g'(u_e)$ is strictly positive, $a > 0$, and the linearized equations are equivalent to (5.14), (5.15) this decay result can be proved by the methods of [9]. For completeness we give a more elementary proof which yields a less sharp estimate on the rate of decay. The idea is to construct a Lyapunov function analogous to those used in the study of the ordinary differential equation (5.3). I would like to thank Professor Dafermos for teaching me this method.

Let $\Phi: H \to \mathbb{R}$ be defined by

$$\Phi(v,w) = \int_\Omega w^2 + |\nabla v|^2 + g'(u_e)v^2 + \alpha vw + \beta(x)v^2\, dx$$

where α is a positive constant and $\beta \in C(\bar{\Omega})$ a positive function to be chosen below. The first restriction on Φ is that $\Phi^{1/2}$ should be equivalent to the norm in H . The positivity of $-\Delta + g'(u_e)$ implies that there is a constant c such that

$$\int_\Omega |\nabla v|^2 + g'(u_e)v^2\, dx \geq c_1||v||^2_{H_1(\Omega)} \quad \forall v \in \overset{\circ}{H}_1(\Omega)$$

Thus to show that

$$\Phi(v,w) \geq c_2||v,w||^2_H \qquad (5.17)$$

it suffices to choose α and β so that

$$\alpha^2 < 4\beta(x) \quad \forall x \in \bar{\Omega} . \qquad (5.18)$$

Next we would like $\Phi(Z)$ to decrease exponentially if Z is a solution of the linearized equations (5.13). If $Z(t) = (v(t),w(t))$ then a tedious computation shows that

$$\frac{d}{dt}\Phi(v(t),w(t)) = -2\alpha\int_\Omega |\nabla v|^2 + g'(u_e)v^2\, dx$$

$$+ \int_\Omega (\alpha - 2a)w^2 + (2\beta - a\alpha)vw\, dx$$

We choose $\beta(x) = \frac{1}{2}a(x)\alpha$ so that the last term vanishes and α is chosen so that

$$\alpha < 2a \quad \forall x \in \Omega \quad . \tag{5.19}$$

With this choice (5.18) holds automatically and there is a positive constant C_3 independent of Z so that

$$\frac{d}{dt}\Phi(Z(t)) \leq -C_3\Phi(Z(t)) \quad \text{so}$$

$$\Phi(Z(t)) \leq e^{-C_3 t}\Phi(Z(0)) \quad \text{for} \quad t \geq 0 \; . \tag{5.20}$$

Since $\Phi^{1/2}$ is equivalent to the norm in H this proves (5.16) and therefore completes the proof of the theorem.

Example. Let $a(x) \equiv \varepsilon > 0$ and $g(u) = \gamma u + u|u|^{r-1}$ where $\gamma > 0, r \geq 2$ and $f \equiv 0$ and $\Omega = \mathbb{R}^3$. The resulting equation is the nonlinear Klein Gordon equation with friction

$$u_{tt} - \Delta u + \gamma u + u|u|^{r-1} = -\varepsilon u_t \quad \text{on} \quad [0,\infty) \times \mathbb{R}^3 \tag{5.21}$$

The equilibrium is $u_e \equiv 0$. Then $g'(u_e) = \gamma$ so $-\Delta + g'(u_e) = -\Delta + \gamma$ which is strictly positive on $H_2(\Omega) \cap \overset{\circ}{H}_1(\Omega)$. We conclude that if $(u(0), u_t(0))$ is sufficiently close to zero in B then there is a unique u continuous on $[0,\infty)$ with values in H_2 and C^1 with values in $\overset{\circ}{H}_1$ which satisfy (5.21) and $||u(t)||_{H_2} + ||u_t(t)||_{H_1}$ decays exponentially as $t \to +\infty$. Note that it is not known whether there is global existence of such smooth solutions to (5.21) when the data is an arbitrary member of B .

§6. DIFFICULTIES WITH HIGH DIMENSIONS AND GENERAL BOUNDARY CONDITIONS

In the previous two sections we studied one problem with Neumann boundary conditions and one with Dirichlet conditions. These two boundary conditions have the desirable property that if w satisfies the boundary condition then so does $g(w)$ (provided $g(0) = 0$ in the Dirichlet case). This property is not shared by more general boundary problems. For example, if one wanted to study the parabolic problem

$$u_t = \Delta u - g(u) \quad \text{on} \quad [0,\infty) \times \Omega \tag{6.1}$$

$$\frac{\partial u}{\partial \nu} = a(x)u \quad \text{on} \quad [0,\infty) \times \partial\Omega \tag{6.2}$$

it would be natural to take

$$B = \{\psi \in H_2(\Omega) : \frac{\partial \psi}{\partial \nu} = a(x)\psi \ \text{on} \ \partial\Omega\} .$$

However, if $J(\psi) = -g(\psi)$ it is not true that J maps B to itself. Thus the theorems of sections two and three do not apply. On a formal level one can still linearize at a solution u to get the linearized equations

$$v_t = \Delta v - g'(u)v \quad \text{on} \quad [0,\infty) \times \Omega$$

$$\frac{\partial v}{\partial \nu} = av \quad \text{on} \quad [0,\infty) \times \partial\Omega$$

and it is more than likely that decay for the linearization at an equilibrium implies stability. It seems to me that the appropriate point of view might be consider the differential equation as defining a flow, on a larger space, in which B is invariant. For nonlinear boundary conditions the larger space may even be a nonlinear submanifold of a Banach space. These problems are wide open.

If one tries to extend the results of sections four and five to higher dimensions similar difficulties arise. One wants to work in a Banach space B with the property that nonlinear maps are well-behaved on B. In §4 and §5 this was achieved by choosing B to be the domain of an appropriate elliptic operator with $B \subset H_2(\Omega) \subset C(\bar{\Omega})$. A natural generalization is to take B to be the domain of a power of such an operator. This idea does not work as well as one might hope. For example, consider the problem of §4. Let

$$B_2 = D(A^2) = \{\psi \in H_4(\Omega): \frac{\partial \psi}{\partial \nu} = \frac{\partial \Delta\psi}{\partial \nu} = 0 \quad \text{on} \quad \partial\Omega\} .$$

If J is defined by $J(\psi) = g(\psi)$ we ask whether J maps $B2$ into itself. If $\partial\Omega$ is not a hyperplane it does not. As an example suppose Ω is the disc, that is, $\Omega = \{x \in \mathbb{R}^2 \mid |x|^2 < 1\}$. Using polar coordinates one finds that $\psi \in H_4$ is in B_2 if and only if

$$\psi_r = \psi_{rrr} - 2\psi_{\sigma\sigma} + \psi_{rr} = 0 \quad \text{when} \quad r = 1 .$$

For such ψ we find that for $r = 1$

$$\frac{\partial}{\partial r}(\Delta g(\psi)) = -2g''(\psi)\psi_\theta^2$$

which need not vanish.

For parabolic problems there is another way out. One may use Sobolev spaces based on L_p. If

$$B = \{\psi \in H_{2,p}(\Omega): \frac{\partial\psi}{\partial\nu} = 0 \quad \text{on} \quad \partial\Omega\} .$$

Then if $p > n/2$, $B \subset C(\overline{\Omega})$ and J maps B to itself nicely. In this way the results of section 4 may be extended to arbitrary dimensions. To carry this out one must prove decay of the linearized equations in $H_{2,p}$. This is done as follows. Choose $N > 0$ so that $\mathcal{D}(A^N) \subset H_{2,p}(\Omega)$ where A is defined as in §4. Then

$$e^{A + dJ_{<u_e>}}: H_{2,p} \to \mathcal{D}(A^N)$$

continuously by virtue of the smoothing properties of the heat operator. Then for $t > 1$ one finds decay in the norm of $\mathcal{D}(A^N)$ which proves decay in $H_{2,p}$.

We next consider the wave equation of §5. With A as in that section we might try

$$B = \mathcal{D}(A^2) = \{(\phi,\psi): \phi \in H_4(\Omega) \cap \mathring{H}_1(\Omega) ,$$

$$\psi \in H_2(\Omega) \cap \mathring{H}_1(\Omega) \text{ , and, } \Delta\phi \in \mathring{H}_1(\Omega)\} .$$

Here $J(\phi,\psi) = (0, g(\phi)-f)$ is a well behaved map of B into itself provided g and f satisfy (5.11) and $n \leq 7$. (One needs $H_4(\overline{\Omega}) \subset C(\overline{\Omega})$). In this way the results of §5 may be extended to dimensions $n \leq 7$. Unfortunately it is not true that J maps $\mathcal{D}(A^3)$ to itself so one reaches an impasse at $n = 8$. For the wave equation, one cannot use Sobolev spaces based on L_p for $p > 2$ since according to a theorem of Littman [7], A does not generate a semigroup on these spaces. Formally, linearization works in all dimension and I fell that there must be a way to make it rigorous. This remains an open problem.

REFERENCES

1. H. Brezis, Monotonicity methods in Hilbert spaces and some applications to nonlinear partial differential equations, in Contributions to Nonlinear Functional Analysis, E. Zarontonello ed., Academic Press, New York, 1971.
2. H. Brezis, Quelque proprieties des operateurs monotones et des semigroup nonlinear, Proc. of Nato Conference Brussels 1975, to appear.
3. C. Dafermos and M. Slemrod, Asymptotic behavior of nonlinear contraction semigroups, Jnl. Funct. Anal., 13(1973), 97-106.
4. J. Diedonné, Foundations of Modern Analysis, Academic Press, New York, 1960.
5. E. Hille and R. Phillips, Functional Analysis and Semigroups, Am. Math. Soc. Collquium Publ Vol. 31, 1957.
6. P.D. Lax and R.S. Phillips, Scattering Theory, Academic Press, New York, 1967.
7. W. Littman, The wave operator and Lp norms, J. Math. Mech. 12(1963), 55-68.
8. P.J.McKenna and J. Rauch, Strongly nonlinear elliptic boundary value problems with kernel, to appear.
9. J.Rauch, Qualitative behavior of dissipative wave equations on bounded domains, Archiv. Rat. Mech. Anal., 1976.
10. J.Rauch, Global existence for the FitzHugh-Nagumo Equations, Comm. P.D.E., (to appear)
11. J. Rauch and J. Smoller, Strongly nonlinear perturbations of nonnegative boundary value problems with kernel, to appear.
12. M.Reed, Abstract Non-Linear Wave Equations, Springer Lecture Notes no 507, Springer-Verlag, New York, 1976.
13. D.Sattinger, Stability of nonlinear hyperbolic equations, Archi. Rat. Mech. Anal., 28(1968) 226-244.
14. W. Strauss, Nonlinear scattering theory, Proceedings Nato Advanced Study Institute June 1973, D. Reidel Publishing Co., Holland.

THE APPROXIMATION OF SEMI-GROUPS OF LINEAR OPERATORS AND THE FINITE ELEMENT METHOD

Teruo Ushijima

Department of Information Mathematics,
The University of Electro-Communications,
Chofu-shi, Tokyo, Japan 182

ABSTRACT

In this talk an approximation theory for semi-groups of linear operators and its application to the numerical analysis of semi-linear heat equation of blow-up type will be discussed.

In §1, the Trotter-Kato's Theorem will be reformulated, and the variable time step approximation will be discussed. In §2, our approximate scheme will be described, and the convergence of approximate solutions to the true solution will be established. In §3 a numerical algorithm for the blow-up problem will be proposed with a justification.

Main part of this talk is a product of the collaboration with Dr. Nakagawa in Tokyo. Details of proofs will be reported in our works [8] and [9].

1. A VARIANT OF TROTTER-KATO'S THEOREM

Let X be a Banach space. The totality of bounded linear operators is denoted by L(X). In this article a C_0-semi-group $T(t) \in L(X)$ $(t \geqq 0)$ is simply called a continuous semi-group. An L(X)-valued step function $T(t)$ $(t \geqq 0)$ is called a discrete semi-group with time unit τ $(\tau > 0)$ if there exists an operator $T(\tau) \in L(X)$ satisfying

$$T(t) = T(\tau)^{[t/\tau]} \qquad \text{for } t \geqq 0$$

where [ ] denotes the Gaussian bracket. The generator of a

Garnir (ed.), Boundary Value Problems for Linear Evolution Partial Equations. 351-362.

discrete semi-group $T(t)$ is defined by

$$A = \tau^{-1}(T(\tau) - 1).$$

A sequence of Banach spaces $\{X_h : h > 0\}$ is said to K-converge (or converge in the sense of Kato) to a Banach space X ($X_h \xrightarrow{K} X$), in short if there exist approximating operators $P_h \in L(X,X_h)$ satisfying the following conditions (K.1) and (K.2):

(K.1) $\sup_{h>0} \|P_h x\| < \infty$ and $\lim_{h\to 0} \|P_h x\| = \|x\|$ for any $x \in X$.

(K.2) For any $x_h \in X_h$ can be expressed as $x_h = P_h x^{(h)}$ with some $x^{(h)} \in X$ satisfying $\| x^{(h)} \| \leq N \| x_h \|$, where N is independent of h.

Now we fix a sequence of Banach spaces $\{X_h\}$ which K-converges to a Banach space X. A sequence $\{x_h \in X_h\}$ is said to K-converge to a point $x \in X$ ($x_h \xrightarrow{K} x$, in short) if $\lim_{h\to 0} \| x_h - P_h x\| = 0$, and sequences $\{x_{\lambda,h} \in x_h\}_{\lambda\in\Lambda}$ are said to K-converge to points $x_\lambda \in X$ uniformly in $\lambda \in \Lambda$ if $\lim_{h\to 0} \| x_{\lambda,h} - P_h x_\lambda \| = 0$ hold uniformly in $\lambda \in \Lambda$. A sequence $\{A_h \in L(X_h)\}$ is said to K-converge to an operator $A \in L(X)$ ($A_h \xrightarrow{K} A$, in short) if $A_h P_h x \xrightarrow{K} Ax$ for any $x \in X$, and sequences $\{A_{\lambda,h} \in L(X_h)\}_{\lambda\in\Lambda}$ are said to K-converge to operators $A_\lambda \in L(X)$ uniformly in $\lambda \in \Lambda$, if $A_{\lambda,h} P_h x \xrightarrow{K} A_\lambda x$ uniformly in $\lambda \in \Lambda$ for any $x \in X$.

Let us fix a continuous semi-group $T(t) \in L(X)$. And let A be its generator. Suppose that there is either a sequence of continuous semi-groups $T_h(t) \in L(X_h)$ or a sequence of discrete semi-groups $T_h(t) \in L(X_h)$ with time unit τ_h. Let A_h be the generator of semi-group $T_h(t)$. When the discrete semi-groups are considered, it is always assumed that $\lim_{h\to 0} \tau_h = 0$.

Consider the following three conditions:

(A) (Consistency). For some complex number λ, there exist $(\lambda - A_h)^{-1} \in L(X_h)$ $(h > 0)$ and $(\lambda - A)^{-1} \in L(X)$ satisfying

$$(\lambda - A_h)^{-1} \xrightarrow{K} (\lambda - A)^{-1}.$$

(B) (Boundedness) For some $T < \infty$,

$$\sup_{h,\ 0\leq t\leq T} \|T_h(t)\| < \infty .$$

(C) (Convergence). For any $T < \infty$

$$T_h(t) \xrightarrow{K} T(t) \text{ uniformly in } t \in [0,T].$$

Then we have the following result.

Theorem 1.1. (A-B-C Theorem). The conditions (A) and (B) hold if and only if the condition (C) holds.

In case $X_h \equiv X$ and $P_h \equiv I$, Theorem 1.1 is a corollary of Trotter-Kato's theory of approximation of semi-groups (Cf. Trotter [10], Chapter IX of Kato[7]). The notion of K-convergence is suggested in [7]. One can easily obtain the proof of Theorem 1 if he modifies Kato's treatment in [7] appropriately. See also the author's work [13].

For the convenience of our purpose, we discuss here a variable step approximation of semi-groups in a restricted situation. Let the operator A generate a continuous semi-group e^{tA} in the space X. Assume that there is a sequence of bounded operators $A_h \in L(X_h)$ satisfying the following conditions:

(1.1) For any $T < \infty$

$$e^{tA_h} \xrightarrow{K} e^{tA}$$

uniformly in $t \in [0,T]$.

(1.2) For any h there is a positive number τ_h such that

$$\| 1 + \tau A_h \| \leq 1 \qquad \text{for any } \tau \leq \tau_h.$$

This condition implies

$$(1.3) \qquad \| e^{tA_h} \| \leq 1 \qquad \text{for any } h \text{ and } t.$$

An infinite sequence $\boldsymbol{\tau}$ of positive numbers:

$$\boldsymbol{\tau} = (\tau_0, \tau_1, \tau_2, \cdots)$$

is said to be a time mesh vector. Fix $T > 0$. For any $h > 0$ choose a time mesh vector $\boldsymbol{\tau}_h$ satisfying that

$$(1.4) \qquad \| \boldsymbol{\tau}_h \|_\infty = \sup_{\tau\in\boldsymbol{\tau}_h} \tau \leq \tau_h$$

and that

$$(1.5) \qquad \| \boldsymbol{\tau}_h \|_1 = \sum_{\tau\in\boldsymbol{\tau}_h} \tau > T.$$

Let us define a family of evolution operators $\{U_h(t,s): 0 \leq s \leq t \leq T\}$ as follows

$$t_{-1} = -\infty,\ t_0 = 0,\ t_{j+1} = t_j + \tau_j \quad (j \geq 0)$$

$$U_h(t,s) = \begin{cases} \prod_{k=j+1}^{n} (1 + \tau_k A_h) & \text{if } t_{n+2} > t \geq t_{n+1} > t_{j+1} \geq s > t_j, \\ 1 & \text{if } t_{n+2} > t \geq s > t_n. \end{cases}$$

Let $\Delta_T = \{(t,s): 0 \leq s \leq t \leq T\}$.

Theorem 1.2. Under the above assumptions, $U_h(t,s)$ K-converges to $e^{(t-s)A}$ uniformly in $(t,s) \in \Delta_T$ if $\| \boldsymbol{\tau}^h \|_\infty \longrightarrow 0$ as $h \longrightarrow 0$.

Remark 1.1. Let $\{\sigma_h: h > 0\}$ be a sequence of positive numbers such that $\lim_{h\to 0} \sigma_h = 0$, and that $\sigma_h \leq \tau_h$. Consider a family $\mu = \{\boldsymbol{\tau}_h: h > 0\}$ of time mesh vectors $\boldsymbol{\tau}_h$ satisfying (1.5) and

$$(1.4)' \qquad \| \boldsymbol{\tau}_h \|_\infty \leq \sigma_h .$$

We regard this family μ as an index, and denote by M the totality of these indices. Then for each $\mu \in M$, Theorem 1.2 holds. It is, however, to be noted that the convergence is uniform with respect to the index $\mu \in M$.

2. THE LUMPED MASS APPROXIMATION OF THE SEMI-LINEAR HEAT EQUATION

We consider the following problem.

$$(E) \begin{cases} \frac{\partial u}{\partial t} = \Delta u + f(u), & x \in \Omega,\ t > 0, \\ u(t,x) = 0, & x \in \Gamma,\ t > 0, \\ u(0,x) = a(x), & x \in \overline{\Omega}, \end{cases}$$

where the set Ω is a bounded open set in $\mathbb{R}^n$ with the smooth boundary Γ. The function $f(u)$ is assumed to be Lipshitz continuous in the variable u, and $a(x)$ to be continuous on $\overline{\Omega}$ vanishing at Γ.

First we impose the following assumptions on the problem (E).

Assumption 1. For some fixed $T < \infty$, there exists one and only one solution $u(t,x)$ of (E) such that

$$(1) \quad u(t,x) \in C([0,T] \times \overline{\Omega}),$$

(2) $u(t,x)$ is continuously differentiable in t and twice continuously differentiable in x for

$$(t,x) \in (0,T) \times \Omega.$$

Assumption 2. There is a sequence, $\{\Omega_h : h > 0\}$, of polyhedral domains contained in Ω such that

(Ω.1) $\Omega_{h1} \supset \Omega_{h2}$ if $h_1 \leq h_2$,

(Ω.2) $\max_{x \in \Gamma_h} \operatorname{dist}(x,\Gamma) \to 0$ as $h \to 0$ where Γ_h is the boundary of Ω_h.

We consider the lumped mass approximation of (E) in the following manner.

A family $\textcircled{T}_h$ of finite numbers of closed nondegenerate n-simplices is said to be a triangulation of the bounded polyhedral domain Ω_h if the closure $\overline{\Omega}_h$ is expressed as

(T.1) $$\overline{\Omega}_h = \bigcup_{T \in \textcircled{T}_h} T$$

such that the interior of any simplex of $\textcircled{T}_h$ is disjoint with that of another simplex of $\textcircled{T}_h$, and such that any one of faces of a simplex is either a face of another simplex of $\textcircled{T}_h$, or else is a portion of the boundary of Ω_h.

Now let us define the notion of the lumped mass region $B = B_b$ corresponding to the nodal point b with respect to the triangulation $\textcircled{T}_h$. Here we say that a point which is a vertex for some $T \in \textcircled{T}_h$ is a nodal point. Let $b_0 = b, b_1, \cdots\cdots, b_n$ be the vertices of some n-simplex T of $\textcircled{T}_h$. Let λ_i be the barycentric coordinate corresponding to the vertex b_i $(0 \leq i \leq n)$. Then the barycentric subdivision B_{bT} of the simplex T corresponding to the point b is defined as follows:

$$B_{bT} = \{x : 1 \geq \frac{\lambda_0(x)}{\lambda_0(x) + \lambda_i(x)} > 1/2 \quad \text{for any } i = 1, 2, \cdots\cdots, n\}.$$

The lumped mass region B_b is the union of the subdivisions B_{bT} of simplices T having the point b as its vertex:

$$B_b = \bigcup_{b \text{ is a vertex of } T} B_{bT}.$$

The linear shape function corresponding to the nodal point b is denoted by $\hat{w}_b(x)$, which coincides with $\lambda_0(x)$ if x is a point of a simplex T having the vertex b as b_0, and equals zero otherwise. The characteristic function of the region B_b is denoted by $\overline{w}_b(x)$. Let us count the interior, and boundary, nodal points of Ω_h as $b_1, b_2, \cdots, b_N$, and $b_{N+1}, b_{N+2}, \cdots, b_{N+M}$, respectively. And we write

$$\hat{w}_j = \hat{w}_{b_j} \qquad \text{and} \qquad \overline{w}_j = \overline{w}_{b_j}.$$

Following to Ciarlet-Raviart [1], the triangulation $\mathcal{T}_h$ is said to be nonnegative if and only if it holds

$$\text{(T.2)} \qquad (\nabla\hat{w}_i, \nabla\hat{w}_j) \leqq 0 \quad \text{for } i \neq j,\ 1 \leqq i \leqq N,\ 1 \leqq j \leqq N+M.$$

For any simplex T, its diameter and the maximum of the diameters of the inscribed spheres of T, are denoted by $h(T)$ and $\rho(T)$ respectively.

<u>Assumption 3.</u> For any $h > 0$, there is a nonnegative triangulation $\mathcal{T}_h$ of Ω_h such that

$$\text{(T.3)} \qquad \max_{T \in \mathcal{T}_h} h(T) \leqq h,$$

and that

$$\text{(T.4)} \qquad \inf_h \min_{T \in \mathcal{T}_h} \frac{\rho(T)}{h(T)} = \gamma > 0.$$

Now we introduce the space $\hat{V}_h$ and $\overline{V}_h$ as an approximation of the space $V = H_0^1(\Omega)$. Namely we have

$$\hat{V}_h = \{\hat{u}_h = \sum_{j=1}^{N} \alpha_j \hat{w}_j\}, \quad \overline{V}_h = \{\overline{u}_h = \sum_{j=1}^{N} \alpha_j \overline{w}_j\}$$

where the scalers $\alpha_j (1 \leqq i \leqq N)$ take arbitrary values. An element of the space $\hat{V}_h$ or $\overline{V}_h$ is considered to be defined on the whole $\overline{\Omega}$ taking zero in the complement of its support. Linear mappings J_h from $\hat{V}_h$ onto $\overline{V}_h$ and K_h from $\overline{V}_h$ onto $\hat{V}_h$ are defined as follows,

$$J_h\hat{u}_h = J_h(\sum_{j=1}^{N} \alpha_j\hat{w}_j) = \sum_{j=1}^{N} \alpha_j\overline{w}_j = \overline{u}_h,$$

$$K_h\overline{u}_h = K_h(\sum_{j=1}^{N} \alpha_j\overline{w}_j) = \sum_{j=1}^{N} \alpha_j\hat{w}_j = \hat{u}_h.$$

Hereafter correspondence $\hat{u}_h \longleftrightarrow \overline{u}_h$ will be frequently used. The orthogonal projections from $L_2(\Omega)$ to $\hat{V}_h$, and to $\overline{V}_h$, are denoted by $\hat{P}_h$, and $\overline{P}_h$, respectively. Let X be the space of real valued continuous functions on $\overline{\Omega}$ vanishing at Γ:

$$X = C_0(\Omega) = \{u \in C(\overline{\Omega}): u(x) = 0 \quad \text{for } x \in \Gamma\}.$$

The interpolation operator $\tilde{P}_h$ from X onto $\hat{V}_h$ is defined as

$$(\tilde{P}_h u)(x) = \sum_{j=1}^{N} u(b_j)\hat{w}_j(x) \quad \text{for } u \in X,$$

and $J_h\tilde{P}_h$ is denoted by P_h.

The function f naturally mapps the space $\bar{V}_h$ into itself. Namely we have for $\bar{u}_h = \sum_{j=1}^{N} \alpha_j \bar{w}_j \in \bar{V}_h$,

$$f(\bar{u}_h) = \sum_{j=1}^{N} f(\alpha_j)\bar{w}_j \in \bar{V}_h.$$

Now we introduce the negative definite self-adjoint operator A_h in $\bar{V}_h$ defined by the formula

$$(A_h\bar{\phi}_h, \bar{\psi}_h)_{L^2(\Omega_h)} = -(\nabla\hat{\phi}_h, \nabla\hat{\psi}_h)_{L^2(\Omega_h)}$$

for any $\bar{\phi}_h, \bar{\psi}_h \in \bar{V}_h$.

Let

$$\tau_h = \min_{1\leq i\leq N} \frac{\| \bar{w}_i \|^2_{L^2(\Omega)}}{\|\nabla\hat{w}_i\|^2_{L^2(\Omega)}}.$$

Now we fix an index set M mentioned in Remark 1.1. For any index $\mu = \{\tau_h: \ h > 0\}$, we have the following explicit approximation of (E).

Find the $\bar{V}_h$-valued function $u_h(t)$ such that

$$(E^{\tau}_h)\begin{cases} u_h(t) = u_h(t_k),\ t_k \leq t < t_{k+1} = t_k + \tau_k, \\ \dfrac{u_h(t_{k+1}) - u_h(t_k)}{\tau_k} = A_h u_h(t_k) + f(u_h(t_k)), \\ u_h(0) = a_h = P_h a. \end{cases}$$

Theorem 2.1. Let $u_h(t,x)$ be the solution of (E^{τ}_h). Then

$$\lim_{h\to 0} \max_{0\leq t\leq T, x \in \Omega_h} | u_h(t,x) - u(t,x)| = 0.$$

This convergence is uniform with respect to $\mu \in M$.

Remark 2.1. Consider the set $C_0(\Omega)$, and $\bar{V}_h$, as a Banach space X, and X_h, respectively, with the maximum norm. Then the sequence of spaces $\{X_h: h > 0\}$ K-converges to the space X with the approximating operators P_h. Let $\{T_t: t \geq 0\}$ be the continuous semi-group in $X = C_0(\Omega)$ corresponding to the heat equation with the Dirichlet boundary condition. The generator A has the bounded inverse $A^{-1} \in L(X)$. With the aid of the result of Ciarlet-Raviart [1], we have $A_h^{-1} \xrightarrow{K} A^{-1}$. On the other hand Fujii [2] established that if $\tau \leq \tau_h$ then $\|(1 + \tau A_h)\|_{L(X_h)} \leqq 1$. This implies $\|e^{tA_h}\|_{L(X_h)} \leqq 1$. By Theorem 1.1, we have $e^{tA_h} \xrightarrow{K} e^{tA}$ uniformly in $t \in [0,T]$. Hence the conditions (1.1) and (1.2) hold.

3. A NUMERICAL ALGORITHM FOR THE BLOW-UP PROBLEM

In the problem (E), we further assume that f is a positive convex function satisfying that for some positive γ and C

$$(3.1) \qquad f(u) \geqq Cu^{1+\gamma} \qquad \text{as } u \to \infty.$$

The initial data a(x) is continuous on $\bar{\Omega}$ vanishing at Γ, the totality of such functions is denoted by $C_0(\bar{\Omega})$. By Kaplan's classical argument [6], the solution u(t,x) tends to infinity at a finite time T for some a(x). This fact is called the blowing-up of solution, and the time T is called the blowing-up time or the finite escape time. Fujita studied extensively this problem in [3], [4] and so forth. There are also some works based on different criteria by other authors, for example, Tsutsumi [11], [12], Ito [5], among others.

Now we provide a numerical method of (E) by making use of the finite element approximation of lumped mass type, based on Kaplan'Fujita's criterion.

3.1 Kaplan-Fujita's criterion

Let λ denote the smallest eigenvalue of $-\Delta$ with the Dirichlet boundary condition, and let $\phi(x)$ denote the eigenfunction associated with λ, $\phi(x)$ being normalized as

$$\begin{cases} \phi(x) > 0, \ x \in \Omega, \\ \int_\Omega \phi(x)dx = 1. \end{cases}$$

Denote by J(t) the inner product of u(t,x) and $\phi(x)$, i.e.,

$$J(t) = (u(t,x),\phi(x))_{L^2(\Omega)} = \int_\Omega u(t,x)\phi(x)dx.$$

<u>Definition 3.1.</u> The classical solution u(t,x) of (E) J-blows up at t = T if and only if

$$\begin{cases} u(t,x) \in C([0,T),C_0(\bar{\Omega})) \text{ satisfies (E)}, \\ \lim_{t\uparrow T} J(t) = \infty. \end{cases}$$

Let J^1 be the largest positive root of the equation of

$$-\lambda J + f(J) = 0.$$

If the equation has no positive roots, then let $J^1 = 0$.

Proposition 3.1. The solution $u(t,x)$ J-blows up at a finite time T if and only if there exists a $t_0 \geqq 0$ such that

$$\begin{cases} u(t,x) \in C([0,t_0],\ C_0(\bar{\Omega})) \text{ satisfies (E)}, \\ J(t_0) > J^1. \end{cases}$$

Corollary 3.2. The blowing-up time T is bounded from above as

$$T \leqq t_0 + \int_{J(t_0)}^{\infty} \frac{dJ}{-\lambda J+f(J)} .$$

3.2 An algorithm for controlling time steps

Proposition 3.3. If $(\nabla\hat{w}_i,\nabla\hat{w}_j)_{L^2(\Omega_h)} \leqq 0$ for $i \neq j$, $1 \leqq i \leqq N$, $1 \leqq J \leqq N+M$, then it holds that the smallest eigenvalue λ_h of $-A_h$ is simple, and that there is the associated eigenfunction $\phi_h(x)$ normalized as $\phi_h(x) \geqq 0 (x \in \Omega_h)$ and $\int_{\Omega_h} \phi_h(x)dx = 1$.

Define $J_h(t)$, the discrete analogue to $J(t)$, by

$$J_h(t) = (u_h(t,x),\ \phi_h(x))_{L^2(\Omega_h)} .$$

Let J_h^1 denote the largest positive root of the equation of

$$-\lambda_h J + f(J) = 0.$$

If the equation has no positive roots, then let $J_h^1 = 0$.

Define τ_h by the formula

$$\tau_h = \min_{1 \leqq i \leqq N} \| \bar{w}_i \|^2 / \| \nabla\hat{w}_i \|^2 .$$

Choose a fixed value of τ which is not greater than τ_h. Then our algorithm for controlling the time step τ_n is given by

$$\text{(3.2)} \quad \begin{cases} \tau_0 = \tau, \text{ and} \\ \tau_n = \begin{cases} \tau & \text{if } J_h(t_{n-1}) \leqq J_h^{1}, \\ \min\left\{\tau, \dfrac{J_h(t_n)-J_h(t_{n-1})}{-\lambda_h J_h(t_n)+f(J_h(t_n))}\right\} & \text{otherwise,} \end{cases} \end{cases}$$

$$\text{for } n = 1, 2, 3, \cdots.$$

Definition 3.2. The solution $u_h(t,x)$ of $(E_h^{\boldsymbol{\tau}})$ where $\boldsymbol{\tau}$ is the time mesh vector obtained by the algorithm described above, J_h-blows up at $t = T_h$ if and only if

$$T_h = \sum_{n=o}^{\infty} \tau_n < \infty .$$

Corollary 3.4. If the solution $u_h(t,x)$ J_h-blows up at finite $t = T_h$, then

$$\lim_{t \to T_h} J_h(t) = \infty.$$

Proposition 3.5. The solution $u_h(t,x)$ J_h-blows up at a finite time T_h if and only if there is a $t_n \geqq 0$ such that

$$J_h(t_n) > J_h^{1}.$$

Corollary 3.6. The blowing-up time T_h is bounded from above as

$$T_h \leqq t_n + \tau_n + \int_{J_h(t_n)}^{\infty} \frac{dJ}{-\lambda_h J + f(J)}.$$

3.3 Convergence of the blowing up time

Theorem 3.1. Assume the following two conditions:

(i) $\lambda_h \to \lambda$ and $\phi_h \to \phi$ in $L^2(\Omega)$ as $h \to 0$.

(ii) Let the solution u of (E) J-blow up at a finite time T. For any $T' < T$ and for any sufficiently small h, there is a solution $u_h(t)$ of $(E_h^{\boldsymbol{\tau}_h})$ for $0 \leqq t \leqq T'$ satisfying $\max_{0 \leqq t \leqq T'} \| u_h(t) - u(t) \|_{L^2(\Omega)} \to 0$ as $h \to 0$. Here $\boldsymbol{\tau}_h$ is the time mesh vector obtained by (3.2).

Then it holds that

$$T_h \to T \text{ as } h \to 0$$

provided that $\|\boldsymbol{\tau}_h\|_\infty \to 0$ as $h \to 0$.

Remark 3.1. Because of Theorem 1.1 and Corollary 3.4, the condition (ii) of Theorem 3.1 follows from the fact that for any fixed $T' < T$ one can choose h_0 in such a way that $u_h(t)$ never blows-up within the interval $[0,T']$ if $h \leq h_0$. This fact is also implied by Theorem 1.1. In fact, let h_0 be such that

$$\max_{0 \leq t \leq T'} \max_{x \varepsilon \Omega} |u_h^\mu(t,x) - u(t,x)| \leq 1$$

for any $\mu \varepsilon M$ and $h \leq h_0$ in the situation of Theorem 1.1. This implies that there is a finite number N satisfying

$$(3.3) \qquad \sup_{0 \leq t \leq T',\ \mu \varepsilon M, h \leq h_0} \| u_h^\mu(t) \|_{L^2(\Omega)} = N < \infty.$$

Assume that there is a solution $u_h(t)$ J_h-blowing up at $t = T_h < T'$. Then there is a mesh point t_n such that $\| u_h(t_n) \| > N$ since $\| \phi_h \| \, \| u_h(t) \| \geq J_h(t) \uparrow \infty$. This contradicts the condition (3.3), since there is a μ containing the time mesh vector in the form

$$\boldsymbol{\tau}_h = (\tau_0, \tau_1, \ldots, \tau_{n-1}, \tau_{n-1}, \ldots)$$

where τ_j, $0 \leq j \leq n-1$, are the mesh lengths determined by our algorithm.

It is seemingly well known that the condition (i) of Theorem 3.1 holds under the Assumptions 2 and 3 in §2.

REFERENCES

[1] Ciarlet, P. G. and P. A. Raviart, Maximum principle and uniform convergence for the finite element method, Computer Methods in Applied Mechanics and Engineering 2, 17-31 (1973).

[2] Fujii, H., Some remarks on finite element analysis of time-dependent field problems, Theory and practice in finite element structural analysis (Proceedings of 1973 Tokyo Seminar on Finite Element Analysis), (Tokyo Univ. Press, Tokyo, 1973).

[3] Fujita, H., On the blowing up of solutions to the Cauchy problem for $u_t = \Delta u + u^{1+\alpha}$, J. Fac. Sci. Univ. Tokyo, 13, 109-124 (1966).

[4] Fujita, H., On some nonexistence and nonuniqueness

theorems for nonlinear parabolic equations, Proc. Symposium in Pure Math., AMS, 18, 105-113 (1970).

[5] Ito, S., On the blowing up of solutions for semi-linear parabolic equations (in Japanese), Sugaku, 18, 44-47 (1966).

[6] Kaplan, S., On the growth of solutions of quasilinear parabolic equations, Comm. Pure Appl. Math., 16, 305-330 (1963).

[7] Kato, T., Perturbation theory for linear operators, (Springer, Berlin, 1966).

[8] Nakagawa, T. and T. Ushijima, Finite element analysis of the semi-linear heat equation of blow-up type. (to appear in Topics in Numerical Analysis III, Academic Press)

[9] Nakagawa, T. and T. Ushijima, On the lumped mass approximation of semi-linear heat equations (in preparation).

[10] Trotter, H. F., Approximation of semi-groups of operators Pacific Jour. Math., 8, 887-919 (1958).

[11] Tsutsumi, M., Existence and nonexistence of global solutions for nonlinear parabolic equations, Publ. RIMS, Kyoto Univ., 8, 211-229 (1972).

[12] Tsutsumi, M., Existence and nonexistence of global solutions of the first boundary value problem for a certain quasilinear parabolic equation, Funkcialaj Ekvacioj, 17, 13-24 (1974).

[13] Ushijima, T., Approximation theory for semi-groups of linear operators and its application to approximation of wave equations, Japanese Journal of Mathematics, 1, 185-224 (1975).

PROPAGATION OF SINGULARITIES FOR HYPERBOLIC MIXED PROBLEMS

S. Wakabayashi

Depertment of Mathematics, University of Tsukuba,
Ibaraki, Japan

1. INTRODUCTION

Duff [4] studied the location and structures of singularities of fundamental solutions for hyperbolic mixed problems with constant coefficients in a quarter-space making use of the method of stationary phase. Deakin [3] treated first order hyperbolic systems by the same method. However, it seems that it is difficult to apply the method to the study of fundamental solutions for more general hyperbolic mixed problems. Matsumura [7] gave an inner estimate of the location of singularities of fundamental solutions which correspond to main reflected waves, making use of the localization method developed by Atiyah, Bott and Gårding [1] and Hörmander [5]. A localization theorem describing the location of singularities of fundamental solutions which correspond to lateral waves was obtained by the author [13] under some restrictive assumptions. In [14] the author proved a localization theorem describing the location of singularities of fundamental solutions which correspond to main reflected waves, lateral waves and boundary waves. Tsuji [12] also studied the same problem in the cases where operators are homogeneous and obtained similar results. On the other hand outer estimates of the location of singularities of fundamental solutions were given in [15] by the same method as in [1] which treated the Cauchy problems.

Microlocal parametrices for hyperbolic mixed problems with variable coefficients were constructed in some cases by using the theory of Fourier integral operators (see [2], [9], [11]). Microlocal parametrices for the Dirichlet problem for second order operators were constructed at diffractive points by Melrose [9] and Taylor [11]. But it seems that it is very difficult to construct

Garnir (ed.), Boundary Value Problems for Linear Evolution Partial Equations. 363-384.

microlocal parametrices at glancing points which are not diffractive. On the other hand there is a question of constructing microlocal parametrices when Lopatinski's determinant has real zeros. Although there are no difficulties in so doing, we can investigate reflection of singularities corresponding to boundary waves by the construction of microlocal parametrices (see [16]).

We summarize now the contents of this note. In §2 we shall give inner and outer estimates of wave front sets of fundamental solutions for hyperbolic mixed problems with constant coefficients in a quarter-space (see [14], [15]). Lateral waves arise from the presence of branch points in reflection coefficients and boundary waves are caused by real zeros of Lopatinski's determinant. The results obtained in §2 will show that the above characterizations of lateral waves and boundary waves are valid. In §3 we shall introduce the results for hyperbolic mixed problems in plane-stratified media which were obtained by Matsumura [8]. In §4 we shall construct microlocal parametrices for hyperbolic mixed problems at non-glancing points in the case where Lopatinski's determinants have real zeros.

2. WAVE FRONT SETS OF FUNDAMENTAL SOLUTIONS

Let R^n denote the n-dimensional euclidean space and write $x'=(x_1, \cdots, x_{n-1})$, $x''=(x_2,\cdots,x_n)$ for the coordinate $x=(x_1,\cdots,x_n)$ in R^n and $\xi'=(\xi_1,\cdots,\xi_{n-1})$, $\xi''=(\xi_2,\cdots,\xi_n)$, $\tilde{\xi}=(\xi,\xi_{n+1})$ for the dual coordinate $\xi=(\xi_1,\cdots,\xi_n)$. We shall also denote by R^n_+ the half-space $\{x=(x',x_n)\in R^n;\ x_n>0\}$ and use the symbol $D=i^{-1}(\partial/\partial x_1,\cdots,\partial/\partial x_n)$. Let $P=P(\xi)$ be a hyperbolic polynomial of order m of n variables ξ with respect to $\theta=(1,0,\cdots,0)$ in R^n in the sense of Gårding, i.e., $P^0(\theta)\neq 0$ and $P(\xi+s\theta)\neq 0$ when ξ is real and Im $s<-\gamma_0$, where P^0 denotes the principal part of P. Moreover we assume that $P^0(0,\cdots,0,1)\neq 0$. In this section we consider the mixed initial-boundary value problem for the hyperbolic operator P(D) in a quarter-space

$$\begin{aligned} &P(D)u(x) = f(x), \quad x\in R^n_+,\ x_1>0,\\ &(D_1^k u)(0,x'') = 0, \quad x_n>0,\ 0\le k\le m-1,\\ &B_j(D)u(x)|_{x_n=0} = 0, \quad x_1>0,\ 1\le j\le \ell. \end{aligned}$$

Here the $B_j(D)$ are partial differential operators with constant coefficients and the number ℓ of boundary conditions is equal to that of the roots with positive imaginary part of the equation $P(\xi'-i\gamma\theta',\lambda)=0$ with respect to λ, where $\gamma>\gamma_0$.

Let us denote by $\Gamma=\Gamma(P,\theta)$ $(\subset R^n)$ the component of the set $\{\xi\varepsilon R^n;\ P^0(\xi)\neq 0\}$ which contains θ and put $\Gamma_0=\{\xi'\varepsilon R^{n-1};\ (\xi',0)\varepsilon\Gamma\}$. When $\xi'\varepsilon R^{n-1}-i\gamma_0\theta'-i\Gamma_0$, we can denote the roots of $P(\xi',\lambda)=0$ with respect to λ by $\lambda_1^+(\xi'),\cdots,\lambda_\ell^+(\xi'),\lambda_1^-(\xi'),\cdots,\lambda_{m-\ell}^-(\xi')$, which are enumerated so that

$$\text{Im }\lambda_k^+(\xi')>0,\quad 1\le k\le \ell,$$

$$\text{Im }\lambda_k^-(\xi')<0,\quad 1\le k\le m-\ell.$$

Put

$$P_+(\xi',\lambda)=\Pi_{j=1}^{\ell}\ (\lambda-\lambda_j^+(\xi')),\quad \xi'\varepsilon R^{n-1}-i\gamma_0\theta'-i\Gamma_0.$$

We now define Lopatinski's determinant for the system $\{P,B_j\}$ by

$$R(\zeta')=\det\ ((2\pi i)^{-1}\oint P_+(\zeta)^{-1}B_j(\zeta)\zeta_n^{k-1}d\zeta_n)_{j,k=1,\cdots,\ell},$$

$$\zeta'\varepsilon R^{n-1}-i\gamma_0\theta'-i\Gamma_0.$$

We assume throughout this section that

(A.1) $P(\xi)=p_1(\xi)^{\nu_1}\cdots p_q(\xi)^{\nu_q}$,

where the $p_j(\xi)$ are distinct strictly hyperbolic polynomials with respect to θ,

(A.2) the system $\{P,B_j\}$ is $\mathcal{E}$-well posed, i.e.,

$$R^0(-i\theta')\neq 0,$$

$$R(\xi'+s\theta')\neq 0\ \text{ for } \xi'\varepsilon R^{n-1} \text{ and Im } s<-\gamma_1,$$

where $R^0(\xi')$ denotes the principal part of $R(\xi')$, i.e.,

$$R(t\xi')=t^{h_0}R^0(\xi')+o(t^{h_0})\quad \text{as } t\to\infty \tag{2.1}$$

(see Sakamoto [10]).

Now we can construct the fundamental solution $G(x,y)$ for $\{P,B_j\}$ which describes the propagation of waves produced by unit impulse given at position $y=(0,y'')$ in R_+^n (see [10], [12]). Write

$$G(x,y)=E(x-y)-F(x,y),\quad x\varepsilon R_+^n,\ x_1>0,\ y=(0,y'')\varepsilon R_+^n,$$

where $E(x)$ is the fundamental solution for the Cauchy problem represented by

$$E(x)=(2\pi)^{-n}\int_{R^n-i\eta}\exp[ix\cdot\zeta]P(\zeta)^{-1}d\zeta,\quad \eta\varepsilon\gamma_0\theta+\Gamma.$$

Then the reflected Riemann function F(x,y) is written in the form

$$F(x,y) = (2\pi)^{-n-1}\int_{R^{n+1}-i\tilde{\eta}} i^{-1}\Sigma_{j,k=1}^{\ell} \exp[i\{(x'-y')\cdot\zeta'-y_n\zeta_n +x_n\zeta_{n+1}\}](R(\zeta')P_+(\zeta',\zeta_{n+1})P(\zeta))^{-1}R_{jk}(\zeta')B_k(\zeta)\zeta_{n+1}^{j-1}d\tilde{\zeta},$$

$$\eta\in\gamma_1\theta+\Gamma,\ \eta'\in\gamma_1\theta'+\Gamma_0,\ \eta_{n+1}=0,$$

where

$$R_{jk}(\zeta') = (k,j)\text{-cofactor of } ((2\pi i)^{-1}\oint P_+(\zeta)^{-1}B_j(\zeta)\zeta_n^{k-1}d\zeta_n).$$

F(x,y) has to be interpreted in the sense of distribution with respect to (x,y) in $R_+^n\times R_+^n$. We put

$$\tilde{F}(x',y_n,x_n) = F(x,0,y_n)$$

and regard $\tilde{F}(x',y_n,x_n)$ as a distribution on $X=R^{n-1}\times R_+^1\times R_+^1$.

2.1 Localization theorem

Put

$$\dot{\Gamma} = \{\xi'\in R^{n-1};\ (\xi',\xi_n)\in\Gamma \text{ for some } \xi_n\in R^1\}.$$

Then $R(\xi')$ is holomorphic in $R^{n-1}-i\gamma_0\theta'-i\dot{\Gamma}$. Let us denote by $\dot{\Sigma}=\Gamma(R,\theta')$ $(\subset R^{n-1})$ the component of the set $\{\xi'\in\dot{\Gamma};\ R^0(-i\xi')\neq 0\}$ which contains θ'. $\dot{\Sigma}$ is an open convex cone and we have

$$R(\xi') \neq 0 \quad \text{for } \xi'\in R^{n-1}-i\gamma_1\theta'-i\dot{\Sigma}.$$

Let $\xi^{0\prime}$ be arbitrarily fixed in $R^{n-1}\setminus\{0\}$ and let $\{j_k\}_{1\le k\le r_1}$ be the set of suffixes so that $p_{j_k}^0(\xi^{0\prime},\mu)=0$ has a real multiple root μ_k. We define $\dot{\Gamma}_{\xi^{0\prime}}$ by

$$\dot{\Gamma}_{\xi^{0\prime}}\times R^1 = \bigcap_{k=1}^{r_1}\Gamma(p_{j_k(\xi^{0\prime},\mu_k)},\theta).^{\dagger}$$

Here p_{ξ^0} is the localization of p at ξ^0 defined by

$$p(\nu^{-1}\xi^0+\eta) = \nu^h p_{\xi^0}(\eta) + o(\nu^h) \quad \text{as } \nu\to 0.$$

Let $p(\xi)$ be a strictly hyperbolic polynomial with respect to θ and assume that $p^0(0,1)\neq 0$, $p(\xi)\neq 0$ for $\xi\in R^n-i\gamma_0\theta-i\Gamma$. Put

† If $r_1=0$, then we put $\dot{\Gamma}_{\xi^{0\prime}}=R^{n-1}$.

$$p(\xi',\lambda;\nu) = p^0(\xi',\lambda)+\nu p^1(\xi',\lambda)+\cdots+\nu^{m'}p^{m'}$$
$$(= \nu^{m'}p(\nu^{-1}\xi',\nu^{-1}\lambda)),$$

where deg $p=m'$. We can assume without loss of generality that $\lambda=0$ is an ℓ'-ple root of $p^0(\xi^{0\prime},\mu)=0$. Let $\lambda(\xi';\nu)$ be a root of $p(\xi',\lambda;\nu)$ such that $\lambda(\xi^{0\prime},0)=0$. Then we have the following

Lemma 2.1. (Lemma 2.5 in [14]) For any compact set K in $R^{n-1}-i\gamma_0\theta'-i\dot{\Gamma}$ and any positive integer N there exists $\varepsilon>0$ such that

$$\lambda(\xi^{0\prime}+\nu\eta';\nu) = \Sigma_{j=1}^{N} c_j(\eta')\nu^{j/\ell'} + O(\nu^{(N+1)/\ell'})$$

if $\eta'\varepsilon K$ and $|\nu|<\varepsilon$. If $\ell'=1$, the $c_j(\eta')$ are polynomials of η', and if $\ell'>1$, the $c_j(\eta')$ are equal to $\Sigma_{\text{finite sum}}(\text{polynomials of }\eta')\times c_1(\eta')^{-n_{jk}}$, where the n_{jk} are integers. In particular,

$$c_1(\eta') = \text{const. } p_{(\xi^{0\prime},0)}(\eta')^{1/\ell'}.$$

Lemma 2.2. (Lemma 2.6 in [14]) For any compact set K in $R^{n-1}-i\gamma_0\theta'-i\dot{\Gamma}$ and any non-negative integer N there exists $\varepsilon>0$ such that if $\eta'\varepsilon K$ and $0<\nu<\varepsilon$,

$$\nu^{h_1}R(\nu^{-1}\xi^{0\prime}+\eta') = \Sigma_{j=0}^{N} Q_j(\eta')\nu^{j/L} + O(\nu^{(N+1)/L}), \qquad (2.2)$$

where $Q_0(\eta')\neq 0$, L is a positive integer and h_1 is a rational number. Moreover the $Q_j(\eta')$ are holomorphic in $R^{n-1}-i\gamma_0\theta'-i\dot{\Gamma}_{\xi^{0\prime}}$ and $Q_0(\eta')$ is equal to the localization $R_{\xi^{0\prime}}(\eta')$ of $R(\xi')$ at $\xi^{0\prime}$.

Let $Q_0^0(\eta')$ be the principal part of $Q_0(\eta')$. We denote by $\dot{\Sigma}_{\xi^{0\prime}}=\Gamma(Q_0,\theta')$ the component of the set $\{\eta'\varepsilon\dot{\Gamma}_{\xi^{0\prime}};\ Q_0^0(-i\eta')\neq 0\}$ which contains θ'.

Lemma 2.3. (Lemma 2.5 in [15]) $\dot{\Sigma}_{\xi^{0\prime}}$ is an open convex cone and

$$Q_0(\eta') \neq 0 \quad \text{for } \eta'\varepsilon R^{n-1}-i\gamma_1\theta'-i\dot{\Sigma}_{\xi^{0\prime}},$$
$$Q_0^0(\eta') \neq 0 \quad \text{for } \eta'\varepsilon R^{n-1}-i\dot{\Sigma}_{\xi^{0\prime}}.$$

Let ξ_{n+1}^0 be arbitrarily fixed in R^1 and let $\{s_k\}_{1\leq k\leq r_0}$ be the set of suffixes so that $p_{s_k}^0(\xi^{0\prime},\xi_{n+1}^0)=0$ and $\partial p_{s_k}^0/\partial\xi_1(\xi^{0\prime},\xi_{n+1}^0)\cdot\partial p_{s_k}^0/\partial\mu(\xi^{0\prime},\mu)|_{\mu=\xi_{n+1}^0}>0$. This implies that ξ_{n+1}^0 is a real simple root of $p^0_{s_k}(\xi^{0\prime},\mu)=0$ which corresponds to a root with positive imaginary part of $p^0_{s_k}(\xi^{0\prime}-i\gamma\theta',\mu)=0$, $\gamma>0$. Define

$$\tilde{\Gamma}_{(\xi^{0\prime},\xi^0_{n+1})} = \cap_{k=1}^{r_0} \{\tilde{\xi}\in R^{n+1};\ (\xi',\xi_{n+1})\in\Gamma(p_{s_k(\xi^{0\prime},\xi^0_{n+1})},\theta)\}.^{\dagger}$$

Let $\tilde{\xi}^0$ be arbitrarily fixed in $R^{n+1}\setminus\{0\}$ and put

$$\Gamma_{\tilde{\xi}^0} = (\Gamma(P_{\xi^0},\theta)\times R^1)\cap\tilde{\Gamma}_{(\xi^{0\prime},\xi^0_{n+1})}\cap(\dot{\Sigma}_{\xi^{0\prime}}\times R^2).$$

Here we put, if $\xi^{0\prime}=0$,

$$\dot{\Gamma}_{\xi^{0\prime}} = \dot{\Gamma},$$

$$\tilde{\Gamma}_{(\xi^{0\prime},\xi^0_{n+1})} = \begin{cases} R^{n+1} & \text{if } \xi^0_{n+1}\neq 0, \\ \{\tilde{\xi}\in R^{n+1};\ (\xi',\xi_{n+1})\in\Gamma(P,\theta)\} & \text{otherwise.} \end{cases}$$

Theorem 2.4. (Theorem 1.1 in [14]) Assume that the conditions (A.1) and (A.2) are satisfied and that $\tilde{\xi}^0\in R^{n+1}$. Then we have

$$t^{N/L}\{t^{p_0}\exp[-it(x'\cdot\xi^{0\prime}-y_n\xi^0_n+x_n\xi^0_{n+1})]\tilde{F}(x',y_n,x_n)$$
$$-\ \Sigma_{j=0}^{N}\ \tilde{F}_{\tilde{\xi}^0,j}(x',y_n,x_n)t^{-j/L}\} \longrightarrow 0 \quad \text{as } t\to\infty, \text{ in } \mathcal{D}'(X),\ N=0,1,\cdots.$$

Moreover we have

$$\cup_{j=0}^{\infty}\ \mathrm{supp}\ \tilde{F}_{\tilde{\xi}^0,j}(x',y_n,x_n)\times\{(\xi^{0\prime},-\xi^0_n,\xi^0_{n+1})\}$$
$$\subset WF(\tilde{F}(x',y_n,x_n)) \quad \text{for } \tilde{\xi}^0\neq 0,$$

and

$$\overline{ch}[\cup_{j=0}^{\infty}\ \mathrm{supp}\ \tilde{F}_{\tilde{\xi}^0,j}(x',y_n,x_n)]\subset\tilde{K}_{\tilde{\xi}^0}, \tag{2.3}$$

where

$$\tilde{K}_{\tilde{\xi}^0} = \{(x',y_n,x_n)\in X;\ x'\cdot\eta'-y_n\eta_n+x_n\eta_{n+1}\geq 0 \text{ for all } \eta\in\Gamma_{\tilde{\xi}^0}\}$$

and the closure in (2.3) is taken in X.

Proof. From Lemmas 2.1-2.3 and Seidenberg's lemma it follows that for any non-negative integer N there exist positive constant a and c such that

$$|t^{N/L}\{t^{p_0}\Sigma^{\ell}_{j,k=1}\ (R(t\xi^{0\prime}+\eta')P_+(t\xi^{0\prime}+\eta',t\xi^0_{n+1}+\eta_{n+1})P(t\xi^0+\eta))^{-1}$$
$$\times R_{jk}(t\xi^{0\prime}+\eta')B_k(t\xi^0+\eta)(t\xi^0_{n+1}+\eta_{n+1})^{j-1}\ -\ \Sigma_{j=0}^{N}\ \hat{F}_{\tilde{\xi}^0,j}(\tilde{\eta})t^{-j/L}\}|$$

---

† If $r_0=0$, then $\tilde{\Gamma}_{(\xi^{0\prime},\xi^0_{n+1})}=R^{n+1}$.

$$\leq at^{-1/L}(1+|\tilde{\eta}|)^c$$

when $\tilde{\eta}\in R^{n+1}-is\tilde{\theta}$ and $t\geq 1$, where s is large enough and $\tilde{\theta}=(\theta,0)\in R^{n+1}$. Here $\hat{F}_{\tilde{\xi}^0,j}(\tilde{\eta})$ are equal to $\Sigma_{\text{finite sum}}$ (polynomial of $\tilde{\eta}$)$\times Q_0(\eta')^{-n_1}$ $\times\Pi_{j=1}^{r_0} p_{s_j}(\xi^{0\prime},\xi^0_{n+1})(\eta',\eta_{n+1})^{-n_2}\times P_{\xi^0}(\eta)^{-n_3}\Pi_{k=1}^{r_1} p_{j_k}(\xi^{0\prime},\mu_k)(\eta)^{n_4/\ell_k}$. Put

$$\tilde{F}_{\tilde{\xi}^0,j}(x',y_n,x_n) = (2\pi)^{-n-1}\int_{R^{n+1}-is\tilde{\theta}} i^{-1}\exp[i\{x'\cdot\zeta'-y_n\zeta_n +x_n\zeta_{n+1}\}]\hat{F}_{\tilde{\xi}^0,j}(\tilde{\zeta})d\tilde{\zeta}.$$

Then we have for $\phi\in C_0^\infty(X)$

$$\langle\exp[-it\{x'\cdot\xi^{0\prime}-y_n\xi^0_n+x_n\xi^0_{n+1}\}]\tilde{F}(x',y_n,x_n),\phi(x',y_n,x_n)\rangle$$
$$= t^{-p_0}\{\Sigma_{j=0}^N \langle\tilde{F}_{\tilde{\xi}^0,j}(x',y_n,x_n),\phi(x',y_n,x_n)\rangle t^{-j/L} + O(t^{-(N+1)/L})\}.$$

Q.E.D.

2.2 Analytic wave front sets

Lemma 2.5. (Lemma 3.2 in [15]) Let $\xi^{0\prime}\in R^{n-1}\setminus\{0\}$ and let M be a compact set in $\Gamma_{\xi^{0\prime}}$. Then there exist a conic neighborhood Δ_1 $(\subset R^{n-1})$ of $\xi^{0\prime}$ and positive numbers C, t_0 such that $P_+(\zeta',\lambda)$ is holomorphic in $(\zeta',\lambda)\in\Lambda\times C^1$, where

$$\Lambda = \{\zeta'=\xi'-it|\xi'|\eta'-i\gamma_0\theta'\in R^{n-1}-iR^{n-1};\ \xi'\in\Delta_1,\ |\xi'|\geq C,\ \eta'\in M \text{ and } 0<t\leq t_0\}.$$

Therefore $R(\zeta')$ and $R_{jk}(\zeta')$ are also holomorphic in Λ.

Lemma 2.6. (Lemma 3.4 in [15]) Let K be a compact set in $R^{n-1}-i\Gamma_{\xi^{0\prime}}$. For any non-negative integer N there exist positive numbers ν_0 and r_0 such that

$$\nu^{h_1}R(\nu^{-1}r\xi^{0\prime}+r\eta') = r^{h_1}\Sigma_{j=0}^N Q_j(r\eta')(\nu r^{-1})^{j/L} + O(r^{h_0}\nu^{(N+1)/L})$$

if $r_0\eta'\in R^{n-1}-i\gamma_0\theta'-i\dot{\Gamma}_{\xi^{0\prime}}$, $\alpha\eta'\in K$ for some $\alpha\in C^1$ $(|\alpha|=1)$, $0<\nu\leq\nu_0$ and $r\geq r_0$, where h_0 and h_1 were defined by (2.1) and (2.2), respectively, and L is a positive integer.

Define the principal parts Q_j^0 of Q_j and rational numbers q_j by

$$Q_j(r\eta') = r^{q_j}\{Q_j^0(\eta') + r^{-1}Q_j^1(\eta') + \cdots\}, \quad Q_j^0(\eta') \not\equiv 0.$$

Moreover it is easy to see that $p_j \equiv h_1 + q_j - j/L$ is an integer and that $p_j \le h_0$. Put

$$p = \max p_j, \quad \alpha = \min_{p=p_j} j. \tag{2.4}$$

Lemma 2.7. (Lemma 3.6 in [15]) There exists the localization $R^0_{\xi^{0\prime}}(\eta')$ of $R^0(\xi')$ at $\xi^{0\prime}$ and

$$R^0_{\xi^{0\prime}}(\eta') = Q^0_\alpha(\eta').$$

Moreover $R^0_{\xi^{0\prime}}(\eta')$ is holomorphic in $R^{n-1} - i\dot{\Gamma}_{\xi^{0\prime}}$.

Let $\dot{\Sigma}^0_{\xi^{0\prime}}$ $(=\Gamma(R^0_{\xi^{0\prime}}, \theta'))$ be the component of the set $\{\eta' \varepsilon \dot{\Gamma}_{\xi^{0\prime}};\ R^0_{\xi^{0\prime}}(-i\eta') \neq 0\}$ which contains θ'. Here we define $\dot{\Sigma}^0_{\xi^{0\prime}} = \dot{\Sigma}$ if $\xi^{0\prime} = 0$. We can also prove that $\dot{\Sigma}^0_{\xi^{0\prime}}$ is an open convex cone and that $R^0_{\xi^{0\prime}}(\eta') \neq 0$ for $\eta' \varepsilon R^{n-1} - i\dot{\Sigma}^0_{\xi^{0\prime}}$.

Lemma 2.8. (Lemma 3.7 in [15]) For any compact set M in $\dot{\Sigma}^0_{\xi^{0\prime}}$, there exist a conic neighborhood Δ_1 $(\subset R^{n-1})$ of $\xi^{0\prime}$ and positive numbers C, t_0 such that

$$R(\xi' - it|\xi'|\eta' - i\gamma_1\theta') \neq 0 \quad \text{if } \eta' \varepsilon M,\ \xi' \varepsilon \Delta_1,\ |\xi'| \ge C \text{ and } 0 < t \le t_0.$$

From the above lemmas we have the following

Theorem 2.9. (Theorem 4.2 in [15]) Under the conditions (A.1) and (A.2) we have

$$(WF(\tilde{F}) \subset)\ WF_A(\tilde{F}(x', y_n, x_n)) \subset \bigcup_{\tilde{\xi} \varepsilon R^{n+1} \setminus \{0\}} \tilde{K}^0_{\tilde{\xi}} \times \{(\xi', -\xi_n, \xi_{n+1})\},$$

where

$$\Gamma^0_{\tilde{\xi}^0} = (\Gamma(P_{\xi^0}, \theta) \times R^1) \cap \tilde{\Gamma}_{(\xi^0, \xi^0_{n+1})} \cap (\dot{\Sigma}^0_{\xi^{0\prime}} \times R^2),$$

$$\tilde{K}^0_{\tilde{\xi}^0} = \{(x', y_n, x_n) \varepsilon X;\ x' \cdot \eta' - y_n\eta_n + x_n\eta_{n+1} \ge 0 \text{ for all } \tilde{\eta} \varepsilon \Gamma^0_{\tilde{\xi}^0}\}.$$

Proof. Let us assume that $(x^{0\prime}, y^0_n, x^0_n, \xi^{0\prime}, -\xi^0_n, \xi^0_{n+1}) \notin \tilde{K}^0_{\tilde{\xi}^0}$. Then from Lemmas 2.5 and 2.8 there exist an open conic neighborhood Δ_1 $(\subset R^{n+1} \setminus \{0\})$ of $(\xi^{0\prime}, -\xi^0_n, \xi^0_{n+1})$, $\tilde{\eta} \varepsilon \Gamma^0_{\tilde{\xi}^0}$, a neighborhood U of $(x^{0\prime}, y^0_n, x^0_n)$, positive numbers δ, C, t_0 and a rational number a such that

$$x' \cdot \eta' - y_n\eta_n + x_n\eta_{n+1} < 0 \quad \text{when } (x', y_n, x_n) \varepsilon U, \tag{2.5}$$

$$|R(\xi'-i(t|\tilde{\xi}|\eta'+\gamma_2\theta'))P_+(\xi'-i(t|\tilde{\xi}|\eta'+\gamma_2\theta'),\xi_{n+1}-it|\tilde{\xi}|\eta_{n+1})$$
$$\times P(\xi'-i(t|\tilde{\xi}|\eta'+\gamma_2\theta'),-\xi_n-it|\tilde{\xi}|\eta_n)| \geq \delta|\tilde{\xi}|^a \tag{2.6}$$

when $\tilde{\xi}\epsilon\Delta_1$, $|\tilde{\xi}|\geq C$, $0\leq t\leq t_0$, where $\gamma_2=\gamma_1+1$. Let Δ ($\subset\subset\Delta_1$) be conic neighborhood of $(\xi^{0\prime},-\xi_n^0,\xi_{n+1}^0)$. Let $\{\phi_N(x',y_n,x_n)\}$ be a bounded sequence in $C_0^\infty(U)$ such that $\phi_N=1$ on a fixed neighborhood of $(x^{0\prime}, y_n^0,x_n^0)$ and

$$|\tilde{D}^\alpha\phi_N| \leq C(CN)^{|\alpha|} \quad \text{for } |\alpha|\leq N,$$

where $\tilde{D}=i^{-1}(\partial/\partial x_1,\cdots,\partial/\partial x_{n-1},\partial/\partial y_n,\partial/\partial x_n)$. Let V_t, $0\leq t\leq t_0$, be the chain $\tilde{\zeta}=(\xi'-i(t\psi(\tilde{\xi})|\tilde{\xi}|\eta'+\gamma_2\theta'),\xi_n+it\psi(\tilde{\xi})|\tilde{\xi}|\eta_n,\xi_{n+1}-it\psi(\tilde{\xi})|\tilde{\xi}|\eta_{n+1})$, $|\tilde{\xi}|\geq C$, where $\psi(\tilde{\xi})\epsilon C^\infty(R^{n+1}\setminus\{0\})$ is positively homogeneous of degree zero and $\psi(\tilde{\xi})=1$ on Δ, supp $\psi\subset\Delta_1$, $0\leq\psi(\tilde{\xi})\leq 1$. From (2.5),(2.6) and Stokes' formula we have

$$\mathcal{F}[\phi_N\tilde{F}](\tilde{\xi}) = \int_{R^{n+1}-i\gamma_2\tilde{\theta}} \hat{\phi}_N(\tilde{\xi}-\tilde{\zeta})\times i^{-1}\Sigma_{j,k=1}^{\ell}(R(\zeta')P_+(\zeta',\zeta_{n+1}))^{-1}$$
$$\times P(\zeta',-\zeta_n)^{-1}R_{jk}(\zeta')B_k(\zeta',-\zeta_n)\zeta_{n+1}^{j-1}d\tilde{\zeta} = \int_\gamma + \int_{V_{t0}} \equiv I_1 + I_2,$$

where γ is a compact chain. It is obvious that

$$|I_1| \leq C(CN)^N(1+|\tilde{\xi}|)^{-N} \quad \text{when } \tilde{\xi}\epsilon\Delta.$$

Since $|\tilde{\zeta}-\tilde{\xi}|\geq\delta(|\tilde{\zeta}|+|\tilde{\xi}|)$ for $\tilde{\xi}\epsilon\Delta$ and $\tilde{\zeta}\epsilon V_{t_0}$, we have

$$|I_2| \leq C(CN)^N(1+|\tilde{\xi}|)^{-N+b} \quad \text{for } \tilde{\xi}\epsilon\Delta,$$

where b is a constant $>n+1$ and $N\geq b$. Q.E.D.

2.3 Some remarks

Let us consider $\tilde{K}_{\tilde{\xi}^0}^0$ and $\tilde{K}_{\tilde{\xi}^0}$.

Lemma 2.10. (Lemma 3.8 in [15]) $\dot{\Sigma}_{\xi^{0\prime}}^0\subset\dot{\Sigma}_{\xi^{0\prime}}$ and, therefore, $\tilde{K}_{\tilde{\xi}^0}^0\supset\tilde{K}_{\tilde{\xi}^0}$.

Lemma 2.11. (Theorem 4.1 in [14]) Assume that each $p_j^0(\xi^{0\prime},\mu)=0$ has no real multiple roots. Then the localization $Q_0(\eta')$ of $R(\xi')$ at $\xi^{0\prime}$ is a hyperbolic polynomial. Moreover $Q_0^0(\eta')$ is equal to the localization of $R^0(\xi')$ at $\xi^{0\prime}$, i.e. α defined by (2.4) is equal to zero, if at least one of the following conditions is satisfied:
(i) The system $\{P(-D),B_j(-D)\}$ satisfies the Lopatinski condition.

(ii) $\xi^{0\prime}\in\partial\dot{\Sigma}$.

Lemma 2.12. Let K be a compact set in $R^{n-1}-i\dot{\Gamma}_{\xi^{0\prime}}$. Then there exists a positive number ν_K such that for $0<\nu\le\nu_K$ and $\eta'\in K$

$$\nu^{-h_0+h_1-\alpha/L}R^0(\xi^{0\prime}+\nu\eta') = Q^0_\alpha(\eta') + O(\nu^{1/L}).$$

From Lemma 2.12 we have the following

Lemma 2.13. Let M be a compact set in $\dot{\Sigma}^0_{\xi^{0\prime}}$. Then there exist a neighborhood U of $\xi^{0\prime}$ and positive number t_0 such that $R^0(\xi')$ is holomorphic in U-iD and $R^0(\xi')\neq 0$ for $\xi'\in U-iD$, where $D=\{t\eta';\ \eta'\in\mathring{M},\ 0<t\le t_0\}$.

Using Lemmas 2.12 and 2.13 we can prove the inner semi-continuity of $\dot{\Sigma}^0_{\xi'}$ and, therefore, $\Gamma^0_{\tilde{\xi}}$.

Theorem 2.14. Let M be a compact set in $\dot{\Sigma}^0_{\xi^{0\prime}}$. Then there exists a neighborhood U of $\xi^{0\prime}$ such that

$$M\subset\dot{\Sigma}^0_{\xi'}\quad\text{for } \xi'\in U.$$

Theorem 2.15. $\bigcup_{\tilde{\xi}\in R^{n+1}\setminus\{0\}}\tilde{K}^0_{\tilde{\xi}}\times\{(\xi',-\xi_n,\xi_{n+1})\}$ is closed in $T^*X\setminus 0$.

In the following example $\dot{\Sigma}_{\xi'}$ does not have the property of inner semi-continuity and $\bigcup_{\tilde{\xi}\in R^{n+1}\setminus\{0\}}\tilde{K}_{\tilde{\xi}}\times\{(\xi',-\xi_n,\xi_{n+1})\}$ is not closed.

Example 2.16. Put n=4 and

$$P(\xi) = (\xi_1^2-\xi_2^2-\xi_3^2-\xi_4^2+a\xi_3)(\xi_1^2-\xi_4^2),\quad a>0,$$

$$B_1(\xi) = 1,\quad B_2(\xi) = (-\xi_1-i\xi_3)\xi_4 - \xi_4^2.$$

Then we have $R(\xi')=i\xi_3+{}^{+}\sqrt{\xi_1^2-\xi_2^2-\xi_3^2+a\xi_3}$. It is obvious that $\{P,B_1,B_2\}$ satisfies the conditions (A.1) and (A.2). Put $\tilde{\xi}^0=(0,0,-1,0,0)$ and $\tilde{\xi}^j=(1/j,i/j,-1,1/j,1/j)$, $j=1,2,\cdots$. Then it is easily seen that

$$\Gamma(P_{\xi^0},\theta) = \{\eta\in R^4;\ \eta_1-\eta_4>0 \text{ and } \eta_1+\eta_4>0\},$$

$$\tilde{\Gamma}_{(\xi^{0\prime},\xi_5^0)} = \{\tilde{\eta}\in R^5;\ \eta_1+\eta_5>0\},$$

$$\dot{\Sigma}_{\xi^{0\prime}} = R^3,$$

$$\Gamma(P_{\xi^j},\theta) = \{\eta\in R^4;\ \eta_1-\eta_4>0\},$$

$$\tilde{\Gamma}_{(\xi^{j'},\xi_5^j)} = \{\eta\in R^5;\ \eta_1+\eta_5>0\},$$

$$\dot{\Sigma}_{\xi^{j'}} = \{\eta\in R^3;\ \eta_1-\eta_2>0\}.$$

Thus $(3,-1,0,1,1)\notin\tilde{K}_{\tilde{\xi}^0}$ and $(3,-1,0,1,1)\in\tilde{K}_{\tilde{\xi}^j}$. This implies that $\bigcup_{\tilde{\xi}\in R^5\setminus\{0\}} \tilde{K}_{\tilde{\xi}}\times\{(\xi',-\xi_4,\xi_5)\}$ is not closed. Using the following lemmas and Lemma 8.3 in [1] after some calculations we can show that

$$\tilde{K}_{\tilde{\xi}} = \bigcup_{j=0}^{\infty} \operatorname{supp} \tilde{F}_{\tilde{\xi},j}(x',y_4,x_4).$$

Lemma 2.17. Let f_j $(j=1,2)$ be non-negative measures and assume that $\operatorname{supp} f_j=C_j$, where C_j is a closed cone included by the set $\{x\in R^n;\ x\cdot\theta>0\}\cup\{0\}$. Then f_1*f_2 is well-defined and a non-negative measure. Moreover $\operatorname{supp}(f_1*f_2)=C_1+C_2$.

Lemma 2.18. Let f_1 be non-negative measure and assume that αf_2 is non-negative measure when $\alpha\in C_0^\infty(R^n)$ and $\alpha(0)=0$ and that $\operatorname{supp} f_j=C_j$, $j=1,2$, are closed cones included by the set $\{x\in R^n,\ x\cdot\theta>0\}\cup\{0\}$. If $x^0\notin C_1$ and $x^0\in C_1+C_2$ then $x^0\in\operatorname{supp}(f_1*f_2)$.

As for Example 2.16 we can show that

$$\begin{aligned}
&\bigcup_{\tilde{\xi}\in R^5\setminus\{0\}}\bigcup_{j=0}^{\infty} \operatorname{supp} \tilde{F}_{\tilde{\xi},j}\times\{(\xi',-\xi_4,\xi_5)\}\\
&=\bigcup_{\tilde{\xi}\in R^5\setminus\{0\}} \tilde{K}_{\tilde{\xi}}\times\{(\xi',-\xi_4,\xi_5)\} \subsetneq WF(\tilde{F}) \subset WF_A(\tilde{F})\\
&\subset\bigcup_{\tilde{\xi}\in R^5\setminus\{0\}} \tilde{K}^0_{\tilde{\xi}}\times\{(\xi',-\xi_4,\xi_5)\}
\end{aligned}$$

and that

$$\overline{ch}[WF(\tilde{F})|_{\tilde{\xi}^0}] = \overline{ch}[WF_A(\tilde{F})|_{\tilde{\xi}^0}] = \tilde{K}^0_{\tilde{\xi}^0} \quad \text{for } \tilde{\xi}^0\neq 0.$$

Moreover Lemma 2.11 implies that $\{P(-D),B_1(-D),B_2(-D)\}$ does not satisfy the Lopatinski condition, which is easily verified. So this example shows that the $\mathcal{E}$-well posedness of $\{P(D),B_j(D)\}$ does not always imply that of $\{P(-D),B_j(-D)\}$.

3. HYPERBOLIC MIXED PROBLEMS IN PLANE-STRATIFIED MEDIA

In this section we shall introduce the results obtained by Matsumura [8]. Let $P_j(\xi)$ be hyperbolic polynomial of order m_j with respect to θ, $j=1,2$. We assume that $P_j^0(0,\cdots,0,1)\neq 0$. Now we consider the hyperbolic mixed problem in plane-stratified media

$$P_1(D)u(x) = f(x), \quad x_1>0, \ 0<x_n<h, \tag{3.1}$$

$$P_2(D)u(x) = f(x), \quad x_1>0, \ x_n>h, \tag{3.2}$$

$$D_1^{j-1}u(0,x'') = g_{1j}(x''), \quad 0<x_n<h, \ 1\le j\le m_1, \tag{3.3}$$
$$D_1^{j-1}u(0,x'') = g_{2j}(x''), \quad x_n>h, \ 1\le j\le m_2,$$

$$Q_j(D)u(x)|_{x_n=0} = k_{0j}(x'), \quad x_1>0, \ 1\le j\le \ell_1, \tag{3.4}$$

$$B_j(D)u(x)|_{x_n=h-0} = C_j(D)u(x)|_{x_n=h+0} + k_j(x'), \quad x_1>0, \ 1\le j\le m-\ell_1+\ell_2. \tag{3.5}$$

Here $Q_j(D)$, $B_j(D)$ and $C_j(D)$ are partial differential operators with constant coefficients and ℓ_j is equal to the number of the roots with positive imaginary part of the $P_j(\xi'-i\gamma\theta',\lambda)=0$ with respect to λ, where $\gamma>\gamma_0$ and γ_0 is sufficiently large.

Put $\Gamma_j=\Gamma(P_j,\theta)$ and $\Gamma_{j0}=\{\xi'\in R^{n-1};\ (\xi',0)\in\Gamma_j\}$. When $\xi'\in R^{n-1}-i\gamma_0\theta'-i\Gamma_{j0}$, we can denote the roots of $P_j(\xi',\lambda)=0$ with respect to λ by $\lambda_{j1}^+(\xi'),\cdots,\lambda_{j\ell_j}^+(\xi'),\lambda_{j1}^-(\xi'),\cdots,\lambda_{jm_j-\ell_j}^-(\xi')$ in the same way as in §2. Put

$$L_1(\zeta') = \left((2\pi i)^{-1}\oint P_{1+}(\zeta)^{-1}Q_j(\zeta)\zeta_n^{k-1}d\zeta_n\right)_{j,k=1,\cdots,\ell_1},$$
$$L_2(\zeta') = \Big((2\pi i)^{-1}\oint P_{1-}(\zeta)^{-1}B_j(\zeta)d\zeta_n,$$
$$\cdots,(2\pi i)^{-1}\oint P_{1-}(\zeta)^{-1}B_j(\zeta)\zeta_n^{m-\ell_1-1}d\zeta_n,$$
$$(2\pi i)^{-1}\oint P_{2+}(\zeta)^{-1}C_j(\zeta)d\zeta_n,$$
$$\cdots,(2\pi i)^{-1}\oint P_{2+}(\zeta)^{-1}C_j(\zeta)\zeta_n^{\ell_2-1}d\zeta_n\Big)_{j\downarrow 1,\cdots,m-\ell_1+\ell_2},$$
$$R_0(\zeta') = \det L_1(\zeta')\cdot\det L_2(\zeta').$$

We note that $\det L_1(\xi')$ is Lopatinski's determinant for the system $\{P_1,Q_j\}$ and that $\det L_2(\xi')$ is Lopatinski's determinant for transmission problem $\{P_1,P_2,B_j,C_j\}$.

Theorem 3.1. In order that the problem (3.1)-(3.5) is $\mathcal{E}$-well posed, it is necessary and sufficient that

$$R_0^0(\theta') \ne 0, \tag{3.6}$$

$$R_0(\xi'-is\theta') \ne 0 \quad \text{for any } \xi'\in R^{n-1},\ s>\gamma_1. \tag{3.7}$$

Here $R_0^0(\xi')$ is the principal part of $R_0(\xi')$.

The necessity of Theorem 3.1 can be proved by the same argument as in [10]. The sufficiency of the theorem will follow from the explicit expressions of the fundamental solution and the Poisson kernels. In this note we shall construct and study only the fundamental solution which describes the propagation of waves produced by unit impulse given at position $y=(0,y'')$, $0<y_n<h$. For we can construct and study in a similar way the Poisson kernels and the fundamental solution which describes the propagation of waves produced by unit impulse given at position $y=(0.y'')$, $h<y_n$.

The fundamental solution for the mixed problem (3.1)-(3.5) is defined as the unique distribution solution $G(x,y)$, whose support is included by $\{x_1\geq 0\}$, of the mixed problem (3.1)-(3.5) with $f=0$, $g_{1j}=\delta_{m_1 j}\delta(x''-y'')$, $g_{2j}=0$, $k_{0j}=0$ and $k_j=0$. Let us consider the case where $0<y_n<h$ and assume that (3.6) and (3.7) hold. Write

$$G(x,y) = \begin{cases} E_1(x-y) - F_1(x,y), & 0<x_n<h, \\ F_2(x,y), & x_n>h, \end{cases}$$

where

$$E_1(x) = (2\pi)^{-n}\int_{R^n-i\eta} \exp[ix\cdot\zeta]P_1(\zeta)^{-1}d\zeta, \quad \eta\in\gamma_0\theta+\Gamma_1.$$

Then $F_1(x,y)$ and $F_2(x,y)$ satisfy the equations

$$P_1(D_x)F_1(x,y) = 0, \quad 0<x_n<h,$$

$$P_2(D_x)F_2(x,y) = 0, \quad x_n>h,$$

$$Q_j(D_x)F_1(x,y)|_{x_n=0} = Q_j(D_x)E_1(x-y)|_{x_n=0}, \quad x_1>0,\ 1\leq j\leq \ell_1,$$

$$B_j(D_x)F_1(x,y)|_{x_n=h} + C_j(D_x)F_2(x,y)|_{x_n=h}$$
$$= B_j(D_x)E_1(x-y)|_{x_n=h}, \quad x_1>0,\ 1\leq j\leq m-\ell_1+\ell_2.$$

Taking formally partial Fourier-Laplace transforms with respect to x' in these equations, we obtain a system of ordinary differential equations in x_n with coefficients depending on the parameter ζ'. Put

$$\hat{F}_1(\zeta',x_n,y) = \Sigma_{j=1}^{\ell_1} \alpha_j^+(\zeta',y)\times(2\pi i)^{-1}\oint \exp[ix_n\zeta_n]P_{1+}(\zeta)^{-1}\zeta_n^{j-1}d\zeta_n$$
$$+ \Sigma_{j=1}^{m-\ell_1} \alpha_j^-(\zeta',y)\times(2\pi i)^{-1}\oint \exp[i(x_n-h)\zeta_n]P_{1-}(\zeta)^{-1}\zeta_n^{j-1}d\zeta_n,$$

$$\hat{F}_2(\zeta',x_n,y) = \Sigma_{j=1}^{\ell_2} \beta_j(\zeta',y)$$
$$\times(2\pi i)^{-1}\oint \exp[i(x_n-h)\zeta_n]P_{2+}(\zeta)^{-1}\zeta_n^{j-1}d\zeta_n,$$

and consider the linear equations for $\alpha_j^{\pm}$ and β_j

$$L(\zeta')^t(\alpha_1^+(\zeta'),\cdots,\alpha_{\ell_1}^+,\alpha_1^-,\cdots,\alpha_{m-\ell_1}^-,\beta_1,\cdots,\beta_{\ell_2})$$
$$= (2\pi)^{-1}{}^t(\int_{-\infty}^{\infty}\exp[-iy\cdot\zeta]P_1(\zeta)^{-1}Q_1(\zeta)d\zeta_n,$$
$$\cdots,\int_{-\infty}^{\infty}\exp[-iy\cdot\zeta]P_1(\zeta)^{-1}Q_{\ell_1}(\zeta)d\zeta_n,$$
$$\int_{-\infty}^{\infty}\exp[-iy\cdot\zeta+ih\zeta_n]P_1(\zeta)^{-1}B_1(\zeta)d\zeta_n,$$
$$\cdots,\int_{-\infty}^{\infty}\exp[-iy\cdot\zeta+ih\zeta_n]P_1(\zeta)^{-1}B_{m-\ell_1+\ell_2}(\zeta)d\zeta_n),$$

where

$$L(\zeta') = \begin{pmatrix} L_1(\zeta') & L_3(\zeta') & 0 \\ L_4(\zeta') & & L_2(\zeta') \end{pmatrix},$$
$$L_3(\zeta') = ((2\pi i)^{-1}\oint \exp[-ih\zeta_n]P_{1-}(\zeta)^{-1}Q_j(\zeta)\zeta_n^{k-1}d\zeta_n),$$
$$j\downarrow 1,\cdots,\ell_1;\ k\rightarrow 1,\cdots,m-\ell_1,$$
$$L_4(\zeta') = ((2\pi i)^{-1}\oint \exp[ih\zeta_n]P_{1+}(\zeta)^{-1}B_j(\zeta)\zeta_n^{k-1}d\zeta_n),$$
$$j\downarrow 1,\cdots,m-\ell_1+\ell_2;\ k\rightarrow 1,\cdots,\ell_1.$$

It easily follows that

$$\det L(\zeta') \equiv R(\zeta') = R_0(\zeta')\cdot\det(I-R_0(\zeta')^{-1}{}^t\mathrm{cof}\ L_2\cdot L_4 \times{}^t\mathrm{cof}\ L_1\cdot(L_3,0)). \tag{3.9}$$

The hyperbolicity of $P_1(\xi)$ implies that there exists a positive number ε such that

$$|\mathrm{Im}\ \lambda_{1j}^{\pm}(\xi'-i\gamma\theta')| > \varepsilon\gamma \quad \text{for } \gamma>2\gamma_0.$$

In fact, we have

$$P_1(\xi-i\eta) \neq 0 \quad \text{if } \eta\in\gamma_0\theta+\Gamma_1 \text{ and } \xi\in R^n,$$

and there exists a positive number ε such that $\gamma\theta+\eta\in\Gamma_1$ if $|\eta|\le 2\varepsilon\gamma$ and $\eta\in R^n$. Thus we have

$$P_1(\xi-i(\gamma_0+\gamma)\theta-i\eta) \neq 0 \quad \text{if } |\eta|\le 2\varepsilon\gamma,$$

that is,

$$P_1(\xi-i\gamma\theta-i\eta) \neq 0 \quad \text{if } |\eta|\le\varepsilon\gamma \text{ and } \gamma>2\gamma_0.$$

From (3.7) and Seidenberg's lemma it follows that there exist positive constants δ and M such that

$$\|R_0(\xi'-i\gamma\theta')^{-1}{}^t\mathrm{cof}\, L_2(\xi'-i\gamma\theta')\cdot L_4{}^t\mathrm{cof}\, L_1\cdot(L_3,0)\| \leq \delta < 1 \tag{3.10}$$

when $\xi'\varepsilon R^{n-1}$ and $\gamma=M\cdot\log(2+|\xi'|)$. Here $\|\cdot\|$ denote the matrix norm. In fact, there exist positive numbers C_0, C_1 and rational numbers α_0, α_1 such that

$$|R_0(\xi'-i\gamma\theta')| \geq C_0(1+|\xi'|+|\gamma|)^{\alpha_0}, \quad \xi'\varepsilon R^{n-1}, \ \gamma\geq\gamma_1+1,$$

$$|\text{each entry of } L_3(\xi'-i\gamma\theta') \text{ and } L_4(\xi'-i\gamma\theta')|$$

$$\leq C_1\exp[-\varepsilon h\gamma](1+|\xi'|+|\gamma|)^{\alpha_1}, \quad \xi'\varepsilon R^{n-1}, \ \gamma>2\gamma_0.$$

By (3.9) and (3.10) we can solve the equations (3.8) when $\zeta'=\xi'-i\gamma\theta'$ and $\gamma=M\cdot\log(2+|\xi'|)$. Let S_M be the chain $\{\zeta=(\xi_1-i\gamma,\xi''); \xi\varepsilon R^n, \gamma=M\cdot\log(2+|\xi'|)\}$. Then $F_1(x,y)$ and $F_2(x,y)$ can be obtained by applying the inverse Fourier-Laplace transformation along S_M to $\hat{F}_1(\zeta',x_n,y)$ and $\hat{F}_2(\zeta',x_n,y)$.

The wave front sets of $F_1(x,y)$ and $F_2(x,y)$ can be estimated by the same argument as in §2 if $P_j(\xi)$ satisfy the condition (A.1).

Lemma 3.2. Let K be an $\ell\times\ell$ matrix. Then

$$\det(I-\lambda K)^{-1} = \Sigma_{j=0}^{\infty}\lambda^j \,\Sigma\, F^{\ell}_{(i_1,k_1),\cdots,(i_j,k_j)}\, K_{i_1k_1}\cdots K_{i_jk_j}$$

when $|\lambda|\|K\|<1$, where $F^{\ell}_{(i_1,k_1),\cdots,(i_j,k_j)}$ is independent of K.

This lemma gives the developments of $F_1(x,y)$ and $F_2(x,y)$. Moreover we can apply the argument in §2 to each term of the developments. We can also see that the supports of each terms of the developments are locally finite. Roughly speaking, we have in the development of $F_1(x,y)$, for example,

$$\int \exp[i\{(x'-y')\cdot\zeta'-iy_n\zeta_n+x_n\zeta_{n+1}+h(\eta_1+\cdots+\eta_s-\eta_{s+1}-\cdots-\eta_{s+r})\}]$$

$$\times f(\zeta,\zeta_{n+1},\eta_1,\cdots,\eta_{s+r})R_0(\zeta')^{-h}(P_1(\zeta)P_{1+}(\zeta',\zeta_{n+1})P_{1+}(\zeta',\eta_1)$$

$$\cdots P_{1+}(\zeta',\eta_s)P_{1-}(\zeta',\eta_{s+1})\cdots P_{1-}(\zeta',\eta_{s+r}))^{-1}d\tilde{\zeta}d\eta_1\cdots d\eta_{s+r}.$$

The explicit developments of $F_1(x,y)$ and $F_2(x,y)$ were given in [8] as concerns wave equations.

4. MICROLOCAL PARAMETRICES IN THE CASE WHERE BOUNDARY WAVES APPEAR

Let $P(x,\xi)$ be a polynomial of order m of ξ variables with C^∞ coefficients and $p(x,\xi)$ its principal part. We assume that $p(x,\xi)$ is a strictly hyperbolic polynomial with respect to θ and $p(x,0,1)=1$. Thus we can write

$$p(x,\xi) = \Pi_{j=1}^{\ell} (\xi_n-\mu_j^+(x,\xi'))\cdot\Pi_{j=1}^{m-\ell} (\xi_n-\mu_j^-(x,\xi')),$$

where the $\mu_j^\pm(x,\xi')$ are continuous in (x,ξ') and

$$\mathrm{Im}\ \mu_j^\pm(x,\xi') \gtrless 0 \quad \text{when } \mathrm{Im}\ \xi_1<0,\ \xi''\varepsilon R^{n-2}.$$

We consider the mixed initial-boundary value problem for hyperbolic operator $P(x,D)$ in a quarter-space

$$P(x,D)u(x) = 0, \quad x\varepsilon R_+^n,\ x_1>0, \tag{4.1}$$

$$D_1^{k-1}u(x)|_{x_1=0} = 0, \quad x_n>0,\ 1\le k\le m, \tag{4.2}$$

$$B_j(x',D)u(x)|_{x_n=0} = \delta_{1j}g_j(x'), \quad x_1>0,\ 1\le j\le \ell. \tag{4.3}$$

Here the $B_j(x',D)$ are boundary operators with C^∞ coefficients.

Now let $(x^{0\prime},\xi^{0\prime})$ be a fixed point in $T^*R^{n-1}\setminus 0$ and put $x^0=(x^{0\prime},0)$. We may assume that the $\mu_j^+(x,\xi')$ are enumerated in the following way:

$$\mathrm{Im}\ \mu_j^+(x^0,\xi^{0\prime}) = 0 \quad \text{for } 1\le j\le\mu,$$

$$\mathrm{Im}\ \mu_j^+(x^0,\xi^{0\prime}) > 0 \quad \text{for } \mu+1\le j\le\ell.$$

Then we put

$$\tilde{L}(x',\xi') = (b_j(x',\mu_j^+(x',0,\xi'),\xi'),\cdots,b_j(x',\mu_\mu^+,\xi'),$$
$$(2\pi i)^{-1}\int_{C_{\xi'}} p(x',0,\xi)^{-1}b_j(x',\xi)d\xi_n,$$
$$\cdots,(2\pi i)^{-1}\int_{C_{\xi'}} p(x',0,\xi)^{-1}b_j(x',\xi)\xi_n^{\ell-\mu-1}d\xi_n)_{j\downarrow 1,\cdots,\ell},$$

where $b_j(x',\xi)$ is the principal part of $B_j(x',\xi)$ and $C_{\xi'}$ is a simple closed curve enclosing only roots $\mu_{\mu+1}^+(x',0,\xi'),\cdots,\mu_\ell^+(x',0,\xi')$ of $p(x',0,\xi',\lambda)=0$, and we define

$$\tilde{R}(x',\xi') = \det \tilde{L}(x',\xi').$$

Remark. It is easy to see that

$$\tilde{R}(x',\xi') = (-1)^{\mu(\ell-\mu)}\Pi_{1\le j<k\le\mu}\,(\mu_j^+(x',0,\xi')-\mu_k^+)$$
$$\times\Pi_{\mu+1\le j\le\ell,\ 1\le k\le m-\ell}\,(\mu_j^+-\mu_k^-)^{-1}R(x',\xi'),$$

where $R(x',\xi')$ is Lopatinski's determinant for the system $\{p,b_j\}$.

We state the assumptions that we impose on $\{p,b_j\}$:

(A.3) $(x^{0},\xi^{0\prime})$ is not a glancing point for p, i.e., $\mu_j^+(x^0,\xi^{0\prime})$, $1\le j\le\mu$, are simple real roots of $p(x^0,\xi^{0\prime},\lambda)=0$.

(A.4) $\tilde{R}(x',\xi') = (\xi_1-\xi_1(x',\xi'''))^{\theta}r(x',\xi')$,

where $\xi_1(x',\xi''')$ and $r(x',\xi')$ are C^∞ functions defined in a conic neighborhood of $(x^{0\prime},\xi^{0\prime})$ in $T^*R^{n-1}\setminus 0$, $\xi_1(x',\xi''')$ is real-valued and homogeneous of degree 1 in ξ''', $\xi_1(x^{0\prime},\xi^{0\prime\prime\prime})=\xi_1^0$, $r(x^{0\prime},\xi^{0\prime})\ne 0$, $\xi'''=(\xi_2,\cdots,\xi_{n-1})$ and θ is a positive integer.

(A.5) There exist $\ell\times\ell$ matrix-valued C^∞ functions $U(x',\xi')$ and $V(x',\xi')$ defined in a conic neighborhood of $(x^{0\prime},\xi^{0\prime})$ in $T^*R^{n-1}\setminus 0$ such that

$$U(x',\xi')\tilde{L}(x',\xi')V(x',\xi') = \begin{pmatrix}(\xi_1-\xi_1(x',\xi'''))I_{\theta'} & 0\\ 0 & \tilde{L}_e(x',\xi')\end{pmatrix},$$

$\det U(x^{0\prime},\xi^{0\prime})\ne 0$, $\det \hat{L}_e(x^{0\prime},\xi^{0\prime})\ne 0$, the (i,j)-entry of U is homogeneous of degree $1-\rho_i-m_j$ and the (i,j)-entry of V is homogeneous of degree ρ_j for $1\le i\le\mu$ and of degree $\rho_j+m+\mu-i$ for $\mu+1\le i\le\ell$, where θ' is a positive integer, $I_{\theta'}$ is the identity matrix of order θ', $\hat{L}_e$ is an $(\ell-\theta')\times(\ell-\theta')$ matrix and $\deg B_j=m_j$.

Remark. (i) If the condition (A.4) with $\theta=1$ is satisfied then the condition (A.5) also holds. In fact, taking $U(x',\xi')=I$ and $V(x',\xi')=r(x',\xi')^{-1}{}^t\mathrm{cof}\,\tilde{L}(x',\xi')$, we have $U\tilde{L}V=(\xi_1-\xi_1(x',\xi'''))I$.
(ii) Suppose that $\mu_j^+(x^0,\xi^{0\prime})$, $1\le j\le\mu'$, are simple roots of $p(x^0,\xi^{0\prime},\lambda)=0$. If $\mathrm{rank}\,(B_j(x^0,\xi^{0\prime},\mu_k^+(x^0,\xi^{0\prime})))_{j\downarrow 1,\cdots,\ell;k\to 1,\cdots,\mu'}=\mu'-\theta$ the condition (A.5) follows from (A.4) (see [6]).

Let Γ be a conic neighborhood of $(x^{0\prime},\xi^{0\prime})$ in $T^*R^{n-1}\setminus 0$ and U a neighborhood of $x^{0\prime}$ in R^{n-1}.

Definition 4.1. A right microlocal parametrix (Poisson operator) for the problem (4.1)-(4.3) at $(x^0,\xi^{0\prime})$ is a triple $\{E_1,\Gamma,[0,\varepsilon)\times U\}$ satisfying the conditions
(i) E_1 is a continuous linear map: $\mathcal{D}'(U)\to C^\infty([0,\varepsilon);\mathcal{D}'(U))$,

(ii) $PE_1(g)\in C^\infty([0,\varepsilon)\times U)$;†

(iii) $B_jE_1(g)|_{x_n=0} - \delta_{1j}g\in C^\infty(U)$, $1\le j\le \ell$, if $WF(g)\subset\Gamma$,

(iv) $E_1(g)|_{x_1<c}$ is smooth if $WF(g)\subset\{x_1\ge c\}$.

There exist a conic neighborhood Γ_0 of $(x^{0\prime},\xi^{0\prime})$ in $T^*R^{n-1}\setminus 0$ and $\tilde\psi(x',\xi')\in C^\infty(\Gamma_0)$ such that $\tilde\psi(x',\xi')$ satisfies the equations

$$\partial_1\tilde\psi(x',\xi') - \xi_1(x',\nabla_{x'''}\tilde\psi(x',\xi')) = \xi_1 - \xi_1(x^{0\prime},\xi'''),$$
$$\tilde\psi(x_1^0,x''',\xi''') = x'''\cdot\xi''',$$

where $\partial_j=\partial_{x_j}=\partial/\partial x_j$ and $\nabla_{x'''}f=(\partial_2 f,\cdots,\partial_{n-1}f)$. Moreover $\tilde\psi(x',\xi')$ is homogeneous of degree 1 in ξ'. Let $\chi(x',y',\xi')$ be a C^∞ function in R^{3n-3} such that $\chi=1$ in $\dot\Gamma_2\cap\{|\xi'|\ge 1\}$ and supp $\chi\subset\dot\Gamma_1$, where Γ_1 $(\subset\subset\Gamma_0)$ and Γ_2 are conic neighborhoods of $(x^{0\prime},\xi^{0\prime})$ in $T^*R^{n-1}\setminus 0$ and

$$\dot\Gamma = \{(x',y',\xi');\ (x',\xi')\in\Gamma \text{ and } (y',\xi')\in\Gamma\}.$$

Since $(\partial^2/\partial x_j\partial\xi_k\tilde\psi(x^{0\prime},\xi^{0\prime}))=I$, it follows that the operator A:

$$\mathcal{D}'(R^n)\ni g(x')\longrightarrow(Ag)(x')=\int\exp[i(\tilde\psi(x',\xi')-\tilde\psi(y',\xi'))]$$
$$\times\chi(x',y',\xi')g(y')dy'd\!\!{}^-\xi'\in\mathcal{D}'(R^{n-1})$$

is a properly supported pseudo-differential operator, if necessary, shrinking Γ_1, where $d\!\!{}^-\xi'=(2\pi)^{-n+1}d\xi'$. A is elliptic in a conic neighborhood of $(x^{0\prime},\xi^{0\prime})$. Thus there is a microlocal parametrix (pseudo-differential operator) B of A at $(x^{0\prime},\xi^{0\prime})$, i.e., there exists a conic neighborhood Γ of $(x^{0\prime},\xi^{0\prime})$ such that $ABg-g\in C^\infty(R^{n-1})$ if $WF(g)\subset\Gamma$.

Let us formally construct a microlocal parametrix for the problem (4.1)-(4.3) at $(x^{0\prime},\xi^{0\prime})$ in the form

$$E_1(g) = \Sigma_{j=1}^{\mu}\int\exp[i\phi_j(x,y',\xi')]a_j(x,y',\xi')(Bg)(y')dy'd\!\!{}^-\xi'$$
$$+\int\exp[i(\tilde\psi(x',\xi')-\tilde\psi(y',\xi'))]a(x,y',\xi')(Bg)(y')dy'd\!\!{}^-\xi',$$
$$a(x,y',\xi') = \Sigma_{j=1}^{\ell-\mu}(2\pi i)^{-1}\int_{C_{\xi'}}\exp[ix_n\xi_n]c_j(x,y',\xi)\xi_n^{j-1}d\xi_n.$$

Then we have

† $x_n\in[0,\varepsilon)$, $x'\in U$.

$$\begin{aligned}PE_1(g) &= \Sigma_{j=1}^{\mu} \int \exp[i\phi_j(x,y',\xi')]\{p(x,\nabla_x\phi_j)\\&+\Sigma_{|\alpha|=1}\, p^{(\alpha)}(x,\nabla_x\phi_j)D^\alpha+S(\phi_j;x)+q(\phi_j;x,D)\}\\&\times a_j(x,y',\xi')(Bg)(y')dy'đ\xi'\\&+\int \exp[i(\tilde\psi(x',\xi')-\tilde\psi(y',\xi'))][\Sigma_{j=1}^{\ell-\mu}\,(2\pi i)^{-1}\\&\times\int_{C_{\xi'}}\{p(x,\nabla_x\psi_0)+\Sigma_{|\alpha|=1}\, p^{(\alpha)}(x,\nabla_x\psi_0)D^\alpha+S(\psi_0;x)\\&+q(\psi_0;x,D)\}c_j(x,y',\xi)\xi_0^{j-1}\exp[ix_n\xi_n]d\xi_n]dy'đ\xi',\end{aligned}\tag{4.4}$$

where $p^{(\alpha)}(x,\zeta)=\partial_\zeta^\alpha p(x,\zeta)$ and $\psi_0(x,\xi)=\tilde\psi(x',\xi')+x_n\xi_n$. Thus $\phi_j(x,y',\xi')$, $1\le j\le\mu$, are determined by the eiconal equations

$$\partial_n\psi_j(x,\xi') = \lambda_j^+(x,\nabla_{x'}\psi_j),\quad \psi_j(x',0,\xi') = \tilde\psi(x',\xi'),\tag{4.5}$$

where $\phi_j(x,y',\xi')=\psi_j(x,\xi')-\psi_j(y',0,\xi')$. We easily see that $\phi_j(x,y',\xi')\in C^\infty([0,\varepsilon)\times\dot\Gamma_0)$ for some $\varepsilon>0$, if necessary, shrinking Γ_0. If $a_j(x,y',\xi')$, $1\le j\le\mu$, can be written as asymptotic sums

$$a_j(x,y',\xi') \sim \Sigma_{\nu=0}^{\infty}\, a_j^\nu(x,y',\xi')$$

in a certain sense, we obtain the transport equations

$$\begin{aligned}&\{\Sigma_{|\alpha|=1}\, p^{(\alpha)}(x,\nabla_x\phi_j)D^\alpha+S(\phi_j;x)\}a_j^\nu(x,y',\xi')\\&+ q(\phi_j;x,D)a_j^{\nu-1}(x,y',\xi') = 0,\\&a_j^{-1}(x,y',\xi') \equiv 0,\quad 1\le j\le\mu,\ \nu=0,1,2,\cdots.\end{aligned}\tag{4.6}$$

(4.6) is an ordinary differential equation for a_j^ν along rays corresponding to (4.5). Thus we can solve the transport equations (4.6) when the boundary values of a_j^ν are given. Put

$$a_j^\nu(x',0,y',\xi') = \tilde a_j^\nu(x',y',\xi')\ \varepsilon C^\infty(\dot\Gamma_0),\quad 1\le j\le\mu,\ \nu=0,1,\cdots.$$

We represent $c_j(x,y',\xi)$ as asymptotic sums

$$c_j(x,y',\xi) \sim \Sigma_{\nu,\tau=0}^{\infty}\, c_j^{\nu\tau}(x,y',\xi).$$

From (4.4) we put

$$\begin{aligned}c_j^{\nu 0}(x,y',\xi) &= \tilde c_j^\nu(x',y',\xi')\rho(x_0)p(x,\nabla_x\psi_0)^{-1},\\c_j^{\nu\tau+1}(x,y',\xi) &= -[\{\Sigma_{|\alpha|=1}\, p^{(\alpha)}(x,\nabla_x\psi_0)D^\alpha+S(\psi_0;x)\}\end{aligned}$$

$$\times c_j^{\nu\tau}(x,y',\xi) + q(\psi_0;x,D)c_j^{\nu-1\tau}(x,y',\xi)]p(x,\nabla_x\psi_0)^{-1},$$

$$c_j^{-1\tau} \equiv 0, \quad 1\le j\le \ell-\mu, \quad \nu,\tau=0,1,2,\cdots,$$

where $\rho(x_0)\varepsilon C^\infty_{(0)}([0,\varepsilon))$, $\rho(x_0)=1$ in a neighborhood of $x_0=0$. From the boundary conditions (4.3) we have

$$\tilde{L}(x',\nabla_{x'}\tilde{\psi}(x',\xi'))\vec{a}^\nu(x',y',\xi') + \Sigma_{j=1}^{n-1} \partial\tilde{L}/\partial\zeta_j D_j\vec{a}^\nu(x',y',\xi')$$

$$+ S(x',\xi')\vec{a}^\nu(x',y',\xi') = \vec{F}^\nu(x',y',\xi'), \quad \nu=0,1,2,\cdots,$$

where $\partial\tilde{L}/\partial\zeta_j=\partial\tilde{L}/\partial\zeta_j(x',\zeta')|_{\zeta'=\nabla_{x'}\tilde{\psi}}$,

$$\vec{a}^\nu(x',y',\xi') = {}^t(\tilde{a}_1^\nu(x',y',\xi'),\cdots,\tilde{a}_\mu^\nu,\tilde{c}_1^\nu(x',y',\xi'),\cdots,\tilde{c}_{\ell-\mu}^\nu),$$

and $\vec{F}^\nu$ is an ℓ-vector depending only on $a_j^{\nu-1}$ and $c_j^{\nu-1\tau}$ (see [16]). Finally we obtain

$$\begin{aligned}E_1(g) = \Sigma_{j=1}^{\mu}[&\int \exp[i\phi_j(x,y',\xi')]a_{0j}(x,y',\xi')(Bg)(y')dy'đ\xi'\\ &+ \int dy'đ\xi'\int_{-\infty}^{0}ds \exp[i\{\phi_j(x,y',\xi')+(\xi_1-\xi_1(x^{0'},\xi'''))s\}]\\ &\times a_{1j}(x,y',\xi',s)(Bg)(y')] + \int \exp[i(\tilde{\psi}(x',\xi')\\ &-\tilde{\psi}(y',\xi'))]e_0(x,y',\xi')(Bg)(y')dy'đ\xi' \qquad (4.7)\\ &+ \int dy'đ\xi'\int_{-\infty}^{0}ds \exp[i\{\tilde{\psi}(x',\xi')-\tilde{\psi}(y',\xi')\\ &+(\xi_1-\xi_1(x^{0'},\xi'''))s\}]e_1(x,y',\xi',s)(Bg)(y'), \quad g\varepsilon\mathcal{D}'(U).\end{aligned}$$

Definition 4.2. Let $x'=x'(t;y',\eta')$ and $\zeta'=\zeta'(t;y',\eta')$ be the solutions of a system of the equations

$$dx'/dt = (1,-\nabla_{\zeta'''}\xi_1(x',\zeta''')),$$

$$d\zeta'/dt = \nabla_{x'}\xi_1(x',\zeta'''),$$

$$x'=y', \ \zeta'=\eta' \text{ and } \eta_1-\xi_1(y',\eta''')=0 \text{ when } t=0.$$

Then the curves $\{(x'(t;y',\eta'),\zeta'(t;y',\eta'))\varepsilon\Gamma_0;\ t\varepsilon R^1\}$ are said to be boundary null-bicharacteristic strips. Let $x=x_j(t;y',\eta')$ and $\zeta=\zeta_j(t;y',\eta')$, $1\le j\le\mu$, be the solutions of a system of the equations

$$dx/dt = (-\nabla_{\zeta'}\mu_j^+(x,\zeta'),1),$$

$$d\zeta/dt = \nabla_x\mu_j^+(x,\zeta'),$$

$$x_n=0, \ x'=y', \ \zeta'=\eta' \text{ and } \zeta_n=\mu_j^+(y',0,\eta') \text{ when } t=0.$$

Then the curves $\{(x_j(t;y',\eta'),\zeta_j(t;y',\eta'));\ t\geq 0\}$, $1\leq j\leq \mu$, are said to be outgoing null-bicharacteristic strips. Further we define

$$C_0(\Gamma_0) = \{(x',\zeta',y',\eta')\in\Gamma_0\times\Gamma_0;\ (x',\zeta')=(y',\eta') \text{ or } x_1>y_1 \text{ and there exists a boundary null-bicharacteristic strip which contains both } (x',\zeta') \text{ and } (y',\eta')\},$$

$$C_j(\Gamma_0) = \{(x,\zeta,y',\eta')\in(T^*((0,\varepsilon)\times U)\setminus 0)\times\Gamma_0;\ \text{there exists a outgoing null-bicharacteristic strip which contains both } (x,\zeta) \text{ and } (y',0,\eta',\mu_j^+(y',0,\eta'))\},\quad 1\leq j\leq \mu.$$

Let us define wave front sets for $u\in C^\infty([0,\varepsilon);\mathcal{D}'(U))$. Since we can regard $u\in C^\infty([0,\varepsilon);\mathcal{D}'(U))$ as an element of $\mathcal{D}'((0,\varepsilon)\times U)$ we can define $WF(u)$ for $u\in C^\infty([0,\varepsilon);\mathcal{D}'(U))$ by regarding u as an element of $\mathcal{D}'((0,\varepsilon)\times U)$.

Definition 4.3. For $u\in C^\infty([0,\varepsilon);\mathcal{D}'(U))$ we say that a point $(x^{1\prime},\xi^{1\prime})$ in $T^*U\setminus 0$ is not in the set $WF_0(u)$ if there exist $\phi\in C_0^\infty(U)$, a conic neighborhood γ_1 of $\xi^{1\prime}$ and a positive constant ε_1 such that $\phi(x^{1\prime})\neq 0$ and

$$|\mathcal{F}_{x'}[\phi(x')D_0^j u(x)](x_0,\xi')| \leq C_{jk}(1+|\xi'|)^{-k}$$

when $\xi'\in\gamma_1$, $x_0\in[0,\varepsilon_1)$ and $j,k=0,1,2,\cdots$.

Theorem 4.4. Assume that the conditions (A.3)-(A.5) are satisfied. Then $\{E_1,\Gamma,[0,\varepsilon)\times U\}$ is a right microlocal parametrix for the problem (4.1)-(4.3) at $(x^0,\xi^{0\prime})$, where the operator E_1 is defined by (4.7) and ε (>0) and U are suitably chosen. Moreover we have

$$WF(E_1(g))\subset \bigcup_{j=1}^{\mu} C_j(\Gamma_0)\circ C_0(\Gamma_0)\circ WF(g),$$

$$WF_0(E_1(g))\subset C_0(\Gamma_0)\circ WF(g) \quad \text{for } g\in\mathcal{D}'(U).$$

Remark. $C_0(\Gamma_0)$ is related to a boundary wave.

We note that we can construct a microlocal parametrix as the composition of a microlocal parametrix for the Dirichlet problem and a microlocal parametrix for the Cauchy problem for a system of pseudo-differential operators on the boundary (see [16]).

REFERENCES

1. Atiyah, M. F., Bott, R. and Gårding, L., Lacunas for hyperbolic differential operators with constant coefficients I, Acta Math., 124 (1970), 109-189.
2. Chazarain, J., Construction de la paramétrix du problème mixte hyperbolique pour l'équation des ondes, C. R. Acad. Sci. Paris, 276 (1973), 1213-1215.
3. Deakin, A. S., Uniform asymptotic expansions for a hyperbolic-boundary problem, Comm. Pure Appl. Math., 24 (1971), 227-252.
4. Duff, G. F. D., On wave fronts, and boundary waves, Comm. Pure Appl. Math., 17 (1964), 189-225.
5. Hörmander, L., On the singularities of solutions of partial differential equations, International Conference of Functional Analysis and Related Topics. Tokyo 1969.
6. Ikawa, M., Problèmes mixtes pas nécessairement L^2-bien posés pour les équations strictement hyperboliques, Osaka J. Math., 12 (1975), 69-115.
7. Matsumura, M., Localization theorem in hyperbolic mixed problems, Proc. Japan Acad., 47 (1971), 115-119.
8. ___________, On the singularities of the Riemann functions of mixed problems for the wave equation in plane-stratified media I, II, Proc. Japan Acad. 52 (1976), 289-295.
9. Melrose, R. B., Microlocal parametrices for diffractive boundary value problems, Duke Math. J., 42 (1975), 605-635.
10. Sakamoto, R., $\mathcal{E}$-well posedness for hyperbolic mixed problems with constant coefficients, J. Math. Kyoto Univ., 14 (1974), 93-118.
11. Taylor, M. S., Grazing rays and reflection of singularities of solutions to wave equations, Comm. Pure Appl Math., 29 (1976), 1-38.
12. Tsuji, M., Fundamental solutions of hyperbolic mixed problems with constant coefficients, Proc. Japan Acad., 51 (1975), 369-373.
13. Wakabayashi, S., Singularities of the Riemann functions of hyperbolic mixed problems in a quarter-space, Proc. Japan Acad., 50 (1974), 821-825.
14. ___________, Singularities of the Riemann functions of hyperbolic mixed problems in a quarter-space, Publ. RIMS, Kyoto Univ., 11 (1976), 417-440.
15. ___________, Analytic wave front sets of the Riemann functions of hyperbolic mixed problems in a quarter-space, Publ. RIMS, Kyoto Univ., 11 (1976), 785-807.
16. ___________, Microlocal parametrices for hyperbolic mixed problems in the case where boundary waves appear, to appear.

SPECTRAL AND ASYMPTOTIC ANALYSIS OF ACOUSTIC WAVE PROPAGATION

Calvin H. Wilcox

Department of Mathematics, University of Utah,
Salt Lake City, Utah, USA

1. INTRODUCTION

Classical theories of acoustic wave propagation provide a wealth of examples of boundary value problems for evolution partial differential equations. These problems may be described categorically as initial-boundary value problems for certain systems of linear hyperbolic partial differential equations with variable coefficients. However, the known existence, uniqueness and regularity theorems for these problems are only a first step toward understanding the structure of the solutions. To obtain a deeper insight it is essential to discover how the nature of the solutions changes with the geometry of the boundary and with the coefficients. An examination of recent scientific literature on acoustics reveals a great variety of physically distinct phenomena. Examples include phenomena associated with acoustic wave propagation in stratified fluids, anisotropic solids such as crystals and man-made composites, open and closed waveguides, periodic media and many others. A theory which treats all of these phenomena on the same footing can provide only the most superficial information about the structure of acoustic waves.

The purpose of these lectures is to present a method for determining the structure of acoustic waves in unbounded media. The method will be explained in the context of four specific classes of propagation problems. No attempt will be made to formulate the most general problem that can be analyzed by the method. Indeed, such a formulation would necessarily be too abstract to be useful. However, it will be clear from the examples that the method is applicable to many other wave propagation problems, both in acoustics and in other areas of physics.

Garnir (ed.), Boundary Value Problems for Linear Evolution Partial Equations. 385-473.

It will be helpful to outline the main steps of the method here before passing to a detailed discussion of specific cases. The method is based on the fact that the states of an acoustic medium which occupies a spatial domain $\Omega \subset R^3$ can be described by the elements of a Hilbert space $\mathcal{H}$ of functions on Ω. The evolution of an acoustic wave in the medium is then described by a curve $t \to u(t,\cdot) \in \mathcal{H}$. Moreover, there is a selfadjoint real positive operator A on $\mathcal{H}$, determined by the geometry of Ω and the physical properties of the medium, such that the evolution of acoustic waves in the medium is governed by the equation

$$\frac{d^2u}{dt^2} + Au = 0 \tag{1.1}$$

It follows that the evolution is given by

$$u(t,\cdot) = \text{Re}\,\{\exp\,(-itA^{1/2})h\} \tag{1.2}$$

where $h \in \mathcal{H}$ characterizes the initial state of the wave.

The spectral theorem may be used to construct the solution operator $\exp\,(-itA^{1/2})$. However, the very generality of this theorem implies that it can give little specific information about the structure of the wave functions $u(t,x)$. Accordingly, the next step in the method is to construct an eigenfunction expansion for A. In each of the cases discussed below A has a purely continuous spectrum and the eigenfunctions are therefore generalized eigenfunctions. They define a complete set of steady-state modes of propagation of the medium and the most general time-dependent acoustic wave in $\mathcal{H}$ can be constructed as a spectral integral over these modes.

The final step in the method is an asymptotic analysis for $t \to \infty$ of the spectral integral representing $u(t,x)$. The result is an asymptotic wave function $u^\infty(t,x)$ which approximates $u(t,x)$ in $\mathcal{H}$ when $t \to \infty$; that is,

$$\lim_{t\to\infty} \| u(t,\cdot) - u^\infty(t,\cdot)\|_{\mathcal{H}} = 0 \tag{1.3}$$

Stronger forms of convergence can also be proved under appropriate supplementary hypotheses about the medium and its initial state.

The result (1.3) offers a fundamental insight into the nature of transient acoustic waves in unbounded media. For it is found in each case that the form of the asymptotic wave function $u^\infty(t,x)$ is determined entirely by the geometry of the domain Ω and the physical characteristics of the medium that fills it. Only the fine structure of $u^\infty(t,x)$ depends on the initial state of the wave. Thus in the simple case of a homogeneous fluid filling R^3, $u^\infty(t,x)$ is a spherical wave:

$$u^{\infty}(t,x) = F(r - t,\theta)/r, \; x = r\theta, \; |\theta| = 1 \tag{1.4}$$

The initial state affects only the shape of the profile $F(\tau,\theta)$. In other cases the form of $u^{\infty}(t,x)$ is entirely different, but in each case the form of $u^{\infty}(t,x)$ is determined solely by the geometry and physical characteristics of the medium. In each case $u^{\infty}(t,x)$ gives the final form of any transient wave in the medium. The details of how the wave is excited have only a secondary effect on the ultimate waveform.

The remainder of these lectures is organized as follows. The fundamental boundary value problems of acoustics are formulated in section 2. The spectral and asymptotic analysis of the four classes of propagation problems is presented in sections 3 through 8. The four classes, which are physically quite different, were chosen to illustrate the flexibility and scope of the method. In each of the four classes there is a special case for which, because of additional symmetry, the eigenfunctions can be constructed explicitly. The remaining cases of the class are then treated as perturbations of the special case. When used in this context, perturbation theory is usually called the steady-state, or time-dependent, theory of scattering. The first class of problems treated below corresponds physically to the scattering of acoustic waves by bounded obstacles immersed in a homogeneous fluid. Mathematically, it is an initial-boundary problem for the d'Alembert equation in an exterior domain $\Omega \subset R^3$ ($R^3-\Omega$ compact). The simple special case where $\Omega = R^3$ is treated in section 3 and the general case in section 4. The second class of problems deals with tubular waveguides. Thus Ω is the union of a bounded domain and a finite number of semi-infinite cylinders. The special case of a single cylinder is treated in section 5 and the general case in section 6. The third class of problems, treated in section 7, deals with acoustic wave propagation in plane stratified fluids filling a half-space. Here the novel feature is the possibility of the trapping of waves by total internal reflection. The fourth and final class of problems, dealing with acoustic waves in crystalline solids, is discussed in section 8. The new feature in this case is the anisotropy which has a profound effect on the form of the asymptotic wave functions.

The results presented below are based primarily on the author's research. Sections 3 and 4 are based on the author's monograph on "Scattering Theory for the d'Alembert Equation in Exterior Domains" [42]. The spectral theory of acoustic wave propagation and scattering in tubular waveguides was developed by C. Goldstein [9-12] and by W. C. Lyford [21,22]. More recently, J. C. Guillot and the author [13] have developed the theory for domains Ω which are the union of a bounded domain and a finite number of cylinders and cones. Sections 5 and 6 present spectral and scattering theory for tubular waveguides following the plan of [13]. Sections 7 and 8 are based on the author's publications [39,40,43,44].

The goal of these lectures is to provide an introduction to the method of spectral and asymptotic analysis of wave propagation. Therefore, the lectures emphasize concepts and results, rather than techniques of proof. Proofs of the results given here may be found in the references listed at the end of the lectures.

2. BOUNDARY VALUE PROBLEMS OF ACOUSTICS

Acoustic waves are the mechanical vibrations of small amplitude that are observed in all forms of matter. The classical equations of acoustics are the linear partial differential equations which govern small perturbations of the equilibrium states of matter. Derivations of these equations from the laws of mechanics, together with a discussion of their range of validity, may be found in [3,4,8,20,31]. In this section the equations and their physical interpretation are reviewed briefly and the principal boundary value problems for them are formulated and discussed. Applications of the equations to particular classes of acoustic wave propagation problems are developed in sections 3 through 8.

The following notation is used throughout the remainder of the lectures. $t \in R$ denotes a time coordinate. $x = (x_1, x_2, x_3) \in R^3$ denote Cartesian coordinates of a point in Euclidean space. $\Omega \subset R^3$ denotes a domain in R^3 and $\partial\Omega$ denotes the boundary of Ω. $\nu = (\nu_1, \nu_2, \nu_3) = \nu(x)$ denotes the unit exterior normal vector to $\partial\Omega$ at points $x \in \partial\Omega$ where it exists. The equations of acoustics are written below in the notation of Cartesian tensor analysis. In particular, the summation convention is used. Acoustic waves in fluids (gases and liquids) and solids are discussed separately. The simpler case of fluids is treated first.

2.1 Acoustic waves in fluids

The case of an inhomogeneous fluid occupying a domain $\Omega \subset R^3$ is considered. The propagation of acoustic waves in such a fluid is governed by two functions of $x \in \Omega$:

$$\rho = \rho(x), \text{ the equilibrium density of the fluid} \tag{2.1}$$

and

$$c = c(x), \text{ the local speed of sound in the fluid} \tag{2.2}$$

The state of the acoustic field in the fluid is determined by

$$v_j = v_j(t,x), \text{ the velocity field of the fluid at time } t \text{ and position } x \tag{2.3}$$

and

$$p = p(t,x), \text{ the pressure field of the fluid at time } t \text{ and position } x \tag{2.4}$$

Moreover, it is assumed that

$$p(t,x) = p_0(x) + u(t,x) \tag{2.5}$$

where $p_0(x)$ is the equilibrium pressure of the fluid and $u(t,x)$ remains small. With this notation the equations satisfied by the acoustic field in the fluid are

$$\frac{\partial v_j}{\partial t} + \frac{1}{\rho(x)} \frac{\partial u}{\partial x_j} = 0, \; j = 1,2,3 \tag{2.6}$$

$$\frac{\partial u}{\partial t} + c^2(x)\, \rho(x) \frac{\partial v_j}{\partial x_j} = 0 \tag{2.7}$$

Elimination of the velocity field gives the single equation

$$\frac{\partial^2 u}{\partial t^2} - c^2(x)\, \rho(x) \frac{\partial}{\partial x_j} \left(\frac{1}{\rho(x)} \frac{\partial u}{\partial x_j} \right) = 0 \tag{2.8}$$

for the pressure increment $u = p - p_0$. Moreover, if u is known then the velocity field v_j can be calculated from (2.6).

The wave equation (2.8) must be supplemented by a boundary condition at the fluid boundary $\partial\Omega$. Two physically distinct cases are considered here. The first case is that of a free boundary $\partial\Omega$. Here the pressure at the boundary is unperturbed; that is,

$$u\big|_{\partial\Omega} = 0 \text{ if } \partial\Omega \text{ is a free boundary} \tag{2.9}$$

This condition is often used to represent an air-water interface in the theory of underwater sound. The second case is that of a rigid boundary $\partial\Omega$. Here the normal component of the fluid velocity must vanish: $\nu_j v_j = 0$ on $\partial\Omega$. It follows from (2.6) that

$$\frac{\partial u}{\partial \nu}\bigg|_{\partial\Omega} = \nu_j \frac{\partial u}{\partial x_j}\bigg|_{\partial\Omega} = 0 \text{ if } \partial\Omega \text{ is a rigid boundary} \tag{2.10}$$

The solvability of the boundary value problems (2.8), (2.9) and (2.8), (2.10) is discussed below, after the discussion of acoustic waves in solids.

2.2 Acoustic waves in solids

The case of an inhomogeneous elastic solid occupying a domain $\Omega \subset R^3$ is considered. The propagation of acoustic waves in such a solid is governed by the following functions of $x \in \Omega$:

$$\rho = \rho(x), \text{ the equilibrium density of the solid} \tag{2.11}$$

and

$$c^{jk}_{\ell m} = c^{jk}_{\ell m}(x), \text{ the stress-strain tensor for the solid} \tag{2.12}$$

The stress-strain tensor must have the symmetry properties [4].

$$c^{jk}_{\ell m} = c^{kj}_{\ell m} = c^{kj}_{m\ell} = c^{m\ell}_{kj} \text{ for all } j,k,\ell,m = 1,2,3 \tag{2.13}$$

It follows that the 81 components $c^{jk}_{\ell m}(x)$ are determined by 21 functions. The state of the acoustic field in the solid is determined by

$$u_j = u_j(t,x), \text{ the displacement field of the solid at time } t \text{ and position } x \tag{2.14}$$

and

$$\sigma_{jk} = \sigma_{jk}(t,x), \text{ the stress tensor field of the solid at time } t \text{ and position } x \tag{2.15}$$

Moreover, the stress tensor field is symmetric:

$$\sigma_{jk} = \sigma_{kj} \text{ for all } j,k = 1,2,3 \tag{2.16}$$

With this notation the equations satisfied by the acoustic field in the solid are

$$\sigma_{jk} = c^{\ell m}_{jk}(x) \frac{\partial u_\ell}{\partial x_m}, \quad j,k = 1,2,3 \tag{2.17}$$

$$\frac{\partial^2 u_j}{\partial t^2} = \frac{1}{\rho(x)} \frac{\partial \sigma_{jk}}{\partial x_k}, \quad j = 1,2,3 \tag{2.18}$$

Elimination of the stress tensor gives the equations

$$\frac{\partial^2 u_j}{\partial t^2} - \frac{1}{\rho(x)} \frac{\partial}{\partial x_k} \left(c^{\ell m}_{jk}(x) \frac{\partial u_\ell}{\partial x_m} \right) = 0, \quad j = 1,2,3 \tag{2.19}$$

for the displacement field u_j. Moreover, if u_j is known then the stress tensor field σ_{jk} can be calculated from (2.17).

The wave equation (2.19) must be supplemented by boundary conditions at the boundary $\partial\Omega$ of the solid. Only the cases of free and rigid boundaries will be considered here. In the first case the normal component of the stress must vanish at the boundary. Hence

$$\sigma_{jk}\nu_k\Big|_{\partial\Omega} = c_{jk}^{\ell m}\frac{\partial u_\ell}{\partial x_m}\nu_k\Big|_{\partial\Omega} = 0 \text{ if } \partial\Omega \text{ is a free boundary} \tag{2.20}$$

In the second case the displacement must vanish at the boundary; that is,

$$u_j\Big|_{\partial\Omega} = 0 \text{ if } \partial\Omega \text{ is a rigid boundary} \tag{2.21}$$

2.3 Energy integrals

One of the most important formal properties of the equations of acoustics is the existence of quadratic energy integrals. The first order system (2.6), (2.7) for acoustic waves in fluids has the quadratic energy density

$$\eta(t,x) = \frac{1}{2}\left\{\rho(x)v_jv_j + \frac{1}{c^2(x)\rho(x)}u^2\right\} \tag{2.22}$$

and corresponding energy integral

$$E(v_1,v_2,v_3,u,K,t) = \int_K \eta(t,x)\,dx \tag{2.23}$$

where $dx = dx_1dx_2dx_3$ denotes Lebesgue measure in R^3. The energy density for the derived field $v_j' = \partial v_j/\partial t$, $u' = \partial u/\partial t$, which also satisfies the field equations (2.6), (2.7), can be written

$$\eta'(t,x) = \frac{1}{2}\left\{\frac{1}{\rho(x)}\frac{\partial u}{\partial x_j}\frac{\partial u}{\partial x_j} + \frac{1}{c^2(x)\rho(x)}\left(\frac{\partial u}{\partial t}\right)^2\right\} \tag{2.24}$$

by (2.6). The integral

$$E(u,K,t) = \int_K \eta'(t,x)\,dx \tag{2.25}$$

is an energy integral for solutions of the scalar wave equation (2.8). The importance of these integrals in the theory of acoustic waves derives from the conservation laws for them. In differential form they state that

$$\frac{\partial\eta(t,x)}{\partial t} = -\frac{\partial}{\partial x_j}(uv_j) \tag{2.26}$$

and

$$\frac{\partial\eta'(t,x)}{\partial t} = \frac{\partial}{\partial x_j}\left(\frac{1}{\rho(x)}\frac{\partial u}{\partial t}\frac{\partial u}{\partial x_j}\right) \tag{2.27}$$

These equations follow immediately from (2.6), (2.7) and the definitions. The integral forms of the conservation laws follow from (2.26), (2.27) and the divergence theorem. They may be written

$$dE(v_1,v_2,v_3,u,K,t)/dt = -\int_{\partial K} u(\nu_j v_j)\,dS \tag{2.28}$$

and

$$dE(u,K,t)/dt = \int_{\partial K} \frac{1}{\rho(x)}\frac{\partial u}{\partial t}\frac{\partial u}{\partial \nu}\,dS \tag{2.29}$$

where $K \subset R^3$ is any domain for which the divergence theorem is valid and dS is the element of area on ∂K. In particular, if $u(t,x)$ is a solution of (2.8) which satisfies (2.9) or (2.10) then (2.29) implies that $dE(u,\Omega,t)/dt = 0$.

The equations for acoustic waves in solids have an analogous quadratic energy integral

$$E(u_1,u_2,u_3,K,t) = \int_K \eta(t,x)\,dx \tag{2.30}$$

with density

$$\eta(t,x) = \frac{1}{2}\left\langle\rho(x)\frac{\partial u_j}{\partial t}\frac{\partial u_j}{\partial t} + c^{\ell m}_{jk}(x)\frac{\partial u_j}{\partial x_k}\frac{\partial u_\ell}{\partial x_m}\right\rangle \tag{2.31}$$

The corresponding conservation law, which follows from (2.19), is

$$\frac{\partial\eta(t,x)}{\partial t} = \frac{\partial}{\partial x_k}\left(c^{\ell m}_{jk}(x)\frac{\partial u_\ell}{\partial x_m}\frac{\partial u_j}{\partial t}\right) \tag{2.32}$$

in differential form and

$$dE(u_1,u_2,u_3,K,t)/dt = \int_{\partial K} (\sigma_{jk}\nu_k)\frac{\partial u_j}{\partial t}\,dS \tag{2.33}$$

in integral form. In particular, solutions of (2.19) which satisfy (2.20) or (2.21) also satisfy $dE(u_1,u_2,u_3,\Omega,t)/dt = 0$.

The preceding remarks emphasize the mathematical relationship of the quadratic energy integrals to the field equations of acoustics. The term "energy" has been used because in certain cases the integrals can be interpreted as the portion of the energy of the acoustic field that is in the set K at time t. This interpretation is not always correct because the linear equations of acoustics are only a first-order approximation to more complicated nonlinear equations and the energy densities defined above are second-order quantities. Hence, it is possible that other second-order terms which were dropped in the linearization should be included in the energy densities. A correct calculation of the energy must begin with the original nonlinear problem. A discussion of these problems may be found in [8,31] for the case of fluids and in [4] for the case of solids.

It is important to realize that the energy integrals defined above play an essential role in the theory of acoustic fields, whether or not they represent the actual physical energy of the fields. Indeed, it was shown in [33] and [34] that the existence of these integrals implies the existence and uniqueness of solutions to the basic initial-boundary value problems for acoustic fields. Moreover, recent work on eigenfunction expansions and scattering theory makes use of Hilbert spaces based on energy integrals. The one indispensible hypothesis that must be made is that the quadratic forms (2.22) or (2.24) and (2.31) be positive definite. For (2.22) and (2.24) this means that

$$\rho(x) > 0 \text{ and } c^2(x) > 0 \text{ for all } x \in \Omega \tag{2.34}$$

In any case, these hypotheses are essential because of the physical interpretation of $\rho(x)$ and $c(x)$. The form (2.31) is positive definite if $\rho(x) > 0$ and

$$c_{jk}^{\ell m}(x)\, \xi_{\ell m}\, \xi_{jk} > 0 \text{ for all } x \in \Omega \text{ and } \xi_{\ell m} = \xi_{m\ell} \neq 0 \tag{2.35}$$

The last condition can also be expressed by means of the well-known determinantal criteria for a quadratic form to be positive definite. It is assumed throughout these lectures that (2.34) and (2.35) are satisfied.

It has been shown that the acoustic fields in both fluids and solids satisfy partial differential equations of the form

$$\frac{\partial^2 u}{\partial t^2} + Au = 0 \tag{2.36}$$

where A is a second order partial differential operator in the space variables. In the case of fluids $u(t,x) \in R$, $Au(t,x) \in R$ and

$$Au = -c^2(x)\ \rho(x) \frac{\partial}{\partial x_j} \left(\frac{1}{\rho(x)} \frac{\partial u}{\partial x_j} \right) \tag{2.37}$$

while in the case of solids $u(t,x) = (u_1(t,x), u_2(t,x), u_3(t,x)) \in R^3$, $Au(t,x) \in R^3$ and

$$(Au)_j = -\frac{1}{\rho(x)} \frac{\partial}{\partial x_k} \left(c_{jk}^{\ell m}(x) \frac{\partial u_\ell}{\partial x_m} \right),\ j = 1,2,3 \tag{2.38}$$

Thus in both cases the evolution of acoustic waves in a medium which fills a domain $\Omega \subset R^3$ is described by the solution of an initial-boundary value problem of the form

$$\frac{\partial^2 u}{\partial t^2} + Au = 0 \text{ for } t > 0,\ x \in \Omega \tag{2.39}$$

$$Bu = 0 \text{ for } t \geq 0,\ x \in \partial\Omega \tag{2.40}$$

$$u(0,x) = f(x) \text{ and } \partial u(0,x)/\partial t = g(x) \text{ for } x \in \Omega \tag{2.41}$$

Here (2.40) represents one of the boundary conditions (2.9), (2.10) in the case of a fluid and (2.20), (2.21) in the case of a solid.

It is interesting to note that the positive definiteness of the energy densities, hypothesized above on physical grounds, implies the hyporbolicity of the equation (2.36). It follows that the initial-boundary value problem (2.39) - (2.41) has compact domains of dependence and influence [6,33]. In physical terms this means that acoustic waves propagate into undisturbed portions of a medium with finite speed.

A simple and rigorous solution theory for the initial-boundary value problem (2.39) - (2.41) can be based on the theory of selfadjoint operators in Hilbert space. This possibility follows from the divergence theorem which implies the formal selfadjointness of the operators A relative to suitable inner products. Indeed, for the operator (2.37) the divergence theorem implies

$$\int_\Omega \overline{Au}\ v\ c^{-2}(x)\ \rho^{-1}(x)dx = \int_\Omega \overline{\frac{\partial u}{\partial x_j}} \frac{\partial v}{\partial x_j} \rho^{-1}(x)dx - \int_{\partial\Omega} \overline{\frac{\partial u}{\partial \nu}}\ v\ \rho^{-1}(x)dS \tag{2.42}$$

and hence

$$\int_\Omega \{\overline{Au}\ v - \overline{u}\ Av\} c^{-2}(x)\ \rho^{-1}(x)dx = \int_{\partial\Omega} \{\overline{u} \frac{\partial v}{\partial \nu} - \overline{\frac{\partial u}{\partial \nu}} v\}\ \rho^{-1}(x)dS \tag{2.43}$$

Thus if an inner product is defined by

$$(u,v) = \int_\Omega \overline{u(x)}\ v(x)\ c^{-2}(x)\ \rho^{-1}(x)dx \tag{2.44}$$

then

$$(Au,v) = (u,Av) \tag{2.45}$$

for all u and v in the domain of A which satisfy the boundary condition (2.9) or (2.10). Moreover, (2.44) defines the inner product in the Hilbert space $\mathcal{H} = L_2(\Omega, c^{-2}(x)\rho^{-1}(x)dx)$ of functions on Ω which are square-integrable with respect to the measure $c^{-2}(x)\rho^{-1}(x)dx$. Hence (2.45) implies that A, acting in the classical sense on functions which satisfy (2.9) or (2.10), is a symmetric operator in $\mathcal{H}$. Moreover, (2.42) implies that

$$(Au,u) = \int_\Omega \overline{\frac{\partial u}{\partial x_j}}\ \frac{\partial u}{\partial x_j}\ \rho^{-1}(x)dx \geq 0 \tag{2.46}$$

for all u in the domain of A. Hence A is positive. It was shown in [42] and [43] how the domain of A could be enlarged to obtain an extension A of A which is selfadjoint and positive in $\mathcal{H}$. The boundary condition (2.9) or (2.10) is incorporated into the definition of the domain of A. Moreover, the construction provides a meaningful generalization of the boundary conditions for arbitrary domains $\Omega \subset R^3$. The precise definitions and results are reviewed in sections 3-7 below.

The operator (2.38) for acoustic waves in solids can be treated similarly. The divergence theorem implies

$$\int_\Omega \overline{(Au)_j}\ v_j\ \rho(x)dx \tag{2.47}$$

$$= \int_\Omega c_{jk}^{\ell m}(x)\ \overline{\frac{\partial u_\ell}{\partial x_m}}\ \frac{\partial v_j}{\partial x_k}\ dx - \int_{\partial\Omega} \overline{\left(c_{jk}^{\ell m}(x)\ \frac{\partial u_\ell}{\partial x_m}\ \nu_k\right)}\ v_j dS$$

It follows that if an inner product is defined by

$$(u,v) = \int_\Omega \overline{u_j(x)}\ v_j(x)\ \rho(x)dx \tag{2.48}$$

then (2.45) holds for all u and v in the domain of A which satisfy the boundary condition (2.20) or (2.21). Moreover, (2.48) defines the inner product in the Hilbert space $\mathcal{H} = L_2(\Omega, C^3, \rho(x)dx)$ of functions from Ω to C^3 which are square integrable with respect to the measure $\rho(x)dx$. Hence A, acting in the classical sense on functions which satisfy (2.20) or (2.21), is a symmetric operator in $\mathcal{H}$. Moreover, (2.47) implies that

$$(\mathcal{A}u,u) = \int_\Omega c_{jk}^{\ell m}(x) \overline{\frac{\partial u_\ell}{\partial x_m}} \frac{\partial u_j}{\partial x_k} dx \geq 0 \tag{2.49}$$

for all u in the domain of $\mathcal{A}$ by the assumed positivity of the energy density, (2.35). It will be shown in section 8 below how the domain of $\mathcal{A}$ can be enlarged to obtain a selfadjoint positive extension A of $\mathcal{A}$.

A Hilbert space $\mathcal{H}$ and selfadjoint positive operator A on $\mathcal{H}$ can be associated with each acoustic wave propagation problem by the method indicated above. A theory of solutions of the initial-boundary value problem (2.39) - (2.41) may then be based on A in the following way. First of all, the problem can be formulated as an initial value problem in $\mathcal{H}$. A function u: $R \to \mathcal{H}$ is sought such that

$$\frac{d^2u}{dt^2} + Au = 0 \text{ for all } t \in R \tag{2.50}$$

$$u(0) = f \text{ and } \frac{du(0)}{dt} = g \text{ in } \mathcal{H} \tag{2.51}$$

The spectral theorem for A:

$$A = \int_0^\infty \lambda \, d\Pi(\lambda) \tag{2.52}$$

and the associated operator calculus make it possible to construct the generalized solution

$$u(t) = (\cos t A^{1/2})f + (A^{-1/2} \sin t A^{1/2})g \tag{2.53}$$

The coefficient operators in (2.53) are bounded and hence u(t) is defined for all f and g in $\mathcal{H}$ and defines a curve in $C(R,\mathcal{H})$, the class of continuous $\mathcal{H}$-valued functions on R. The differentiability properties of u(t) depend on those of f and g. Two cases will be mentioned.

2.4 Solutions in $\mathcal{H}$

If $f \in \mathcal{H}$ and $g \in \mathcal{H}$ then u(t) is continuous in $\mathcal{H}$ and u(0) = f. However, u(t) will not in general be differentiable, and hence (2.50) and the second initial condition need not hold. In this case u(t) coincides with the "generalized solution in $\mathcal{H}$" which was defined and studied by M. Vishik and O. A. Ladyzhenskaya [32].

2.5 Solutions with finite energy

If $f \in D(A^{1/2})$ and $g \in \mathcal{H}$ then u is in the class

$$C^1(R,\mathcal{H}) \cap C(R,D(A^{1/2})) \tag{2.54}$$

This follows easily from (2.53) and the spectral theorem. Hence, u satisfies (2.51) but (2.50) need not hold. In this case u(t) coincides with the "solution with finite energy" which, for arbitrary domains Ω, was defined and studied by the author in [33,34, 42]. The existence and uniqueness of solutions with finite energy was proved in [33,34].

3. PROPAGATION IN HOMOGENEOUS FLUIDS

Propagation in an unlimited homogeneous fluid is analyzed in this section. In the notation of section 2 this is the special case where $\Omega = R^3$ and $\rho(x) = \rho$ and $c(x) = c$ are constant for all $x \in R^3$. It will be enough to treat the case $c = 1$ since the general case can be reduced to this one by the change of variable $ct \to t$. With these simplifications the wave equation (2.8) reduces to the d'Alembert equation

$$\frac{\partial^2 u}{\partial t^2} - \left(\frac{\partial^2 u}{\partial x_1^2} + \frac{\partial^2 u}{\partial x_2^2} + \frac{\partial^2 u}{\partial x_3^2}\right) = 0 \tag{3.1}$$

and the propagation problem is simply the Cauchy problem for (3.1). The spectral and asymptotic analysis of solutions in $L_2(R^3)$ of (3.1) was developed in detail in [42]. Only the principal concepts and results are reviewed here.

The operator in $L_2(R^3)$ defined by $Au = -(\partial^2 u/\partial x_1^2 + \partial^2 u/\partial x_2^2 + \partial^2 u/\partial x_3^2)$ acting in the domain $D(A) = \mathcal{D}(R^3)$, the L. Schwartz space of testing functions, is known to be essentially selfadjoint [18]. Thus A has a unique selfadjoint extension in $L_2(R^3)$ which will be denoted here by A_0. This operator may be defined by

$$D(A_0) = L_2(R^3) \cap \left\{u: \frac{\partial^2 u}{\partial x_1^2} + \frac{\partial^2 u}{\partial x_2^2} + \frac{\partial^2 u}{\partial x_3^2} \in L_2(R^3)\right\} \tag{3.2}$$

and

$$A_0 u = -\left(\frac{\partial^2 u}{\partial x_1^2} + \frac{\partial^2 u}{\partial x_2^2} + \frac{\partial^2 u}{\partial x_3^2}\right) \quad \text{for all } u \in D(A_0) \tag{3.3}$$

where the derivatives are interpreted in the sense of Schwartz's theory of distributions. A_0 is known to be non-negative and it is obviously real; that is

$$A_0 \overline{u} = \overline{A_0 u} \quad \text{for all } u \in D(A_0) \tag{3.4}$$

where the bar denotes the complex conjugate.

The d'Alembert equation (3.1) will be interpreted as the equation

$$\frac{d^2 u}{dt^2} + A_0 u = 0 \tag{3.5}$$

for an $L_2(R^3)$-valued function. Hence the solution in $L_2(R^3)$ of the Cauchy problem can be written

$$u(t) = (\cos t A_0^{1/2}) f + (A_0^{-1/2} \sin t A_0^{1/2}) g \tag{3.6}$$

where $u(0) = f$ and $du(0)/dt = g$ are in $L_2(R^3)$. If it is assumed that $f(x)$ and $g(x)$ are real-valued and

$$f \in L_2(R^3), \quad g \in D(A_0^{-1/2}) \tag{3.7}$$

then it follows from (3.4) that

$$u(t,x) = \mathrm{Re}\, \{v(t,x)\} \tag{3.8}$$

where

$$v(t,\cdot) = \exp(-itA_0^{1/2}) h \tag{3.9}$$

and

$$h = f + i A_0^{-1/2} g \in L_2(R^3) \tag{3.10}$$

In what follows attention is restricted to this case.

An eigenfunction expansion for A_0 may be based on the Plancherel theory of the Fourier transform in $L_2(R^3)$. If

$$w_0(x,p) = \frac{1}{(2\pi)^{3/2}} \exp(i\, x \cdot p), \quad p \in R^3 \tag{3.11}$$

where $x \cdot p = x_1 p_1 + x_2 p_2 + x_3 p_3$ then the main results of the theory state that for all $f \in L_2(R^3)$ the following limits exist

$$\left.\begin{aligned} \hat{f}(p) \equiv (\Phi_0 f)(p) &= L_2(R^3)\text{-}\lim_{M\to\infty} \int_{|x|\le M} \overline{w_0(x,p)}\, f(x)\, dx \\ f(x) = (\Phi_0^* \hat{f})(x) &= L_2(R^3)\text{-}\lim_{M\to\infty} \int_{|p|\le M} w_0(x,p)\, \hat{f}(p)\, dp \end{aligned}\right\} \tag{3.12}$$

and Φ_0: $L_2(R^3) \to L_2(R^3)$ is unitary. These relations will often be written in the symbolic form

$$\hat{f}(p) = \int_{R^3} \overline{w_0(x,p)}\, f(x)dx, \quad f(x) = \int_{R^3} w_0(x,p)\, \hat{f}(p)dp \tag{3.13}$$

but must be interpreted in the sense (3.12). The utility of the Fourier transform is due to the fact that if f and $\partial f/\partial x_j$ are in $L_2(R^3)$ then

$$\left(\Phi_0 \frac{\partial f}{\partial x_j}\right)(p) = ip_j\, \hat{f}(p),\ j = 1,2,3 \tag{3.14}$$

In particular, it follows that

$$D(A_0) = L_2(R^3) \cap \{u: \ |p|^2\hat{u}(p) \in L_2(R^3)\} \tag{3.15}$$

A_0 has the spectral representation

$$A_0 = \int_0^\infty \lambda\ d\Pi_0(\lambda) \tag{3.16}$$

with spectral family $\{\Pi_0(\lambda)\}$ defined by

$$\Pi_0(\lambda)f(x) = \int_{|p|\le\sqrt{\lambda}} w_0(x,p)\ \hat{f}(p)dp,\ \lambda \ge 0 \tag{3.17}$$

It follows that A_0 is an absolutely continuous operator [18,42] whose spectrum is the interval $[0,\infty)$.

The above results imply that Φ_0 defines a spectral representation for A_0 and functions of A_0. In particular, if $\Psi(\lambda)$ is any bounded Lebesgue-measurable function of $\lambda \ge 0$ then

$$\Phi_0\Psi(A_0)f(p) = \Psi(|p|^2)\hat{f}(p) \tag{3.18}$$

These results imply that the wave function v(t,x) defined by (3.9) has the representation

$$v(t,x) = \int_{R^3} w_0(x,p)\ \exp\ (-it|p|)\ \hat{h}(p)dp \tag{3.19}$$

The function $w_0(x,p)$ is a generalized eigenfunction for A_0. This means that $w_0(\cdot,p)$ is locally in $D(A_0)$; i.e., $\phi w_0(\cdot,p) \in D(A_0)$ for every $\phi \in \mathcal{D}(R^3)$ and

$$A_0 w_0(\cdot,p) = \lambda\ w_0(\cdot,p),\ \lambda = |p|^2 \tag{3.20}$$

The functions

$$w_0(x,p)\ \exp\ (-it|p|) = \frac{1}{(2\pi)^{3/2}} \exp\ \{i(x\cdot p - t|p|)\} \tag{3.21}$$

are solutions of the d'Alembert equation which represent plane waves propagating in the direction of the vector $p \in R^3$. Hence, (3.19) is a representation of a localized acoustic wave as a superposition of the elementary waves (3.21).

The spectral integral (3.19) is the starting point for the asymptotic analysis of the behavior for $t \to \infty$ of solutions in $L_2(R^3)$ of the d'Alembert equation. It is convenient to begin the analysis with the special case where $\hat{h}$ is in the class

$$\mathcal{D}_0(R^3) = \mathcal{D}(R^3) \cap \{\hat{h}:\ \hat{h}(p) \equiv 0 \text{ for } |p| \leq a,\ a = a(\hat{h}) > 0\} \tag{3.22}$$

The analysis will then be extended to the general case by using the easily verified fact that $\mathcal{D}_0(R^3)$ is dense in $L_2(R^3)$.

If $\hat{h} \in \mathcal{D}_0(R^3)$ and the support of $\hat{h}$ satisfies

$$\text{supp}\ \hat{h} \subset \{p:\ \ 0 < a \leq |p| \leq b\} \tag{3.23}$$

then the spectral integral (3.19) converges both in $L_2(R^3)$ and pointwise to $v(t,x)$ and

$$v(t,x) = \frac{1}{(2\pi)^{3/2}} \int_{a\leq|p|\leq b} \exp\ \{i(x\cdot p - t|p|)\}\hat{h}(p)\,dp \tag{3.24}$$

To find the behavior of $v(t,\cdot) \in L_2(R^3)$ for $t \to \infty$ introduce spherical coordinates for p:

$$p = \rho\omega,\ \rho \geq 0,\ \omega \in S^2,\ dp = \rho^2 d\rho d\omega \tag{3.25}$$

where S^2 represents the unit sphere in R^3 with center at the origin and $d\omega$ is the element of area on S^2. This gives the representation

$$v(t,x) = \frac{1}{(2\pi)^{3/2}} \int_a^b e^{-it\rho}\ V(x,\rho)\ \rho^2 d\rho \tag{3.26}$$

where

$$V(x,\rho) = \int_{S^2} e^{i\rho x\cdot\omega}\ \hat{h}(\rho\omega)\,d\omega \tag{3.27}$$

The asymptotic behavior of $V(x,\rho)$ for $|x| \to \infty$ will be calculated and used to find the behavior of $v(t,x)$ for $t \to \infty$. Application of the method of stationary phase [2,23] to (3.27) with $x = r\theta$, $r \geq 0$, $\theta \in S^2$ implies that if

$$V(x,\rho) = \left(\frac{2\pi}{i\rho r}\right) e^{i\rho r}\, \hat{h}(\rho\theta) + \left(\frac{2\pi}{-i\rho r}\right) e^{-i\rho r}\, \hat{h}(-\rho\theta) + q_0(x,\rho) \tag{3.28}$$

then there exists a constant $M_0 = M_0(\hat{h})$ such that

$$|q_0(x,\rho)| \leq M_0/r^2 \text{ for all } r > 0,\ a \leq \rho \leq b \text{ and } \theta \in S^2 \tag{3.29}$$

Substituting (3.28) into (3.26) gives

$$v(t,x) = G(r-t,\theta)/r + G'(r+t,\theta)/r + q_1(t,x) \tag{3.30}$$

where $G(\tau,\theta)$ and $G'(\tau,\theta)$ are the functions of $\tau \in R$ and $\theta \in S^2$ defined by

$$G(\tau,\theta) = \frac{1}{(2\pi)^{1/2}} \int_a^b e^{i\tau\rho}\, \hat{h}(\rho\theta)(-i\rho)\,d\rho \tag{3.31}$$

and

$$G'(\tau,\theta) = \frac{1}{(2\pi)^{1/2}} \int_{-b}^{-a} e^{i\tau\rho}\, \hat{h}(\rho\theta)(-i\rho)\,d\rho \tag{3.32}$$

Moreover, the estimate (3.29) implies that $q_1(t,x)$ satisfies

$$|q_1(t,x)| \leq M_1/r^2 \text{ for all } r > 0,\ t \in R \text{ and } \theta \in S^2 \tag{3.33}$$

where $M_1 = M_1(\hat{h}) = (2\pi)^{-3/2}(b^3 - a^3)\, M_0(\hat{h})/3$.

The principal result of this section states that

$$v^\infty(t,x) = G(r-t,\theta)/r,\ x = r\theta \tag{3.34}$$

is an asymptotic wave function for $v(t,x)$ in $L_2(R^3)$; that is,

$$\lim_{t\to\infty} \| v(t,\cdot) - v^\infty(t,\cdot)\|_{L_2(R^3)} = 0 \tag{3.35}$$

Before indicating a proof it is necessary to complete the statement of the theorem by defining the profile G for arbitrary $h \in L_2(R^3)$. When $\hat{h} \in \mathcal{D}_0(R^3)$, G is defined by (3.31) and a simple calculation gives

$$\|G\|^2_{L_2(R\times S^2)} = \int_{a\leq|p|\leq b} |\hat{h}(p)|^2 dp = \|\hat{h}\|^2_{L_2(R^3)} = \|h\|^2_{L_2(R^3)} \tag{3.36}$$

Hence the correspondence

$$h \to G = \Theta h \in L_2(R \times S^2) \tag{3.37}$$

can be extended to all $h \in L_2(R^3)$ by completion. Another method, based on the Plancherel theory in $L_2(R, L_2(S^2))$ is given in [42]. It is not difficult to verify by constructing Θ^{-1} that

$$\Theta: \; L_2(R^3) \to L_2(R \times S^2) \text{ is unitary} \tag{3.38}$$

A similar extension of the definition (3.32) of G' may be made.

A proof of (3.35) will now be outlined. Note first that the function $G'(r+t,\theta)/r$ tends to zero in $L_2(R^3)$ when $t \to \infty$. This follows from the simple calculation

$$\begin{aligned} \int_{R^3} |G'(r+t,\theta)/r|^2 dx &= \int_0^\infty \int_{S^2} |G'(r+t,\theta)|^2 d\theta dr \\ &= \int_t^\infty \int_{S^2} |G'(r,\theta)|^2 d\theta dr \end{aligned} \tag{3.39}$$

and the fact that $G' \in L_2(R \times S^2)$. The proof that, in (3.30), $q_1(t,\cdot) \to 0$ in $L_2(R^3)$ when $t \to \infty$ is based on the following lemma.

3.1 Convergence lemma

Let $\Omega \subset R^3$ be an unbounded domain and let $u(t,x)$ have the properties

$$u(t,\cdot) \in L_2(\Omega) \text{ for every } t > t_0 \tag{3.40}$$

$$\lim_{t\to\infty} \|u(t,\cdot)\|_{L_2(K\cap\Omega)} = 0 \text{ for every compact } K \subset R^3 \tag{3.41}$$

$$|u(t,x)| \le M/|x|^2 \text{ for every } |x| > r_0 \tag{3.42}$$

where t_0, r_0 and M are constants. Then

$$\lim_{t\to\infty} \|u(t,\cdot)\|_{L_2(\Omega)} = 0 \tag{3.43}$$

Only the case $\Omega = R^3$ of the lemma is needed here. The more general case is used in section 4. A simple proof of the lemma is given in [42].

The proof of (3.35) for the case $\hat{h} \in \mathcal{D}_0(R^3)$ may be completed by applying the lemma to $u(t,x) = q_1(t,x)$. (3.33) states that q_1

satisfies (3.42) while (3.40) and (3.41) follow from (3.30). To verify (3.41) note that $G'(r+t,\theta)/r$ satisfies it by (3.39). Moreover, if $K \subset \{x: |x| \leq R\}$ then by direct calculation

$$\int_K |G(r-t,\theta)/r|^2 dx \leq \int_{|x|\leq R} |G(r-t,\theta)/r|^2 dx$$
$$= \int_0^R \int_{S^2} |G(r-t,\theta)|^2 d\theta dr = \int_{-t}^{R-t} \int_{S^2} |G(r,\theta)|^2 d\theta dr \qquad (3.44)$$

The last integral tends to zero when $t \to \infty$ because $G \in L_2(R \times S^2)$. Finally, $v(t,x)$ satisfies (3.41). When $\hat{h} \in \mathcal{D}_0(R^3)$ this can be verified directly from (3.24) by an integration by parts.

The proof of (3.35) indicated above is valid when $\hat{h} \in \mathcal{D}_0(R^3)$. To prove (3.35) for general $h \in L_2(R^3)$ note that

$$v(t,\cdot) = U_0(t)h \text{ where } U_0(t) = \exp(-itA_0^{1/2}) \qquad (3.45)$$

is unitary. In particular,

$$\|U_0(t)\| = 1 \text{ for all } t \in R \qquad (3.46)$$

Similarly, if $U_0^\infty(t): L_2(R^3) \to L_2(R^3)$ is defined by

$$v^\infty(t,\cdot) = U_0^\infty(t)h \qquad (3.47)$$

then it follows from (3.44) and (3.36) that $U_0^\infty(t)$ is contractive:

$$\|U_0^\infty(t)\| \leq 1 \text{ for all } t \in R \qquad (3.48)$$

The general case of (3.35) now follows from the special case $\hat{h} \in \mathcal{D}_0(R^3)$, the density of $\mathcal{D}_0(R^3)$ in $L_2(R^3)$ and the estimates (3.46) and (3.48). The details are given in [42].

The real part of the asymptotic wave function (3.34) is another function of the same form. Hence, (3.8) and (3.35) imply a similar result for the solution in $L_2(R^3)$ of the Cauchy problem. The result may be formulated as follows.

3.2 Theorem

Let f and g be real-valued functions such that $f \in L_2(R^3)$ and $g \in D(A_0^{-1/2})$. Let $u(t,x)$ be the corresponding solution in $L_2(R^3)$ of the d'Alembert equation given by (3.6). Define the asymptotic wave function

$$u^{\infty}(t,x) = \frac{F(r-t,\theta)}{r}, \quad x = r\theta \tag{3.49}$$

where

$$F(\tau,\theta) = \text{Re}\,\{G(\tau,\theta)\} \tag{3.50}$$

and

$$G = \Theta h = \Theta(f + iA_0^{-1/2}g) \tag{3.51}$$

Then

$$\lim_{t\to\infty} \|u(t,\cdot) - u^{\infty}(t,\cdot)\|_{L_2(R^3)} = 0 \tag{3.52}$$

Stronger forms of convergence than (3.52) can also be proved under suitable hypotheses on the initial state. In particular, convergence in energy holds if the initial state has finite energy. A result of this type is formulated at the end of section 4 for the more general case of an initial-boundary value problem for the d'Alembert equation in an exterior domain.

4. SCATTERING BY OBSTACLES IN HOMOGENEOUS FLUIDS

The scattering of localized acoustic waves by bounded rigid obstacles immersed in an unlimited homogeneous fluid is analyzed in this section. The corresponding boundary value problem is

$$\frac{\partial^2 u}{\partial t^2} - \left(\frac{\partial^2 u}{\partial x_1^2} + \frac{\partial^2 u}{\partial x_2^2} + \frac{\partial^2 u}{\partial x_3^2}\right) = 0 \text{ for } t > 0,\ x \in \Omega \tag{4.1}$$

$$\frac{\partial u}{\partial \nu} = 0 \text{ for } t \geq 0,\ x \in \partial\Omega \tag{4.2}$$

$$u(0,x) = f(x) \text{ and } \partial u(0,x)/\partial t = g(x) \text{ for } x \in \Omega \tag{4.3}$$

where $\Omega \subset R^3$ is an exterior domain (i.e., $\Gamma = R^3 - \Omega$ is compact). This problem will be treated as a perturbation of the Cauchy problem of section 3.

A formulation of the initial-boundary value problem (4.1) - (4.3) which is applicable to arbitrary domains $\Omega \subset R^3$ was given by the author in [33,42]. That work provides the starting point for the analysis of this section and sections 5 and 6. The principal definitions and results are summarized here briefly.

The formulation makes use of the Hilbert space $L_2(\Omega)$ and the following subsets of $L_2(\Omega)$.

$$L_2^1(\Omega) = L_2(\Omega) \cap \{u: \partial u/\partial x_j \in L_2(\Omega) \text{ for } j = 1,2,3\} \tag{4.4}$$

$$L_2(\Delta,\Omega) = L_2(\Omega) \cap \{u: \Delta u \in L_2(\Omega)\} \tag{4.5}$$

$$L_2^1(\Delta,\Omega) = L_2^1(\Omega) \cap L_2(\Delta,\Omega) \tag{4.6}$$

where $\Delta u = \partial^2u/\partial x_1^2 + \partial^2u/\partial x_2^2 + \partial^2u/\partial x_3^2$ denotes the Laplacian of u. The derivatives in these definitions are to be interpreted in the sense of the theory of distributions. The sets (4.4), (4.5) and (4.6) are linear subsets of $L_2(\Omega)$. Moreover, they are Hilbert spaces with inner products meaningful for arbitrary domains Ω. Moreover, it reduces to (4.2)

$$(u,v)_1 = (u,v) + \sum_{j=1}^{3} (\partial u/\partial x_j, \partial v/\partial x_j) \tag{4.7}$$

$$(u,v)_\Delta = (u,v) + (\Delta u, \Delta v) \tag{4.8}$$

$$(u,v)_{1,\Delta} = (u,v)_1 + (\Delta u, \Delta v) \tag{4.9}$$

respectively, where (u,v) is the inner product in $L_2(\Omega)$.

4.1 Definition

A function $u \in L_2^1(\Delta,\Omega)$ is said to satisfy the generalized Neumann condition for Ω if and only if

$$(\Delta u, v) + \sum_{j=1}^{3} (\partial u/\partial x_j, \partial v/\partial x_j) = 0 \text{ for all } v \in L_2^1(\Omega) \tag{4.10}$$

Note that (4.10) defines a closed subspace

$$L_2^N(\Delta,\Omega) = L_2^1(\Delta,\Omega) \cap \{u: u \text{ satisfies } (4.10)\} \tag{4.11}$$

in the Hilbert space $L_2^1(\Delta,\Omega)$. The condition "$u \in L_2^N(\Delta,\Omega)$" is a generalization of the Neumann boundary condition (4.2). It is meaningful for arbirary domains Ω. Moreover, it reduces to (4.2) whenever $\partial\Omega$ is sufficiently smooth (see [42,p.41] for a discussion).

The construction of solutions of the initial-boundary value problem (4.1) - (4.3) given below is based on the linear operator $A = A(\Omega)$ in $L_2(\Omega)$ defined by

$$D(A) = L_2^N(\Delta,\Omega) \tag{4.12}$$

$$Au = -\Delta u \text{ for all } u \in D(A) \tag{4.13}$$

The utility of this operator is based on the following theorem which is proved in [42].

4.2 Theorem

A is a selfadjoint real positive operator in $L_2(\Omega)$. Moreover, $D(A^{1/2}) = L_2^1(\Omega)$ and

$$\|A^{1/2}u\|^2 = \sum_{j=1}^{3} \|\partial u/\partial x_j\|^2 \text{ for all } u \in D(A^{1/2}) \tag{4.14}$$

The operator A may be used to construct "solutions in $L_2(\Omega)$" and "solutions with finite energy" of (4.1) - (4.3), as described in section 2. The solution in $L_2(\Omega)$ will be considered here. As in section 3, if $f \in L_2(\Omega)$ and $g \in D(A^{-1/2})$ then

$$u(t,x) = \text{Re}\,\{v(t,x)\} \tag{4.15}$$

where

$$v(t,\cdot) = \exp\,(-itA^{1/2})h, \quad h = f + iA^{-1/2}g \tag{4.16}$$

The properties of the operator A stated in the theorem above are valid for arbitrary domains $\Omega \subset R^3$. It was shown in [42] that if Ω is an exterior domain then A has a continuous spectrum. Moreover, if Ω has the local compactness property (defined below) then there exist eigenfunction expansions for A in terms of generalized eigenfunctions which are perturbations of the plane wave eigenfunctions of section 3. In the remainder of this section the eigenfunction expansions are described and used to analyze the structure of solutions of the scattering problem (4.1) - (4.3). The principal result of the analysis states that the behavior of the acoustic field for large times is described by an asymptotic wave function of exactly the same form (3.49) as when there is no obstacle. The only effect of an obstacle is to modify the wave profile $F(\tau,\theta)$. Moreover, a procedure is given for calculating the modified profile when the obstacle and the initial state are known.

4.3 Distorted plane wave eigenfunctions

Two families of generalized eigenfunctions of A, denoted by $w_+(x,p)$ and $w_-(x,p)$ respectively, were defined in [42]. They are perturbations of the plane wave eigenfunctions $w_0(x,p)$ and have the form

$$w_\pm(x,p) = w_0(x,p) + w'_\pm(x,p), \quad p \in R^3 \tag{4.17}$$

where $w'_+(x,p)$ and $w'_-(x,p)$ may be interpreted as secondary fields which are produced when the obstacle $\Gamma = R^3 - \Omega$ is irradiated by the plane wave $w_0(x,p)$. Mathematically, $w_+(x,p)$ and $w_-(x,p)$ must satisfy

$$(\Delta + |p|^2)\, w_\pm(x,p) = 0 \text{ for } x \in \Omega \tag{4.18}$$

$$\frac{\partial w_\pm(x,p)}{\partial \nu} = 0 \text{ for } x \in \partial\Omega \tag{4.19}$$

However, they are not completely determined by these conditions. Instead, $w_+(x,p)$ is determined by (4.18), (4.19) and the condition that $w_+'(x,p)$ should describe an outgoing secondary wave. This is implied by the Sommerfeld condition for outgoing waves:

$$\left.\begin{aligned} &\frac{\partial w_+'(x,p)}{\partial |x|} - i|p|\, w_+'(x,p) = o(|x|^{-1}), \quad |x| \to \infty \\ &w_+'(x,p) = O(|x|^{-1}), \quad |x| \to \infty \end{aligned}\right\} \tag{4.20}$$

Similarly, $w_-(x,p)$ is determined by (4.18), (4.19) and the condition that $w_-'(x,p)$ should describe an incoming secondary wave, which is implied by the Sommerfeld condition for incoming waves:

$$\left.\begin{aligned} &\frac{\partial w_-'(x,p)}{\partial |x|} + i|p|\, w_-'(x,p) = o(|x|^{-1}), \quad |x| \to \infty \\ &w_-'(x,p) = O(|x|^{-1}), \quad |x| \to \infty \end{aligned}\right\} \tag{4.21}$$

Of course, if $\partial\Omega$ is not smooth then the boundary condition (4.19) must be understood in the generalized sense of (4.10). A technical difficulty is caused by the fact that $w_\pm(\cdot,p)$ cannot be in $D(A) = L_2^N(\Delta,\Omega)$ because the spectrum of A is continuous. This is overcome by requiring that

$$\phi w_\pm(\cdot,p) \in L_2^N(\Delta,\Omega) \tag{4.22}$$

for all $\phi \in \mathcal{D}(R^3)$ such that $\phi(x) \equiv 1$ in a neighborhood of $\partial\Omega$. Generalized eigenfunctions with these properties will be called "distorted plane waves," following T. Ikebe [16].

The uniqueness of distorted plane waves satisfying (4.18), (4.20) or (4.21) and (4.22) was proved in [42] for arbitrary exterior domains. However, to prove their existence it was necessary to impose a condition on $\partial\Omega$. To define it let

$$\Omega_R = \Omega \cap \{x: \ |x| < R\} \tag{4.23}$$

$$L_2^{\ell oc}(\overline{\Omega}) = \{u: \ u \in L_2(\Omega_R) \text{ for every } R > 0\} \tag{4.24}$$

$$L_2^{1,\ell oc}(\overline{\Omega}) = L_2^{\ell oc}(\overline{\Omega}) \cap \{u: \partial u/\partial x_j \in L_2^{\ell oc}(\overline{\Omega}) \text{ for } j = 1,2,3\} \tag{4.25}$$

and define the

4.4 Local compactness property

A domain $\Omega \subset R^3$ is said to have the local compactness property if and only if for each set $S \subset L^{1,\ell oc}(\overline{\Omega})$ and each $R > 0$ the condition

$$\|u\|_{L_2^1(\Omega_R)} \leq C(R) \text{ for all } u \in S \tag{4.26}$$

implies that S is precompact in $L_2(\Omega_R)$; i.e., every sequence $\{u_n\}$ in S which satisfies (4.26) has a subsequence which converges in $L_2(\Omega_R)$. The class of domains with the local compactness property will be denoted by LC.

The local compactness property is known to hold for large classes of domains. S. Agmon has proved it for domains with the "segment property" [1]. A generalization of the segment property, called the "finite tiling property" was given by the author in [42]. As an application of this condition it can be shown that the local compactness property holds for the many simple, but non-smooth, boundaries that arise in applications, such as polyhedra, finite sections of cylinders, cones, spheres, disks, etc. The following existence theorem was proved in [42].

4.5 Theorem

Let $\Omega \subset R^3$ be an exterior domain such that $\Omega \in LC$. Then for each $p \in R^3$ there exists a unique outgoing distorted plane wave $w_+(x,p)$ and a unique incoming distorted plane wave $w_-(x,p)$.

The outgoing (resp. incoming) property of $w_+'(x,p)$ (resp. $w_-'(x,p)$) is made explicit by the following corollary.

4.6 Corollary

Under the same hypotheses there exist functions $T_\pm(\theta,p) \in C^\infty(S^2 \times \{R^3 - 0\})$ such that

$$w_\pm'(x,p) = \frac{e^{\pm i|p|r}}{r} T_\pm(\theta,p) + w_\pm''(x,p), \quad x = r\theta \tag{4.27}$$

where

$$w''_{\pm}(x,p) = O(r^{-2}), \quad r \to \infty \tag{4.28}$$

uniformly for $\theta = x/r \in S^2$ and p in any compact subset of $R^3 - \{0\}$.

In acoustics the functions $T_+(\theta,p)$ and $T_-(\theta,p)$ are called the far-field amplitudes of the distorted plane waves.

4.7 The eigenfunction expansion theorem

Each of the families $\{w_+(\cdot,p): p \in R^3\}$ and $\{w_-(\cdot,p): p \in R^3\}$ defines a complete set of generalized eigenfunctions of A in the sense described by the following theorems.

4.8 Theorem

For each $f \in L_2(\Omega)$ the following limits exist

$$\left.\begin{aligned} \hat{f}_{\pm}(p) &= L_2(R^3)\text{-}\lim_{M\to\infty} \int_{\Omega_M} \overline{w_{\pm}(x,p)}\, f(x)dx \\ f(x) &= L_2(\Omega)\text{-}\lim_{M\to\infty} \int_{|p|\le M} w_{\pm}(x,p)\, \hat{f}_{\pm}(p)dp \end{aligned}\right\} \tag{4.29}$$

where $\Omega_M = \Omega \cap \{x: |x| < M\}$. Moreover, the operators $\Phi_{\pm}: L_2(\Omega) \to L_2(R^3)$ defined by

$$\Phi_{\pm} f = \hat{f}_{\pm} \tag{4.30}$$

are unitary.

The relations (4.29) will usually be written in the symbolic form

$$\hat{f}_{\pm}(p) = \int_{\Omega} \overline{w_{\pm}(x,p)}\, f(x)dx, \quad f(x) = \int_{R^3} w_{\pm}(x,p)\, \hat{f}_{\pm}(p)dp \tag{4.31}$$

but must be understood in the sense of (4.29).

4.9 Theorem

If $\{\Pi(\lambda)\}$ denotes the spectral family of A:

$$A = \int_0^{\infty} \lambda \, d\Pi(\lambda) \tag{4.32}$$

then $\Pi(\lambda)$ has the eigenfunction expansions

$$\Pi(\lambda)\ f(x) = \int_{|p|\leq\sqrt{\lambda}} w_{\pm}(x,p)\ \hat{f}_{\pm}(p)dp, \quad \lambda \geq 0 \tag{4.33}$$

In particular, A is an absolutely continuous operator whose spectrum is the interval $[0,\infty)$.

The last result implies that Φ_+ and Φ_- define spectral representations for A in the sense of the following corollary.

4.10 Corollary

If $\Psi(\lambda)$ is a bounded Lebesgue-measurable function of $\lambda \geq 0$ then for all $f \in L_2(\Omega)$

$$\Phi_{\pm}\Psi(A)f(p) = \Psi(|p|^2)\ \hat{f}_{\pm}(p) \tag{4.34}$$

These results provide a complete generalization of the Plancherel theory to exterior domains $\Omega \in LC$.

4.11 The eigenfunction expansions and scattering theory

The results stated above imply that the wave functions

$$v(t,\cdot) = \exp\ (-itA^{1/2})h, \quad h \in L_2(\Omega) \tag{4.35}$$

have the spectral integral representations

$$v(t,x) = \int_{R^3} w_{\pm}(x,p)\ \exp\ (-it|p|)\ \hat{h}_{\pm}(p)dp \tag{4.36}$$

Note that (4.36) defines two representations, corresponding to $w_+(x,p)$ and $w_-(x,p)$. They will be called the outgoing and incoming representations, respectively.

The representations (4.36) and the results of section 3 will now be used to derive the asymptotic behavior of $v(t,x)$ for $t \to \infty$. To begin consider an initial state $h \in L_2(\Omega)$ such that

$$\hat{h}_- \in \mathcal{D}_0(R^3) \tag{4.37}$$

Such states are dense in $L_2(\Omega)$ because $\mathcal{D}_0(R^3)$ is dense in $L_2(R^3)$ and Φ_-: $L_2(\Omega) \to L_2(R^3)$ is unitary. The wave function corresponding to (4.37) is

$$v(t,x) = \int_{R^3} w_-(x,p) \exp(-it|p|) \hat{h}_-(p)dp \tag{4.38}$$

where the integral converges both pointwise and in $L_2(\Omega)$ to $v(t,x)$. To discover the behavior of $v(t,x)$ for $t \to \infty$ substitute the decompositions (4.17) and (4.27) for $w_-(x,p)$ into (4.38) and write

$$v(t,x) = v_0(t,x) + v'(t,x) + v''(t,x) \tag{4.39}$$

where

$$v_0(t,x) = \int_{R^3} w_0(x,p) \exp(-it|p|) \hat{h}_-(p)dp \tag{4.40}$$

$$v'(t,x) = \frac{1}{r} \int_{R^3} \exp\{-i|p|(r+t)\} T_-(\theta,p) \hat{h}_-(p)dp \tag{4.41}$$

and

$$v''(t,x) = \int_{R^3} w''_-(x,p) \exp(-it|p|) \hat{h}_-(p)dp \tag{4.42}$$

Note that $v_0(t,x)$ is a solution in $L_2(R^3)$ of the d'Alembert equation. Indeed, $\hat{h}_- = \Phi_- h = \Phi_0(\Phi_0^*\Phi_- h) = \hat{h}_0$ where

$$h_0 = \Phi_0^*\Phi_- h \in L_2(R^3) \tag{4.43}$$

and

$$v_0(t,\cdot) = \int_{R^3} w_0(\cdot,p) \exp(-it|p|) \hat{h}_0(p)dp = \exp(-itA_0^{1/2})h_0 \tag{4.44}$$

Thus $v_0(t,x)$ represents a wave in an unlimited fluid containing no obstacles. It will be shown that $v(t,x)$ is asymptotically equal to this wave when $t \to \infty$; i.e.,

$$\lim_{t\to\infty} \| v(t,\cdot) - v_0(t,\cdot)\|_{L_2(\Omega)} = 0 \tag{4.45}$$

To see this note that, in (4.39), $v'(t,x)$ has the form

$$v'(t,x) = G'(r+t,\theta)/r \tag{4.46}$$

It was shown in section 3 that such functions tend to zero in $L_2(R^3)$ when $t \to \infty$ (see (3.39)). It is easy to check that (4.37) implies that $G' \in L_2(R \times S^2)$. Finally, condition (4.28) for $w''_-(x,p)$ implies that the term $v''(t,x)$ in (4.39) satisfies

$$|v''(t,x)| \leq M/|x|^2 \text{ for all } |x| > 0 \text{ and } t \in R \tag{4.47}$$

with a suitable constant M. Hence, the convergence lemma of section 3, applied to $v'' = v - v_0 - v'$ implies (4.45) if $v''(t,x)$ satisfies the local decay condition (3.41). For $v'(t,x)$ this condition follows from (4.46). For $v(t,x)$ and $v_0(t,x)$ it follows from the local compactness property. A proof may be found in [42]. Thus (4.45) is established for all $\hat{h}_- \in \mathcal{D}_0(R^3)$. The main result of this section is the

4.12 Theorem

For all $h \in L_2(\Omega)$ if $v(t,\cdot) = \exp(-itA^{1/2})h$ and $v_0(t,\cdot) = \exp(-itA_0^{1/2})(\Phi_0^*\Phi_-)h$ then

$$\lim_{t\to\infty} \| v(t,\cdot) - v_0(t,\cdot)\|_{L_2(\Omega)} = 0 \tag{4.48}$$

This result follows immediately from the special case (4.37) proved above, the density of $\mathcal{D}_0(R^3)$ in $L_2(R^3)$ and the unitarity of the operators $\exp(-itA^{1/2})$, $\exp(-itA_0^{1/2})$, Φ_0 and Φ_-.

4.13 Corollary

If J_Ω: $L_2(\Omega) \to L_2(R^3)$ is defined by $J_\Omega u(x) = u(x)$ for all $x \in \Omega$ and $J_\Omega u(x) = 0$ for all $x \in R^3 - \Omega$ then the strong limit

$$W_+ = W_+(A_0^{1/2}, A^{1/2}, J_\Omega) = \operatorname*{s-lim}_{t\to\infty} \exp(itA_0^{1/2}) J_\Omega \exp(-itA^{1/2}) \tag{4.49}$$

exists in $L_2(\Omega)$ and W_+: $L_2(\Omega) \to L_2(R^3)$ is given by

$$W_+ = \Phi_0^*\Phi_- \tag{4.50}$$

In particular, W_+ is unitary.

The operator W_+ is the wave operator for the pair $A_0^{1/2}$, $A^{1/2}$ in the sense of the time dependent theory of scattering. The equivalence of (4.48) and (4.50) is proved in [42].

4.14 Asymptotic wave functions in $L_2(\Omega)$

The wave function in $L_2(R^3)$ defined by

$$v_0(t,\cdot) = \exp(-itA_0^{1/2})h_0, \quad h_0 = \Phi_0^*\Phi_- h \tag{4.51}$$

has an asymptotic wave function in $L_2(R^3)$, by the results of section 3; i.e.,

$$\lim_{t\to\infty} \| v_0(t,\cdot) - v^\infty(t,\cdot)\|_{L_2(R^3)} = 0 \tag{4.52}$$

where

$$v^\infty(t,x) = G(r-t,\theta)/r, \quad x = r\theta \tag{4.53}$$

and

$$G = \Theta h_0 = \Theta\Phi_0^*\Phi_- h \tag{4.54}$$

Equations (4.48), (4.52) and the triangle inequality imply the

4.15 Theorem

For each $h \in L_2(\Omega)$ the wave function $v^\infty(t,\cdot)$ defined by (4.53), (4.54) is an asymptotic wave function in $L_2(\Omega)$ for $v(t,\cdot) = \exp(-itA^{1/2})h$; that is,

$$\lim_{t\to\infty} \| v(t,\cdot) - v^\infty(t,\cdot)\|_{L_2(\Omega)} = 0 \tag{4.55}$$

4.16 Corollary

The profile of the asymptotic wave function is given by

$$G(\tau,\theta) = \frac{1}{(2\pi)^{1/2}} \int_0^\infty e^{i\tau\rho}\, \hat{h}_-(\rho\theta)(-i\rho)\,d\rho \tag{4.56}$$

where the integral converges in $L_2(R \times S^2)$.

This follows immediately from (4.54) and (3.31). Note that the only difference between the asymptotic wave functions for R^3 and those for Ω is that $\hat{h} = \Phi_0 h$ is replaced by $\hat{h}_- = \Phi_- h$.

4.17 Asymptotic energy distributions

If the initial state $h \in L_2(\Omega)$ has derivatives in $L_2(\Omega)$ then the corresponding profile G and asymptotic wave function $v^\infty(t,x)$ will have corresponding derivatives. In particular, the following result was proved in [42].

4.18 Corollary

If $\partial h(x)/\partial x_j \in L_2(\Omega)$ for $j = 1,2,3$ then $\partial v(t,x)/\partial t$ and $\partial v(t,x)/\partial x_j$ are in $L_2(\Omega)$ for all $t \in R$ and $j = 1,2,3$ and

$$\left.\begin{aligned} &\lim_{t\to\infty} \| \partial v(t,\cdot)/\partial t - v_0^\infty(t,\cdot)\|_{L_2(\Omega)} = 0 \\ &\lim_{t\to\infty} \| \partial v(t,\cdot)/\partial x_j - v_j^\infty(t,\cdot)\|_{L_2(\Omega)} = 0, \ j = 1,2,3 \end{aligned}\right\} \quad (4.57)$$

where

$$v_k^\infty(t,x) = G_k(r - t,\theta)/r, \ k = 0,1,2,3 \quad (4.58)$$

$$G_0(\tau,\theta) = -\partial G(\tau,\theta)/\partial\tau \quad (4.59)$$

$$G_j(\tau,\theta) = -G_0(\tau,\theta)\theta_j, \ j = 1,2,3 \quad (4.60)$$

and $G(\tau,\theta)$ is given by (4.56).

The energy integral for a homogeneous fluid is given by

$$E(u,K,t) = \frac{1}{2} \int_K \left< \left\{ \left(\frac{\partial u(t,x)}{\partial t}\right)^2 + \sum_{j=1}^3 \left(\frac{\partial u(t,x)}{\partial x_j}\right)^2 \right\} \right> dx \quad (4.61)$$

if $\rho = 1$, $c = 1$. The last corollary implies that if $u(t,x) = \mathrm{Re}\,\{v(t,x)\}$ is a solution with finite energy in Ω then the energy in any measurable cone

$$C = \{x = r\theta: \ r > 0, \ \theta \in C_0 \subset S^2\} \quad (4.62)$$

has a limit as $t \to \infty$ which can be calculated from the initial state $u(0,x) = f(x)$, $\partial u(0,x)/\partial t = g(x)$. The following result was proved in [42].

4.19 Theorem

If $f \in L_2^1(\Omega)$, $g \in L_2(\Omega)$ and if C is any measurable cone in R^3 then

$$\lim_{t\to\infty} E(u,C \cap \Omega,t) = \frac{1}{2} \int_C \left| |p|\hat{f}_-(p) + i\hat{g}_-(p)\right|^2 dp \quad (4.63)$$

5. PROPAGATION IN UNIFORM TUBULAR WAVEGUIDES

The propagation and scattering of localized acoustic waves is simple and compound tubular waveguides with rigid walls, and filled with a homogeneous fluid, is analyzed in this section and the next. The simplest case is the uniform semi-infinite cylinder, closed by a plane wall perpendicular to the axis. Other special

Figure 1. Uniform semi-infinite cylindrical waveguide.

cases which are of interest in applied acoustics include the cylindrical waveguide terminated by a resonator, the tubular

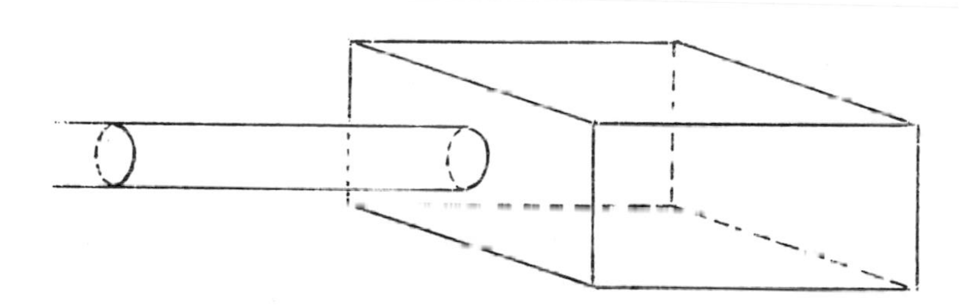

Figure 2. Cylindrical waveguide terminated by a resonator.

waveguide with a bend, or elbow, coupled cylindrical waveguides with different cross-sections, the T-joint in a waveguide, uniform waveguides containing an iris, waveguides containing obstacles, and many others.

The most general compound tubular waveguide considered here is described by a domain $\Omega \subset R^3$ of the form

$$\Omega = \Omega_0 \cup S_1 \cup S_2 \cup \cdots \cup S_m \tag{5.1}$$

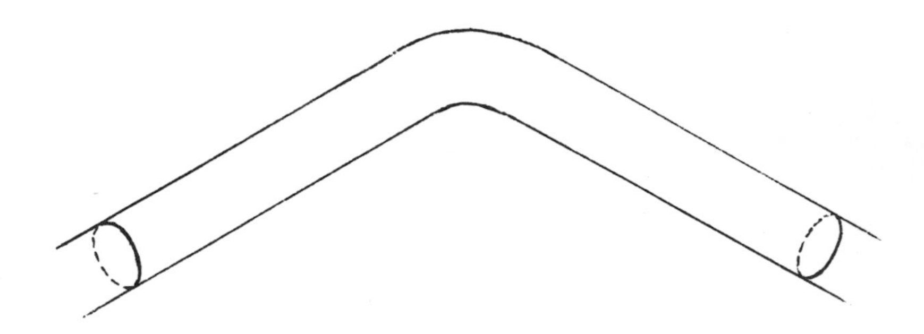

Figure 3. Waveguide with elbow.

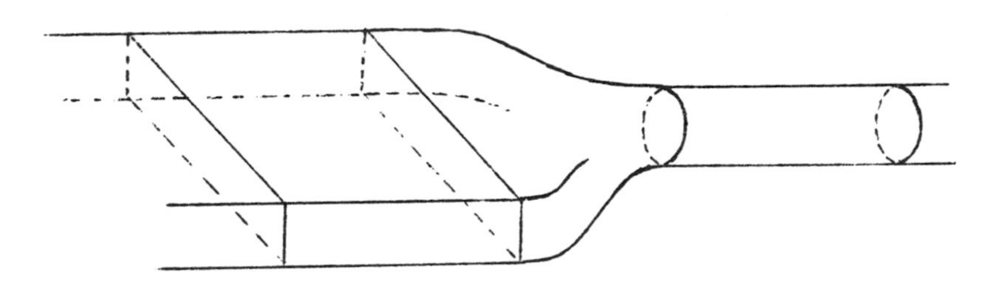

Figure 4. Coupled waveguides.

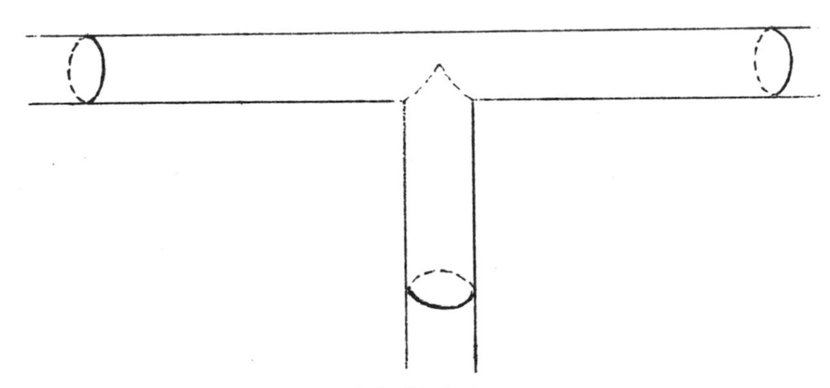

Figure 5. Waveguide with T-joint.

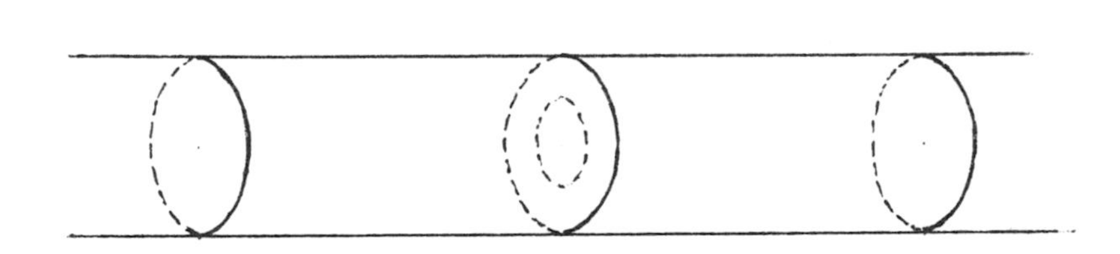

Figure 6. Waveguide with iris.

where Ω_0 is a bounded domain and $S_1, S_2, \cdots, S_m$ are disjoint uniform semi-infinite cylinders. If Ω is a waveguide with rigid walls, filled with a homogeneous fluid, the corresponding boundary value problem is again problem (4.1) - (4.3), but for a domain with the structure (5.1). Hence, the Hilbert space formulation of (4.1) - (4.3) given at the beginning of section 4, which is valid for arbitrary domains $\Omega \subset R^3$, provides a starting point for the analysis of the waveguide problems. The remainder of this section presents the spectral and asymptotic analysis of acoustic waves in a uniform semi-infinite cylindrical waveguide. The general case (5.1) is analyzed in section 6.

5.1 The uniform semi-infinite cylinder

It will be convenient to use coordinates

$$(x_1, x_2, y) \equiv (x, y) \in R^3 \tag{5.2}$$

such that the y-axis lies in the waveguide. With this choice the waveguide may be described by a domain of the form

$$S = \{(x,y): \quad x \in G \text{ and } y > 0\} \tag{5.3}$$

where $G \subset R^2$ defines the waveguide cross section. It will be assumed that G is bounded and that $S \in LC$.

The spectral analysis of the operator $A = A(S)$, acting in $L_2(S)$, will be based on the spectral analysis of $A(G)$ acting in $L_2(G)$. It can be shown that the hypothesis $S \in LC$ implies that $G \in LC$ as a domain in R^2. This property and the boundedness of

G imply that A(G) has a discrete spectrum with eigenvalues

$$0 = \lambda_0 \leq \lambda_1 \leq \lambda_2 \leq \cdots \tag{5.4}$$

such that

$$\lim_{j\to\infty} \lambda_j = \infty \tag{5.5}$$

Each eigenvalue has finite multiplicity and it is assumed that in the enumeration (5.4) each eigenvalue is repeated according to its multiplicity. There exists a corresponding orthonormal set $\{\phi_j(x)\}$ of eigenfunctions which is complete in $L_2(G)$. Each ϕ_j satisfies $\phi_j \in D(A(G)) = L_2^N(\Delta,G)$ and $A(G)\phi_j = \lambda_j\phi_j$. Formally, the $\phi_j(x)$ are solutions of the eigenvalue problem

$$\left.\begin{aligned} &\frac{\partial^2\phi}{\partial x_1^2} + \frac{\partial^2\phi}{\partial x_2^2} + \lambda\phi = 0 \text{ for } x \in G \\ &\frac{\partial\phi}{\partial\nu} = 0 \text{ for } x \in \partial G \end{aligned}\right\} \tag{5.6}$$

Of course, if ∂G is not smooth then the boundary condition is the generalized Neumann condition defined in section 4. It is known that the first eigenvalue $\lambda_0 = 0$ is simple with normalized eigenfunction

$$\phi_0(x) = \frac{1}{|G|^{1/2}} = \text{const.} \tag{5.7}$$

where $|G|$ is the Lebesgue measure of G.

5.2 The eigenfunction expansion

The eigenfunctions of A may be constructed by separation of variables. From a more sophisticated point of view, A is a sum of tensor products

$$A = A(G) \otimes 1 + 1 \otimes A(R_+) \tag{5.8}$$

where $R_+ = \{y: \; y > 0\}$. It follows that the eigenfunctions of A are products of eigenfunctions of A(G) and $A(R_+)$. The spectral analysis of $A(R_+)$ is given by the Fourier cosine transform in $L_2(R_+)$:

$$\hat{f}(p) = L_2(R_+)\text{-}\lim_{M\to\infty} \left(\frac{2}{\pi}\right)^{1/2} \int_0^M \cos py\, f(y)dy \tag{5.9}$$

$$f(y) = L_2(R_+)\text{-}\lim_{M\to\infty} \left(\frac{2}{\pi}\right)^{1/2} \int_0^M \cos py\, \hat{f}(p)dp \tag{5.10}$$

$$\|\hat{f}\|_{L_2(R_+)} = \|f\|_{L_2(R_+)} \tag{5.11}$$

It follows that a complete normalized family of generalized eigenfunctions for A is defined by

$$w_j(x,y,p) = \left(\frac{2}{\pi}\right)^{1/2} \cos py\ \phi_j(x),\ p \in R_+,\ j = 0,1,2,\cdots \tag{5.12}$$

The Plancherel theory for $A(G)$ and $A(R_+)$, quoted above, implies that

$$\hat{f}_j(p) = L_2(R_+)\text{-}\lim_{M\to\infty} \int_0^M \int_G \overline{w_j(x,y,p)}\ f(x,y)dxdy \tag{5.13}$$

exists for all $f \in L_2(S)$, and the operator Φ_j: $L_2(S) \to L_2(R_+)$ defined by $\Phi_j f = \hat{f}_j$ has range $\Phi_j L_2(S) = L_2(R_+)$. Moreover

$$\|f\|^2_{L_2(S)} = \sum_{j=0}^{\infty} \|\hat{f}_j\|^2_{L_2(R_+)} \tag{5.14}$$

and

$$f(x,y) = L_2(S)\text{-}\lim_{M,N\to\infty} \sum_{j=0}^{N} \int_0^M w_j(x,y,p)\ \hat{f}_j(p)dp \tag{5.15}$$

The relations (5.13) and (5.15) are frequently written in the more concise symbolic form

$$\hat{f}_j(p) = \int_S \overline{w_j(x,y,p)}\ f(x,y)dxdy \tag{5.16}$$

and

$$f(x,y) = \sum_{j=0}^{\infty} \int_{R_+} w_j(x,y,p)\ \hat{f}_j(p)dp \tag{5.17}$$

but must be understood in the sense of (5.13) and (5.15).

Note that, formally, $\hat{f}_j(p)$ is just the $L_2(S)$ inner product of $f(x,y)$ and the eigenfunction (5.12). For a more detailed discussion of this expansion see [21].

The generalized eigenfunctions (5.12) are locally in $D(A)$ and satisfy

$$Aw_j(\cdot,\cdot,p) \equiv -\Delta w_j(\cdot,\cdot,p) = (p^2 + \lambda_j)w_j(\cdot,\cdot,p) \tag{5.18}$$

This fact and the Plancherel theory imply the following construction of the spectral family of A.

5.3 Theorem

If $\{\Pi(\lambda), \lambda \geq 0\}$ denotes the spectral family of $A = A(S)$ then $\Pi(\lambda)$ has the eigenfunction expansion

$$\begin{aligned} \Pi(\lambda)\ f(x,y) &= \sum_{\lambda_j \leq \lambda} \int_0^{\sqrt{\lambda-\lambda_j}} w_j(x,y,p)\ \hat{f}_j(p)dp \\ &= \sum_{\lambda_j \leq \lambda} \left[\int_0^{\sqrt{\lambda-\lambda_j}} \left(\frac{2}{\pi}\right)^{1/2} \cos py\ \hat{f}_j(p)dp \right] \phi_j(x) \end{aligned} \tag{5.19}$$

for all $\lambda \geq 0$. In particular, A is an absolutely continuous operator whose spectrum is the interval $[\lambda_0,\infty) = [0,\infty)$.

Note that the sum in (5.19) is actually finite by (5.5). (5.19) implies that the eigenfunction expansion (5.17) defines a spectral representation for A in the sense of the following corollary.

5.4 Corollary

If $\Psi(\lambda)$ is any bounded Lebesgue-measurable function of $\lambda \geq 0$ then for all $f \in L_2(S)$

$$\Psi(A)\ f(x,y) = L_2(S)\text{-}\lim_{M,N\to\infty} \sum_{j=0}^{N} \int_0^M w_j(x,y,p)\ \Psi(p^2+\lambda_j)\ \hat{f}_j(p)dp \tag{5.20}$$

The eigenfunction expansion (5.17) defines a decomposition of the Hilbert space $L_2(S)$. To describe its properties let $f \in L_2(S)$ and define

$$\begin{aligned} P_j\ f(x,y) &= \left(\int_G \overline{\phi_j(x')}\ f(x',y)dx' \right) \phi_j(x) \\ &= f_j(y)\ \phi_j(x),\ j = 0,1,2,\cdots \end{aligned} \tag{5.21}$$

where

$$f_j(y) = \int_G \overline{\phi_j(x')}\, f(x',y)dx', \quad j = 0,1,2,\cdots \tag{5.22}$$

The orthonormality of $\{\phi_j\}$ in $L_2(G)$ implies that $\{P_j: j=0,1,2,\cdots\}$ defines a complete family of orthogonal projections in $L_2(S)$:

$$P_j^* = P_j, \; P_jP_k = \delta_{jk}P_k \text{ for } j,k = 0,1,2,\cdots \tag{5.23}$$

and

$$\sum_{j=0}^{\infty} P_j = 1 \tag{5.24}$$

Moreover, a simple calculation gives

$$\Psi(A)\, P_jf = P_j\Psi(A)f = \int_{R_+} w_j(\cdot,\cdot,p)\Psi(p^2+\lambda_j)\hat{f}_j(p)dp \tag{5.25}$$

for $j = 0,1,2,\cdots$. In particular,

$$P_jf(x,y) = \int_{R_+} w_j(x,y,p)\; \hat{f}_j(p)dp \tag{5.26}$$

An equivalent operator-theoretic representation is $P_j = \Phi_j^*\Phi_j$. If

$$\mathcal{H}_j = P_jL_2(S) = \{f(x,y) = f_j(y)\phi_j(x): \; f_j \in L_2(R_+)\} \tag{5.27}$$

then (5.23) - (5.25) imply the

5.5 Corollary

The direct sum decomposition

$$L_2(S) = \sum_{j=0}^{\infty} \oplus \mathcal{H}_j \tag{5.28}$$

is a reducing decomposition for A.

Note that each $\mathcal{H}_j$ is isomorphic to $L_2(R_+)$ under the mapping $f(x,y) \to f_j(y)$ defined by (5.22).

5.6 Solutions in $L_2(S)$ of the propagation problem

Only the case where $f \in L_2(S)$ and $g \in D(A^{-1/2})$ will be discussed. As in sections 3 and 4, the solution in $L_2(S)$ of the propagation problem (4.1) - (4.3) has the form

$$u(t,x,y) = \mathrm{Re}\,\{v(t,x,y)\} \tag{5.29}$$

where

$$v(t,\cdot,\cdot) = \exp\,(-it\,A^{1/2})h, \quad h = f + iA^{-1/2}g \in L_2(S) \tag{5.30}$$

The decomposition (5.28) implies that

$$v(t,x,y) = \sum_{j=0}^{\infty} v_j(t,x,y) \text{ in } L_2(S) \tag{5.31}$$

where

$$v_j(t,x,y) = P_j v(t,x,y) = v_j(t,y)\phi_j(x) \in L_2(S) \tag{5.32}$$

with

$$v_j(t,y) = \left(\frac{2}{\pi}\right)^{1/2} \int_{R_+} \cos py\; e^{-it\omega_j(p)}\, \hat{h}_j(p)dp \tag{5.33}$$

and

$$\omega_j(p) = (p^2 + \lambda_j)^{1/2} \geq \lambda_j^{1/2} \geq 0 \tag{5.34}$$

In the theory of waveguides (5.31) is called a modal decomposition and the partial waves $v_j(t,x,y)$ are called waveguide modes. $v_j(t,x,y)$ will be said to be in mode j of the waveguide S. In particular, mode 0

$$v_0(t,x,y) = v_0(t,y)/|G|^{1/2} \tag{5.35}$$

will be called the fundamental mode of S. It is not difficult to show that

$$\begin{aligned} u_0(t,y) &= \mathrm{Re}\,\{v_0(t,y)\} \\ &= \frac{1}{2}\,\{f_0(y-t) + f_0(y+t)\} + \frac{1}{2}\int_{y-t}^{y+t} g_0(y')dy' \end{aligned} \tag{5.36}$$

where $f_0(-y) = f_0(y)$ and $g_0(-y) = g_0(y)$. Note that the modal waves propagate independently in the sense that different modes are orthogonal in $L_2(S)$ for all t.

The spectral representation (5.31), (5.32), (5.33) will now be used to study the asymptotic behavior for $t \to \infty$ of solutions in $L_2(S)$. Because of the independence of the modes it will be enough to study the individual modal waves (5.33). The substitution $2 \cos py = \exp(ipy) + \exp(-ipy)$ gives the decomposition

$$v_j(t,y) = v_j^+(t,y) + v_j^-(t,y) \tag{5.37}$$

where

$$v_j^+(t,y) = v_j^-(t,-y) = \frac{1}{(2\pi)^{1/2}} \int_{R_+} e^{i(yp - t\omega_j(p))} \hat{h}_j(p)\,dp \tag{5.38}$$

and the integral converges in $L_2(R_+)$ (and in $L_2(R)$) for each $h_j \in L_2(R_+)$. The special case of the fundamental mode is discussed first.

5.7 Asymptotic wave functions for the fundamental mode

This case is closely related to that of section 3, since $\omega_0(p) = p$ for all $p \in R_+$. Thus

$$v_0^+(t,y) = \frac{1}{(2\pi)^{1/2}} \int_{R_+} e^{i(y-t)p} \hat{h}_0(p)\,dp = G_0(y-t) \tag{5.39}$$

where

$$G_0(y) = \frac{1}{(2\pi)^{1/2}} \int_{R_+} e^{iyp} \hat{h}_0(p)\,dp \in L_2(R) \tag{5.40}$$

Moreover, it is easy to verify by direct calculation that $v_0^-(t,y) = v_0^+(t,-y) = G_0(-y-t) \to 0$ in $L_2(R_+)$ when $t \to \infty$. Thus

$$v_0^\infty(t,y) = G_0(y-t) \tag{5.41}$$

is an asymptotic wave function for $v_0(t,y)$ in $L_2(R_+)$:

$$\lim_{t\to\infty} \| v_0(t,\cdot) - v_0^\infty(t,\cdot)\|_{L_2(R_+)} = 0 \tag{5.42}$$

for all $\hat{h}_0 \in L_2(R_+)$.

For the higher order modes $j \geq 1$ the functions $\omega_j(p) = (p^2 + \lambda_j)^{1/2}$ with $\lambda_j > 0$. For these cases the spectral integrals (5.38) all have the same form, differing only in the value of λ_j and the function $\hat{h}_j \in L_2(R_+)$. The asymptotic behavior of these integrals may be determined by the method of stationary phase, as follows.

5.8 Application of the method of stationary phase

Consider the wave function defined by

$$v(t,y,\lambda,h) = (2\pi)^{-1/2} \int_{R_+} \exp\{i(yp - t\omega(p,\lambda))\}h(p)dp \quad (5.43)$$

where

$$\omega(p,\lambda) = (p^2+\lambda)^{1/2} \geq \lambda^{1/2} > 0 \quad (5.44)$$

and

$$h \in L_2(R_+) \quad (5.45)$$

The phase function

$$\theta(p,\lambda,y,t) = yp - t\omega(p,\lambda) \quad (5.46)$$

is stationary with respect to p if and only if

$$\partial\theta(p,\lambda,y,t)/\partial p = y - t\partial\omega(p,\lambda)/\partial p = 0 \quad (5.47)$$

or

$$\frac{y}{t} = \frac{\partial\omega(p,\lambda)}{\partial p} \equiv U(p,\lambda) = \frac{p}{(p^2+\lambda)^{1/2}} \quad (5.48)$$

The function $U(p,\lambda)$ defined by (5.48) is the group velocity [5] for the wave function (5.43). Note that

$$\frac{\partial U(p,\lambda)}{\partial p} = \frac{\partial^2\omega(p,\lambda)}{\partial p^2} = \frac{\lambda}{(p^2+\lambda)^{3/2}} > 0 \quad (5.49)$$

and hence $U(p,\lambda)$ is a monotone increasing function of p. Moreover,

$$0 \leq U(p,\lambda) < 1 \text{ for all } p \geq 0 \text{ and } \lambda > 0 \quad (5.50)$$

Hence for $t > 0$ equation (5.48) has the unique solution

$$p = \left(\frac{(y/t)^2}{1-(y/t)^2}\lambda\right)^{1/2} = \left(\frac{y^2\lambda}{t^2-y^2}\right)^{1/2} \geq 0 \quad (5.51)$$

if

$$0 \leq y/t < 1 \quad (5.52)$$

and has no solution for other positive values of t. The principle of stationary phase asserts that for large values of y^2+t^2 the stationary point (5.51) will make a contribution

$$v^{\infty}(t,y,\lambda,h) = \chi\left(\frac{y}{t}\right) \frac{e^{i(yp-t\omega(p,\lambda)-\pi/4)}}{(t\partial U(p,\lambda)/\partial p)^{1/2}} h(p), \qquad (5.53)$$

$$p = \left(\frac{y^2\lambda}{t^2 - y^2}\right)^{1/2}$$

to the integral (5.43), where $\chi(y/t)$ is the characteristic function of the set (5.52). More precisely, if $h \in \mathcal{D}(R_+)$ then the following error estimate is known [2,23].

5.9 Theorem

Let $h \in \mathcal{D}(R_+)$ and define the remainder $q(t,y,\lambda,h)$ by

$$v(t,y,\lambda,h) = v^{\infty}(t,y,\lambda,h) + q(t,y,\lambda,h) \qquad (5.54)$$

Then there exists a constant $C = C(\lambda,h)$ such that

$$|q(t,y,\lambda,h)| \leq C/(y^2 + t^2)^{3/4} \quad \text{for all } y \in R \text{ and } t > 0 \qquad (5.55)$$

It follows from (5.54) and (5.55), by direct integration, that for all $h \in \mathcal{D}(R_+)$

$$\lim_{t\to\infty} \| v(t,\cdot,\lambda,h) - v^{\infty}(t,\cdot,\lambda,h)\|_{L_2(R_+)} = 0 \qquad (5.56)$$

The stationary phase method is not applicable to (5.43) when $\lambda = 0$. However, the results for this case are described by the same equations if

$$v^{\infty}(t,y,0,h) = (2\pi)^{-1/2} \int_{R_+} \exp(iyp)\, h(p)dp \qquad (5.57)$$

With this notation, (5.56) with $\lambda = 0$ is equivalent to (5.42).

The estimate (5.55) implies (5.56) for all $h \in \mathcal{D}(R_+)$. For more general $h \in L_2(R_+)$ the estimate (5.55) may not hold. Nevertheless, the following results hold.

5.10 Theorem

For all $\lambda \geq 0$ and all $h \in L_2(R_+)$

$$v^{\infty}(t,\cdot,\lambda,h) \in L_2(R_+) \text{ for all } t \neq 0 \qquad (5.58)$$

$$t \to v^{\infty}(t,\cdot,\lambda,h) \in L_2(R_+) \text{ is continuous for all } t \neq 0 \qquad (5.59)$$

$$\| v^{\infty}(t,\cdot,\lambda,h)\|_{L_2(R_+)} \leq \|h\|_{L_2(R_+)} \text{ for all } t \neq 0 \qquad (5.60)$$

Moreover, the relation (5.56) holds for all $h \in L_2(R_+)$.

Properties (5.58) - (5.60) follow from the definitions (5.53), (5.57) by direct integration. Moreover, the validity of (5.56) for all $h \in L_2(R_+)$ follows from the special case $h \in \mathcal{D}(R_+)$, the density of $\mathcal{D}(R_+)$ in $L_2(R_+)$ and the uniform boundedness in t of $\|v(t,\cdot,\lambda,h)\|_{L_2(R_+)}$ and $\|v^{\infty}(t,\cdot,\lambda,h)\|_{L_2(R_+)}$. More detailed proofs may be found in [22,40].

5.11 Asymptotic wave functions for the higher order modes

Define the modal asymptotic wave functions by

$$v_j^{\infty}(t,y) = v^{\infty}(t,y,\lambda_j,\hat{h}_j), \quad j = 0,1,2,\cdots \qquad (5.61)$$

Then (5.38), (5.43) and (5.56) imply

$$\lim_{t\to\infty} \| v_j^{+}(t,\cdot) - v_j^{\infty}(t,\cdot)\|_{L_2(R_+)} = 0, \quad j = 0,1,2,\cdots \qquad (5.62)$$

Moreover, (5.38) for v_j^- and (5.43) imply

$$\begin{aligned} v_j^{-}(t,y) &= v(t,-y,\lambda_j,\hat{h}_j) \\ &= (2\pi)^{-1/2} \int_{R_+} \exp\{-i(yp + t\omega(p,\lambda_j)\}\, \hat{h}_j(p)\,dp \end{aligned} \qquad (5.63)$$

The stationary phase method, applied to (5.63), implies that

$$\lim_{t\to\infty} \| v_j^{-}(t,\cdot)\|_{L_2(R_+)} = 0 \qquad (5.64)$$

because the phase $yp + t\omega(p,\lambda_j)$ in (5.63) has no stationary points when $y \geq 0$ and $t > 0$. Combining (5.37), (5.62) and (5.64) gives

$$\lim_{t\to\infty} \| v_j(t,\cdot) - v_j^{\infty}(t,\cdot)\|_{L_2(R_+)} = 0 \qquad (5.65)$$

for all $\hat{h}_j \in L_2(R_+)$ and $j = 0,1,2,\cdots$. The results and the decomposition (5.31), (5.32) imply the

5.12 Asymptotic convergence theorem

For all $h \in L_2(S)$ define

$$v^{\infty}(t,x,y) = \sum_{j=0}^{\infty} v_j^{\infty}(t,y)\,\phi_j(x), \quad (x,y) \in S \tag{5.66}$$

Then

$$v(t,\cdot,\cdot) \in L_2(S) \text{ for all } t \neq 0 \tag{5.67}$$

$$t \to v(t,\cdot,\cdot) \in L_2(S) \text{ is continuous for all } t \neq 0 \tag{5.68}$$

$$\| v^{\infty}(t,\cdot,\cdot)\|_{L_2(S)} \leq \| h\|_{L_2(S)} \text{ for all } t \neq 0 \tag{5.69}$$

and

$$\lim_{t\to\infty} \| v(t,\cdot,\cdot) - v^{\infty}(t,\cdot,\cdot)\|_{L_2(S)} = 0 \tag{5.70}$$

The proof of this result will be outlined. First, note that the convergence in $L_2(S)$ of the series in (5.66) follows from the orthogonality of its terms in $L_2(S)$, (5.60) which implies

$$\| v_j^{\infty}(t,\cdot)\phi_j\|_{L_2(S)} = \| v_j^{\infty}(t,\cdot)\|_{L_2(R_+)} \leq \|\hat{h}_j\|_{L_2(R_+)} \tag{5.71}$$

for all $t \neq 0$ and (see (5.14))

$$\| h\|^2_{L_2(S)} = \sum_{j=0}^{\infty} \|\hat{h}_j\|^2_{L_2(R_+)} < \infty \tag{5.72}$$

Properties (5.68) and (5.69) follow from (5.59) and (5.60), applied to $v_j^{\infty}(t,y)$. Finally, to verify (5.70) note that for $j = 0,1,2,\cdots$

$$\begin{aligned}\| v_j(t,\cdot) - v_j^{\infty}(t,\cdot)\|_{L_2(R_+)} &\leq \| v_j(t,\cdot)\|_{L_2(R_+)} + \| v_j^{\infty}(t,\cdot)\|_{L_2(R_+)} \\ &\leq 2\|\hat{h}_j\|_{L_2(R_+)}\end{aligned} \tag{5.73}$$

for all $t \neq 0$. It follows that

$$\| v(t,\cdot,\cdot) - v^{\infty}(t,\cdot,\cdot)\|^2_{L_2(S)} = \sum_{j=0}^{\infty} \| v_j(t,\cdot) - v_j^{\infty}(t,\cdot)\|^2_{L_2(R_+)} \tag{5.74}$$

$$\leq \sum_{j=0}^{N} \|v_j(t,\cdot) - v_j^\infty(t,\cdot)\|^2_{L_2(R_+)} + 4 \sum_{j=N+1}^{\infty} \|\hat{h}_j\|^2_{L_2(R_+)} \quad (5.74\text{ Cont})$$

for $N = 0,1,2,\cdots$. Fixing N and making $t \to \infty$ gives, by (5.65)

$$\overline{\lim_{t\to\infty}} \|v(t,\cdot,\cdot) - v^\infty(t,\cdot,\cdot)\|^2_{L_2(S)} \leq 4 \sum_{j=N+1}^{\infty} \|\hat{h}_j\|^2_{L_2(R_+)} \quad (5.75)$$

for $N = 0,1,2,\cdots$. Thus (5.70) follows from (5.72) and (5.75).

If $f \in L_2^1(S) = D(A^{1/2})$ and $g \in L_2(S)$ then the same method can be used to show convergence in energy:

$$\lim_{t\to\infty} E(u - u^\infty, S, t) = 0 \quad (5.76)$$

where $u^\infty(t,x,y) = \text{Re}\,\{v^\infty(t,x,y)\}$ but the details will not be recorded here.

6. SCATTERING BY OBSTACLES AND JUNCTIONS IN TUBULAR WAVEGUIDES

The analysis of section 5 is extended to compound tubular waveguides in this section. The mathematical problem is the initial-boundary value problem (4.1) - (4.3) for an unbounded domain $\Omega \subset R^3$ of the form

$$\Omega = \Omega_0 \cup S_1 \cup \cdots \cup S_m \quad (6.1)$$

where Ω_0 is a bounded domain and $S_1,\cdots,S_m$ are disjoint uniform semi-infinite cylinders. Examples include waveguides of the types described at the beginning of section 5 and many others. It will be assumed that $\Omega \in LC$.

6.1 Notation

It will be convenient to think of R^3 as a 3-dimensional differentiable manifold. The generic point of R^3 will be denoted by q. A special Cartesian coordinate system

$$(x_1^\alpha, x_2^\alpha, y^\alpha) \equiv (x^\alpha, y^\alpha) \in R^3 \quad (6.2)$$

may be associated with each semi-infinite cylinder S_α $(\alpha = 1,\cdots,m)$ in such a way that

$$S_\alpha = \{q \in R^3:\ x^\alpha(q) \in G_\alpha \text{ and } y^\alpha(q) > 0\} \quad (6.3)$$

where $G_\alpha \subset R$ is a bounded domain. The assumption that $\Omega \in LC$ implies that $G_\alpha \in LC$ for $\alpha = 1,\cdots,m$ and hence that each $A(G_\alpha)$ has a discrete spectrum with eigenvalues

$$0 = \lambda_{\alpha 0} \le \lambda_{\alpha 1} \le \lambda_{\alpha 2} \le \cdots \tag{6.4}$$

such that

$$\lim_{\ell\to\infty} \lambda_{\alpha\ell} = \infty \tag{6.5}$$

and corresponding eigenfunctions

$$\phi_{\alpha 0}(x^\alpha) = 1/|G_\alpha|^{1/2},\ \phi_{\alpha 1}(x^\alpha),\ \phi_{\alpha 2}(x^\alpha),\cdots \tag{6.6}$$

which form a complete orthonormal sequence in $L_2(G_\alpha)$.

6.2 Solutions of $Aw = \lambda w$ in S_α

Suppose that w is locally in $D(A)$; i.e., $\phi w \in D(A)$ for every $\phi \in \mathcal{D}(R^3)$. Then the completeness of the eigenfunctions (6.6) implies that

$$w(q) = \sum_{\ell=0}^{\infty} w_{\alpha\ell}(y^\alpha)\ \phi_{\alpha\ell}(x^\alpha) \text{ for all } q \in S_\alpha \tag{6.7}$$

where $x^\alpha = x^\alpha(q)$, $y^\alpha = y^\alpha(q)$. Moreover, if

$$Aw = \lambda w \text{ in } S_\alpha \tag{6.8}$$

then the coefficients $w_{\alpha\ell}(y^\alpha)$ will satisfy

$$w''_{\alpha\ell}(y^\alpha) + (\lambda - \lambda_{\alpha\ell})\ w_{\alpha\ell}(y^\alpha) = 0 \text{ for all } y^\alpha > 0 \tag{6.9}$$

In particular, if it is assumed that

$$\lambda \ne \lambda_{\alpha\ell};\ \alpha = 1,\cdots,m;\ \ell = 0,1,2,\cdots \tag{6.10}$$

then

$$w_{\alpha\ell}(y^\alpha) = C^+_{\alpha\ell} \exp\{i\sqrt{\lambda-\lambda_{\alpha\ell}}\ y^\alpha\} + C^-_{\alpha\ell} \exp\{-i\sqrt{\lambda-\lambda_{\alpha\ell}}\ y^\alpha\} \tag{6.11}$$

where, for definiteness, $\mu^{1/2} > 0$ for $\mu > 0$ and

$$\sqrt{\lambda - \lambda_{\alpha\ell}} = \begin{cases} (\lambda - \lambda_{\alpha\ell})^{1/2} & \text{for } \lambda > \lambda_{\alpha\ell} \\ i(\lambda_{\alpha\ell} - \lambda)^{1/2} & \text{for } \lambda < \lambda_{\alpha\ell} \end{cases} \tag{6.12}$$

6.3 Eigenfunctions of A and non-propagating modes

It was discovered by F. Rellich [27] that the operators A for waveguide regions of the form (6.1) may have a point spectrum. A point $\lambda \in R$ is in the point spectrum of A if and only if there is a non-zero function $w \in D(A)$ such that $Aw = \lambda w$. In particular, the requirement that $w \in L_2(\Omega)$ implies that in the expansions (6.7), (6.11) the coefficients $C^+_{\alpha\ell} = C^-_{\alpha\ell} = 0$ for $\lambda > \lambda_{\alpha\ell}$ and $C^-_{\alpha\ell} = 0$ for $\lambda < \lambda_{\alpha\ell}$. Thus any eigenfunction of A must have the form

$$w(q) = \sum_{\{\ell:\lambda<\lambda_{\alpha\ell}\}} C^+_{\alpha\ell} \exp\{-(\lambda_{\alpha\ell} - \lambda)^{1/2} y^\alpha\} \phi_{\alpha\ell}(x^\alpha) \qquad (6.13)$$

for all $q \in S_\alpha$. In particular, the eigenfunctions are exponentially damped in each cylinder S_α.

D. S. Jones [17] has shown that the point spectrum of A is a discrete subset of $(0,\infty)$; i.e., each eigenvalue has finite multiplicity and each finite subinterval of $(0,\infty)$ contains at most a finite number of eigenvalues. Thus if the point spectrum of A is not empty then there exists an M such that $1 \le M \le \infty$ and $\lambda_{(n)}$, $1 \le n < M$, is an enumeration of the eigenvalues of A, each repeated according to its multiplicity. It may be assumed that

$$0 < \lambda_{(n)} \le \lambda_{(n+1)} \text{ for } 1 \le n < n+1 < M \qquad (6.14)$$

The corresponding eigenfunctions will be denoted by $w_{(n)}$. The subspace spanned by $\{w_{(n)}: 1 \le n < M\}$ will be denoted by $\mathcal{H}^p(A)$ and called the subspace of discontinuity of A [18]. Thus

$$\mathcal{H}^p(A) = \{w = \sum_{1\le n<M} c_n w_{(n)}: \sum_{1\le n<M} |c_n|^2 < \infty\} \qquad (6.15)$$

It is known that

$$L_2(\Omega) = \mathcal{H}^p(A) \oplus \mathcal{H}^c(A) \qquad (6.16)$$

where $\mathcal{H}^c(A)$, the orthogonal complement of $\mathcal{H}^p(A)$ in $L_2(\Omega)$, is that largest subspace of $L_2(\Omega)$ on which the spectral measure of A is continuous. $\mathcal{H}^c(A)$ is called the subspace of continuity of A [18]. Moreover, (6.16) is a reducing decomposition for A [18].

If the initial state of an acoustic field in Ω satisfies $u(0,\cdot) = f \in \mathcal{H}^p(A)$ and $\partial u(0,\cdot)/\partial t = g \in \mathcal{H}^p(A)$ then $h = f + iA^{-1/2}g \in \mathcal{H}^p(A)$ and hence

$$v(t,q) = \exp(-itA^{1/2})\, h(q) = \sum_{1 \le n < M} c_{(n)} \exp(-it\lambda_{(n)}^{1/2})\, w_{(n)}(q) \tag{6.17}$$

It follows that the energy of the acoustic field $u(t,q) = \mathrm{Re}\{v(t,q)\}$ in any bounded portion of Ω is an oscillatory function of t. In particular, there is no propagation of energy in the cylinders S_α. For this reason the eigenfunctions $w_{(n)}(q)$ are called non-propagating modes of the waveguide. By contrast, it is shown below that for fields with initial state in $\mathcal{H}^c(A)$ the energy in every bounded portion of Ω tends to zero when $t \to \infty$ and hence all the energy propagates outward in the cylinders S_α.

6.4 Generalized eigenfunctions of A

The operator A has two families of generalized eigenfunctions, analogous to the functions $w_+(x,p)$ and $w_-(x,p)$ of section 4, each of which spans the subspace $\mathcal{H}^c(A)$. The structure and properties of these functions are described next.

Consider a single term in the expansion (6.7) for the cylinder S_α. It has the form (cf. (6.11))

$$w_{\alpha\ell}(q) = \left(C_{\alpha\ell}^{+} \exp\{i\sqrt{\lambda - \lambda_{\alpha\ell}}\; y^\alpha\} + C_{\alpha\ell}^{-} \exp\{-i\sqrt{\lambda - \lambda_{\alpha\ell}}\; y^\alpha\}\right) \phi_{\alpha\ell}(x^\alpha) \tag{6.18}$$

where $q \leftrightarrow (x^\alpha, y^\alpha)$. Assume that $\lambda > \lambda_{\alpha\ell}$, so that (6.18) represents a propagating mode in S_α, and write

$$p = (\lambda - \lambda_{\alpha\ell})^{1/2} > 0 \tag{6.19}$$

and

$$\lambda^{1/2} \equiv \omega_{\alpha\ell}(p) = (p^2 + \lambda_{\alpha\ell})^{1/2} > \lambda_{\alpha\ell}^{1/2} \tag{6.20}$$

If one associates a time-dependence $\exp\{-i\lambda^{1/2}t\} = \exp\{-i\omega_{\alpha\ell}(p)t\}$ with (6.18), as in the spectral representation of $v(t,\cdot) = \exp(-itA^{1/2})h$, then

$$w_{\alpha\ell}(q) \exp\{-i\omega_{\alpha\ell}(p)t\} = C_{\alpha\ell}^{+} \exp\{i(py^\alpha - \omega_{\alpha\ell}(p)t)\}\phi_{\alpha\ell}(x^\alpha) + C_{\alpha\ell}^{-} \exp\{-i(py^\alpha + \omega_{\alpha\ell}(p)t)\}\phi_{\alpha\ell}(x^\alpha) \tag{6.21}$$

is the sum of an outgoing wave in S_α, with coefficient $C^+_{\alpha\ell}$, and an incoming wave with coefficient $C^-_{\alpha\ell}$. For this reason, a solution of (6.8) of the form

$$C^+_{\alpha\ell} \exp (ipy^\alpha) \ \phi_{\alpha\ell}(x^\alpha) \tag{6.22}$$

will be called an "outgoing" wave in S_α in mode ℓ, while a solution of the form

$$C^-_{\alpha\ell} \exp (-ipy^\alpha) \ \phi_{\alpha\ell}(x^\alpha) \tag{6.23}$$

will be called an "incoming" wave in S_α in mode ℓ. Note that this terminology is based on the convention that the time-dependence is $\exp (-i\omega_{\alpha\ell}(p)t)$, as in (6.21). If a time-dependence $\exp (i\omega_{\alpha\rho}(p)t)$ were used it would be necessary to interchange the terms "outgoing" and "incoming."

In the case of the uniform semi-infinite cylinder of section 5, m is equal to 1 and the generalized eigenfunctions have the form

$$\begin{aligned} w_\ell(x,y,p) &= \frac{1}{(2\pi)^{1/2}} \exp (ipy) \ \phi_\ell(x) \\ &+ \frac{1}{(2\pi)^{1/2}} \exp (-ipy) \ \phi_\ell(x) \end{aligned} \tag{6.24}$$

Thus they are the sum of an incoming and an outgoing wave in mode ℓ, with equal amplitudes and phases. This symmetry is due to the symmetry of the waveguide. In the general case of a compound waveguide (6.1) it is possible to prescribe the amplitudes and phases of the incoming (resp., outgoing) waves in each cylinder S_α and mode ℓ. The amplitudes and phases of the outgoing (resp., incoming) waves in each cylinder S_β and mode m are thereby determined. The most useful generalized eigenfunctions are those that have an incoming (resp., outgoing) wave of prescribed amplitude and phase in a single prescribed cylinder S_α and mode ℓ. They may be described as follows.

6.5 Definition

The mode (α,ℓ)-outgoing eigenfunction for Ω is the function $w^+_{\alpha\ell}(q,p)$ defined by the properties

$$w^+_{\alpha\ell}(\cdot,p) \text{ is locally in } D(A) \tag{6.25}$$

$$(A - \omega^2_{\alpha\ell}(p)) \ w^+_{\alpha\ell}(q,p) \equiv -(\Delta + \omega^2_{\alpha\ell}(p))w^+_{\alpha\ell}(q,p) = 0 \tag{6.26}$$

for all $q \in \Omega$ and

$$w^{+}_{\alpha\ell}(q,p) = \frac{\delta_{\alpha\beta}}{(2\pi)^{1/2}} \exp(-ipy^{\alpha})\, \phi_{\alpha\ell}(x^{\alpha}) + \sum_{m=0}^{\infty} C^{+}_{\alpha\ell,\beta m}(p) \exp\{i\sqrt{p^2+\lambda_{\alpha\ell}-\lambda_{\beta m}}\, y^{\beta}\}\phi_{\beta m}(x^{\beta}) \qquad (6.27)$$

for all $q \in S_\beta$ ($\beta = 1,2,\cdots,m$). Similarly, the mode (α,ℓ)-incoming eigenfunction for Ω is the function $w^{-}_{\alpha\ell}(q,p)$ defined by the properties that $w^{-}_{\alpha\ell}(\cdot,p)$ is locally in $D(A)$, $(\Delta + \omega^2_{\alpha\ell}(p))w^{-}_{\alpha\ell}(q,p) = 0$ for all $q \in \Omega$ and, for $q \in S_\beta$ ($\beta = 1,2,\cdots,m$)

$$w^{-}_{\alpha\ell}(q,p) = \frac{\delta_{\alpha\beta}}{(2\pi)^{1/2}} \exp(ipy^{\alpha})\, \phi_{\alpha\ell}(x^{\alpha}) + \sum_{m=0}^{\infty} C^{-}_{\alpha\ell,\beta m}(p) \exp\{-i\sqrt{p^2+\lambda_{\alpha\ell}-\lambda_{\beta m}}\, y^{\beta}\}\phi_{\beta m}(x^{\beta}) \qquad (6.28)$$

where $-i\sqrt{p^2+\lambda_{\alpha\ell}-\lambda_{\beta m}} < 0$ for $\lambda_{\beta m} > \lambda_{\alpha\ell} + p^2$.

The eigenfunction $w^{+}_{\alpha\ell}(q,p)$ may be interpreted physically as the steady-state acoustic field in the waveguide Ω due to a single incoming wave (6.23) in cylinder S_α and mode ℓ, with amplitude and phase defined by $C^{-}_{\alpha\ell}(p) = 1/(2\pi)^{1/2}$, and no incoming waves in the other cylinders or in the other modes of cylinder S_α. The amplitudes and phases of the corresponding outgoing waves are defined by the coefficients $C^{+}_{\alpha\ell,\beta m}(p)$ which are determined by the incident wave and the geometry of Ω. Note that, in general, an incoming wave in mode (α,ℓ) will produce outgoing waves in all the cylinders and modes; i.e., scattering produces coupling among the cylinders and modes.

The form of the exponential which multiplies $\phi_{\beta m}(x^{\beta})$ in (6.27) is determined by the requirement (6.26). Note that the sum in (6.27) includes propagating modes with $\lambda_{\beta m} < \lambda_{\alpha\ell} + p^2$ and modes "beyond cutoff" with $\lambda_{\beta m} > \lambda_{\alpha\ell} + p^2$. The latter decrease exponentially when $y^\beta \to \infty$.

The eigenfunctions $w^{-}_{\alpha\ell}(q,p)$ have an interpretation analogous to that of $w^{+}_{\alpha\ell}(q,p)$, but with "outgoing" and "incoming" interchanged. It is easy to verify from the defining conditions that the two families satisfy the relation

$$w^{-}_{\alpha\ell}(q,p) = \overline{w^{+}_{\alpha\ell}(q,p)} \qquad (6.29)$$

The case of the uniform semi-infinite cylinder is a very special case in which m = 1 (so that no index α is needed) and

$$w_\ell^+(q,p) = w_\ell^-(q,p) = w_\ell(x,y,p), \quad q \Leftrightarrow (x,y) \tag{6.30}$$

(see (6.24)). Moreover, in this case the symmetry implies that there is no coupling between different modes:

$$c_{\ell,m}^{\pm}(p) = \delta_{\ell,m}/(2\pi)^{1/2} \tag{6.31}$$

Existence and uniqueness theorems for the eigenfunctions $w_{\alpha\ell}^{\pm}(q,p)$ were proved in [21]. The following notation will be used to formulate them.

$$Z_{\alpha\ell}(\Omega) = \{p \in R_+ : \; p^2 + \lambda_{\alpha\ell} \in \sigma_p(A)\} \tag{6.32}$$

where $\sigma_p(A) = \{\lambda = \lambda_{(n)} : \; 1 \le n < M\}$. Similarly,

$$Z_{\alpha\ell}(G_\beta) = \{p \in R_+ : \; p^2 + \lambda_{\alpha\ell} \in \sigma(A(G_\beta))\} \tag{6.33}$$

where $\sigma(A(G_\beta)) = \{\lambda = \lambda_{\beta\ell} : \; \ell = 1,2,\cdots\}$. Finally

$$Z_{\alpha\ell} = Z_{\alpha\ell}(\Omega) \cup \bigcup_{\beta=1}^{m} Z_{\alpha\ell}(G_\beta) \tag{6.34}$$

and

$$Z = \bigcup_{\alpha=1}^{m} \bigcup_{\ell=1}^{\infty} Z_{\alpha\ell} \tag{6.35}$$

Note that the information on the spectra of $A(G_\alpha)$ and A given above implies that each of these sets is a denumerable subset of R_+. The results of [21] imply the following theorem.

6.6 Theorem

Let $\Omega \in LC$ be a waveguide domain of the form (6.1). Then for each $p \in R_+ - Z$, each $\alpha = 1,\cdots,m$ and each $\ell = 1,2,\cdots$ the eigenfunctions $w_{\alpha\ell}^+(\cdot,p)$ and $w_{\alpha\ell}^-(\cdot,p)$ exist and are unique.

6.7 The eigenfunction expansion theorem

The families $\{w_{\alpha\ell}^+(\cdot,p) : \; p \in R_+ - Z; \; \alpha = 1,\cdots,m; \; \ell = 0,1,2,\cdots\}$ and $\{w_{\alpha\ell}^-(\cdot,p) : \; p \in R_+ - Z; \; \alpha = 1,\cdots,m; \; \ell = 0,1,2,\cdots\}$ define two complete sets of generalized eigenfunctions for the part of A in the subspace of continuity $\mathcal{H}^c(A)$. The eigenfunction expansions, which are of the Plancherel type described in the preceding sections, may be formulated as follows.

6.8 Theorem

Define

$$S_{\alpha,M} = \{q \in R^3 : x^\alpha(q) \in G_\alpha \text{ and } 0 < y^\alpha(q) < M\} \tag{6.36}$$

and

$$\Omega_M = \Omega_0 \cup S_{1,M} \cup \cdots \cup S_{m,M} \tag{6.37}$$

Then for all $f \in L_2(\Omega)$ the limits

$$\hat{f}_{\alpha\ell}(p) = L_2(R_+)\text{-}\lim_{M\to\infty} \int_{\Omega_M} \overline{w^{\pm}_{\alpha\ell}(q,p)}\, f(q)\, dV_q \tag{6.38}$$

exist, where dV_q is the element of Lebesgue measure in R^3. Moreover, the operators $\Phi^{\pm}_{\alpha\ell}$: $L_2(\Omega) \to L_2(R_+)$ defined by $\Phi^{\pm}_{\alpha\ell} f = \hat{f}^{\pm}_{\alpha\ell}$ have range $L_2(R_+)$ and, if P^c denotes the orthogonal projection of $L_2(\Omega)$ onto $\mathcal{H}^c(A)$ then

$$\| P^c f \|^2_{L_2(S)} = \sum_{\alpha=1}^{m} \sum_{\ell=0}^{\infty} \| \hat{f}^{\pm}_{\alpha\ell} \|^2_{L_2(R_+)} \tag{6.39}$$

for all $f \in L_2(\Omega)$, and

$$P^c f(q) = L_2(\Omega)\text{-}\lim_{M,N\to\infty} \sum_{\alpha=1}^{m} \sum_{\ell=0}^{N} \int_0^M w^{\pm}_{\alpha\ell}(q,p)\, \hat{f}^{\pm}_{\alpha\ell}(p)\, dp \tag{6.40}$$

The relations (6.38) and (6.40) will be written in the symbolic form

$$\hat{f}^{\pm}_{\alpha\ell}(p) = \int_{\Omega} \overline{w^{\pm}_{\alpha\ell}(q,p)}\, f(q)\, dV_q \tag{6.41}$$

$$P^c f(q) = \sum_{\alpha=1}^{m} \sum_{\ell=0}^{\infty} \int_{R_+} w^{\pm}_{\alpha\ell}(q,p)\, \hat{f}^{\pm}_{\alpha\ell}(p)\, dp \tag{6.42}$$

but they must be understood in the sense of (6.38), (6.40). The following corollaries are almost immediate; see [13,21].

6.9 Corollary

For each $f \in L_2(\Omega)$ the limits

$$f^{\pm}_{\alpha\ell}(q) = L_2(\Omega)\text{-}\lim_{M\to\infty} \int_0^M w^{\pm}_{\alpha\ell}(q,p)\, \hat{f}^{\pm}_{\alpha\ell}(p)\,dp \tag{6.43}$$

exist and

$$(f^{\pm}_{\alpha\ell}, f^{\pm}_{\beta m})_{L_2(\Omega)} = 0 \text{ whenever } (\alpha,\ell) \neq (\beta,m) \tag{6.44}$$

Moreover,

$$P^c f = \sum_{\alpha=1}^{m} \sum_{\ell=1}^{\infty} f^{\pm}_{\alpha\ell} \tag{6.45}$$

6.10 Corollary

Define

$$\mathcal{H}^{\pm}_{\alpha\ell} = \{f^{\pm}_{\alpha\ell} \in L_2(\Omega): \quad f \in L_2(\Omega)\} \tag{6.46}$$

Then each $\mathcal{H}^{\pm}_{\alpha\ell}$ is a closed subspace of $\mathcal{H}^c(A)$, $\mathcal{H}^{\pm}_{\alpha\ell}$ and $\mathcal{H}^{\pm}_{\beta m}$ are orthogonal whenever $(\alpha,\ell) \neq (\beta,m)$ and

$$\mathcal{H}^c(A) = \sum_{\alpha=1}^{m} \sum_{\ell=1}^{\infty} \oplus\, \mathcal{H}^{+}_{\alpha\ell} = \sum_{\alpha=1}^{m} \sum_{\ell=1}^{\infty} \oplus\, \mathcal{H}^{-}_{\alpha\ell} \tag{6.47}$$

The eigenfunction expansions (6.40) provide the following construction of the spectral family of A in $\mathcal{H}^c(A)$.

6.11 Theorem

If $\{\Pi(\lambda): \lambda \geq 0\}$ denotes the spectral family of A then

$$\Pi(\lambda)P^c f(q) = \sum_{\alpha=1}^{m} \sum_{\lambda_{\alpha\ell}\leq\lambda} \int_0^{\sqrt{\lambda-\lambda_{\alpha\ell}}} w^{\pm}_{\alpha\ell}(q,p)\, \hat{f}^{\pm}_{\alpha\ell}(p)\,dp \tag{6.48}$$

for all $\lambda \geq 0$. In particular, AP^c is an absolutely continuous operator whose spectrum is $[0,\infty)$.

Note that the sums in (6.48) are actually finite because $\lambda_{\alpha\ell} \to \infty$ when $\ell \to \infty$. (6.48) implies that the eigenfunction expansions (6.40) define spectral representations for A in the sense of the following corollary.

6.12 Corollary

If $\Psi(\lambda)$ is any bounded Lebesgue-measurable function of $\lambda \geq 0$ then for all $f \in \mathcal{H}^c(A) = P^c L_2(\Omega)$

$$\Psi(A)f(q) = \sum_{\alpha=1}^{m} \sum_{\ell=1}^{\infty} \int_{R_+} w^{\pm}_{\alpha\ell}(q,p)\ \Psi(p^2 + \lambda_{\alpha\ell})\hat{f}^{\pm}_{\alpha\ell}(p)dp \quad (6.49)$$

It follows from (6.47) and (6.49) that the eigenfunction expansions (6.40) define reducing decompositions of $\mathcal{H}^c(A)$. More precisely, the following generalization of the results of section 5 is valid [13,21].

6.13 Corollary

The operator $P^{\pm}_{\alpha\ell}$ defined by $P^{\pm}_{\alpha\ell}f = f^{\pm}_{\alpha\ell}$ is an orthogonal projections of $L_2(\Omega)$ onto $\mathcal{H}^{\pm}_{\alpha\ell}$ and

$$P^c = \sum_{\alpha=1}^{m} \sum_{\ell=1}^{\infty} P^{\pm}_{\alpha\ell} \quad (6.50)$$

Moreover,

$$P^{\pm}_{\alpha\ell}\Pi(\lambda) = \Pi(\lambda)P^{\pm}_{\alpha\ell} \text{ for all } \lambda \geq 0 \quad (6.51)$$

and hence (6.47) defines reducing decompositions for AP^c.

The surjectivity of $\Phi^{\pm}_{\alpha\ell}$: $L_2(\Omega) \to L_2(R_+)$, the definition of $P^{\pm}_{\alpha\ell}$ and (6.43) imply that for all $\alpha = 1,\cdots,m$ and $\ell = 1,2,\cdots$

$$P^{\pm}_{\alpha\ell} = \Phi^{\pm *}_{\alpha\ell}\ \Phi^{\pm}_{\alpha\ell}, \quad \Phi^{\pm}_{\alpha\ell}\ \Phi^{\pm *}_{\alpha\ell} = 1 \quad (6.52)$$

In particular the eigenfunction mappings $\Phi^{\pm}_{\alpha\ell}$ are partial isometries [18] with initial sets $\mathcal{H}^{\pm}_{\alpha\ell}$ and final sets $L_2(R_+)$.

6.14 Solution in $\mathcal{H}^c(A)$ of the propagation problem

Only the case where $f \in L_2(\Omega)$ and $g \in D(A^{-1/2})$ will be discussed here. For such initial states it follows, just as in sections 3, 4 and 5, that the solution in $L_2(\Omega)$ of (4.1) - (4.3) has the form

$$u(t,q) = \text{Re}\,\{v(t,q)\} \tag{6.53}$$

where

$$v(t,\cdot) = \exp\,(-itA(\Omega)^{1/2})h, \quad h = f + iA(\Omega)^{-1/2}g \in L_2(\Omega) \tag{6.54}$$

Moreover, the case where $h \in \mathcal{H}^p(A)$ was discussed above. Hence, only the case where $h \in \mathcal{H}^c(A)$ remains to be analyzed. In this case $v(t,\cdot) \in \mathcal{H}^c(A)$ for all $t \in R$ and (6.47) implies that $v(t,q)$ has the decompositions

$$v(t,q) = \sum_{\alpha=1}^{m} \sum_{\ell=1}^{\infty} v_{\alpha\ell}^{\pm}(t,q) \text{ in } \mathcal{H}^c(A) \tag{6.55}$$

where

$$\begin{aligned} v_{\alpha\ell}^{\pm}(t,q) &= \exp\,(-itA(\Omega)^{1/2})h_{\alpha\ell}^{\pm} \\ &= \int_{R_+} w_{\alpha\ell}^{\pm}(q,p)\,\exp\,(-it\omega_{\alpha\ell}(p))\,\hat{h}_{\alpha\ell}^{\pm}(p)dp \end{aligned} \tag{6.56}$$

and $\omega_{\alpha\ell}(p)$ is given by (6.20). The two decompositions defined by (6.55), (6.56) will be called modal decompositions, in analogy with the simple case of section 5, and the partial wave $v_{\alpha\ell}^{\pm}(t,q)$ will be said to be in mode $(\pm,\alpha,\ell)$ of the compound waveguide Ω. Note that for the uniform semi-infinite cylinder of section 5 the $(+,\ell)$ and $(-,\ell)$ modes coincide (see (6.30)).

6.15 Transiency of waves in $\mathcal{H}^c(A)$

The absolute continuity of the operator A in the subspace $\mathcal{H}^c(A)$ implies that all waves in $\mathcal{H}^c(A)$ are transient in the sense of the following theorem [38,42].

6.16 Theorem

If $\Omega \in LC$ is a waveguide domain (6.1) then for every $h \in \mathcal{H}^c(A)$ and every compact set $K \subset R^3$

$$\lim_{t\to\infty} \| \exp(-itA^{1/2})h \|_{L_2(K\cap\Omega)} = 0 \tag{6.57}$$

Thus the decomposition $L_2(\Omega) = \mathcal{H}^p(A) \oplus \mathcal{H}^c(A)$ splits every $h \in L_2(\Omega)$ into a sum of a non-propagating and a propagating state. In particular the partial waves

$$v^{\pm}_{\alpha\ell}(t,\cdot) = \exp(-itA^{1/2})\, P^{\pm}_{\alpha\ell} h \in \mathcal{H}^c(A) \tag{6.58}$$

and hence (6.57) with $K = \overline{\Omega_0}$ implies

$$\lim_{t\to\infty} \| v^{\pm}_{\alpha\ell}(t,\cdot) \|_{L_2(\Omega_0)} = 0 \tag{6.59}$$

for $\alpha = 1,2,\cdots,m$ and $\ell = 0,1,2,\cdots$. Thus waves in $\mathcal{H}^c(A)$ ultimately propagate into the cylinders S_α. The eigenfunction expansion for A will now be used to calculate the asymptotic form of these waves.

6.17 Asymptotic wave functions

Let $h \in \mathcal{H}^c(A)$ and consider the representation

$$v(t,\cdot) = \exp(-itA^{1/2})h = \sum_{\alpha=1}^{m} \sum_{\ell=0}^{\infty} v^{-}_{\alpha\ell}(t,\cdot) \tag{6.60}$$

defined by the incoming eigenfunctions $w^{-}_{\alpha\ell}(q,p)$. Substituting the development (6.28) for $w^{-}_{\alpha\ell}(q,p)$ in S_β into the integral (6.56) for $v^{-}_{\alpha\ell}(t,q)$ gives the representation

$$v^{-}_{\alpha\ell}(t,q) = \delta_{\alpha\beta} v(t,y^\alpha,\lambda_{\alpha\ell},\hat{h}^{-}_{\alpha\ell})\phi_{\alpha\ell}(x^\alpha) + v'_{\alpha\ell}(t,q), \quad q \in S_\beta \tag{6.61}$$

where $v(t,y,\lambda,h)$ is defined by (5.43) and

$$v'_{\alpha\ell}(t,q) = \sum_{m=0}^{\infty} v'_{\alpha\ell,\beta m}(t,y^\beta)\, \phi_{\beta m}(x^\beta), \quad q \in S_\beta \tag{6.62}$$

with

$$v'_{\alpha\ell,\beta m}(t,y) = \int_0^{\sqrt{\lambda_{\beta m}-\lambda_{\alpha\ell}}} \exp\left(-\sqrt{\lambda_{\beta m}-\lambda_{\alpha\ell}-p^2}\ y\right) \times \exp\left(-it\omega_{\alpha\ell}(p)\right) C^-_{\alpha\ell,\beta m}(p)\hat{h}^-_{\alpha\ell}(p)dp$$
$$+ \int_{\sqrt{\lambda_{\beta m}-\lambda_{\alpha\ell}}}^{\infty} \exp\left\{-i\left(\sqrt{p^2+\lambda_{\alpha\ell}-\lambda_{\beta m}}\ y + t\omega_{\alpha\ell}(p)\right)\right\} \times C^-_{\alpha\ell,\beta m}(p)\ \hat{h}^-_{\alpha\ell}(p)dp \tag{6.63}$$

This equation defines $v'_{\alpha\ell,\beta m}$ for the case where $\lambda_{\beta m} > \lambda_{\alpha\ell}$. In the case where $\lambda_{\beta m} \le \lambda_{\alpha\ell}$ the first integral is absent and the second has the lower limit zero. The method of stationary phase may be applied to show that $v'_{\alpha\ell,\beta m}(t,y)$ satisfies an estimate of the form (5.55) for $y \ge 0$, $t \ge 0$ because the integrals in (6.63) have no points of stationary phase in this region. It follows that

$$\lim_{t\to\infty} \| v'_{\alpha\ell}(t,\cdot)\|_{L_2(S_\beta)} = 0, \quad \beta = 1,2,\cdots,m \tag{6.64}$$

The proof may be based on a convergence lemma like that of section 3. The details will not be given here. See [22] for a more complete discussion.

In (6.61) the term $v(t,y^\alpha,\lambda_{\alpha\ell},\hat{h}^-_{\alpha\ell})$ has the form (5.43) studied in section 5. Hence, if $v^\infty(t,y,\lambda,h)$ is defined as in that section and

$$v^\infty_{\alpha\ell}(t,q) = v^\infty(t,y^\alpha,\lambda_{\alpha\ell},\hat{h}^-_{\alpha\ell})\ \phi_{\alpha\ell}(x^\alpha), \quad q \in S_\alpha \tag{6.65}$$

then (6.64) and the results of section 5 imply

$$\lim_{t\to\infty} \| v^-_{\alpha\ell}(t,\cdot) - \delta_{\alpha\beta} v^\infty_{\alpha\ell}(t,\cdot)\|_{L_2(S_\beta)} = 0, \quad \beta = 1,2,\cdots,m \tag{6.66}$$

In particular, $v^-_{\alpha\ell}(t,\cdot) \to 0$ in $L_2(S_\beta)$ for $t \to \infty$ and all $\beta \ne \alpha$; i.e., $v^-_{\alpha\ell}(t,\cdot)$ is asymptotically concentrated entirely in S_α.

The asymptotic wave functions for $v(t,q)$ will be defined by

$$v^{\infty}(t,q) = \sum_{\alpha=1}^{m} \sum_{\ell=0}^{\infty} \chi_{\alpha}(q)\, v_{\alpha\ell}^{\infty}(t,q), \quad q \in \Omega \tag{6.67}$$

where $\chi_{\alpha}(q)$ denotes the characteristic function of S_{α}. Note that the terms in this sum are orthogonal in $L_2(\Omega)$ by (6.65). The decompositions (6.60), (6.67) and the convergence results (6.59) and (6.66) imply the

6.18 Theorem

If $\Omega \in LC$ is a waveguide domain (6.1) then for every $h \in \mathcal{H}^c(A)$ the wave function $v(t,\cdot) = \exp(-itA^{1/2})h$ satisfies

$$\lim_{t\to\infty} \| v(t,\ ) - v^{\infty}(t,\cdot)\|_{L_2(\Omega)} = 0 \tag{6.68}$$

The proof is essentially the same as for the special case described in section 5. The convergence in energy, when h has finite energy, can be proved by the same methods.

7. PROPAGATION IN PLANE STRATIFIED FLUIDS

The propagation of localized acoustic waves in a plane stratified fluid which fills a half-space is analyzed in this section. The asymptotic wave functions for such media are shown to be the sum of an asymptotic free (hemispherical) wave and an asymptotic guided wave which propagates parallel to the boundary. This structure, which is intermediate between that of a homogeneous fluid and that of a tubular waveguide, is called an open waveguide in the physical literature.

7.1 Plane stratified fluids

An inhomogeneous fluid will be said to be plane stratified if the local sound speed $c(x)$ and density $\rho(x)$ are functions of a single Cartesian coordinate. This condition can be written

$$\left.\begin{aligned} c(x_1,x_2,x_3) &= c(x_3) \\ \rho(x_1,x_2,x_3) &= \rho(x_3) \end{aligned}\right\} \tag{7.1}$$

with a suitable numbering of the coordinates. It will be convenient to denote x_3 by a single letter and write

$$x = (x_1, x_2) \in R^2, \ y = x_3 \in R, \ (x,y) \in R^3 \tag{7.2}$$

This notation is used in the remainder of this section.

7.2 Propagation in a stratified fluid with a free surface

A stratified fluid filling a half-space

$$R_+^3 = \{(x,y): \ x \in R^2 \text{ and } y > 0\} \tag{7.3}$$

is often used as a model in the study of acoustic wave propagation in oceans and deep lakes. If the surface $\{(x,0): \ x \in R^2\}$ is free the corresponding initial-boundary value problem is (see section 2)

$$\frac{\partial^2 u}{\partial t^2} - c^2(y)\rho(y) \frac{\partial}{\partial x_j} \left(\frac{1}{\rho(y)} \frac{\partial u}{\partial x_j}\right) = 0 \text{ for } t > 0, \quad (x,y) \in R_+^3 \tag{7.4}$$

$$u(t,x,0) = 0 \text{ for } t \geq 0, \ x \in R^2 \tag{7.5}$$

$$u(0,x,y) = f(x,y) \text{ and } \partial u(0,x,y)/\partial t = g(x,y) \quad \text{for } (x,y) \in R_+^3 \tag{7.6}$$

where in (7.4) j is summed from 1 to 3 and $x_3 = y$. $c(y)$ and $\rho(y)$ are assumed to be Lebesgue measurable on $R_+ = \{y: \ y > 0\}$ and to satisfy

$$\left.\begin{array}{l} 0 < c_1 \leq c(y) \leq c_2 < \infty \\ 0 < \rho_1 \leq \rho(y) \leq \rho_2 < \infty \end{array}\right\} \text{ for all } y \in R_+ \tag{7.7}$$

where c_1, c_2, ρ_1 and ρ_2 are suitable constants.

7.3 Hilbert space formulation

The operator

$$Au = -c^2(y)\rho(y) \frac{\partial}{\partial x_j} \left(\frac{1}{\rho(y)} \frac{\partial u}{\partial x_j}\right) \tag{7.8}$$

was shown in section 2 to be formally selfadjoint with respect to the inner product

$$(u,v) = \int_{R_+^3} \overline{u(x,y)}\, v(x,y)\, c^{-2}(y)\, \rho^{-1}(y)\,dxdy \tag{7.9}$$

where dxdy denotes integration with respect to Lebesgue measure on R_+^3. The corresponding Hilbert space is

$$\mathcal{H} = L_2(R_+^3, c^{-2}(y)\, \rho^{-1}(y)\,dxdy) \tag{7.10}$$

The solution of the initial-boundary value problem (7.4) - (7.6) given below is based on a selfadjoint realization of A in $\mathcal{H}$. To define it let $\mathcal{D}(R_+^3)$ denote the Schwartz space of R_+^3 and $\mathcal{D}'(R_+^3)$ the dual space of all distributions on R_+^3. The Lebesgue space $L_2(R_+^3)$ can be regarded as a linear subspace of $\mathcal{D}'(R_+^3)$. Note that $L_2(R_+^3)$ and $\mathcal{H}$ are equivalent Hilbert spaces by (7.7). Let

$$L_2^1(R_+^3) = L_2(R_+^3) \cap \{u: \ \partial u/\partial x_j \in L_2(R_+^3),\ j = 1,2,3\} \tag{7.11}$$

denote the first Sobolev space of R_+^3. It is a Hilbert space with inner product

$$(u,v)_1 = (u,v)_0 + \sum_{j=1}^{3} (\partial u/\partial x_j, \partial v/\partial x_j)_0 \tag{7.12}$$

where $(u,v)_0$ is the inner product in $L_2(R_+^3)$. $\mathcal{D}(R_+^3)$ defines a linear subset of $L_2^1(R_+^3)$ and hence

$$L_2^{1,0}(R_+^3) = \text{closure of } \mathcal{D}(R_+^3) \text{ in } L_2^1(R_+^3) \tag{7.13}$$

is a closed linear subspace of $L_2^1(R_+^3)$. It is known that all the functions in $L_2^{1,0}(R_+^3)$ satisfy the Dirichlet boundary condition (7.5) as elements of $L_2(R^2)$; see [19] and [43,Cor. 2.7].

A realization of the operator A in $\mathcal{H}$ will be defined by

$$D(A) = L_2^{1,0}(R_+^3) \cap \{u: \ Au \in \mathcal{H}\} \tag{7.14}$$

and

$$Au = Au \text{ for all } u \in D(A) \tag{7.15}$$

To interpret the condition $Au \in \mathcal{H}$ in (7.14) note that if $u \in L_2^1(R_+^3)$ then $\rho^{-1}(y)\partial u/\partial x_j \in L_2(R_+^3)$ for j = 1,2,3. The second derivative in (7.8) may therefore be interpreted in the sense of $\mathcal{D}'(R_+^3)$. Thus $Au \in \mathcal{D}'(R_+^3)$ and the condition $Au \in \mathcal{H}$ is meaningful.

The selfadjointness of A in $\mathcal{H}$ may be proved by the method of [43,§2]. Another proof may be based on the theory of sesquilinear forms in Hilbert space [18,Ch. 6]. These methods imply the

7.4 Theorem

A is a selfadjoint real positive operator in $\mathcal{H}$. Every $u \in D(A)$ satisfies the Dirichlet boundary condition (7.5) as an element of $L_2(R^2)$. Moreover, $D(A^{1/2}) = L_2^{1,0}(R_+^3)$ and

$$\|A^{1/2}u\|^2 = \sum_{j=1}^{3} \int_{R_+^3} |\partial u/\partial x_j|^2 \rho^{-1}(y)\,dxdy \quad \text{for all } u \in D(A^{1/2}) \tag{7.16}$$

The operator A may be used to construct "solutions in $\mathcal{H}$" and "solutions with finite energy" of (7.4) - (7.6), as described in section 2. The detailed analysis of the structure of these solutions will again depend on the construction of an eigenfunction expansion for A. For simplicity, the construction will be described here for a special choice of the functions c(y) and ρ(y). Nevertheless, the results obtained are typical of a large class of stratified fluids.

7.5 The Pekeris model

This name will be used for the stratified fluid defined by

$$c(y) = \begin{cases} c_1, & 0 \le y < h \\ c_2, & y \ge h \end{cases} \tag{7.17}$$

$$\rho(y) = \begin{cases} \rho_1, & 0 \le y < h \\ \rho_2, & y \ge h \end{cases} \tag{7.18}$$

where c_1, c_2, ρ_1, ρ_2 and h are positive constants. This model was used by C. L. Pekeris [26] in his study of acoustic wave propagation in shallow water. The model represents a layer of water with depth h, sound speed c_1 and density ρ_1 which overlays a bottom, such as sand or mud, with sound speed c_2 and density ρ_2.

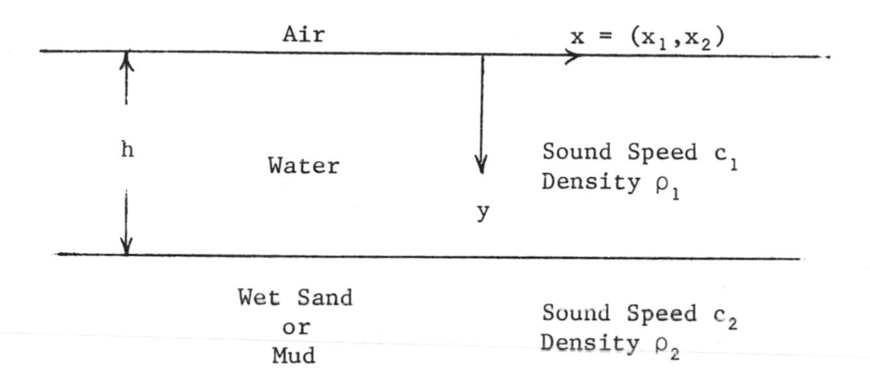

Figure 7. The Pekeris Model

The most interesting case occurs when

$$c_1 < c_2 \tag{7.19}$$

and this condition is assumed to be satisfied in what follows. A detailed study of the Pekeris operator was given by the author in [43]. Here the main results of [43] are reviewed and used to derive the asymptotic wave functions for the Pekeris model.

7.6 Eigenfunctions of A

It was shown in [43] that A has a pure continuous spectrum and a complete family of generalized eigenfunctions was constructed. These functions $w(x,y)$ are characterized by the following properties

$$w \text{ is locally in } D(A) \tag{7.20}$$

$$Aw = \lambda w \text{ for some } \lambda \geq 0 \tag{7.21}$$

$$w(x,y) \text{ is bounded in } R^3_+ \tag{7.22}$$

$$w(x,y) = (2\pi)^{-1} e^{ip\cdot x} w(y), \; p \in R^2 \tag{7.23}$$

where, in (7.23), $w(y)$ is independent of x. The eigenfunctions are of two types, called free wave eigenfunctions and guided wave eigenfunctions. Their definitions and physical interpretations follow.

7.7 Free wave eigenfunctions

These functions exist when the eigenvalue λ satisfies

$$\lambda > c_2^2|p|^2 > c_1^2|p|^2, \quad |p|^2 = p_1^2 + p_2^2 \tag{7.24}$$

To define them let

$$\xi = (\lambda/c_2^2 - |p|^2)^{1/2} > 0, \quad \eta = (\lambda/c_1^2 - |p|^2)^{1/2} > 0 \tag{7.25}$$

and

$$w_0(x,y,p,\lambda) = (2\pi)^{-1} e^{ip\cdot x} w_0(y,p,\lambda) \tag{7.26}$$

where

$$w_0(y,p,\lambda) = a(p,\lambda) \begin{cases} \sin \eta y & , 0 < y < h \\ \gamma_+(\xi,\eta)\, e^{i\xi(y-h)} + \gamma_-(\xi,\eta) e^{-i\xi(y-h)}, & y > h \end{cases} \tag{7.27}$$

with

$$\gamma_\pm(\xi,\eta) = \frac{1}{2}\left(\sin \eta h \mp \frac{\rho_2}{\rho_1}\frac{i\eta}{\xi} \cos \eta h\right) \tag{7.28}$$

In (7.27), $a(p,\lambda)$ is a positive normalizing constant. It was shown in [43] that the eigenfunction expansion takes its simplest form when

$$a(p,\lambda) = \rho_2^{1/2}/2(\pi\xi)^{1/2}|\gamma_+(\xi,\eta)| \tag{7.29}$$

In physical terms, the eigenfunction $w_0(x,y,p,\lambda)$ represents an acoustic field with time dependence $\exp(-it\lambda^{1/2})$ which is the sum of two plane waves in each layer. It may be interpreted as a plane wave which propagates in the region $y > h$, is refracted at $y = h$, reflected at $y = 0$ and refracted again at $y = h$; see Figure 8 where the propagation directions are indicated. It can be verified that Snell's law of refraction is satisfied at $y = h$ and the law of reflection is satisfied at $y = 0$.

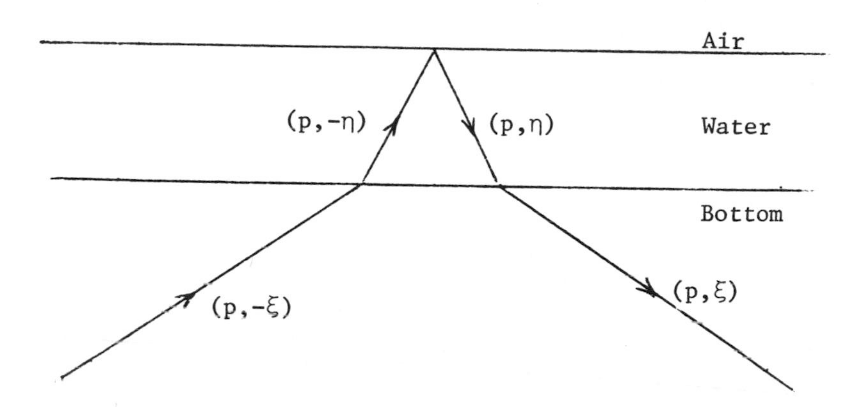

Figure 8. Ray diagram for free wave eigenfunction

7.8 The dispersion relation

For values of λ which satisfy

$$c_1^2|p|^2 < \lambda < c_2^2|p|^2 \tag{7.30}$$

the function $w_0(x,y,p,\lambda)$ defined by (7.25) - (7.28) still satisfies conditions (7.20), (7.21) and (7.23). However, (7.30) implies that ξ is pure imaginary, say

$$\xi = i\xi', \; \xi' = (|p|^2 - \lambda/c_2^2)^{1/2} > 0 \tag{7.31}$$

while η is still real and positive. It follows that $w_0(x,y,p,\lambda)$ satisfies the boundedness condition (7.22) if and only if

$$\gamma_-(i\xi',\eta) = 0 \tag{7.32}$$

or, by (7.28),

$$\xi' = -\frac{\rho_2}{\rho_1}\eta \operatorname{ctn} \eta h \tag{7.33}$$

For λ and $|p|$ which satisfy (7.30), (7.33) is equivalent to the sequence of equations

$$h\eta = (k - \tfrac{1}{2})\pi + \tan^{-1}\left(\frac{\rho_1\xi'}{\rho_2\eta}\right), \quad k = 1,2,\cdots \tag{7.34}$$

where $|\tan^{-1}\alpha| < \pi/2$. Each equation (7.34) defines a functional relation between $|p|$ and λ or, equivalently, between $|p|$ and

$$\omega = \lambda^{1/2} \tag{7.35}$$

The solutions, which will be denoted by

$$\lambda = \lambda_k(|p|), \; \omega = \omega_k(|p|) = \lambda_k(|p|)^{1/2} \tag{7.36}$$

represent a relation between the wave number $|p|$ of the plane waves in $w_0(x,y,p,\lambda)$ and the corresponding frequencies ω. Such relations are called dispersion relations in the theory of wave motion. The relations (7.34), (7.36) were analyzed in [43] and found to have the following properties.

7.9 Properties of $\omega_k(|p|)$

For each $k = 1,2,3,\cdots$ define

$$p_1 = \pi c_1/2h(c_2^2 - c_1^2)^{1/2}, \quad p_k = (2k-1)p_1 \tag{7.37}$$

Then

$$\omega_k(|p|) \text{ is analytic and } \omega_k'(|p|) > 0 \text{ for } |p| \geq p_k \tag{7.38}$$

$$c_1|p| \leq \omega_k(|p|) \leq c_2|p| \text{ for } |p| \geq p_k \tag{7.39}$$

$$\omega_k(p_k) = c_2 p_k, \; \omega_k'(p_k) = c_2 \tag{7.40}$$

$$\omega_k(|p|) \sim c_1|p| \text{ for } |p| \to \infty \tag{7.41}$$

Moreover, an explicit parametric representation of the dispersion curves (7.36) was given in [43].

7.10 Guided wave eigenfunctions

The functions $w_k(x,y,p) = w_0(x,y,p,\lambda_k(|p|))$ are, by construction, the solutions of (7.20) - (7.23) for eigenvalues which satisfy (7.30). It was shown in [43] that there are no solutions of (7.20) - (7.23) when $\lambda < c_1^2|p|^2$. The functions $w_k(x,y,p)$ have the form

$$w_k(x,y,p) = (2\pi)^{-1} e^{ip\cdot x} w_k(y,p) \tag{7.42}$$

where

$$w_k(y,p) = a_k(p) \begin{cases} \sin \eta_k(|p|)y & , 0 < y < h \\ \sin \eta_k(|p|)h \, e^{-\xi_k'(|p|)(y-h)} & , \ y > h \end{cases} \tag{7.43}$$

with

$$\left.\begin{aligned} \eta_k(|p|) &= (\lambda_k(|p|)/c_1^2 - |p|^2)^{1/2}, \\ \xi_k'(|p|) &= (|p|^2 - \lambda_k(|p|)/c_2^2)^{1/2} \end{aligned}\right\} \tag{7.44}$$

In (7.43), $a_k(p)$ is a positive constant which is determined by the condition

$$\int_0^\infty |w_k(y,p)|^2 \, c^{-2}(y) \, \rho^{-1}(y)dy = 1 \tag{7.45}$$

In physical terms, the eigenfunction $w_k(x,y,p)$ represents an acoustic field with time dependence $\exp(-it\omega_k(|p|))$ which corresponds to a plane wave which is trapped in the layer $0 \leq y \leq h$ by reflection at $y = 0$ and total internal reflection at the interface $y = h$. In the layer $y > h$ the field is exponentially damped in the y-direction and propagates strictly in the horizontal direction p; see Figure 9 where the propagation directions are indicated.

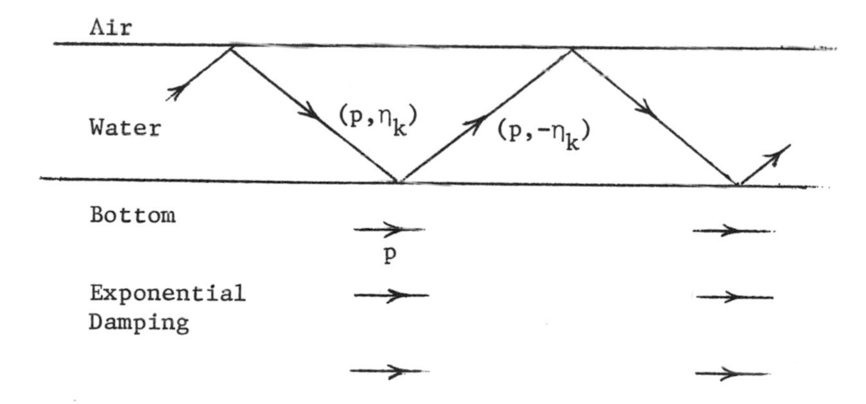

Figure 9. Ray diagram for guided wave eigenfunctions

7.11 The eigenfunction expansion

The free wave eigenfunctions $w_0(x,y,p,\lambda)$ are parameterized by the region

$$\Omega_0 = \{(p,\lambda): \quad p \in R^2 \text{ and } c_2^2|p|^2 < \lambda\} \subset R^3 \tag{7.46}$$

Similarly, the guided wave eigenfunctions $w_k(x,y,p)$ are parameterized by the regions

$$\Omega_k = \{p: \quad |p| > p_k\}, \ k = 1,2,\cdots \tag{7.47}$$

The following expansion theorem was proved in [43]. First of all, for every $f \in \mathcal{H}$ the limits

$$\hat{f}_0(p,\lambda) = L_2(\Omega_0)\text{-}\lim_{M\to\infty} \int_0^M \int_{|x|\le M} \overline{w_0(x,y,p,\lambda)}\, f(x,y)c^{-2}(y)\rho^{-1}(y)\,dxdy \tag{7.48}$$

and

$$\hat{f}_k(p) = L_2(\Omega_k)\text{-}\lim_{M\to\infty} \int_0^M \int_{|x|\le M} \overline{w_k(x,y,p)}\, f(x,y)c^{-2}(y)\rho^{-1}(y)\,dxdy, \quad k = 1,2,\cdots \tag{7.49}$$

exist and satisfy the Parseval relation

$$\|f\|_{\mathcal{H}}^2 = \sum_{k=0}^{\infty} \|\hat{f}_k\|_{L_2(\Omega_k)}^2 \tag{7.50}$$

Moreover, if

$$\left.\begin{aligned} \Omega_0^M &= \{(p,\lambda): \quad p \in R^2 \text{ and } c_2^2|p|^2 < \lambda < M\} \\ \Omega_k^M &= \{p: \quad p_k < |p| < M\}, \ k = 1,2,\cdots \end{aligned}\right\} \tag{7.51}$$

then the limits

$$f_0(x,y) = \mathcal{H}\text{-}\lim_{M\to\infty} \int_{\Omega_0^M} w_0(x,y,p,\lambda)\, \hat{f}_0(p,\lambda)\,dp\,d\lambda \tag{7.52}$$

and

$$f_k(x,y) = \mathcal{H}\text{-}\lim_{M\to\infty} \int_{\Omega_k^M} w_k(x,y,p)\, \hat{f}_k(p)\,dp,\ k = 1,2,\cdots \tag{7.53}$$

exists and satisfy

$$f(x,y) = \mathcal{H}\text{-}\lim_{M\to\infty} \sum_{k=0}^{M} f_k(x,y) \tag{7.54}$$

The relations (7.48), (7.49), (7.52), (7.53) and (7.54) will also be written in the following more concise symbolic forms, in analogy with the notation of previous sections.

$$\hat{f}_0(p,\lambda) = \int_{R_+^3} \overline{w_0(x,y,p,\lambda)}\, f(x,y)\, c^{-2}(y)\, \rho^{-1}(y)\,dx\,dy \tag{7.55}$$

$$\hat{f}_k(p) = \int_{R_+^3} \overline{w_k(x,y,p)}\, f(x,y)\, c^{-2}(y)\, \rho^{-1}(y)\,dx\,dy, \quad k = 1,2,\cdots \tag{7.56}$$

$$f_0(x,y) = \int_{\Omega_0} w_0(x,y,p,\lambda)\, \hat{f}_0(p,\lambda)\,dp\,d\lambda \tag{7.57}$$

$$f_k(x,y) = \int_{\Omega_k} w_k(x,y,p)\, \hat{f}_k(p)\,dp,\ k = 1,2,\cdots \tag{7.58}$$

$$f(x,y) = \sum_{k=0}^{\infty} f_k(x,y) \tag{7.59}$$

Equations (7.55) - (7.59) are the eigenfunction expansion for A and show the completeness of the generalized eigenfunctions defined above. The representation is a spectral representation for A in the sense that, for every $f \in D(A)$,

$$(Af)\hat{}_0(p,\lambda) = \lambda\, \hat{f}_0(p,\lambda) \tag{7.60}$$

and

$$(Af)\hat{}_k(p) = \lambda_k(|p|)\ \hat{f}_k(p), \quad k = 1,2,\cdots \tag{7.61}$$

The representation (7.55) - (7.59) defines a modal decomposition for the Pekeris model. It was shown in [43] that if

$$\mathcal{H}_k = \{f_k: \ f \in \mathcal{H}\} \subset \mathcal{H}, \quad k = 0,1,2,\cdots \tag{7.62}$$

then each $\mathcal{H}_k$ is a closed subspace, $\mathcal{H}_k$ and $\mathcal{H}_\ell$ are orthogonal for $k \neq \ell$ and

$$\mathcal{H} = \sum_{k=0}^{\infty} \oplus \ \mathcal{H}_k \tag{7.63}$$

Moreover, it was shown that (7.60), (7.62) imply that (7.63) reduces A. In fact, more was shown in [43]; namely that

$$\Phi_k f = \hat{f}_k \in L_2(\Omega_k), \quad k = 0,1,2,\cdots \tag{7.64}$$

defines an operator

$$\Phi_k: \ \mathcal{H} \to L_2(\Omega_k), \quad k = 0,1,2,\cdots \tag{7.65}$$

which is a partial isometry with initial set $\mathcal{H}_k$ and final set $L_2(\Omega_k)$; i.e.,

$$\Phi_k^* \ \Phi_k = P_k, \ \Phi_k \ \Phi_k^* = 1, \quad k = 0,1,2,\cdots \tag{7.66}$$

where P_k is the orthogonal projection of $\mathcal{H}$ onto $\mathcal{H}_k$.

7.12 Solution in $\mathcal{H}$ of the propagation problem

Attention will again be restricted to the case where $f \in \mathcal{H}$ and $g \in D(A^{-1/2})$ so that the solution in $\mathcal{H}$ has the form

$$u(t,x,y) = \text{Re}\ \{v(t,x,y)\} \tag{7.67}$$

with

$$v(t,\cdot,\cdot) = \exp\ (-itA^{1/2})h, \quad h = f + iA^{-1/2}g \in \mathcal{H} \tag{7.68}$$

The modal decomposition of v(t,x,y) is

$$v(t,x,y) = \sum_{k=0}^{\infty} v_k(t,x,y) \tag{7.69}$$

where

$$v_0(t,x,y) = \int_{\Omega_0} w_0(x,y,p,\lambda) \exp(-it\lambda^{1/2}) \hat{h}_0(p,\lambda)\, dp\, d\lambda \tag{7.70}$$

and

$$v_k(t,x,y) = \int_{\Omega_k} w_k(x,y,p) \exp(-it\omega_k(|p|))\hat{h}_k(p)\, dp, \quad k = 1,2,\cdots \tag{7.71}$$

Moreover, the modal waves $v_k(t,x,y)$ are independent in the sense that they are orthogonal in $\mathcal{H}$ for every $t \in R$ because (7.63) is a reducing decomposition of A. Asymptotic wave functions for each mode will now be calculated beginning with the guided modes v_k, $k \geq 1$.

7.13 Asymptotic wave functions for the guided modes ($k \geq 1$)

If the representation (7.42) for the eigenfunctions $w_k(x,y,p)$ is substituted into (7.71) the spectral integrals takes the form

$$v_k(t,x,y) = \frac{1}{2\pi} \int_{\Omega_k} \exp\{i(x\cdot p - t\omega_k(|p|))\} w_k(y,p)\hat{h}_k(p)\, dp, \quad k = 1,2,\cdots \tag{7.72}$$

where $w_k(y,p)$ is defined by (7.43). The behavior for large t of these integrals will be calculated by the method of stationary phase. In the present case the integral is a double integral ($\Omega_k \subset R^2$) and the phase function

$$\theta_k(p,x,t) = x \cdot p - t\omega_k(|p|) \tag{7.73}$$

is stationary with respect to p if and only if

$$\frac{\partial\theta_k(p,x,t)}{\partial p_j} = x_j - t\omega_k'(|p|)\frac{p_j}{|p|} = 0, \quad j = 1,2 \tag{7.74}$$

In particular, the number and distribution of the stationary points is determined by the group speed function for the kth guided mode:

$$U_k(|p|) = \omega_k'(|p|), \quad |p| \geq p_k \tag{7.75}$$

The defining relation (7.34) for $\omega_k(|p|)$ implies the following

7.14 Properties of $U_k(|p|)$

For each $k = 1,2,3,\cdots$ there exists a unique $p_k^A \geq p_k$ where $U_k'(p_k^A) = 0$ and $p_k^A > p_k$. Moreover,

$$0 < U_k^A \equiv U_k(p_k^A) \leq U_k(|p|) \leq c_2 \text{ for all } |p| \geq p_k \tag{7.76}$$

$$U_k'(|p|) < 0 \text{ for } p_k \leq |p| < p_k^A \text{ and } U_k'(|p|) > 0 \text{ for } |p| > p_k^A \tag{7.77}$$

$$\lim_{|p|\to p_k} U_k(|p|) = c_2, \quad \lim_{|p|\to\infty} U_k(|p|) = c_1 \tag{7.78}$$

These properties are indicated in Figure 10.

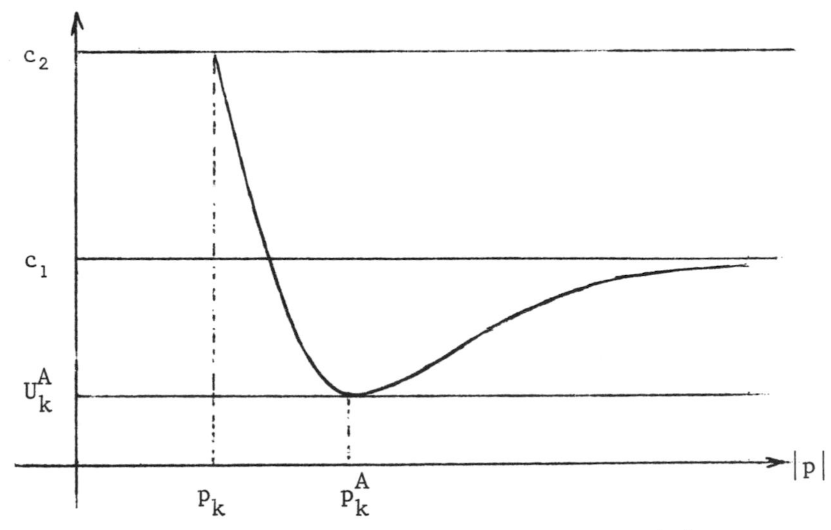

Figure 10. The Group Speed Curve $q = U_k(|p|)$

The stationary points of $\theta_k(p,x,t)$ are defined by (7.74). This may be written in (2-dimensional) vector notation

$$x/t = U_k(|p|)p/|p| \tag{7.79}$$

This is equivalent to the conditions

$$U_k(|p|) = |x|/t \tag{7.80}$$

and

$$p \text{ is parallel to } x \text{ and in the same direction} \tag{7.81}$$

since $U_k(|p|) > 0$ and t is assumed to be positive. Conditions (7.80) and (7.81) determine $|p|$ and $p/|p|$, respectively. In particular, it is clear from Figure 10 that

$$\left.\begin{array}{l} \text{For } |x| > c_2 t \text{ and } |x| < U_k^A t \text{ there are no points of stationary phase} \\ \text{For } c_1 t \le |x| \le c_2 t \text{ and } |x| = U_k^A t \text{ there is one point of stationary phase} \\ \text{For } U_k^A t < |x| < c_1 t \text{ there are two points of stationary phase} \end{array}\right\} \tag{7.82}$$

According to the method of stationary phase each stationary point where $\det(\partial^2\theta_k/\partial p_i\partial p_j) \ne 0$ (regular stationary point) contributes a term

$$\frac{\exp\left(i\theta_k(p,x,t) + i\frac{\pi}{4}\operatorname{sgn}(\partial^2\theta_k/\partial p_i\partial p_j)\right)}{|\det(\partial^2\theta_k/\partial p_i\partial p_j)|^{1/2}}\, w_k(y,p)\hat{h}_k(p) \tag{7.83}$$

to the asymptotic expansion of the integral (7.72), where sgn and det denote the signature and determinant, respectively, of the Gramian matrix $(\partial^2\theta_k/\partial p_i\partial p_j)$. A short calculation shows that the eigenvalues of the Gramian for (7.73) are $-tU_k'(|p|)$ and $-tU_k(|p|)/|p|$ and hence for $t > 0$

$$\operatorname{sgn}(\partial^2\theta_k/\partial p_i\partial p_j) = -1 - \operatorname{sgn} U_k'(|p|) \tag{7.84}$$

$$\det(\partial^2\theta_k/\partial p_i\partial p_j) = t^2 U_k(|p|)U_k'(|p|)/|p| \tag{7.85}$$

In particular, the stationary points are regular when $|p| \neq p_k^A$. Substitution of (7.73), (7.84) and (7.85) into (7.83) gives the function

$$v_k^\infty(t,x,y,p) = \frac{|p|^{1/2} \exp\{i(|x|\,|p| - t\omega_k(|p|) - \frac{\pi}{4} - \frac{\pi}{4} \operatorname{sgn} U_k'(|p|))\}}{t\,\{U_k(|p|)\,|U_k'(|p|)\}^{1/2}}\, w_k(y,p)\hat{h}_k(p) \tag{7.86}$$

To find the asymptotic wave function for $v_k(t,x,y)$ it is necessary to solve (7.79) for p and substitute in (7.86). The result may be described by means of the two inverse functions of $U_k(|p|)$ which may be defined as follows: see Figure 10.

$$\left.\begin{aligned} |p| &= P_k^f(q) \Leftrightarrow U_k(|p|) = q \text{ and } p_k \leq |p| \leq p_k^A \\ |p| &= P_k^s(q) \Leftrightarrow U_k(|p|) = q \text{ and } |p| \geq p_k^A \end{aligned}\right\} \tag{7.87}$$

It is clear from the discussion of $U_k(|p|)$ that P_k^f and P_k^s are analytic functions, P_k^f maps $\{q: U_k^A \leq q \leq c_2\}$ onto $\{|p|: p_k \leq |p| \leq p_k^A\}$ and P_k^s maps $\{q: U_k^A \leq q \leq c_1\}$ onto $\{|p|: |p| \geq p_k\}$.

The asymptotic behavior of $v_k(t,x,y)$ can now be described. The point of stationary phase $|p| = P_k^f(|x|/t)$ makes a contribution

$$v_k^{\infty,f}(t,x,y) = \chi_k^f(t,x)\, v_k^\infty(t,x,y,P_k^f(|x|/t)x/|x|) \tag{7.88}$$

where

$$\chi_k^f(t,x) \text{ is the characteristic function of } \{(t,x):\ U_k^A \leq |x|/t \leq c_2\} \tag{7.89}$$

Similarly, the point of stationary phase $|p| = P_k^s(|x|/t)$ makes a contribution

$$v_k^{\infty,s}(t,x,y) = \chi_k^s(t,x)\, v_k^\infty(t,x,y,P_k^s(|x|/t)x/|x|) \tag{7.90}$$

where

$\chi_k^s(t,x)$ is the characteristic function of

$$\{(t,x):\ U_k^A \leq |x|/t \leq c_1\} \tag{7.91}$$

The functions $v_k^{\infty,f}$ are called the "fast waves" because they describe waves which arrive at points (x,y) at times $t = |x|/c_2$ corresponding to the speed $c(y) = c_2$ of waves in the "fast" medium filling $y > h$. Similarly, the functions $v_k^{\infty,s}$ are called the "slow waves" because they describe waves which arrive at (x,y) at times $t = |x|/c_1$ corresponding to the speed $c(y) = c_1$ of waves in the "slow" medium filling $0 < y < h$. Finally, the total asymptotic wave function is the sum

$$v_k^{\infty}(t,x,y) = v_k^{\infty,f}(t,x,y) + v_k^{\infty,s}(t,x,y) \tag{7.92}$$

The following convergence theorem was proved in [40] by the method outlined in section 5.

7.15 Theorem

Let $h \in \mathcal{H}$. Then for each $k \geq 1$, $v_k^{\infty}(t,\cdot,\cdot) \in \mathcal{H}$ for all $t > 0$ and $t \to v_k^{\infty}(t,\cdot,\cdot) \in \mathcal{H}$ is continuous. Moreover, $v_k^{\infty}(t,\cdot,\cdot)$ is an asymptotic wave function for the modal wave $v_k(t,\cdot,\cdot) = \exp(-itA^{1/2})\, P_k h$; i.e.,

$$\lim_{t\to\infty} \| v_k(t,\cdot,\cdot) - v_k^{\infty}(t,\cdot,\cdot)\|_{\mathcal{H}} = 0 \tag{7.93}$$

The same methods were used in [40] to prove convergence in the energy norm when h has finite energy.

Note that $v_k^{\infty}(t,x,y)$ represents a guided wave which propagates radially outward in horizontal planes y = const. and is exponentially damped in the vertical coordinate y. This is evident from the defining equations (7.86), (7.88), (7.90) and (7.92).

7.16 Asymptotic wave functions for the free mode

It will now be shown that the free mode wave function $v_0(t,x,y)$ is asymptotically equal in $\mathcal{H}$ to a free wave propagating with speed c_2 in the half-space $y \geq h$. To this end note that

$$v_0(t,x,y) = \frac{1}{2\pi} \int_{\Omega_0} \exp\{i(x\cdot p - t\lambda^{1/2})\}\, w_0(y,p,\lambda)\,\hat{h}_0(p,\lambda)\,dp\,d\lambda \tag{7.94}$$

where $\hat{h}_0 \in L_2(\Omega_0)$. The representation (7.27) for $w_0(y,p,\lambda)$ implies that

$$v_0(t,x,y) = v_0^+(t,x,y-h) + v_0^-(t,x,y-h), \quad y > h \tag{7.95}$$

where

$$v_0^+(t,x,y) = \frac{1}{2\pi} \int_{\Omega_0} e^{i(x\cdot p+y\xi-t\lambda^{1/2})} a(p,\lambda)\ \gamma_+(\xi,\eta)\hat{h}_0(p,\lambda)dpd\lambda \tag{7.96}$$

and

$$v_0^-(t,x,y) = \frac{1}{2\pi} \int_{\Omega_0} e^{i(x\cdot p-y\xi-t\lambda^{1/2})} a(p,\lambda)\ \gamma_-(\xi,\eta)\hat{h}_0(p,\lambda)dpd\lambda \tag{7.97}$$

The change of variables

$$(p,\lambda) \to (p,q), \quad q = \xi = (\lambda/c_2^2 - |p|^2)^{1/2} \tag{7.98}$$

in (7.96) gives

$$v_0^+(t,x,y) = \frac{1}{(2\pi)^{3/2}} \int_{q\geq 0} e^{i(x\cdot p+yq-t\omega(p,q))}\hat{h}(p,q)dpdq \tag{7.99}$$

where

$$\hat{h}(p,q) = c_2^2\rho_2^{1/2}(2|q|)^{1/2}(\gamma_+(\xi,\eta)/|\gamma_+(\xi,\eta)|)\hat{h}_0(p,\lambda) \tag{7.100}$$

and

$$\lambda = \lambda(p,q) = \omega(p,q)^2 = c_2^2(|p|^2 + q^2) \tag{7.101}$$

Similarly, the change of variables

$$(p,\lambda) \to (p,q), \quad q = -\xi = -(\lambda/c_2^2 - |p|^2)^{1/2} \tag{7.102}$$

in (7.97) gives

$$v_0^-(t,x,y) = \frac{1}{(2\pi)^{3/2}} \int_{q\leq 0} e^{i(x\cdot p+yq-t\omega(p,q))}\hat{h}(p,q)dpdq \tag{7.103}$$

where $\hat{h}$ is defined by (7.100). Adding (7.99) and (7.103) and using (7.95) shows that

$$v_0(t,x,y+h) = \frac{1}{(2\pi)^{3/2}} \int_{R^3} e^{i(x\cdot p+yq-t\omega(p,q))}\hat{h}(p,q)\,dpdq \tag{7.104}$$

for all $y \geq 0$. Moreover, (7.100) implies that $\hat{h} \in L_2(R^3)$. Thus (7.104) and (7.101) imply that in the half-space $y \geq h$ $v_0(t,x,y)$ coincides with a solution in $L_2(R^3)$ of the d'Alembert equation with propagation speed c_2. Now, the results of section 3 imply that the right-hand side of (7.104) has an asymptotic wave function in $L_2(R^3)$; say

$$w^\infty(t,x,y) = G(r-c_2t,\theta)/r, \quad r^2 = |x|^2 + y^2, \quad \theta = (x,y)/r \tag{7.105}$$

It follows that if

$$v_0^\infty(t,x,y) = \begin{cases} w^\infty(t,x,y-h), & y \geq h \\ 0, & 0 \leq y < h \end{cases} \tag{7.106}$$

then

$$\lim_{t\to\infty} \| v_0(t,\cdot,\cdot) - v_0^\infty(t,\cdot,\cdot)\|_{\mathcal{H}} = 0 \tag{7.107}$$

A proof may be found in [40]. This paper also contains a proof of convergence in the energy norm, when h has finite energy, and applications of these results to the calculation of asymptotic energy distributions in stratified fluids.

7.17 Other cases

The case of the symmetric Epstein profile, defined by

$$c^{-2}(y) = c_0^{-2}\,\mathrm{sech}^2(y/H) + c_\infty^{-2}\tanh^2(y/H) \tag{7.108}$$

and $\rho(y) \equiv 1$ was studied by the author in [41] where eigenfunction expansions and asymptotic wave functions are derived. Eigenfunction expansions for the case of the general Epstein profile

$$c^{-2}(y) = K\cosh^2(y/H) + L\tanh(y/H) + M \tag{7.109}$$

and $\rho(y) \equiv 1$ have been given by J. C. Guillot and the author [14, 15]. Asymptotic wave functions for this case are currently being studied by Y. Dermenjian, J. C. Guillot and the author. Preliminary studies show that the results given above for the Pekeris model are valid for a large class of profiles $c(y)$, $\rho(y)$. The essential hypotheses, apart from the boundedness (7.7), are that $c(y)$ should have a global minimum at some finite point and that $c(y)$ should tend to a limit at infinity sufficiently rapidly. If $c(y)$ does not have a minimum then there are no guided waves. However, these results have not yet been proved in this generality.

8. PROPAGATION IN CRYSTALS

Acoustic wave propagation in an unlimited homogeneous crystal is analyzed in this section. The analysis is similar to that for homogeneous fluids given in section 3. The principal new feature is the influence of anisotropy on the structure of the asymptotic wave functions.

A homogeneous crystal is characterized by a constant density $\rho(x) = \rho$ and stress-strain tensor $c_{\ell m}^{jk}(x) = c_{\ell m}^{jk}$. It will suffice to consider the case $\rho = 1$. Thus the propagation problem reduces in this case to the Cauchy problem for the system

$$\frac{\partial^2 u_j}{\partial t^2} - c_{jk}^{\ell m} \frac{\partial^2 u_\ell}{\partial x_k \partial x_m} = 0, \quad j = 1,2,3 \tag{8.1}$$

where the constants $c_{jk}^{\ell m}$ satisfy (2.13) and (2.35).

8.1 Hilbert space formulation

It was shown in section 2 that the differential operator A defined by

$$(Au)_j = - c_{jk}^{\ell m} \frac{\partial^2 u_\ell}{\partial x_k \partial x_m}, \quad j = 1,2,3 \tag{8.2}$$

is formally selfadjoint in the Hilbert space $\mathcal{H} = L_2(R^3, C^3)$ with inner product

$$(u,v) = \int_{R^3} \overline{u_j(x)}\, v_j(x)\, dx \tag{8.3}$$

In fact, the operator $\mathcal{A}$ in $\mathcal{H}$ with domain $D(\mathcal{A}) = \mathcal{D}(R^3)$ is essentially selfadjoint and its unique selfadjoint extension is the operator A defined by

$$D(A) = \mathcal{H} \cap \{u: \ \mathcal{A}u \in \mathcal{H}\} \tag{8.4}$$

$$Au = \mathcal{A}u \text{ for all } u \in D(A) \tag{8.5}$$

It is easy to verify, using the Plancherel theory of the Fourier transform, the following

8.2 Theorem

A is a selfadjoint, real positive operator in $\mathcal{H}$.

It follows, as in preceding sections, that the Cauchy problem for (8.1) has a solution in $\mathcal{H}$ of the form

$$u_j(t,x) = \operatorname{Re} \{v_j(t,x)\} \tag{8.6}$$

where

$$v(t,\cdot) = \exp(-itA^{1/2})h, \quad h = f + iA^{-1/2}g \in \mathcal{H} \tag{8.7}$$

whenever the Cauchy data $u(0,x) = f(x)$ and $\partial u(0,x)/\partial t = g(x)$ satisfy $f \in \mathcal{H}$, $g \in D(A^{-1/2})$.

8.3 Fourier analysis of A

The Plancherel theory of the Fourier transform $\Phi_0: L_2(R^3) \to L_2(R^3)$ was defined and used in section 3; see (3.12). It may be extended immediately to $\mathcal{H} = L_2(R^3, C^3)$ by defining

$$\Phi_0 u = \Phi_0(u_1, u_2, u_3) = (\Phi_0 u_1, \Phi_0 u_2, \Phi_0 u_3) \tag{8.8}$$

and Φ_0 is also unitary in $\mathcal{H}$. Property (3.14) implies that the operator $\Phi_0 A \Phi_0^*$ corresponds to multiplication by the 3×3 matrix valued function

$$A(p) = (A_{j\ell}(p)) = (c_{jk}^{\ell m} p_k p_m), \quad p \in R^3 \tag{8.9}$$

Thus

$$A = \Phi_0^* \, A(\cdot) \, \Phi_0 \tag{8.10}$$

Moreover, conditions (2.13) and (2.35) imply that A(p) is a real Hermitian positive definite matrix for all $p \in R^3 - \{0\}$. The spectral analysis of A will be based on (8.10). The analysis begins with

8.4 Spectral analysis of A(p)

The eigenvalues of A(p) are the roots μ of the characteristic polynomial

$$\det\,(\mu 1 - A(p)) = 0 \tag{8.11}$$

The Hermitian positive definiteness of A(p) implies that the roots are real and positive for all $p \in R^3 - \{0\}$. They may be uniquely defined as functions of p by enumerating them according to their magnitudes:

$$0 \le \mu_1(p) \le \mu_2(p) \le \mu_3(p) \text{ for all } p \in R^3 \tag{8.12}$$

A result of T. Kato [18] implies (see also [37])

$$\mu_j\colon\ R^3 \to R \text{ is continuous, } j = 1,2,3 \tag{8.13}$$

Equation (8.11) implies that $\mu_j(p)$ is homogeneous of degree 2

$$\mu_j(\alpha p) = \alpha^2 \mu_j(p) \text{ for all } \alpha \in R \text{ and } p \in R^3 \tag{8.14}$$

The functions

$$\lambda_j(p) = \sqrt{\mu_j(p)},\ p \in R^3,\ j = 1,2,3 \tag{8.15}$$

are also needed below. A detailed study of these functions has been made by the author in connection with a formulation of elasticity theory in terms of first order symmetric hyperbolic systems; see [29,35,37,44]. A number of results from these papers are quoted and used below.

It was shown in [44] that there exists a homogeneous polynomial $\mathcal{O}(p) \not\equiv 0$ such that the points $p \in R^3$ where two or more roots $\mu_j(p)$ coincide are contained in the cone

$$Z = \{p \in R^3 : \mathcal{O}(p) = 0\} \tag{8.16}$$

Thus

$$0 < \mu_1(p) < \mu_2(p) < \mu_3(p) \text{ for all } p \in R^3 - Z \tag{8.17}$$

It follows that

$$\mu_j(p) \text{ is analytic on } R^3 - Z,\ j = 1,2,3 \tag{8.18}$$

The orthogonal projection of C^3 onto the eigenspace for $\mu_j(p)$ is given by [18]

$$\hat{P}_j(p) = -\frac{1}{2\pi i}\int_{\gamma_j(p)} (A(p) - zI)^{-1}dz,\ j = 1,2,3 \tag{8.19}$$

where

$$\gamma_j(p) = \{z: |z - \mu_j(p)| = c_j(p)\},\ j = 1,2,3 \tag{8.20}$$

and the radii $c_j(p)$ are chosen so small that the 3 circles $\gamma_j(p)$ are disjoint. This is possible for all $p \in R^3 - Z$ by (8.17). The matrix valued functions $\hat{P}_j$ so defined can be shown to have the following properties [18,44]:

$$\hat{P}_j(p) \text{ is analytic on } R^3 - Z,\ j = 1,2,3 \tag{8.21}$$

$$\hat{P}_j(\alpha p) = \hat{P}_j(p) \text{ for all } \alpha \neq 0 \tag{8.22}$$

$$\hat{P}_j(p)^* = \hat{P}_j(p),\ \hat{P}_j(p)\,\hat{P}_k(p) = \delta_{jk}\hat{P}_k(p) \text{ for } p \in R^3 - Z \tag{8.23}$$

$$\sum_{j=1}^{3} \hat{P}_j(p) = 1 \text{ for } p \in R^3 - Z \tag{8.24}$$

$$A(p)\,\hat{P}_j(p) = \mu_j(p)\,\hat{P}_j(p) \text{ for } p \in R^3 - Z,\ j = 1,2,3 \tag{8.25}$$

The last two properties imply that the projections $\hat{P}_j(p)$ define a spectral representation for $A(p)$; i.e.,

$$A(p) = \sum_{j=1}^{3} \mu_j(p)\, \hat{P}_j(p) \text{ for } p \in R^3 - Z \tag{8.26}$$

8.5 Spectral analysis of A

The representations (8.10) and (8.26) provide a complete spectral analysis of A. In particular, it follows that A is an absolutely continuous operator whose spectrum is $[0,\infty)$ (cf. [36, 44]). Moreover, if $\Psi(\mu)$ is any bounded Lebesgue-measurable function of $\mu \geq 0$ then

$$\Psi(A) = \Phi_0^* \sum_{k=1}^{3} \Psi(\mu_k(\cdot))\, \hat{P}_k(\cdot)\, \Phi_0 \tag{8.27}$$

8.6 Solution in $\mathcal{H}$ of the Cauchy problem

Application of (8.27) to the solution in $\mathcal{H}$ (8.7) yields the representation

$$v(t,x) = \sum_{k=1}^{3} v_k(t,x) \tag{8.28}$$

where

$$v_k(t,x) = \frac{1}{(2\pi)^{3/2}} \int_{R^3} e^{i(x\cdot p - t\lambda_k(p))} \hat{P}_k(p)\hat{h}(p)\,dp \tag{8.29}$$

and $\lambda_k(p) = \sqrt{\mu_k(p)}$. Of course, the integral in (8.29) converges in $\mathcal{H}$, in the sense of the Plancherel theory, rather than pointwise. Equations (8.28), (8.29) represent solutions in $\mathcal{H}$ of (8.1) as a superposition of solutions

$$e^{i(x\cdot p - t\lambda_k(p))} \hat{P}_k(p)\hat{h}(p) \tag{8.30}$$

This may be interpreted as a plane wave which propagates in the crystal with direction $p/|p|$, wave number $|p|$ and frequency

$$\omega = \lambda_k(p) \tag{8.31}$$

The polarization of the wave is determined by $\hat{P}_k(p)$. The corresponding generalized eigenfunctions of A are the matrix plane waves [30]

$$w_k(x,p) = \frac{1}{(2\pi)^{3/2}} \exp(ix\cdot p)\, \hat{P}_k(p) \tag{8.32}$$

8.7 The dispersion relation, phase and group velocities

The dispersion relation between the frequency ω and wave vector p of plane waves in the crystal is (8.31) or, by (8.15) and (8.11)

$$\det(\omega^2 1 - A(p)) = 0 \tag{8.33}$$

The phase velocity for (8.1) is

$$v_{ph}(p) = \frac{\omega}{|p|} \cdot \frac{p}{|p|} = \frac{\lambda_k(p)}{|p|^2}\, p = \lambda_k\left(\frac{p}{|p|}\right)\frac{p}{|p|} \tag{8.34}$$

by the homogeneity of $\lambda_k(p)$. The group velocity for (8.1) is

$$v_g(p) = \nabla_p\omega = \nabla_p\lambda_k(p) \tag{8.35}$$

The medium is said to be isotropic if $v_{ph}(p)$ and $v_g(p)$ have the same direction for all $p \in R^3 - \{0\}$. Otherwise it is said to be anisotropic. It is easy to verify that the medium is isotropic if and only if $\lambda_k(p)$ is a function of $|p|$ alone. In this case $\lambda_k(p) = c_k|p|$ and $v_{ph}(p) = v_g(p) = c_k p/|p|$.

The phase and group speeds for (8.1) are the magnitudes of the corresponding velocities. Thus

$$\left.\begin{aligned} c_{ph}(p) &= |v_{ph}(p)| = \lambda_k(p/|p|) \\ c_g(p) &= |v_g(p)| = \nabla_p\lambda_k(p) \end{aligned}\right\} \tag{8.36}$$

Note that both are homogeneous of degree zero in p and hence depend only on the direction of propagation $p/|p|$. The anisotropy of the medium characterized by (8.1) can be visualized by means of

8.8 The slowness surface S

This is the real algebraic variety defined by

$$S = \{p \in R^3: \ \det (1 - A(p)) = 0\} \tag{8.37}$$

It is clear from the definition of the $\lambda_k(p)$ that

$$S = \bigcup_{k=1}^{3} S_k \tag{8.38}$$

where

$$S_k = \{p \in R^3: \ \lambda_k(p) = 1\} \tag{8.39}$$

or, by (8.36) and the homogeneity of $\lambda_k(p)$,

$$S_k = \{p \in R^3: \ |p| c_k(p) = 1\} \tag{8.40}$$

Thus $p \in S$ if and only if $|p|$ is the reciprocal of a phase speed for the direction p. Note that the slowness surface of an isotropic medium is a set of concentric spheres with centers at the origin.

The properties of slowness surfaces were studied in [35] and [44]. In particular, the following properties were established

$$S_k \text{ is continuous and star-shaped with respect to } 0 \tag{8.41}$$

As an algebraic variety, S will in general have singular points and these are precisely the set

$$Z_S' = \{p \in S: \ p \in S_j \cap S_k \text{ for some } j \neq k\} \tag{8.42}$$

Hence

$$S_k - Z_S', \ k = 1,2,3, \text{ are disjoint and analytic} \tag{8.43}$$

8.9 The wave surface W

The variation of the phase speed with direction is represented by the slowness surface S. Similarly, the variation of the group speed is represented by the wave surface W. W may be defined as the polar reciprocal of S with respect to the unit sphere. This means that

$$W = \{x \in R^3: \quad x \cdot p = 1 \text{ is a tangent plane to } S\} \tag{8.44}$$

It is known that W is a real algebraic variety whose degree is the class number of S [7,28]. Moreover, the relation of S and W is symmetric: S is also the polar reciprocal of W. It is clear that if

$$N(p) = \text{the } \underline{\text{set}} \text{ of all exterior unit normals to } S \text{ at } p \tag{8.45}$$

then

$$W = \{x = (p \cdot N(p))^{-1} N(p): \quad p \in S\} \tag{8.46}$$

Now the group velocity $v_g(p) = \nabla_p \lambda_k(p)$ is normal to S at each $p \in S - Z_S^!$. Moreover, $p \cdot \nabla_p \lambda_k(p) = \lambda_k(p) = 1$ for such points p by (8.39) and the homogeneity of $\lambda_k(p)$. Hence

$$\{x = v_g(p) \equiv \nabla_p \lambda_k(p): \quad p \in S - Z_S^!\} \subset W \tag{8.47}$$

for k = 1,2,3.

8.10 The polar reciprocal map T: $S \to W$

This is the map defined in (8.46); i.e.,

$$T(p) = (p \cdot N(p))^{-1} N(p) \text{ for all } p \in S \tag{8.48}$$

As indicated above, N(p) is not, in general, single valued. It follows that T may be neither single-valued nor injective. However, it was shown in [49] that if

$$\left.\begin{array}{l} Z_S^! = \text{set of singular points of } S \\ \\ Z_W^! = \text{set of singular points of } W \end{array}\right\} \tag{8.49}$$

$$Z_S'' = T^{-1} Z_W', \quad Z_W'' = T Z_S^! \tag{8.50}$$

$$Z_S = Z_S^! \cup Z_S'', \quad Z_W = Z_W^! \cup Z_W'' \tag{8.51}$$

then Z_S and Z_W are sub-varieties of dimension ≤ 1 and

$$T \text{ is bijective and analytic from } S - Z_S \text{ to } W - Z_W \tag{8.52}$$

8.11 Examples

The equation (8.37) for the slowness surface of a crystal contains 21 independent parameters in the most general case (triclinic crystals). Hence a great variety of slowness surfaces are possible. Crystal symmetries may reduce the number of parameters. The slowness surfaces of the various symmetry classes have been studied by many authors. Thorough discussions and examples may be found in [3] and [24] where specific numerical information on the stress-strain tensors of real crystals may also be found. Here two examples will be described briefly to show the kind of surfaces that may occur.

Cubic crystals. In this case symmetry reduces the number of independent parameters to 3 and the equation for S can be written [24]

$$\sum_{j=1}^{3} \frac{p_j^2}{a-b|p|^2-cp_j^2} = 1 \tag{8.53}$$

Of course, the positive definiteness of $c_{jk}^{\ell m}$ imposes certain numerical restrictions on a, b and c. Equation (8.53) represents a surface of degree 6 which is irreducible except for special parameter values.

Hexagonal crystals. In this case symmetry reduces the number of independent parameters to 5. Moreover, S is necessarily a surface of revolution and reduces to two components whose equations can be written [24]

$$a^2(p_1^2 + p_2^2) + b^2p_3^2 = 1 \tag{8.54}$$

$$\frac{p_1^2+p_2^2}{c^2-d^2|p|^2+e(p_1^2+p_2^2)} + \frac{p_3^2}{c^2-d^2|p|^2+fp_3^2} = 1 \tag{8.55}$$

(where a, b, c, d, e and f can be expressed in terms of 5 independent parameters). The two equations have degrees 2 and 4, respectively. These surfaces of revolution can be visualized from their traces on the p_1,p_3-plane; see [24,p.99] for a graph of such an S and the corresponding W. It is seen that in the example Z_S' consists of 2 circles and 2 points lying in S while Z_W' consists of 8 circles and 2 points lying in W.

8.12 Asymptotic wave functions for crystals

It was shown in [44] that the equations (8.1) for acoustic waves in crystals have asymptotic wave functions of the form

$$v^{\infty}(t,x) = \sum_{\alpha=1}^{\nu(\theta)} F(x\cdot s^{(\alpha)}(\theta) - t, s^{(\alpha)}(\theta))/|x|, \quad x = |x|\theta \tag{8.56}$$

where

$$s^{(\alpha)}(\theta) \in S, \ \alpha = 1,2,\cdots,\nu(\theta) \tag{8.57}$$

is the solution set of the equation

$$N(s) = \theta \tag{8.58}$$

Thus $s^{(\alpha)}$ defines the multivalued inverse of the Gauss map N of S. The principal properties of $v^{\infty}(t,x)$ are described by the following theorem whose proof is contained in [44].

8.13 Theorem

For each $h \in \mathcal{H}$ there exists a unique $F: R \times S \to C^3$ such that

$$v^{\infty}(t,\cdot) \in \mathcal{H} \text{ for all } t \in R \tag{8.59}$$

$$t \to v^{\infty}(t,\cdot) \in \mathcal{H} \text{ is continuous for all } t \in R \tag{8.60}$$

$$\| v^{\infty}(t,\cdot)\|_{\mathcal{H}} \leq C\| h\|_{\mathcal{H}} \text{ where } C \text{ is independent of } h \text{ and } t \tag{8.61}$$

Finally, v^{∞} is an asymptotic wave function for $v(t,\cdot) = \exp(-itA^{1/2})h$:

$$\lim_{t\to\infty} \| v(t,\cdot) - v^{\infty}(t,\cdot)\|_{\mathcal{H}} = 0 \tag{8.62}$$

Moreover, explicit constructions of $s^{(\alpha)}(\theta)$ and $F(\tau,s)$ are given in [49]. In the present case they take the following form.

8.14 Construction of $s^{(\alpha)}(\theta)$

The construction consists of two steps.

$x^{(\alpha)}(\theta)$, $\alpha = 1,\cdots,\nu(\theta)$ is the intersection of $W - Z_W$ and the ray from 0 along θ (8.63)

$$s^{(\alpha)}(\theta) = T^{-1}x^{(\alpha)}(\theta) \in S - Z_S \tag{8.64}$$

Note that this defines $s^{(\alpha)}(\theta)$ for all θ outside of the null set

$$Z^0 = \{\theta: \quad x = |x|\theta \in Z_W\} \subset S^2 = \{\theta: \quad |\theta| = 1\} \tag{8.65}$$

8.15 Construction of $F(\tau,s)$

F is calculated from $h = v(0,\cdot) \in \mathcal{H}$ by the rule

$$F(\tau,s) = (2\pi)^{-1/2} \quad \Psi(s) \int_0^\infty e^{i\tau\lambda} \, \hat{h}(\lambda s)\lambda d\lambda \tag{8.66}$$

where

$$\Psi(s) = \psi(s)|K(s)|^{-1/2} \quad |T(s)|^{-1} \hat{P}(s) \tag{8.67}$$

$$\psi(s) = \exp\{i \tfrac{\pi}{4} (p^-(s) - p^+(s))\} \tag{8.68}$$

$p^\pm(s)$ = the number of principal curvatures of S at s which are $\gtrless 0$. (8.69)

$K(s)$ = Gaussian curvature of S at s (8.70)

$\hat{P}(s)$ = orthogonal projection of C^3 onto the eigenspace for the eigenvalue $\mu = 1$ of $A(s)$ $(s \in S)$ (8.71)

It is shown in [44] that $\Psi(s)$ is defined for all $s \in S - Z_S$. In particular, the parabolic points of S lie in Z_S. The integral for F need not converge pointwise, but it converges in the Hilbert space $\mathcal{H}(S)$ with norm defined by

$$\|F\|^2_{\mathcal{H}(S)} = \int_0^\infty \int_S |F(\tau,s)|^2 \, |K(s)T(s)| \, dS d\tau \qquad (8.72)$$

Moreover, the operator Θ: $\mathcal{H} \to \mathcal{H}(S)$ defined by $\Theta h = F$ is an isometry.

8.16 Propagation in non-uniform crystals

The method developed in [42] and section 4 can be applied to local perturbations of uniform crystals. Eigenfunction expansions for non-uniform crystals, and more general systems, have been given by G. Nenciu [25].

REFERENCES

1. S. Agmon, Lectures on Elliptic Boundary Value Problems, New York: Van Nostrand, 1965.
2. O. Arena and W. Littman, 'Farfield' behavior of solutions to partial differential equations, Ann. Scuola Norm. Sup. Pisa 26, 807-827 (1972).
3. B. A. Auld, Acoustic Fields and Waves in Solids, Vols. I and II, New York: J. Wiley and Sons, 1973.
4. L. Brillouin, Les Tenseurs en Mécanique et en Elasticité, New York: Dover, 1946.
5. L. Brillouin, Wave Propagation and Group Velocity, New York: Academic Press, 1960.
6. R. Courant and D. Hilbert, Methods of Mathematical Physics, Vol. 2, New York: Interscience Publishers, 1962.
7. G. F. D. Duff, The Cauchy problem for elastic waves in an anisotropic medium, Phil. Trans. Roy. Soc. London, Series A 252, 249-273 (1960).
8. F. G. Friedlander, Sound Pulses, Cambridge: Cambridge University Press, 1958.
9. C. Goldstein, Eigenfunction expansions associated with the Laplacian for certain domains with infinite boundaries, Trans. Amer. Math. Soc. 135, 1-50 (1969).
10. C. Goldstein, The singularities of the S-matrix and Green's function associated with perturbations of $-\Delta$ acting in a cylinder, Bull. Amer. Math. Soc. 79, 1303-1307 (1973).
11. C. Goldstein, Meromorphic continuation of the S-matrix for the operator $-\Delta$ acting in a cylinder, Proc. Amer. Math. Soc. 42, 555-562 (1974).
12. C. Goldstein, Analytic perturbations of the operator $-\Delta$, J. Math. Ann. Appl. 25, 128-148 (1969).
13. J. C. Guillot and C. H. Wilcox, Théorie spectrale du laplacian dans des ouverts coniques et cylindriques non bornés, C. R. Acad. Sci. Paris 282, 1171-1174 (1976).

14. J. C. Guillot and C. H. Wilcox, Théorie spectrale de l'opérateur d'Epstein (théorie des guides d'onde ouverts), C. R. Acad. Sci. Paris 281, 399-402 (1975).
15. J. C. Guillot and C. H. Wilcox, Spectral analysis of the Epstein operator, ONR Technical Summary Rept. #27, University of Utah, August 1975.
16. T. Ikebe, Eigenfunction expansions associated with the Schrödinger operator and their application to scattering theory, Arch. Rational Mech. Anal. 5, 1-34 (1960).
17. D. S. Jones, The eigenvalues of $\nabla^2 u + \lambda u = 0$ when the boundary conditions are given on semi-infinite domains, Proc. Cambridge Phil. Soc. 49, 668-684 (1953).
18. T. Kato, Perturbation Theory for Linear Operators, New York: Springer, 1966.
19. J. L. Lions and E. Magenes, Non-Homogeneous Boundary Value Problems and Applications, New York: Springer, 1972.
20. A. E. H. Love, A Treatise on the Mathematical Theory of Elasticity, 4th Edition, New York: Dover, 1944.
21. W. C. Lyford, Spectral analysis of the Laplacian in domains with cylinders, Math. Ann. 218, 229-251 (1975).
22. W. C. Lyford, Asymptotic energy propagation and scattering of waves in waveguides with cylinders, Math. Ann. 219, 193-212 (1976).
23. M. Matsumura, Asymptotic behavior at infinity for Green's functions of first order systems with characteristics of nonuniform multiplicity, Publ. R.I.M.S., Kyoto Univ., Ser. A (to appear).
24. M. J. P. Musgrave, Crystal Acoustics, San Francisco: Holden-Day, 1970.
25. G. Nenciu, Eigenfunction expansions for wave propagation problems in classical physics, preprint, 1975.
26. C. L. Pekeris, Theory of propagation of explosive sound in shallow water, Geol. Soc. Am., Memoir 27 (1948).
27. F. Rellich, Das Eigenwertproblem von $\Delta u + \lambda u = 0$ in Halbröhren, in Studies and Essays Presented to R. Courant, New York: Interscience, 1948.
28. G. Salmon, Analytic Geometry of Three Dimensions, Vol. 2, Fifth Edition, Dublin, 1914. Reprinted New York: Chelsea Publishing Co., 1965.
29. J. R. Schulenberger and C. H. Wilcox, A coerciveness inequality for a class of nonelliptic operators of constant deficit, Ann. Mat. Pura Appl. 92, 77-84 (1972).
30. J. R. Schulenberger and C. H. Wilcox, Eigenfunction expansions and scattering theory for wave propagation problems of classical physics, Arch. Rational Mech. Anal. 46, 280-320 (1972).
31. E. Skudrzyk, The Foundations of Acoustics, New York: Springer, 1971.
32. M. Vishik and O. A. Ladyzhenskaya, Boundary value problems for partial differential equations and certain classes of operator equations, Uspehi Mat. Nauk. (N.S.) 11, 41-97 (1956) [Russian]. A.M.S. Translations, Series II, 10, 223-281 (1958).

33. C. H. Wilcox, Initial-boundary value problems for linear hyperbolic partial differential equations of the second order, Arch. Rational Mech. Anal. 10, 361-400 (1962).
34. C. H. Wilcox, The domain of dependence inequality for symmetric hyperbolic systems, Bull. Amer. Math. Soc. 70, 149-154 (1964).
35. C. H. Wilcox, Wave operators and asymptotic solutions of wave propagation problems of classical physics, Arch. Rational Mech. Anal. 22, 37-78 (1966).
36. C. H. Wilcox, Transient wave propagation in homogeneous anisotropic media, Arch. Rational Mech. Anal. 37, 323-343 (1970).
37. C. H. Wilcox, Measurable eigenvectors for Hermitian matrix-valued polynomials, J. Math. Anal. Appl. 40, 12-19 (1972).
38. C. H. Wilcox, Scattering states and wave operators in the abstract theory of scattering, J. Functional Anal. 12, 257-274 (1973).
39. C. H. Wilcox, Spectral analysis of the Laplacian with a discontinuous coefficient, Proc. of C.I.M.E. Conf. on Spectral Analysis - Sept. 1973, 233-253, Rome: Edizioni Cremonesa, 1974.
40. C. H. Wilcox, Transient electromagnetic wave propagation in a dielectric waveguide, Proc. of Conf. on the Mathematical Theory of Electromagnetism, Istituto Nazionale di Alta Matematica, Rome, 1974.
41. C. H. Wilcox, Transient acoustic wave propagation in a symmetric Epstein duct, ONR Technical Summary Rept. #25, University of Utah, May 1974.
42. C. H. Wilcox, Scattering Theory for the d'Alembert Equation in Exterior Domains, Lecture Notes in Mathematics, Vol. 442, New York: Springer, 1975.
43. C. H. Wilcox, Spectral analysis of the Pekeris operator in the theory of acoustic wave propagation in shallow water, Arch. Rational Mech. Anal. 60, 259-300 (1976).
44. C. H. Wilcox, Asymptotic wave functions and energy distributions in strongly propagative anisotropic media, ONR Technical Rept. #28, University of Utah, March 1976.